Geographic Information Systems in Petroleum Exploration and Development

Edited by
Timothy C. Coburn
and
Jeffrey M. Yarus

AAPG Computer Applications in Geology, No. 4

Published by
The American Association of Petroleum Geologists
Tulsa, Oklahoma, U.S.A.
Printed in the U.S.A.

Printed in the U.S.A.
Published May 2000

ISBN: 0-89181-703-4

Association Editor: Neil F. Hurley
Geoscience Director: Robert C. Millspaugh
Publications Manager: Kenneth M. Wolgemuth
Managing Editor, Publications: Anne H. Thomas
Production: Custom Editorial Productions, Inc., Cincinnati, Ohio

This and other AAPG publications are available from:

The AAPG Bookstore
P.O. Box 979
Tulsa, OK 74101-0979
Telephone: 1-918-584-2555 or 1-800-364-AAPG (USA)
Fax: 1-918-560-2652 or 1-800-898-2274 (USA)
http://bookstore.aapg.org

Geological Society Publishing House
Unit 7, Brassmill Enterprise Centre
Brassmill Lane, Bath, U.K.
BA1 3JN
Tel +44-1225-445046
Fax +44-1225-442836
www.geolsoc.org.uk

Australian Mineral Foundation
AMF Bookshop
63 Conyngham Street
Glenside, South Australia 5065
Australia
Tel. +61-8-8379-0444
Fax +61-8-8379-4634
www.amf.com.au/amf

Affiliated East-West Press Private Ltd.
G-1/16 Ansari Road Darya Ganj
New Delhi 110 002
India
Tel +91 11 3279113
Fax +91 11 3260538
e-mail: affiliat@nda.vsnl.n

Preface

The idea about compiling this text was originally broached in a conversation we had in late 1995 in conjunction with our participation in the activities of the Geologic Computing Committee of the American Association of Petroleum Geologists (AAPG). A similar book, *Stochastic Modeling and Geostatistics: Principles, Methods, and Case Studies*, had recently been published by AAPG, and we felt that it had been generally well received. Owing to our mutual interest and professional activities in the quantitative geosciences, we recognized there were several other computer-based technologies that were beginning to have a significant impact on the upstream segment of the petroleum industry. Seeing the potential in these technologies, we felt it was important to encourage their proliferation; and that an appropriate mechanism for doing so would be to publish additional volumes as part of AAPG's Computer Applications series. GIS was one of those technologies that were beginning to explode on the scene, and we felt it should be the next topic for a book. After discussing our ideas with the Committee and with the AAPG staff, we got the "green light" to proceed; and the rest, as they say, is history!

No doubt many readers will agree that GIS—particularly in the context of petroleum exploration and development—is one of those technologies that is hard "to get one's arms around." At least part of the reason is the absence of a standard definition of just what a GIS is and what it's supposed to do. Consequently, taking this indefiniteness into account, we felt the petroleum industry would be best served by structuring the book to be as inclusive as possible in terms of the products and applications represented. As a result, the book is quite diverse in terms of topics, approaches, and philosophies, as well as in terms of the geological settings and operational aspects that are addressed. There is a good mix of theory, applications, resources, and practical advice, along with some historical perspectives and a sprinkling of futurism.

While embracing diversity, we still felt it necessary to impose some editorial consistency. For example, the acronym GIS has been used in various contexts in both the singular and plural sense, but for purposes of this particular text, we have taken it to be singular. Similarly, we have chosen the word "data" to be plural. In the editing process, we have made every effort to strictly enforce these and other rules, but clearly with an effort of this size, such a task is nearly impossible to accomplish. In addition, we selected a particular presentation style that we have maintained as rigorously as possible, but disparities obviously still remain. Finally, we have worked diligently to eliminate spelling and grammatical mistakes in an effort to make the delivery as clear and precise as possible, all the while acknowledging the likelihood that some have gone undetected.

Throughout the text there are numerous references to vendors, suppliers, products, and resources that are of a proprietary nature. To the greatest degree possible, we have attempted to appropriately credit all such enterprises in an appendix, and we offer our apologies in advance to those who may not have received proper acknowledgment. Reference to specific commercial or government entities by various authors, or mention of products or services available from such organizations, does not necessarily represent our views or opinions, or those of the publisher, nor does it in any way constitute an endorsement.

During the life of this project, our governing objectives have been fairly straightforward: we wanted to more fully communicate information about GIS to petroleum professionals and managers, we wanted to underscore its applicability as an effective geoscience tool, and we wanted to encourage and broaden its use within the industry. The degree to which we have achieved these objectives is, of course, yet to be debated and determined. Our personal belief is that the resulting text will be an extremely useful resource and guide to individuals and companies seeking to incorporate GIS into their own projects and working environments. It is our hope that readers will develop a greater appreciation for what GIS is and what it can do, and that they will use this understanding to achieve greater success in all their petroleum exploration and development endeavors.

Timothy C. Coburn, Abilene, TX, and Lawrence, KS
Jeffrey M. Yarus, Houston, TX

Dedication

This book is dedicated to Dr. R. Charles Ivey—friend, teacher, mentor, entrepreneur. As former President and CEO of Petroleum Information/Dwights, Charles empowered the petroleum industry with knowledge about itself. In an unheralded, persistent way, he brought integration and synthesis to information resources that were often fragmented and sometimes contradictory; and he pointed the industry toward data management solutions that will stand the test of time well into the 21st century. With this dedication, we honor him for his countless contributions to students and business people alike.

AAPG
wishes to thank the following
for their generous contributions
to

Geographic Information Systems in Petroleum Exploration and Development

❖

ExxonMobil Exploration

❖

Geotech Computer Systems

❖

Plains Resources

❖

Contributions are applied against the production costs of publication, thus directly reducing the book's purchase price and making the volume available to a greater audience.

Acknowledgments

A project of this nature cannot be successfully completed without the voluntary contributions of many individuals. The authors of the individual chapters, of course, have given the most; and without their hard work, patience, and willingness to accept criticism, there would be no book at all. Others have given selflessly of their time and support to help us bring this project to fruition, including the staffs of the American Association of Petroleum Geologists and of Custom Editorial Productions. We are particularly grateful to Ken Wolgemuth and Anne Thomas of AAPG for their patient and sensitive direction of our efforts.

We are also indebted to our employers who have provided us nurturing environments in which to work, technology with which to experiment, and industry-wide challenges that have enlarged our collective experiences and knowledge base. Special thanks go to John Davis and the staff of the Mathematical Geology Section of the Kansas Geological Survey for access to the hardware and software that was necessary to complete the final draft, and for their assistance in using those resources.

In addition, we gratefully acknowledge the tireless service of the following individuals who provided peer reviews to ensure that the final product would be of the highest quality:

- Greg Slutz, Texaco
- Jennifer Dungan, NASA Ames Research
- Seiichi Nagihara, University of Houston
- Umberto Fracassi, Università degli Studi di Firenze, Italy
- Doug Peters, Peters Geosciences
- Neil Hurley, Colorado School of Mines

Finally and foremost, we owe the utmost to our wives and our families, who have had to endure our absence and diminished attention throughout the lengthy time in which we have pursued this endeavor. We publicly honor their unfailing patience and love, and their willingness to humor us when we hear the call to chase these professional rabbits!

From Jeffrey: Thank you Mary, David, Jordan, and Rachel.

From Tim: Thank you Nancy, Claire, and Kevin.

Table of Contents

Preface iii

INTRODUCTION 1

Chapter 1 3
Geographic Information Systems in Petroleum Exploration and Development: An Introductory Essay
J. M. Yarus and T. C. Coburn

TECHNOLOGICAL FOUNDATIONS 7

Chapter 2 9
Geographic Information Systems (GIS) Technology: A Brief History, Trends, and Probable Future
J. K. Berry

Chapter 3 17
An Overview of GIS in the Geosciences
G. Bonham-Carter

Chapter 4 27
Geoscientific Modeling: Past, Present, and Future
A. K. Turner

Chapter 5 37
GIS Technology—A Vendor Perspective of its History and Prospects in Petroleum Exploration and Production
G. Wade

PRACTICAL REALITIES 47

Chapter 6 49
A Conversation with Petroleum Information
T. C. Coburn, J. M. Yarus, and P. H. Stark

Chapter 7 61
On the Implementation of GIS for Petroleum Exploration and Development: Issues and Perspectives
T. C. Coburn

CASE STUDIES 69

Chapter 8 71
Building a Regional Study in GIS
C. R. Mazzo and W. A. Burroughs

Chapter 9 121
GIS-Based Analysis of Fluid Migration in Sedimentary Basins
G. Paulus

Chapter 10 137
The Gap Mountain Prospect: A New Approach to Conceptual Geological Modeling
D. R. Jefferis, M. S. Connelly, and S. Pack

Chapter 11 151
Evaluating International Acquisitions Using a Spatial Database System—A Low-Cost Approach
A. Haertlein and D. Rich

Chapter 12 173
Use of Geographic Information Systems in Hydrocarbon Resource Assessment and Opportunity Analysis
K. C. Hood, B. C. South, F. D. Walton, O. D. Baldwin, and W. A. Burroughs

Chapter 13 187
Conducting a Field Study with GIS: Port Hudson Field, The Tuscaloosa Trend, East Baton Rouge Parish, Louisiana
K. A. Barrell

Chapter 14 195
Waterflood Optimization Using GIS Techniques in Citronelle Field, Mobile County, Alabama, U.S.A.
L. E. Safley and M. L. Fowler

Chapter 15 205
Reservoir Characterization of a Stratigraphically and Structurally Complex Gulf of Suez Oil Field Using Three-Dimensional Modeling and GIS Techniques
R. A. Day, K. Talaat, and K. S. Hoffman

SUPPORTING ROLES AND APPLICATIONS 213

Chapter 16 215
Integrated Geo-Systems—A Computational Environment for Integrated Management, Analysis, and Presentation of Petroleum Industry Data
K. A. Khan

Chapter 17 227
Using the World Wide Web to Manage Exploration and Production Team Data
C. Fried

Chapter 18 237
GIS—The Exploration and Exploitation Tool
K. A. Barrell

Chapter 19 249
The USGS GeoExplorer Project: Using New GIS Technology for Scientific Data Publishing
C. E. Barnwell

GIS RESOURCES 261

Chapter 20 263
GIS—Public Domain and Inexpensive Software
B. Gibbs

Chapter 21 281
Bibliographic Resources for GIS in Petroleum Exploration and Development
P. J. Yarka and T. C. Coburn

LOOKING AHEAD 293

Chapter 22 295
GIS and Geostatistics—A Marriage of Technologies
T. C. Coburn and J. M. Yarus

Appendices 299

Index 309

About the Editors

Timothy C. Coburn

Timothy C. Coburn was born in 1951 in Houston, Texas, and graduated from Conroe High School in Conroe, Texas, in 1969. He completed a B.S. degree in mathematics at Abilene Christian University in 1973, with minors in physics and education. While at ACU he was the recipient of an undergraduate research grant from Research Corporation. Coburn earned an M.S. in statistics at Oklahoma State University in 1975, and after a two-year hiatus to work as an applied statistician with the federal government, he returned to OSU and completed a Ph.D. in statistics (with a minor in industrial engineering) in 1980.

Coburn subsequently joined the Applied Statistics Group at Phillips Petroleum Company's research center in Bartlesville, Oklahoma. While at Phillips, he was involved in a variety of projects; but most notably, he was a long-term participant in the Prudhoe Bay Equity Redetermination on behalf of the Mobil-Phillips-Chevron consortium. For much of this time he was assigned to Phillips' Western Division Drilling and Production Office in Denver, Colorado. In 1985, Coburn joined the staff of Marathon Oil Company's Petroleum Technology Center in Littleton, Colorado. After serving two years as Senior Scientist in the Geophysical Technology Department in the Exploration Research Division, he was appointed Supervisor of Quantitative Geosciences, a position that he held until 1994. In this role he championed the implementation of many new quantitative technologies throughout Marathon's worldwide exploration and development activities, including geographic information systems, geostatistical methods, three-dimensional visualization, and related techniques.

Throughout his tenure at Marathon, Coburn was actively involved in many exploration, exploitation, and reservoir characterization projects. In addition, he chaired and served on various corporate committees related to data management, computing, and quantitative technology, managed the in-house training program in these disciplines, served as the corporate liaison to various external research programs, and provided ongoing statistical consultation and support throughout the company in his role as Principal Consulting Statistician. He was subsequently appointed supervisor of mapping and database in the Exploration Services organization at Marathon's Headquarters Office in Houston, Texas, where he was responsible for the development and maintenance of corporate databases, and of computer mapping systems and services. In 1995, Coburn returned to the Denver area where he joined the National Renewable Energy Laboratory to serve as Senior Statistician. In addition to his statistical consulting activities, he served in various roles as Team Leader and Project Manager, with his activities being principally focused on the use of natural gas as a transportation fuel.

In 1997, Coburn returned to Abilene Christian University as Associate Professor and University Statistician in the Department of Mathematics and Computer Science, where he currently teaches courses in applied statistics and mathematics, and engages in statistical consulting with researchers throughout the university. He served as Visiting Research Scientist in the Mathematical

Geology Section of the Kansas Geological Survey during the summer of 1999. Coburn has an extensive publication record focused on applications of statistics, quantitative methods, computational techniques, database issues, and spatial analysis, particularly in the petroleum arena. He is active in the Section on Physical and Engineering Statistics of the American Statistical Association, the International Association of Mathematical Geology (for which he serves on the Editorial Board of Natural Resources Research), and the American Association of Petroleum Geologists (for which he serves on the Publications and Geologic Computing Committees). He has provided numerous technical and peer reviews on behalf of scholarly journals and organizations, and he was a long-time member of the U.S. Energy Information Administration's Energy Statistics Committee.

Jeffrey M. Yarus

Jeffrey M. Yarus is executive vice president, and an owner, of Knowledge Reservoir, Inc., a Houston-based company. Immediately prior to assuming his current responsibilities, he served as vice president of Roxar, Inc. (formerly Smedvig Technologies), a Norwegian-based petroleum services company. Yarus was born in 1951 in Cleveland, Ohio. He began his formal studies in geology at the College of Wooster in Ohio, and earned a B.A. degree with honors, after having spent a year abroad studying geology at the University of Durham, England. Yarus began his graduate studies in geology at Michigan State University and subsequently earned M.Sc. and Ph.D. degrees at the University of South Carolina, studying under Robert Erhlich. It was at Michigan State that Yarus first developed an interest in computer mapping and numerical and statistical analysis of geoscientific data. Yarus joined Amoco Production Company in New Orleans, Louisiana, in 1977 as a production geologist for the Gulf Coast region. In 1980, he left Amoco and moved to Denver, Colorado, where he established Diversified Resources, an independent company involved in oil and gas exploration and production in the Rockies, mid-continent, and Appalachian regions of the United States. In 1988, Yarus joined Marathon Oil Company's Petroleum Technology Center in Littleton, Colorado, where he played a major role in instituting desktop computer mapping and geostatistical technology, and in providing training and consulting in these areas for the entire company. He subsequently served as manager of Technical Services and Support for Geographix, Inc., in Denver, Colorado, and as manager of Reservoir Characterization for Geomath, Inc. (a division of Beicip–Franlab), in Houston, Texas. In both these settings, he worked closely with the Center of Geostatistics in Fontainebleau, France, and the Norwegian Computing Center in Trondheim, Norway, assisting in the practical implementation of geostatistical techniques. Yarus is well known throughout the petroleum industry for his teaching and consulting experience. He has worked on reservoirs in many different areas of the world, and has provided technical services and training for numerous domestic and international companies. He has been an adjunct professor at the University of Colorado in Denver, and a visiting professor at the Colorado School of Mines. Yarus is actively involved in AAPG affairs and has published many technical papers and abstracts. He is co-author of the 1994 text published by AAPG entitled *Stochastic Modeling and Geostatistics: Principles, Methods, and Case Studies*.

Introduction

GIS is a computer-based technology for managing, visualizing, and analyzing spatially oriented data. It is a technology having a particular focus and unique capabilities; and it is one that is becoming increasingly important to the success of many scientific and commercial endeavors. With myriad current and prospective applications, GIS occupies a particularly strategic position within the ever-evolving complex of information technology. It lies at the heart of much innovation pertaining to the enterprise of data delivery that is so significant to today's global economy.

The growing significance of GIS is nowhere more evident than in the upstream sector of the petroleum industry. Since the late 1980s, many exploration and development organizations within petroleum companies of all kinds—major, independent, domestic, international—have considered adopting GIS in one form or another. While some have vigorously pursued the technology from its earliest beginnings, anticipating its potential, others have taken a more reserved approach to its implementation. Irrespective of the pathway, GIS is gaining prominence as an effective means for handling the multifaceted geoscience data that are so critical to the discovery and development of oil and gas resources.

The purpose of this book, then, is to affirm the viability of GIS as an exploration and development tool and to substantiate its importance to the petroleum industry. To this end, the editors have compiled a broad and varied collection of papers that retrospectively assess the development of GIS from the geoscience viewpoint, detail the variety of situations in which it has been applied, and evaluate its present and future utility. This extensive array of material encompasses fundamental notions, novel implementation strategies, unique applications, and futuristic predictions pertaining to the efficacy of GIS in petroleum geoscience that readers of all persuasions should find enlightening.

In the introductory chapter—**Geographic Information Systems in Petroleum Exploration and Development: An Introductory Essay**—the editors initiate this lively discourse with their own perspectives on the technology. This essay recounts a bit of the evolution of GIS in oil and gas exploration, and it addresses the very essence of this technology—its character and composition—that often eludes definition. More importantly, it sets the stage for the broader discussion of techniques, applications, resources, and practical considerations that follows— a discussion aimed at facilitating a more thorough understanding of the impact of GIS throughout the petroleum exploration and development arena.

Yarus, J. M., and T. C. Coburn, 2000, Geographic information systems in petroleum exploration and development: an introductory essay: *in* T. C. Coburn and J. M. Yarus, eds., Geographic information systems in petroleum exploration and development: AAPG Computer Applications in Geology, No. 4, p. 3–6.

Chapter 1

Geographic Information Systems in Petroleum Exploration and Development: An Introductory Essay

J. M. Yarus
Roxar, Inc.
Houston, Texas, U.S.A.

T. C. Coburn
Department of Mathematics and Computer Science, Abilene Christian University
Abilene, Texas, U.S.A.
and
Mathematical Geology Section, Kansas Geological Survey
Lawrence, Kansas, U.S.A.

ABSTRACT

In this opening chapter, the reader is introduced to the contents and organization of the book and to some of the important concepts that characterize the technology of geographic information systems (GIS), particularly as it relates to petroleum exploration and development. Specifically, the spatial functionality of GIS is discussed, along with the vitality of its role in integrated project organization and administration. The authors present some philosophical considerations concerning the practice of GIS and point to its evolving applicability in the upstream sector of the oil and gas industry. A short historical perspective is provided, along with additional commentary highlighting the past, present, and prospective status of GIS as a mainstream geoscience technology.

Perhaps as long as geoscientists have been geoscientists, and explorationists have been explorationists, they have yearned for an ability to assimilate information from various sources and of various vintages in a way that would allow the profusion of data to more definitively inform their decision making. If that information could be combined in some way in a single all-encompassing map, then all the better, because geoscientists are the quintessential map makers. The quest for this ability to combine, assimilate, integrate, and distill information has led geoscientists down many pathways. Some of those pursuits have been absolute dead ends—leaving folks scratching their foreheads in bewilderment—and some have been virtual roundabouts catapulting people off in quasi-technical directions in which they really didn't intend to go.

Fortunately for the petroleum industry, successful technological advances in seemingly unrelated disciplines have often proved equally constructive in geoscientific endeavors. That these technologies have proved successful in other settings has helped provide direction through the sometimes unrelenting maze of new ideas, tools, and speculative methodologies. It has also helped to stimulate thinking about new and exciting ways to address, and perhaps resolve, some age-old problems. The technology of geographic information systems (GIS) is one of these kinds of advances.

This text is about GIS technology and its contemporary applications in the geosciences, particularly in petroleum exploration and development. As a formal method of scientific investigation, GIS is perhaps better known and more often practiced in disciplines such as community planning, environmental remediation, demography, and natural resource management; however, as in the case of many other technical developments, the recent, rapid expansion in computational capabilities has pushed GIS to the forefront of almost every avenue of science, engineering, business, and government (Rathfeeder, 1989). The penetration of GIS into non-traditional arenas is nowhere more evident than in the petroleum industry (Alderman et al., 1994), its success in many other venues having been extensively recorded in both the popular and scientific literature (*Computerworld*, 1996; *Electronic Business*, 1990; Exter, 1992; Felleman, 1990; Goodchild and Palladino, 1995; Grupe, 1990; Wilkie, 1991).

This rush to seize upon the GIS opportunity has been fueled by a desire to address real-life situations from a spatial perspective because almost every aspect of the world is geographic in nature. Prior to the advent of GIS, there was no easy, efficient means to incorporate spatial notions into planning, problem-solving, experimentation, and commercial endeavors. With GIS, spatial thinking is easily accommodated and the benefits of spatial transactions are readily realized. Because it involves some of the most spatially oriented disciplines and situations in all of science and engineering, the petroleum industry has perhaps provided one of the most natural environments in which GIS could evolve and flourish.

GIS is certainly not new to the oil and gas business, particularly not to exploration and development, although it has admittedly been given more emphasis in the past few years. Perhaps the first oil company to invest in a GIS for exploration and development purposes was Standard Oil of Ohio (Sohio) (now a part of British Petroleum), when it began work in the early 1980s on a system to facilitate its activities in Prudhoe Bay field. Other major oil companies, like Tenneco (Calkins, 1990), followed suit and invested heavily in GIS technology in the mid to late 1980s; however, not all of these systems were necessarily oriented toward upstream activities. The Paleomap Project of Malcom Ross and Christopher Scotese was one of the earliest uses of GIS technology to manage and display exploration data (Ross and Scotese (1997) contains one of the latest discussions of the Paleomap Project).

Despite these efforts, GIS didn't really catch on in the upstream segment of the petroleum industry until the early 1990s. In fact, the first reference to a petroleum geology application of GIS in the *Bibliography and Index of Geology*, published by the American Geological Institute (AGI), was in 1992, approximately 10 years later than when the professional organizations of other disciplines began to reference the technology (there are also earlier references to the use of GIS in other geoscience applications in AGI publications, such as environmental geology, hydrology and groundwater assessment, and evaluation of geologic hazards).

In the past few years there has been considerably more activity, largely centering on the incorporation of GIS concepts into major software packages by service companies and data vendors. Many of these packages provide GIS-related links to other exploration-oriented capabilities, such as 3D seismic interpretation, geostatistical analysis, and 3D geocellular modeling. In 1994, the Geological Society of America (GSA) devoted a major segment of its annual conference to applications of GIS, although many of the presentations did not specifically target oil and gas investigations. Since 1997 or so, several annual conferences have been devoted solely to petroleum applications of GIS, including "GIS for the Oil & Gas Industry Conference and Exhibition," "International Conference on Petroleum Data Integration and Management," and "International Conference on Exploration and Production Data Management." Unfortunately, none of these events have primarily targeted upstream activities either. Finally, Environmental Systems Research Institute (ESRI), the vendor of Arc/Info and related GIS software, maintains a petroleum department and conducts an annual Petroleum Users Group meeting, but again, exploration and development are not necessarily the main focus.

From a review of petroleum industry activities, it is clear that GIS software and capabilities are readily available, and that they are being fairly widely used. Yet, use of the technology is not outwardly visible, and there is a pervasive perception that the technology is somehow immature or inappropriate for exploration and development purposes. For whatever reason, GIS is certainly not afforded the kind of status that seismic interpretation or well log analysis enjoys, yet it has the same economic potential that these more well-known technologies do. In fact, it is hard to find a formal GIS group within any of the major oil companies today—particularly in the exploration or exploitation divisions—or to even identify small corners within those companies in which it is actively, though perhaps informally, used. This is not the case, however, in many state geological surveys (e.g., West Virginia Geological Survey, Illinois Geological Survey, Kansas Geological Survey), where the usage and visibility of GIS is much greater. The U.S. Geological Survey has also been an active promoter of GIS technology.

Consequently, the principal objective of this book is to raise the awareness of GIS within the petroleum industry, principally in exploration and development, and to increase its visibility among managers and professionals alike. The intent is to give life and credence to this technology, which is already permeating the industry and making substantial contributions to corporate bottom lines. A second objective is to effectively underscore the technological and economic benefits of integration—integration of data, integration of staff (in teams), integration of disciplines, integration of geoscientific interpretations—that GIS is uniquely capable of facilitating. Integration is the hallmark of much of the petroleum industry's present emphasis on improved efficiency, but without technology to accomplish such integration, it may be no more than just another passing fad.

Part of the reason GIS is not more visible, at least outwardly, in exploration and development organizations may be a definitional problem. Lots of industry folks are really not sure exactly what GIS is or does, and so, to avoid any confusion, and perhaps some personal and organizational embarrassment, they have not promoted it either in name or in function. There are lots of other reasons, as well. Some individuals simply may not see or understand why GIS is different from ordinary computer mapping; they may think it is too complex or intricate for their organizations to become involved with; they may consider it too computer-intensive and to really belong in the computing department; they may feel that it is a tool to be used more in the social sciences and that it really does not have a place in geoscience; or they may just be taking a wait-and-see attitude. Additional discussion along these lines is provided in subsequent chapters of this book (in particular, see the essay by Coburn that presents a management perspective on GIS implementation).

Returning to the problem of definitions, it would be easy to say that GIS technology is simply a way to make a better map. Although not many folks could argue with such a claim, it would be a severe restriction to impose on such a robust and rapidly developing technology. GIS has evolved to the point that, in addition to its spatial data handling and analysis capabilities, it has come to be associated with a broader, more powerful range of data-oriented functionality. In fact, it is the capabilities of GIS to integrate, manage, and visualize diverse data that are so important to the contemporary success of entities like the petroleum industry. The papers by Berry, Turner, and Bonham-Carter in this volume all address various aspects of these evolving capabilities.

Asking "What exactly is GIS?" can initiate a provocative discussion among practitioners who perceive themselves as purists. In the petroleum context, is a GIS simply a type of graphical database that provides information to other disciplines? Or, is it a separate discipline, like reservoir characterization or petrophysics? Is it a computer program that produces improved displays by overlaying maps depicting different attributes? To be a *true* GIS, must it have the ability to launch sophisticated queries on and between layers of information and to report the results in a graphical context? Does the use of the term "GIS" imply a certain data set size, type of computer system, or even a specific application or set of applications? Is buffering or some other spatial analysis function required? These questions demand responses, but simple answers are not always forthcoming.

Most objective investigations of GIS systems and applications discover that a wide range of definitions and specifications are in use. In the petroleum industry, there is actually considerable tolerance as to what a GIS comprises, despite the fact that many products clearly do not conform to the most basic of specifications. To attract the widest audience possible, there has consequently been no attempt in this book to limit the meaning of GIS to a specific set of criteria. Indeed, map-making and data integration may be the only common threads. The authors of the chapters use the acronym "GIS" to mean a variety of different things, ranging from conventional computer mapping software to complex data management and spatial analysis systems, and there has likewise been no effort to standardize the terminology or to classify products as to their degree of GIS-likeness.

The goal is simply to bring the technology to the forefront and to encourage its proliferation; hence, this book specifically emphasizes the mapping and visualization aspects of GIS because those are the characteristics with which geologists, geophysicists, petrophysicists, and engineers are most familiar and in which they are most interested. Further, the material presented here is not so much concerned about traditional database management (roles, mechanisms, etc.), even though administration of data is certainly a part of every GIS project. Finally, as has been suggested, the text heavily focuses on the integration aspects of GIS in the context of performing petroleum exploration and development activities. The ideas of integration are woven throughout almost every chapter in this volume (in particular, note the chapter by Barrell entitled "GIS: The Exploration and Exploitation Tool").

There are many topics that the book does not cover. Specifically, it does not address the relationship of GIS and remote sensing, even though remote sensing is often used at the front end of rank exploration projects (particularly in the international arena). Some authors have nonetheless noted that GIS can be effectively used to merge satellite images with other kinds of geological and geophysical data. Because remote sensing is more of a science in its own right, the decision was made to leave a formal discussion of its relationship to GIS to another time and place. Readers are urged, however, to pursue the many excellent resource materials that are available on the topic (e.g., Legg, 1994; Williams, 1995). Other subjects, such as the use of GIS in petroleum environmental assessment or pipeline/facilities siting, are also not specifically addressed, although some authors have mentioned these applications as well. The focus is truly on exploration and development activities.

From a presentation standpoint, the intent of this book is to address the interests and concerns of a broad spectrum of readers and enthusiasts from the petroleum industry and the geoscience disciplines; however, as noted, it is clearly not possible to present every aspect of the technology in a single volume, and hence some prioritization on coverage has been imposed. The text is organized into seven sections, of which the first one, *"Introduction,"* contains the present discussion. The second section of the book, *"Technological Foundations,"* is a collection of chapters that more formally addresses the definitions of GIS, discusses the historical development of the technology relative to the geosciences, and postulates about future innovations and applications. *"Practical Realities,"* the third section, focuses on what is really involved in implementing the technology in a petroleum exploration and development environment and what it takes to initiate, execute, and follow through on a GIS project. A view of GIS from the management perspective is also contained in this section. The heart of the text comes in the fourth section, *"Case Studies."* These chapters relate the factual experiences of individuals who have applied GIS to real problems in exploration, exploitation, resource assessment, reservoir characterization, and other upstream aspects of the petroleum industry. The collection of case studies spans a wide range of geographic regions, both domestically and internationally, and includes work from both major oil companies and independent operators. *"Supporting Roles and Applications,"* the fifth section, describes uses of GIS that might at first seem somewhat peripheral to rank exploration, yet that are extremely important and relevant in the overall scheme of things. Aspects of computing, publishing, data management, and team-oriented project administration are all covered in these chapters. The sixth section, *"Looking Ahead,"* consists of a single chapter that provides a futuristic assessment of how GIS technology might become more fully integrated with other technologies such as geostatistics. The final section, *"GIS Resources,"* is provided for individuals who want to get started using GIS or who want to learn more about the technology. It specifically includes chapters that address other reading materials and the availability of software. A synopsis of the contents of each section will no doubt be important to the reader; and so additional details about each of the chapters and about the topics included in each segment are presented in summary remarks, which are provided at the beginning of each section.

GIS is an exciting, rapidly evolving technology, one that has the potential to revolutionize the way petroleum exploration and development is accomplished. Several authors in this volume point to current developments in GIS functionality, as well as additional ideas that are on the horizon, that may substantially impact the way explorationists and their engineering partners go about finding, characterizing, and developing oil and gas reserves (see, for example, the chapters by Turner, Wade, and Coburn and Yarus). We hope that the contents of this book will challenge and motivate the readers to contemplate the use of GIS in contexts and situations they have not previously considered, and to more fully embrace its concepts and strengths to bring about improved understanding of subsurface phenomena. The book will have been successful if it results in a larger number of GIS projects, a more thorough understanding and appreciation of GIS functionality, and more widespread dissemination of the technology throughout the industry.

REFERENCES CITED

Alderman, W., C. Iglehart, and M. Hollenbeck, 1994, The petroleum and computer industries share an amazing history of growth. Earth Observation Magazine, vol. 3, no. 5, pp. 42-44.

Calkins, J. F., 1990, Tenneco's GIS: a rough start. Geobyte, vol. 5, no. 3, pp. 31-35.

Computerworld, 1996, Mapping it out: customers are using geographic information systems to improve customer service, 1996. Computerworld, vol. 30, no. 24, p. 71.

Electronic Business, 1990, You don't need a road map to find the growing GIS market. Electronic Business, vol. 16, no. 9, p. 109.

Exter, T. G., 1992, The next step is called GIS. American Geographer (American Demographics, Desk Reference Series, No. 2), pp. 2-5.

Felleman, J., 1990, There's a GIS in your future. Government Information Quarterly, vol. 7, no. 3, p. 261-267.

Goodchild, M. F., and S. D. Palladino, 1995, Geographic information systems as a tool in science and technology education. Speculations in Science and Technology, vol. 18, p. 278.

Grupe, F. H., 1990, Geographic information systems: an emerging component of decision support. The Journal of Information Systems Management, vol. 7, p. 74-78.

Legg, C., 1994, Remote Sensing and Geographic Information Systems: Geological Mapping, Mineral Exploration and Mining. New York: John Wiley and Sons, 172 p.

Rathfeeder, J., 1989, These maps can find oil—or sell burgers. Business Week, March 13, 1989, p. 134.

Ross, M. I., and C. R. Scotese, 1997, Integrating ArcView and the PaleoGIS into an earth system history visualization tool. Proceedings, Annual ESRI Petroleum Users Group Meeting. Redlands, CA: Environmental Systems Research Institute.

Wilkie, T., 1991, Geographic information systems: digital technology with a future. Computers & Society, vol. 21, pp. 28-32.

Williams, J., 1995, Geographic information from space: Processing and Applications of Geocoded Satellite Images. New York: John Wiley & Sons, 222 p.

Technological Foundations

The first section of this book is devoted to the technological foundations of geographic information systems (GIS)—particularly from the standpoint of geoscientific investigation. It is a section that defines the technology by examining its earliest beginnings, establishing its current state of application, and surveying its future growth and development.

Readers will discover among the pages of these chapters the collective experience of some true GIS pioneers—individuals who have devoted their careers to the evolution of GIS and related technologies, and who have nurtured its expansion in the geoscientific arena. The section is rich in wisdom, reflection, and futurism.

The initial chapter comes from Joe Berry, who was part of GIS from the very start, and who has witnessed the technology unfold and proliferate to almost every aspect of scientific endeavor. His paper, **Geographic Information Systems (GIS) Technology: A Brief History, Trends, and Probable Future**, provides an initial perspective from the broader world of GIS juxtaposed against geoscientific investigation. In **An Overview of GIS in the Geosciences**, Graeme Bonham-Carter details the capabilities and functions of GIS, particularly those of a spatial-analytical nature, and describes how they can be most effectively used in geoscience investigations. From his vantage point at the Geological Survey of Canada, Bonham-Carter has actively promoted geoscientific application of GIS for many years through extensive writing, practice, and collaboration. Keith Turner has devoted an entire career at the Colorado School of Mines to the management and analysis of spatially-oriented geoscientific data, and has been engaged in innovation and application of GIS and related technologies throughout the world. In **Geoscientific Modeling: Past, Present, and Future**, Turner provides a uniquely geological perspective and discusses how petroleum exploration and development of the future might be accomplished. Geoff Wade, as Manager of Petroleum GIS at Environmental Systems Research Institute (ESRI), has observed much of the behind-the-scenes activity as GIS technology has evolved, and he has witnessed both the successes and failures of oil and gas companies as they have striven to implement this technology. In his paper entitled **GIS Technology—A Vendor Perspective of its History and Prospects in Petroleum Exploration and Development**, Wade presents an altogether different outlook on past experiences with GIS and the path that its future capabilities and technical development are likely to take.

Berry, J. K., 2000, Geographic information systems, (GIS) technology: A brief history, trends, and probable future, *in* T. C. Coburn and J. M. Yarus, eds., Geographic information systems in petroleum exploration and development: AAPG Computer Applications in Geology, No. 4, p. 9–16.

Chapter 2

Geographic Information Systems (GIS) Technology: A Brief History, Trends, and Probable Future

Joseph K. Berry
Berry and Associates/Spatial Information Systems, Inc.
Ft. Collins, Colorado, U.S.A.

ABSTRACT

The technology of geographic information systems (GIS) has rapidly evolved from its early beginnings, moving from the early stages of computer mapping to the more elaborate and informative notions of spatial database management, spatial modeling, and spatial dialog and reasoning. Having penetrated many disciplines, the influence of this technology is poised for further expansion as the concepts of data mining, predictive modeling, and dynamic simulation are incorporated. The geoscience arena—particularly the upstream operations of the petroleum industry—provides many opportunities for successful deployment of these new GIS ideas and capabilities.

INTRODUCTION

Like other resource and environmental disciplines, geologic exploration is inherently spatial. These data are particularly complex because they require two descriptors, namely, the precise location of what is being described and a clear description of its physical characteristics. For many years, explorers produced manually drafted maps that served to link the "*where* is *what*" descriptors. With an emphasis on accurate location of physical features, early maps helped explorers chart unexplored territory.

Today, these early attributes of maps have evolved from exploratory guides to physical space into scientific and management tools for exploring spatial relationships. This new perspective marks a turning point in the use of maps, from one emphasizing physical descriptions of geographic space, to one of interpreting mapped data and communicating spatially based decision factors. This movement from "*where* is *what*" to "*so what* and *why*" has changed the purpose for which maps are used.

Whereas most of the early applications in the oil and gas industry sprang from statistical analysis, early geographic information systems (GIS) technology focused on management, query, and display of mapped data. As GIS technology matures, a growing interest in the statistical analysis of these data attracts GIS users to geostatistical procedures (Berry, 1993, 1995b). As the oil and gas applications mature, geostatistics users borrow more from the spatial database management heritage and spatial analysis functions of generic GIS. The growing similarities among current applications are blurring the distinction between the two, once dissimilar, approaches. An understanding of the evolutionary stages of GIS technology, its current expression, and probable trends is essential for all professionals in the earth sciences.

EVOLUTIONARY STAGES

Since the 1960s, the decision-making process has become increasingly quantitative, and mathematical models have become commonplace. Prior to the computerized map, most spatial analyses were severely limited by their manual processing procedures. GIS technology provides the means for both efficient handling of voluminous data and effective spatial analysis capabilities (Carter, 1989; Coppock and Rhind, 1991). From this perspective, GIS is rooted in the digital nature of the computerized map.

Computer Mapping

The early 1970s saw computer mapping automate the map drafting process (McHarg, 1969; Steinitz et al., 1976; Berry and Ripple, 1994). The points, lines, and areas defining geographic features on a map are represented as an organized set of *X-Y* coordinates. These data drive pen plotters that can rapidly redraw the connections at a variety of colors, scales, and projections. The map image itself is the focus of this automated cartography.

The pioneering work during this period established many of the underlying concepts and procedures of modern GIS technology (Muehrcke and Muehrcke, 1980; Robertson et al., 1982). An obvious advantage of computer mapping is the ability to change a portion of a map and quickly redraft the entire area. Updates to resource maps, such as those representing regional faulting, which previously could take several days, can be done in a few hours. The less obvious advantage is the radical change in the format of mapped data, from analog inked lines on paper to digital values stored on disk.

Spatial Database Management

During the early 1980s, the changes in format and computer environment of mapped data were used. Spatial database management systems (SDBMS) were developed that linked computer mapping capabilities with traditional database management capabilities (Burrough, 1987; Sheppard, 1991). In these systems, identification numbers are assigned to each geographic feature, such as a reservoir or field. For example, a user is able to point to any location on a map and instantly retrieve information about that location. Alternatively, a user can specify a set of conditions, such as a specific vegetation and soil combination, and all locations meeting the criteria of the geographic search are displayed as a map.

During the early development of GIS, two alternative data structures for encoding maps were debated (Piwowar, 1990; Pueker and Christman, 1990). The *vector* data model closely mimics the manual drafting process by representing map features as a set of lines that, in turn, are stored as a series of *X-Y* coordinates. An alternative structure, termed *"raster,"* establishes an imaginary reference grid over a project area, then stores resource information for each cell in the grid. Early debates in the GIS community attempted to determine the universally best data structure. The relative advantages and disadvantages of both were viewed in a competitive manner that failed to recognize the overall strengths of a GIS approach encompassing both formats.

By the mid-1980s, the general consensus within the GIS community was that the nature of the data and the desired processing determine the appropriate data structure. The realization that dual data structures were needed had a significant impact on geographic information systems. From one perspective, maps form sharp boundaries that are best represented as lines. Property ownership, seismic lines, and road networks are examples where the lines are real and the data are certain. Other maps, such as soil, groundwater flow, and slope maps, are abstract characterizations of terrain conditions. The placement of lines identifying these conditions are subject to judgment, statistical analysis of field data, and broad classification of continuous spatial distributions. From this perspective, the sharp boundary implied by a line is artificial and the data itself is based on expert opinion or probabilistic estimates.

This era of rapidly increasing demand for mapped data focused attention on data availability, accuracy, and standards, as well as data structure issues. Hardware vendors continued to improve digitizing equipment, with manual digitizing tablets giving way to automated scanners at many GIS facilities. A new industry for map encoding and database design emerged, and a marketplace for the sales of digital map products emerged. Regional, national, and international organizations began addressing the necessary standards for digital maps to ensure compatibility among systems. This period saw GIS database development move from being expensed as individual project costs to a corporate investment in a comprehensive information resource.

THE CURRENT FRONTIER

GIS Modeling

As the technology continued its evolution, the emphasis turned from descriptive "geo-query" searches of existing databases to prescriptive analysis of mapped data. For the most part, the earlier eras of GIS concentrated on automating traditional mapping practices. If a user had to repeatedly overlay several maps on a light table, an analogous procedure was developed within the GIS. Similarly, if repeated distance and bearing calculations were needed, systems were programmed with a mathematical solution. The result of this effort was GIS functionality that mimicked the manual procedures in a user's daily activities. The value of these systems was the savings gained by automating tedious and repetitive operations.

By the mid-1980s, the bulk of the geo-query operations was available in most GIS systems, and a comprehensive theory of spatial analysis began to emerge. The dominant feature of this theory is that spatial information is represented numerically, rather than in analog fashion as inked lines on a map. These digital maps are frequently conceptualized as a set of "floating maps" with a common registration, allowing the computer to "look" down and across the stack of digital maps. The spatial relationships of the data can be summarized (database geo-queries) or mathematically manipulated (analytic processing). Because of the analog nature of traditional map sheets, manual analytic techniques are limited in their quantitative processing. Digital representation, however, makes a wealth of quantitative (as well as qualitative) processing possible. The application of this new modeling theory to the management of earth-related data is revolutionary. Its application takes two forms—spatial statistics and spatial analysis.

Spatial statistics (a more general term for geostatistics encompassing both discrete and continuous data) has been used by scientists in various disciplines for many years in characterizing the geographic distribution, or spatial pattern, of data (Ripley, 1981; Myers, 1988; Cressie, 1991, 1993; Cressie and Ver Hoef, 1993). The statistics describe the spatial variation in the data, rather than assuming a typical response is everywhere. For example, water saturation measurements can be made at several wells within an oil field. Traditionally, these data are analyzed for a single value (the average saturation) to characterize the field. Spatial statistics, on the other hand, uses both well locations and the recorded measurements to generate a map of relative water saturation throughout the entire field.

More recently, spatial statistics has evolved from descriptive to predictive to optimization models. To use an agricultural analogy, for example, GIS modeling can be used to investigate the spatial relationships between crop yield and soil nutrients (Berry, 1996). The global positioning system (GPS) continuously locates a harvester in an agricultural field (Leick, 1990) and, for each second, an onboard data card stores the geographic position and yield/moisture of the grain flow. The result is a yield map composed of tens of thousands of sample points throughout a field. Soil samples are analyzed for nutrient levels, such as phosphorous and potassium, then spatially interpolated into maps tracking the spatial patterns of the variations (Burgess and Webster, 1980; Webster and Burgess, 1980; Lam, 1983). Predictive techniques from simple regression to knowledge-based modeling are used to relate the dependent (yield) and independent (nutrients) mapped variables. The derived relationship can be used to determine the optimal fertilizer rates throughout the field.

Additional ideas based on the agricultural analogy are informative. Traditional "whole-field" management, for example, involves a similar analysis to that described, except that field averages are used to derive a single application rate for the entire area. The result is that most parts of highly variable fields receive either too much or too little fertilizer. Precision farming, in contrast, uses site-specific management involving a "prescription map" derived by spatial statistics. Prescription maps can be used in combination with GPS units mounted on spray rigs to dynamically adjust the blend of nutrients applied to fields.

Many other applications, from retail market forecasting to facility siting, are using spatial statistics to relate mapped variables, so the extension into the geoscience arena (and the petroleum industry in general) is a natural one. The environmental sciences, for example, have a rich heritage in the quantitative expression of their systems. Spatial statistics provides a new set of tools for explaining spatially induced variance, that is, variations in geographic space rather than numeric space. From this perspective, the maps in a GIS represent the spatial distributions of mapped variables. In traditional mathematical terms, each map is a variable, each location is a case, and each map value is a measurement. The GIS provides a consistent spatial registration of the numbers. The full impact of this "map-ematical" treatment of maps is yet to be determined. The application of such concepts as spatial correlation, statistical filters, map uncertainty, and error propagation await their translation from other disciplines.

Spatial analysis, however, has a rapidly growing number of current applications (Maguire et al., 1991; Goodchild et al., 1993; Ripple, 1994). For example, a forest manager can characterize timber supply by considering the relative skidding and log-hauling accessibility of harvesting parcels. Wildlife managers can consider such factors as proximity to roads and relative housing density to map human activity and incorporate this information into habitat delineation. Landscape planners can generate visual exposure maps for alternative sites for a proposed facility to sensitive viewing locations, such as recreational areas and scenic overlooks. Soil scientists can identify areas with high sediment loading potential based on proximity to streams and intervening terrain slope, vegetative cover, and soil type. Groundwater and atmospheric scientists can simulate the complex movement of a release as it responds to environmental factors affecting its flow through geographic space. Similarly, petroleum reservoir managers can locate untapped drainage areas in oil fields by considering the spatial relationships among petrophysical properties.

Just as spatial statistics has evolved by extending concepts of conventional statistics, a mathematics supporting spatial analysis has been developed (Unwin, 1981; Berry, 1987a; Goodchild, 1987). This "map algebra" uses sequential processing of spatial operators to perform complex analyses (Berry, 1987b; Tomlin, 1990). It is similar to traditional algebra in that primitive operations (e.g., add, subtract, exponentiate) are logically sequenced on variables to form equations; however, in map algebra, entire maps composed of thousands or millions of numbers represent the variables of the spatial equation.

The change in lead concentration in an aquifer, for example, can be estimated in a simple and straightforward manner by evaluating the algebraic expression

$$\%change = [(new\ value - old\ value)/old\ value] * 100$$

using the average values for two time periods. Taking the map algebra approach, however, involves replacing the simple averages with spatially interpolated maps based on the same sets of field data used to calculate the averages. The *%change* equation is evaluated at each map location, resulting in a map of percent change.

Areas of unusual change can then be identified by evaluating the standard normal variable (SNV) expression

SNV = [(new_%change_average)/ %change_standard_deviation]* 100

This normalizes the map of changes in lead concentration, with areas of statistically unusual increase having SNV values over 100. These potentially hazardous areas can be overlaid on demographic maps to determine the level of environmental risk.

Most of the traditional mathematical capabilities, plus an extensive set of advanced map processing operations, are available in modern GIS software. Users can add, subtract, multiply, divide, exponentiate, root, log, cosine, differentiate, and even integrate maps. After all, maps in a GIS are just organized sets of numbers; however, with this "map-ematics," the spatial coincidence and juxtapositioning of values among and within maps create new operations, such as effective distance, optimal path routing, visual exposure density, and landscape diversity, shape, and pattern.

As an illustration, consider that distance is traditionally defined as the shortest straight line between two points. Both a ruler (analog tool) and the Pythagorean theorem (mathematical tool) adhere to this strict definition; however, the simple definition of distance is rarely sufficient for most applications, including applications in the geosciences. Often " . . . between two points" must be expanded to ". . . among a set of points" to account for proximity, such as buffers around drilling locations; ". . . straight line" has to be expanded to ". . . not necessarily straight lines" because nothing in the real world moves in a straight line (even light bends in the atmosphere). In a GIS, the concept of movement replaces distance by introducing the location of absolute and relative barriers into the calculations (Muller, 1982; Elridge and Jones, 1991). An effective oil field development buffer, for example, reaches out around a producing well capturing an appropriate amount of the subsurface matrix, instead of simply extending out a fixed distance regardless of rock quality.

Another example of advanced spatial analysis tools involves landscape analysis. The ability to quantify landscape structure (topography, or relief, in the geologic setting) is a prerequisite to the study of landscape function and change. For this reason, considerable emphasis has been placed on the development of landscape metrics (Turner, 1990; McGarigal and Marks, 1995). Many of these relationships are derived through analysis of the shape, pattern, and arrangement of landscape elements spatially depicted as patches (individual polygons), classes of related patches (polygons of the same type/condition), and entire landscape mosaics (all polygons). A convexity index compares each patch's perimeter to its area, with an increase in perimeter per unit area indicating more irregularly shaped units (geobodies). The mean proximity index indicates the average distance between the patches within a class as a measure of the relative dispersion. One aspect of the fractal dimension of a landscape assesses the proportion of edge versus interior of all patches, summarizing whether the mosaic is primarily composed of simple shapes (circular, elongated, or square-like) or complex shapes with convoluted, plane-filling perimeters. These, plus myriad other landscape indices, can be used to track the fragmentation induced by exploration drilling and to relate the changes to impacts on oil field development.

This GIS modeling toolbox is rapidly expanding. A detailed discussion of all of the statistical and analysis tools is beyond the scope of this presentation. It suffices to note that GIS technology is not simply automating traditional exploration approaches, but radically changing geoscience. It is not just a faster mapper, nor merely an easier entry to traditional databases. Its new tools and modeling approach to geoscience information combine to extend record-keeping systems and decision-making models into effective decision support systems (Parent and Church, 1989; Densham, 1991; Pereira and Duckstein, 1993).

Spatial Reasoning and Dialog

GIS is at a threshold that is pushing beyond mapping, management, and modeling to spatial reasoning and dialogue. In the past, analysis models have focused on management options that are technically optimal—the scientific solution. Yet in reality, another set of perspectives must be considered—the social solution. This final sieve of management alternatives most often confounds resource decision-making. It uses elusive measures, such as human values, attitudes, beliefs, judgment, trust, and understanding. These are not the usual quantitative measures amenable to computer algorithms and traditional decision-making models.

The step from technically feasible to socially acceptable options is not so much an increase in scientific and econometric modeling, as it is communication (Calkins, 1991; Epstein, 1991; King and Kraemer, 1993; Medyckyj-Scott and Hernshaw, 1993). Basic to effective communication is involvement of interested parties throughout the decision-making process. This new participatory environment has two main elements: consensus building and conflict resolution. Consensus building involves technically driven communication and occurs during the alternative formulation phase. It involves the resource specialist's translation of the various considerations identified by a decision team into a spatial model. Once completed, the model is executed under a wide variety of conditions and the differences in outcome are noted.

From this perspective, a single map rendering of an oil field's geology is not the objective. It is how the field development plan changes as different scenarios are considered that becomes information for decision-making. "What if avoidance of visual exposure is more important than optimizing production in siting a new well?" "How and why does the proposed site change, if at all?" Answers to such analytic queries focus attention on the effects of differing perspectives. Often, seemingly divergent philosophical views result in only slightly different map views. This realization, coupled with active involvement in the decision-making process, often leads to group consensus.

If consensus is not obtained, conflict resolution is necessary. Such socially driven communication occurs during the decision formulation phase. It involves the creation of a "conflicts map," which compares the outcomes from two or more competing uses. Each area being considered is assigned a numeric code describing the conflict over the location. An area might be identified as ideal for both a production facility and a wildlife preserve. Because these alternatives are mutually exclusive, a single use must be assigned. The assignment, however, involves a holistic perspective that simultaneously considers the assignments of all other locations in a project area.

Traditional scientific approaches are rarely effective in addressing conflict resolution. Most are deterministic models involving a succession, or cascade, of individual area assignments. The final result is strongly biased by the ordering of area consideration, mathematical assumptions, and the assignment of discrete model parameters. Even if a scientific solution is reached, it is viewed with suspicion by the layperson. Modern resource information systems provide an alternative approach involving human rationalization and trade-offs. This process involves statements such as, "If you let me drill in this area, I will let you set aside that one as a wildlife preserve." The statement is followed by a persuasive argument and group discussion. The dialogue is far from mathematical optimization, but it is often closer to an effective decision. It uses the information system to focus discussion away from broad philosophical positions to a specific project area and its unique distribution of conditions and potential uses.

TRENDS, DIRECTIONS, AND CHALLENGES

Clearly, what began in the 1960s as a cartographer's tool has quickly evolved into a revolution in many disciplines (Thomas and Huggett, 1980; Goodchild et al., 1992; Berry, 1994; Maguire and Dangermond, 1994; Ottens, 1994; Rix, 1994). As general users become more directly engaged, the nature of GIS applications changes. Early applications emphasized mapping and spatial database management. Increasingly, new applications are emphasizing modeling of the interrelationships among mapped variables. Most of these applications involve cartographic modeling, which employs GIS operations that mimic manual map processing, such as map overlay and simple buffering around features. The latest wave of applications concentrates on GIS modeling, which employs spatial statistics and advanced analytical operations. They can be grouped into three categories: (1) data mining, (2) predictive modeling, and (3) dynamic simulation.

Technological Advances

Data mining uses the GIS to discover relationships among mapped variables. For example, a map of gas production can be statistically compared to maps of driving variables, such as porosity, permeability, and rock type. If a strong spatial correlation (coincidence) is identified for a certain combination of driving variables, this information can be used to direct management action to areas of declining production. Another form of data mining is the derivation of empirical models. For example, the geographic distribution of hydrogen sulfide in a basin can be interpolated from production samples taken at local wells. Areas of unusually high concentrations (more than one standard deviation above the average) can be isolated. If a time series of samples are considered and the maps of the high concentrations are animated, then the contamination will appear to move through the basin, forming an empirical subsurface transport model. From such animations it should be possible to estimate the levels of hydrogen sulfide in the next drilling location being considered. Data investigation and visualization will increasingly extend beyond the perspective of traditional map renderings.

Most *predictive modeling* is currently non-spatial. Environmental data, for example, are collected by sampling large areas, then using these data to solve a mathematical model to establish a prediction equation linking the variable of interest to other more easily obtainable quantities. The model is applied by collecting data on the driving variables for another area or period in time, reducing the measurements to typical values (arithmetic averages), then evaluating the prediction equation.

Another example involves the derivation of a prediction equation for the amount of water encroachment as a result of production. Encroachment is defined in terms of porosity, permeability, and other petrophysical properties (independent variables), possibly with the "sweet spots" experiencing the highest level of encroachment. A traditional non-spatial approach ignores the geographic distribution of variation in field-collected data by assuming that the average well and average geology are everywhere. The predicted encroachment (dependent variable) is a single value for the entire area, extended within a range of probable error (standard deviation). In a mathematical sense, the non-spatial approach assumes that both the dependent and independent variables are randomly, or uniformly, distributed throughout a project area and that the variables are spatially independent. Both parts of the assumption are diametrically opposed to geological theory and evidence. Most geoscientific phenomena exhibit considerable spatial dependency and coalesce into niches responding to a variety of physical factors.

The GIS modeling solution spatially interpolates the data into maps of each variable, then solves the equation for all locations in space. The coincidence of the spatial patterns of the variables are preserved, thereby relaxing the unrealistic assumptions of random/uniform geographic distribution and spatial independence. The direct consideration of the spatial patterns and coincidence among mapped data will increasingly refine geoscience predictions and management actions, making them more responsive to the unique conditions in a project area.

Dynamic simulation allows the user to interact with a GIS model. If model parameters are systematically modified and the induced changes in the final map tracked, then the behavior of the model can be investigated. This "sensitivity analysis" identifies the relative importance of each mapped variable within the context of the unique geoscience/geographic setting to which it is applied. In the water encroachment example, the equation may be extremely sensitive to porosity; however, in a field with a maximum porosity of only 20%, permeability might be identified as the dominant variable. A less disciplined use of dynamic simulation enables a GIS to act like a spatial spreadsheet and to address "what if" questions (Berry, 1991, 1995a). Such queries address natural curiosity as much as they provide insights into system sensitivities. Both simulation versions aid decision-makers in understanding the linkages among the variables and help identify critical ranges. The use of dynamic simulation will increasingly involve decision-makers in the analysis (throughput) phase of exploration and development strategy.

Technology Versus Science

In many respects, the emerging applications of data mining, predictive modeling, and dynamic simulation have the technological cart in front of the scientific horse. GIS can storehouse tremendous volumes of descriptive data and overlay myriad maps to test their coincidence. It has powerful tools for expressing the spatial interactions among mapped variables; however, there is a chasm between GIS technology and applied science. The bulk of scientific knowledge lacks spatial specificity in the relationships among variables. Now that a tool can characterize spatial relationships (cart), the understanding of its expression in complex systems (horse) becomes the void.

An example of the above notion involves the ability of GIS to characterize the changes in the relative amount of edge in an oil field by computing a set of fractal dimension maps. This, and over 60 other landscape analysis indices, allows tracking of changes in feature structure, but the impact of the changes on external factors (such as an expansion of oil field limits into environmentally sensitive areas) is beyond current scientific knowledge. Similarly, a GIS can characterize the effective sediment loading distance from streams as a function of slope, vegetative cover, and soil type. It is common sense that areas with a stable soil on gentle, densely vegetated intervening slopes are effectively farther away from a stream than areas of unstable soils with steep, sparsely vegetated intervening slopes; however, are (or were, in the case of subsurface investigations) effective sediment loading distances related to the survival of fish? Neighborhood variability statistics allow us to track the diversity, interspersion, and juxtapositioning of various subsurface coverages in and around the Arctic National Wildlife Refuge (ANWR), but how are these statistics translated into management decisions about caribou herd populations?

The mechanics of GIS in integrating multiple phenomena are well established. The functionality needed to relate the spatial relationships among mapped variables is in place. What is lacking is the scientific knowledge to exploit these capabilities. Until recently, GIS was thought of as a manager's technology focused on inventory and record-keeping. Even the early scientific applications used it as an electronic planimeter to aggregate data over large areas for input into traditional, non-spatial models. The future will see a new era of scientific research in which spatial analysis plays an integral part and its results are expressed in GIS modeling terms. The opportunity to have both the scientific and managerial communities using the same technology is unprecedented. Until then, however, frustrated managers will use the analytical power of GIS to construct their own models based on common (and uncommon) sense.

The direct engagement of general users will increasingly question the traditional concepts of a map and its use. To more effectively portray the landscape of a subsurface horizon, for example, GIS must step beyond its classical disciplines. Traditional concepts of a map (1) distort reality of a three-dimensional world into a two-dimensional abstraction, (2) selectively characterize just a few elements from the actual complexity of the spatial reality, and (3) attempt to portray geological gradients and conceptual abstractions as distinct spatial objects. The imposition of one's historical concept of a map constructed of inked lines, shading, and symbols thwarts the exploitation of the full potential of mapped data expressed in digital form.

The concepts of synergism, cumulative effects, and system management within the natural resources communities are pushing at the envelope of the ability of GIS to characterize a unified landscape. Historically, system models have required a discrete piecemeal approach; however, a unified landscape is by nature a holistic phenomenon. As an illustration, consider how a hiking trail that maximizes observation of vegetative diversity might be identified. An atomistic approach would begin at the trailhead, test the neighborhood around each location, and step to a different region of interest whenever possible. This approach, however, could commit one to a monotonous path after the first few diverse steps. If, however, a few seemingly suboptimal steps where made at the start it might have lead to a much more interesting route.

A holistic modeling approach requires the assimilation of an entire system at the onset of an analysis. Conventional mapping and GIS modeling approaches characterize the "landscape" in an atomistic fashion. GIS can benefit from the advancements in holistic modeling made by the emerging disciplines of artificial intelligence, chaos theory, and fuzzy logic. These approaches attempt to account for inference, abrupt changes, and uncertainty in spatial data. Instead of a deterministic solution with a single map portrayal, they establish the sidebars of system response. If applied to GIS, these emerging map-ematical techniques might provide a more realistic description of a system whose whole is greater than the sum of its individual parts. The recognition that many spatial phenomena do not always conform to the cartographer's representation of predictable, discrete objects, coupled with these new mathematical approaches, will change mapping paradigms as much as they change data processing.

Equally important is the recognition of perception as an additional element of a landscape. Each individual has a unique set of educational, observational, and interpersonal experiences that form his or her perspective of a landscape. The ability to map these considerations requires a closer marriage between GIS and disciplines not traditionally related to this technology. As the future of GIS unfolds, maps will be viewed less as static descriptions of portions of the earth and more as an active process accounting for inherent variability in perception, as well as spatial descriptors. To move from tool development to a true discipline, GIS needs an infusion of ideas from a wealth of neo-related disciplines not traditionally thought of as its bedfellows.

CONCLUSION

Early information systems relied on physical storage of data and manual processing. With the advent of the computer, most of these data and procedures have been automated during the past three decades. Commensurate with the digital map, geoscience and resource information processing increasingly has become more quantitative. Systems analysis techniques developed links between descriptive data to the mix of management actions that maximize a set of objectives. This mathematical approach to geoscience investigation has been both stimulated and facilitated by modern information systems technology. The digital nature of mapped data in these systems provides a wealth of new analysis operations and an unprecedented ability to spatially model complex earth-related issues. The full impact of the new data form and analytical capabilities is yet to be realized.

It is certain, however, that tomorrow's GIS will build on the cognitive basis, as well as the spatial databases and analytical operations, of the technology. This new view pushes GIS and geostatistical processing beyond data mapping, management, and modeling to spatial reasoning and dialogue focusing on the communication of ideas. In a sense, the allied technologies extend the analytical toolbox to a social sandbox, where alternative perspectives are constructed and discussed, and common knowledge and wisdom are distilled.

REFERENCES CITED

Berry, J. K., 1987a, Fundamental operations in computer-assisted map analysis. International Journal of Geographical Information Systems, vol. 1, p. 119–136.

Berry, J. K., 1987b, A mathematical structure for analyzing maps. Journal of Environmental Management, vol. 11, p. 317–325.

Berry, J. K., 1991, GIS in island resource planning: a case study in map analysis, *in* Geographic Information Systems: Principles and Application, Vol. 2, D. J. Maguire, M. F. Goodchild, and D. W. Rhind, eds., Essex: Longman Scientific and Technical Press, p. 285–295.

Berry, J. K., 1993, Beyond Mapping: Concepts, Algorithms and Issues in GIS. Arlington Heights, IL: Adams Business Media, GIS World Books, 246 p.

Berry, J. K., 1994, A brief history and probable future of GIS in natural resources. The Compiler, vol. 12, no. 1, p. 8–10.

Berry, J. K., 1995a, Characterizing and interacting with GIS model logic, in The AGI Source Book for Geographic Systems. London: The Association for Geographic Information, p. 57–62.

Berry, J. K., 1995b, Spatial Reasoning for Effective GIS. Arlington Heights, IL: Adams Business Media, GIS World Books, 208 p.

Berry, J. K., 1996, Inside the GIS toolbox: the big picture of precision farming. agINNOVATOR, vol. 4, no. 2, p. 4.

Berry, J. K., and W. Ripple, 1994, Emergence and role of GIS in natural resource information systems, *in* The GIS Applications Book: Examples in Natural Resources, W. Ripple, ed., Falls Church, VA: American Society of Photogrammetry and Remote Sensing (ASPRS), p. 3–20.

Burgess, T., and R. Webster, 1980, Optimal interpolation and isarithmic mapping of soil properties: the semi-variogram and punctual kriging. Journal of Soil Science, vol. 31, p. 315–331.

Burrough, P. A., 1986, Principles of Geographical Information Systems for Land Resources Assessment. Oxford: Oxford University Press, 193 p.

Calkins, H. W., 1991, GIS and public policy, *in* Geographic Information Systems: Principles and Applications, Vol. 2, D.J. Maguire, M.F. Goodchild, and D.W. Rhind, eds., Essex: Longman Scientific and Technical Press, p. 233–45.

Carter, J. R., 1989, On defining geographic information systems, *in* Fundamentals of Geographic Information Systems: A Compendium, W. Ripple, ed., Falls Church, VA: American Society of Photogrammetry and Remote Sensing (ASPRS), p. 3–9.

Coppock, J., and D. Rhind, 1991, The history of GIS, *in* Geographic Information Systems: Principles and Applications, Vol. 1, D.J. Maguire, M.F. Goodchild, and D.W. Rhind, eds., Essex: Longman Scientific and Technical Press, p. 45–54.

Cressie, N., 1991, Statistics for Spatial Data. New York: John Wiley & Sons, 900 p.

Cressie, N., 1993, Geostatistics: a tool for environmental modelers, *in* Environmental Modeling with GIS M. F. Goodchild, B. O. Parks, and L. T. Steyaert, eds., Oxford: Oxford University Press, pp. 414–421.

Cressie, N., and J. M. Ver Hoef, 1993, Spatial statistical analysis of environmental and ecological data, *in* Environmental Modeling with GIS, M. F. Goodchild, B. O. Parks, and L. T. Steyaert, eds., Oxford: Oxford University Press, pp. 404–413.

Densham, P. J., 1991, Spatial decision support systems,

in Geographic Information Systems: Principles and Applications, Vol., 1 D.J. Maguire, M.F. Goodchild, and D.W. Rhind, eds., Essex: Longman Scientific and Technical Press, p. 403–12.

Elridge, J., and J. Jones, 1991, Warped space: a geography of distance decay. Professional Geographer, vol. 43, p. 500–511.

Epstein, E. F., 1991, Legal aspects of GIS, *in* Geographic Information Systems: Principles and Applications, Vol. 1, D.J. Maguire, M.F. Goodchild, and D.W. Rhind, eds., Essex: Longman Scientific and Technical Press, p. 489–502.

Goodchild, M. F., 1987, A spatial analytical perspective on geographical information systems. International Journal of Geographical Information Systems, vol. 1, p. 327–334.

Goodchild, M. F., R. Haining, and S. Wise, 1992, Integrating GIS and spatial data analysis: problems and possibilities. International Journal of Geographical Information Systems, vol. 6, p. 407–23.

Goodchild, M. F., B. O. Parks, and L. T. Steyaert, eds., 1993, Environmental Modeling with GIS. Oxford: Oxford University Press, 488 p.

King, J. L., and K. L. Kraemer, 1993, Models, facts and the policy process: the political ecology of estimated truth, *in* Environmental Modeling with GIS, M. F. Goodchild, B. O. Parks, and L. T. Steyaert, eds., Oxford: Oxford University Press, p. 353–360.

Lam, N. S., 1983, Spatial interpolation methods: a review. American Cartographer, vol. 10, p. 129–149.

Leick, A., 1990, GPS Satellite Surveying. New York: John Wiley & Sons, 352 p.

Maguire, J. D., M. F. Goodchild, and D. W. Rhind, eds., 1991, Geographic Information Systems: Principles and Applications, Vol. 1 & 2. Essex: Longman Scientific and Technical Press, 1096 p.

Maguire, D. J., and J. Dangermond, 1994, Future of GIS technology, *in* The AGI Source Book for Geographic Systems. London: The Association for Geographic Information, p. 113–20.

McGarigal, K., and B. Marks, 1995, FRAGSTATS: Spatial Pattern Analysis Program for Quantifying Landscape Structure. Portland, OR: U.S. Department of Agriculture, U.S. Forest Service, Pacific Northwest Research Station, General Technical Report PNW-GTR-351, 122 p.

McHarg, I. L., 1969, Design with Nature. Garden City, NJ: Doubleday/Natural History Press, 197 p.

Medyckyj-Scott, D., and H. M. Hernshaw, eds., 1993, Human Factors in Geographical Information Systems. London: Belhaven Press, 266 p.

Myers, D. E., 1988, Multivariate geostatistics for environmental monitoring. Sciences de la Terra, vol. 27, p. 411–427.

Muehrcke, P. C., and J. O. Muehrcke, 1998, Map Use: Reading, Analysis and Interpretation, 4th ed. Madison, WS: J.P. Publications, 647 p.

Muller, J. C., 1982, Non-Euclidean geographic spaces: mapping functional distances. Geographical Analysis, vol. 14, p. 189–203.

Ottens, H., 1994, Relevant trends for geographic information handling. Geo-Info Systems, vol. 4, no. 8, p. 23.

Parent , P., and R. Church, 1989, Evolution of geographic information systems as decision making tools, *in* Fundamentals of Geographic Information Systems: A Compendium, W. Ripple, ed., Falls Church, VA: American Society of Photogrammetry and Remote Sensing (ASPRS), p. 9–18.

Pereira, J. M., and L. Duckstein, 1993, A multiple criteria decision-making approach to GIS-based land suitability evaluation. International Journal of Geographical Information Systems, vol. 7, p. 407–424.

Pueker, T., and N. Christman, 1990, Cartographic data structures. American Cartographer, vol. 2, p. 55–69.

Piwowar, J. M., 1990, Integration of spatial data in vector and raster formats in a geographic information system. International Journal of Geographic Information Systems, vol. 4, p. 429–44.

Ripley, B. D., 1981, Spatial Statistics. New York: John Wiley & Sons, 252 p.

Ripple, W., ed., 1994, The GIS Applications Book: Examples in Natural Resources. Falls Church, VA: American Society of Photogrammetry and Remote Sensing (ASPRS), 373 p.

Rix, D., 1994, Recent trends in GIS technology, *in* The AGI Source Book for Geographic Systems. London: The Association for Geographic Information, p. 25–39.

Robinson, A. H., J. L. Morrison, P. C. Muehrcke, A. J. Kimerling, and S. C. Guptill, 1995, Elements of Cartography, 6th ed. New York: John Wiley & Sons, 674 p.

Shepherd, J. D., 1991, Information integration and GIS, *in* Geographical Information Systems: Principles and Applications, Vol. 1, D. J. Maguire, M. F. Goodchild, and D.W. Rhind, eds., Essex: Longman Scientific and Technical Press, p. 337–360.

Steinitz, C. F., P. Parker, and L. Jordan, 1976, Hand drawn overlays: their history and prospective uses. Landscape Architecture, vol. 66, p. 444–455.

Thomas, R., and R. Huggett, 1980, Modelling in Geography: A Mathematical Approach. London: Harper and Row, 338 p.

Tomlin, C. D., 1990, Geographic Information Systems and Cartographic Modeling. Englewood Cliffs, NJ: Prentice Hall, 249 p.

Turner, M. G., 1990, Spatial and temporal analysis of landscape patterns. Landscape Ecology, vol. 4, p. 21–30.

Unwin, D., 1981. Introductory Spatial Analysis. London: Methuen, 212 p.

Webster, R., and T. Burgess, 1980, Optimal interpolation and isarithmic mapping of soil properties: changing drift and universal kriging. Journal of Soil Science, vol. 31, p. 505–524.

Bonham-Carter, G., 2000, An overview of GIS in the geosciences, *in* T. C. Coburn and J. M. Yarus, eds., Geographic information systems in petroleum exploration and development: AAPG Computer Applications in Geology, No. 4, p. 17–26.

Chapter 3

An Overview of GIS in the Geosciences

Graeme Bonham-Carter
Mineral Resources Division
Geological Survey of Canada
Ottawa, Ontario, Canada

ABSTRACT

Geographic information systems (GIS) are now firmly entrenched as routine tools in many branches of geoscience. GIS technology is part of a spectrum of software systems dealing with spatial data. The core activities for which GIS is particularly valuable are data organization, visualization and search of spatial data, combining data layers, analyzing spatial relationships, and decision support. The core functions of GIS include spatial data capture, management of multi-layer data sets in a database, exploring spatial data visually and interactively, transformation of spatial objects, and analyzing maps singly, in pairs, and in multiple combinations. From a management perspective, GIS projects have the beneficial effect of integrating people and ideas, as well as data. From a scientific perspective, current GIS practice does not reach the full potential of spatial data analysis.

INTRODUCTION

Although geologists have been using computers to analyze spatial data for over 30 years, developments in hardware and software in the 1980s resulted in the emergence of a type of spatial analysis software known as geographic information systems (GIS). As the name suggests, these computer programs are large and complex, dealing with a variety of spatial data types and providing a broad range of functionality within a single package. The market for GIS grew rapidly in the late 1980s and early 1990s because it quickly became apparent that such systems could handle virtually any type of geographical data and could provide a range of functionality for spatial data handling and analysis that previously had required a combination of software to achieve. Although the principal users of GIS were (and are) those who deal in all aspects of how land at the earth's surface is used (government, industry, and academia), GIS applications appeared in a myriad of diverse fields, such as military planning, business, marketing, and ecology. Geologists were perhaps slow to realize the potential of GIS, partly because most commercial systems deal only with two spatial dimensions, and partly because alternative computing systems for spatial data were strongly embedded in the geological workplace prior to GIS development. For example, before GIS became well known, the petroleum industry already had a considerable investment in sophisticated digital systems for management and analysis of geological data (e.g., well data systems, systems for seismic data, and systems for stratigraphic analysis) with a strong influence of surface modeling by contouring. Nevertheless, many geological organizations (including petroleum companies) began to realize the advantages of GIS (both 2-D and 3-D systems), and during the 1990s this technology established a significant presence in many branches of geoscience.

One of the features of modern GIS products is that they have been developed by the commercial sector.

Although there are some examples of systems developed by universities and government, the GIS market is now dominated by commercial systems. In consequence, the most successful systems have tended to be those that cater to the needs of disciplines having the most abundant users, and geology has not been a major influence on system design. Most GIS products deal with 2-D rather than 3-D data, and requirements for geological information systems (such as cross-sections and stratigraphic relationships) have not played a strong role in influencing the most popular commercial developments. Nevertheless, both 2-D and 3-D GIS capabilities now play an important role in many aspects of geological activity. In particular, government geological surveys that hitherto published spatial data as paper maps now use GIS to produce maps digitally and to distribute spatial databases in digital form, allowing others to make their own maps and to perform customized spatial analyses. The widespread use of the Internet has facilitated the distribution of spatial data, and the demand for software tools for visualizing and analyzing map data is continuing to increase.

This chapter reviews GIS from several viewpoints, including its place in the spatial data handling software spectrum, the primary purposes of GIS, and the core functionality that is characteristic of this technology. It touches on some issues of current interest to geologists.

THE SPATIAL ANALYSIS SOFTWARE SPECTRUM

The software for spatial data analysis consists of a variety of program types. It is important to distinguish between them, and to realize the differences in functionality because at a superficial level, many systems appear similar with closely overlapping functions (see Table 1). The precursors to, or roots of, the modern GIS consist of three well-entrenched types of computer software: computer-aided design systems (CAD), image processing (IP) systems, and database management systems (DBMS). CAD systems use vector data structures, consisting of point, line, and polygon graphical objects, which were initially developed for engineering drawings. They do not generally handle raster (gridded) data, nor are they generally linked to database management systems for handling object attributes; nevertheless, they have powerful visualization and transformation functions that are similar to those used in GIS. Image processing systems, on the other hand, were developed initially for display and analysis of satellite (and medical) images. IP systems are now regularly used for processing raster images of any type with a sophisticated capability for enhancing, visualizing, and analyzing raster data. Although IP systems often have some capability to display vectors, they are not designed for the analysis of vector data, nor for the analysis of the associated attribute tables. The third member of this triumvirate, DBMS, is not designed principally for spatial entities, but it has sophisticated tools for organizing and searching large databases. GIS combines the functionality of these three types of software and provides functions for data viewing, data transformation, and data analysis. However, CAD, IP, and DBMS have all continued to develop in parallel with GIS and have important market niches of their own.

An important class of spatial data analysis software familiar to petroleum geologists contains the packages for surface modeling and contouring (e.g., Tipper, 1979; Jones et al., 1986; Hamilton and Jones, 1992) widely available long before GIS became popular. Generally, these are designed to access point data (and associated attribute tables) and to post the data values, grid and contour selected point attributes, and visualize the resulting surfaces in a variety of ways, such as hill-shading and perspective views. Surface modeling functions are present in full GIS packages but may or may not be available in desktop versions. Gridding and contouring may be done by a variety of methods, including triangulation, inverse distance weighting, spline fitting, kriging, and many others (e.g., Watson, 1992).

The requirements of geological mappers working with computers in the field have been satisfied by specialized systems such as FieldLog (Brodaric and Fyon, 1989; Brodaric, 1993). FieldLog is based on a combination of CAD and DBMS software and provides facilities for tablet digitizing of field stations (or direct GPS entry of station coordinates) and for entry of attribute data. Data entry is accomplished through a sophisticated database interface using standard geological terminology. The system is designed for customization, which is essential for adaptation to the needs of different organizations. Export formats allow transfer of data to popular commercial systems.

There is a distinct class of GIS software specifically designed to handle 3-D geological data (e.g., Houlding, 1994). A prime example is the GOCAD system (Mallet, 1991), or systems such as Stratamodel (Denver and Phillips, 1990). These systems employ a variety of design principles, using both vector and raster data structures, but in general they are not as strong as 2-D systems for dealing with attribute data. They have sophisticated capabilities for input and editing of 3-D data and for visualization. Three-dimensional systems are particularly important to the petroleum industry, as well as to the mineral industry for mine-scale projects.

Software for viewing 2-D spatial data, also known as desktop GIS or desktop mapping systems (e.g., ArcView, MapInfo), provides the user with the ability to make custom views of spatial data (on-screen or in hardcopy) and to do a limited variety of data analysis functions. These software packages excel in providing easily understood linkages between spatial objects and associated attribute tables. In general, these systems have many of the popular functions of a full GIS and have easy-to-use, menu-driven user interfaces, but they are usually unsuitable for GIS database builders because of their restricted functionality. They are ideal, however, for many users of GIS data, including most non-specialists and managers, because these

Table 1. Summary of Types of Software for Spatial Analysis

Name	Characteristics and Function
Computer-aided design (CAD)	Design and display of engineering drawings; vector formats
Image processing (IP)	Raster image processing
Database management systems (DBMS)	Management of entities and attributes; often relational, but also hierarchical and network
Surface modeling software	Gridding, contouring, and display of points and associated attributes
Field-based geographic information system (GIS)	Software for use by field mappers
3-D modeling	Generation, editing, and visualization of 3-D data
Desktop GIS	Visualization and analysis (limited) of GIS data
Statistical packages	Univariate and multivariate statistical analysis
Full GIS	Some of all the previous functions, plus additional GIS functions; may or may not have 3-D functionality
Other software not in a GIS, but used in spatial data processing	Special plotting, custom analysis, geostatistics, mathematical morphology, fractal and multi-fractal analysis, correlation of well log values, neural networks, geophysical data processing, hydrodynamic modeling, basin simulation, etc.

systems have a relatively short learning curve. They can also serve as platforms for building special-purpose applications, using scripting languages (e.g., the Avenue language used with ArcView). Desktop mapping systems dominate the sales in the general GIS market in terms of the number of units sold.

Most GIS products have some limited capability with which to perform descriptive statistical analysis, but they rely on linkages to specialized statistical packages for more sophisticated functions. Many conventional statistical packages have good graphics that can produce displays (such as scatter plots of symbolized data in geographical coordinates) based on limited overlap with GIS capabilities. These packages may be entirely adequate for some statistical users, particularly for individuals dealing with spatial data occurring in point form. Some GIS packages have modules specifically designed to interface with statistical packages (e.g., Arc/Info with S-Plus), but usually such linkages are made in ad hoc fashion via common exchange formats.

Full GIS functionality (a full-fledged GIS) is discussed in greater detail in the next section of this chapter. GIS has a long learning curve, and few users master the whole range of potential functionality.

Many other types of software that deal with spatial data are not specifically mentioned here; for example, software for geostatistics, mathematical morphology, fractal and multifractal analysis, log analysis, hydrodynamic modeling, basin simulation, terrain analysis, geophysical data processing, and many other geomathematical applications. Sometimes, commercial GIS products may be linked loosely to packages such as these on an ad hoc basis. The integration of software systems is a topic outside the scope of this discussion, but it is a subject that is relevant to GIS because the technology may evolve to incorporate such software integration capabilities in the near future.

CORE GIS ACTIVITIES

Given the particular needs of the geoscientist, and the petroleum geologist in particular, it is pertinent to review typical GIS activities (encountered during actual projects) and the purposes they serve. A discussion of GIS activities helps to answer questions about the advantages of using GIS, as opposed to an alternative software mix employed by oil companies, to handle and analyze spatial data. Table 2 summarizes six core activities, with a brief explanation of the purpose served by each one. These activities are (1) data organization (involving data modeling, data compilation, and database construction), (2) data visualization (producing data views and maps, and graphically evaluating spatial patterns), (3) data search (what is here? or where does this occur?), (4) combination (integration) of diverse data types (e.g., what is the relationship of satellite linears to mapped faults?), (5) data analysis (e.g., how large is this area? or is proximity to structural domes related to producing wells?), and (6) prediction, particularly to support decision-making that depends on spatial information (e.g., based on multiple factors, which areas are the most favorable for exploration?). The sequence of these activities also is significant, and the order in the table roughly follows the order in which the activities are undertaken in a particular project. Database building, data visualization, and data search are core activities central to all GIS projects. Data combination, analysis, and prediction are central to many projects, but the degree of sophistication differs considerably depending on project goals and on the personnel and resources available.

Table 2. Core GIS Activities, with a Brief Explanation

Name of Activity	Explanation
Data organization	Data modeling, data collection and digitizing, data compilation and editing; database construction and maintenance
Data visualization	Interactive display and hardcopy views of data (2-D, 3-D, zoom, pan, symbolization, cartographic design, statistical plots, hotlinked views)
Data search	Interactive search for attributes of objects at elected locations; selected objects and attributes displayed on map views
Data combination	Overlay operations between maps in vector or raster formats; generation of new attribute tables of combined objects
Data analysis	Transformations and measurements of objects on data layers to characterize and extract meaning
Data prediction for decision support	Simple or complex analysis of multiple layers for prediction using data-driven (generally statistical) or expert-driven methods; decision support based on multiple factors

Data Organization

A spatial database lies at the heart of every GIS project, and the organization of data according to a well-documented data model is an essential step in the process. In geology, spatial data are represented digitally both in vector form (points, lines, and polygons occurring on maps) and in raster form (grid cells with values of a continuous spatial variable).

GIS projects involve pulling together data from a variety of sources: field data from mapping or instrumental measurements, digitized paper maps (or well logs), and information extracted from previously stored databases or data collections (tapes, CDs, Internet, etc.). The primary purpose of a project may be to build and maintain a spatial database, such as a well database as used in the oil industry. Government geological surveys are mainly in the business of collecting, organizing, and distributing geoscientific data. In these instances, the final product may be a CD or database on the Internet. GIS has been particularly important in the collection, organization, and distribution of geological map data. Field mapping systems (e.g., FieldLog described by Brodaric, 1992) now rely heavily on GIS-like mapping software, usually linking vector and attribute database functionality with facilities for GPS and field data capture. The database is the source of all the subsequent GIS visualizations and analyses.

GIS projects involve compilation of spatial data from a variety of sources: paper maps in various cartographic projections, images, and tables of attribute data of many kinds. GIS tools are invaluable for compilation tasks. The efficient management of GIS projects, particularly large ones, requires teamwork and well-defined tasks and objectives. Data models (as discussed later) are essential for defining how data are to be organized. An important part of ensuring the integrity of a spatial database is making sure that the various layers of information are properly registered so that overlap relationships are correct. Often, the expression "building a GIS" is used to refer to the development of a spatial database using GIS tools, and for later use by those equipped with GIS facilities.

Visualization

Visualization includes the display of spatial data on computer monitors, as well as the production of hardcopy products. The data displays can take the form of maps in addition to other graphics, such as cross-sections, 3-D cutaways, scatter diagrams, and other statistical plots. Traditionally, paper maps were made by cartographers using manual methods. The map was a static interpretation of field data. It could not be modified or reinterpreted, nor could the symbolization, scale, and projection be changed without the expense of creating a new cartographic product. Furthermore, a distinction could not easily be made between what was actually observed and what was interpreted, and the map could represent only a small fraction of the actual known data.

The map, nevertheless, is invaluable as a visual aid to understanding patterns and for examining spatial relationships between spatial objects and their attributes. Thus, visualization is often the most widely used GIS function and the single characteristic most frequently associated with the technology. Among its visualization capabilities, GIS allows the user to select data, to zoom and pan, to change colors and patterns, and to label and create legends. Many systems can now produce cartographic products that aesthetically match those of hand-produced maps. Graphical displays quickly created on a computer monitor can be saved as display files and reproduced as page diagrams, 35 mm slides, or overheads, or they can be included in digital displays for use in oral presentations.

Spatial Query

Spatial query can take two forms: an on-screen selection of a location (or area) to determine its characteristics from the database, or a search of the database for all those locations where a particular attribute (or combination of attributes) occurs. For example, given a display of wells on a map showing property boundaries, one may be interested in determining the characteristics of a particular well (identified by its location on the screen). Such characteristics may include the elevation of the collar; whether it is a producer, past producer, or dry hole; how close it is to a structural dome; whether it intersects a particular formation; and so on. Alternatively, one may wish to display all those wells having a specified set of characteristics against the backdrop of a satellite image with roads superimposed for location. Whereas the visualization function reveals patterns, spatial query is used to identify the specific details about particular occurrences of spatial objects.

Combination of Spatial Layers

The primary purpose of many GIS projects is to combine different map layers together in various ways for purposes of revealing or confirming spatial relationships in the data. The ability to combine map layers visually, or in spatial search operations, or to generate new layers by mathematical or Boolean operators is the GIS equivalent of overlaying maps on a light-table.

Spatial Analysis

Data analysis is the process of deriving meaning from data. Spatial data analysis can mean the analysis of data that are spatially distributed without regard to spatial location (except for spatial coincidence). Suppose, for example, that a grid of sampling points is superimposed on a series of spatial data layers, and the characteristics of each layer are recorded as a field (column) in a data table whose rows correspond to sample points. If the resulting data table is analyzed using univariate or multivariate statistics, then the result would certainly constitute one type of spatial data analysis. If, however, the same data table includes spatial coordinates for each sampling point and the analysis makes specific use of the coordinate data, then it constitutes a different type of spatial analysis wherein the locations of data items play a role in the computational algorithm. Some well-known examples of this kind of spatial analysis include contouring methods, trend surface analysis, the use of variograms and spatial covariance diagrams, kriging and cokriging, spatial filtering, and fractal and multifractal analysis. Spatial analysis may also incorporate the attributes of adjacency and containment, the so-called topological attributes of spatial objects. For example, skarn deposits often occur in impure limestones near the contact with intrusions. By extracting the contact between these two rock types, the distance distribution of skarn deposits to the nearest contact may be tested to determine whether the deposits occur closer than would be expected due to chance. Finally, spatial analysis may involve the characterization of the shape of objects or of the texture of surfaces.

Spatial analysis in GIS is often exploratory in nature, seeking to uncover associations in the data that are otherwise unknown. For example, one might be seeking spatial associations between producing wells and combinations of geological factors, or one might be trying to explain the occurrence of a geochemical anomaly in a stream sediment survey as a function of the distribution of rock types in the associated catchment basins.

Prediction

The ultimate purpose of GIS may be to provide spatial predictions, which are often made to support decisions. Some well-known examples include oil, gas, and mineral exploration, landslides forecasting, and assessment of the effects of oil spills or pollution. Decisions are based on mathematical models of various types; for example, Boolean rules, simple subjectively weighted map combinations, fuzzy membership functions, statistical regression, expert systems, and neural networks.

Some people would argue that there is no need to regard spatial analysis and mathematical modeling as part of a GIS because these functions may be done in software packages external to the GIS. From this perspective, the GIS is merely a tool for creating, editing, transforming, and visualizing information from the database. It principally acts as a spatial data server that interfaces with other specialized information processing and modeling packages. The results of data analyses may subsequently be returned to the GIS for visualization, incorporation in the database, and further processing. The alternative view, however, is that a GIS should incorporate analytical functionality as an integral part of the system because the tight coupling of analysis and data manipulation takes full advantage of the data structures and processing tools.

This discussion provides a broad introduction to the principal GIS activities and capabilities. The next section identifies some of the characteristics of a modern GIS in terms of its functional components.

GIS COMPONENTS

Certain key characteristics are common to every GIS, although they may be implemented differently in particular systems. At least seven functional components can be recognized:

- a spatial database
- data capture, geo-referencing, and editing capabilities
- data search and visualization capabilities
- transformations that can be applied to spatial data
- the ability to analyze single maps
- the ability to analyze map pairs
- the ability to analyze multiple maps

These fundamental components are discussed more extensively in Bonham-Carter (1994), but are briefly introduced here.

Spatial Database

A GIS deals with spatial data organized into layers. Each layer consists of data about some theme and extends over a region of interest. These layers have no stratigraphic connotation. Land ownership parcels, terrain elevation, soil type, depth to a formation top, distribution of lineaments, distribution of wells, rivers, and chloride content of drinking water from water wells are all examples of data layers. The concept of a GIS data layer is useful for separating 2-D spatial data into distinct groups, and although a layer is not a natural concept for dealing with 3-D data, one may indeed talk about 3-D themes (e.g., distribution of permeability in a sedimentary basin).

Spatial data are organized in a GIS either as vectors, rasters, or a combination of the two types. Spatial objects such as points, lines, and polygons are held in vector format. A data layer may consist of point objects (e.g., locations of wells), lines (e.g., seismic network), or polygons (e.g., field outlines). These are treated as discrete spatial objects, and each object type is usually associated with a table of attributes. Attributes may describe spatial characteristics (e.g., geographic coordinates, size and shape, and topological information) or thematic characteristics (e.g., lithology, age, composition, and name).

Rasters are grids containing the values of some spatially distributed variable at node points or cell centers. These values may be real numbers (particularly for continuous variables measured on an interval or ratio scale of measurement), and they may be integers or characters, depending on whether the quantity of interest was measured on an ordinal or categorical (nominal) measurement scale. Integer values may act as pointers to attribute tables, although many raster layers are not associated with an attribute table [e.g., a single band of a satellite image, or a digital elevation model (DEM) value].

The way in which topological information is structured is a key feature of a full-fledged GIS. In GIS parlance, a topological attribute pertains to the adjacency and containment relationships of polygons or to the connectedness of lines. Topological relationships are not affected by translation, scale, or shearing operations. Thus, the USA map polygon is adjacent to the Pacific ocean map polygon, and the Great Salt Lake polygon is fully contained by the state of Utah polygon. These relationships are not affected by geometric transformations such as changes in projection. Topological relationships built into a spatial database can make some types of processing very efficient. For example, a search for all lines that represent a surface contact between limestone and granite requires only that a search be made of the table that holds these topological relationships,whereas locating this information in a database with no topological structure requires each line to be searched with respect to all possible neighboring areas.

A key concept for the organization of data is that of a data model. A data model is a schema for explaining how the data elements are to be structured in the database. Many databases are structured according to relational principles into tables linked by key fields (e.g., Bisland, 1989). Relational databases greatly improve the efficiency of editing, and they provide a structure that facilitates an understanding of how the data are organized. One of the current research issues pertaining to geoscientific applications of GIS involves the development of data models that are suitable for geological maps. The U.S. Geological Survey (USGS) and the U.S. state surveys, along with Canadian participation, have developed a data model that forms the basis for a digital map standard (U.S. Geological Survey, 1996). Although the technical issues of developing good geological map models are certainly not insurmountable, it is difficult to get all geological organizations to agree on a universal standard. Without such a standard, however, the use of GIS to produce geological maps is limited, particularly with regard to regional-scale compilations involving the combination of data from many mapping organizations. Models for well data, as used by the petroleum companies, are better developed than models for geological maps (e.g., Turner, 1990).

Data Capture, Geo-Referencing, and Editing

A major cost of building GIS databases involves the digital capture of analog documents. Printed maps must be either scanned or digitized on a tablet. Spatial objects must then be edited and tagged with attributes and linked to an attribute table. A structured database (i.e., one in which discrete objects and their associated attributes are clearly identified) is much more valuable for subsequent analysis than an unstructured scan (in a raster format, for example). Although a scanned image may have value as a backdrop against which other layers are displayed, a structured data layer can be searched, analyzed, symbolized in different ways, and used to generate selected features. For example, a road network showing all classes of roads together may be almost useless in a regional compilation study, without the ability to separate roads by size and importance.

GIS provides not only tools for digitizing but also tools for editing, building topology, and most important, geo-referencing. Data from maps and images for which the cartographic projection is completely known (e.g., the ellipsoid for defining the Earth's shape and the projection parameters, such as central meridian, standard parallels, and so on) can be straightforwardly transformed into a new working projection that is suitable for the study at hand. Maps and images based on unknown projections, however, must be geometrically rectified using control points. Modules for projection changes and rectification are standard parts of GIS.

Data Search and Visualization

The graphical capabilities of GIS enable the production of maps that meet cartographic standards. They also provide improved efficiency in moving data from the field mappers to a published product. The experience of the Geological Survey of Canada, which may be typical, is that GIS methodology applied to field operations, database construction, and map production has doubled map output since 1991. The turnaround time has been reduced from two years to less than three months despite a 20% reduction in cartographic staff (J. Broome, 1997, personal communication). Desktop mapping systems excel at producing on-screen displays, which can be designed for poster- and page-size hardcopy.

Data structures for efficient handling of spatial information have made interactive search and selection of data fast and efficient. Hotlink facilities, available in many systems, allow a spatial view and attribute table to be opened side-by-side. Identification of objects on the map result in their being highlighted in the table and vice-versa. In some systems, this capability is extended through the addition of an aggregation function. Aggregation permits the characteristics of groups of objects (selected by polygon, circle, rectangle, etc.) to be statistically summarized (mean, median, count, largest, etc).

Transformations

Another useful function of GIS is its ability to perform transformations between object types. A simple example is conversion of a raster image of a continuous variable that is classified into discrete class intervals to vector format, resulting in a set of nested polygons. Suppose the continuous variable of interest is a DEM. The DEM is now held in two separate data structures, one being raster and the other being vector. Such a situation allows the polygon boundaries (lines) to be superimposed on layers containing other objects (e.g., geological map, location of producing wells), which is useful for display; it also allows the processing to be accomplished using the data structure that is most suitable. A DEM, for example, may exist in a raster, vector, or triangulated irregular network (TIN) structure, but the calculation of its slope is easiest in the context of the raster structure. Some examples of the more useful GIS transformations can be illustrated in a "to-and-from" table (see Table 3).

Most of these transformations are self-explanatory, but the reader can refer to Bonham-Carter (1994) for a more detailed discussion. However, one function, the resampling of attributes of polygons at a fixed set of sampling points (either on a grid or an irregular distribution), perhaps needs further explanation. This function involves the addition of one or more new fields to a point attribute table, depending on the attributes of polygons underlying the points. For example, geological map polygon attributes may be appended as a new field to the attribute table of a set of wells using this function, thereby creating a linkage between well location and the geology. The polygon overlay function is really an extension of the same idea, whereby a new set of polygons is produced by the intersection of two previous polygon sets, and the subsequent merging of the attribute tables attached to each new polygon is defined by the combined attributes.

The next three sections briefly describe some spatial analysis functions of GIS.

Analysis of Single Maps

One of the most useful and frequently employed GIS operations is to reclassify a map, usually generalizing it for a particular purpose. For example, given a complex geological map of a region, where the map attribute table contains lithologic and age attributes associated with the map units, it may be desirable to classify it by (1) age alone, (2) lithology alone, or (3) specific combinations of the two (such as "all Devonian granites" or "Paleozoic carbonates").

Another popular operation on single map layers is to classify or select objects based on size, shape, or topological attributes. As an illustration, this would allow the user to remove all lakes smaller than $1 km^2$ or to change the color of all granites where they are in contact with limestones.

Most systems have the built-in capacity to perform raster image processing, such as high and low pass filtering, visibility analysis, and, possibly, spectral analysis in the spatial domain. However, the more specialized analyses used in geophysical data processing, such as reduction-to-pole and upward/downward continuation, are not present in a general-purpose GIS.

Analysis of Map Pairs

As in ordinary statistics, an objective of spatial analysis is often to explore for correlations in the data among pairs of variables, in this instance pairs of data layers. This is not simply a question of calculating correlation coefficients, however, because there may be spatial associations between layers having different object structures. For example, one might be interested in whether producing wells were spatially associated with lineaments or with lineament intersections. The wells are point objects and the lineaments are line objects, so one can pose the question, "Was this particular set of points generated by a process that is independent of the one that generated the set of lines?"

One way to examine such a question is to use the following sequence of GIS operations:

(1) For each member of the point set, calculate the distance to the nearest line.
(2) Buffer the lines at successive distance intervals to produce a map of distance to the nearest line.
(3) Generate a table (or plot) showing the cumulative area and cumulative number of points for each distance interval.
(4) Use a Kolmogorov-Smirnov test to determine whether the null hypothesis that the two distributions are independent can be rejected at a selected level of confidence.

Step 4, and possibly step 3, cannot be performed directly in the GIS and may require special programming

Table 3. Some Elementary GIS Transformations

From/To	Points	Lines	Polygons
Points	Gridding	Contouring	Thiessen polygons
Lines	Intersecting lines	Smoothing lines	Buffering lines
Polygons	Resampling at points	Medial axis	Polygon overlay

in a scripting language or exporting the table (step 3) to a statistical package.

The GIS procedure of overlaying two map layers together produces an attribute table that is often the source of data for analyzing map pairs. This is also a key operation in the analysis of multiple map layers together. Map overlay operations using Boolean or mathematical operators can be used to generate derived map layers. For example, a map showing only those locations where the slope is less than 1% and the soil type is not clay may be useful for evaluating slope stability. Other approaches can be taken, some of which are discussed in Bonham-Carter (1994).

Analysis of Multiple Maps

Simultaneous operations on many map layers are achieved in various ways depending on the GIS design. The process of carrying out a sequence of operations on maps is often known as map or cartographic modeling (e.g., Tomlin, 1990). In fact, the operations are carried out not only on the values at some location in a map layer (e.g., as a pixel value in a raster) but also on attributes associated with the objects at that location. A map or cartographic model is implemented with a map algebra that takes various forms depending on the system (Berry, this volume). Common to all systems is the concept of a sequence of operations (the map model) that defines an algorithm, and this algorithm is applied either pixel-by-pixel in a raster environment, or polygon-by-polygon in a vector environment.

An important GIS concept is that of an overlay of multiple layers. This leads to production of an attribute table in which the entities (rows) are polygon or pixel objects and the columns (fields) are the layers being combined. Each row can be thought of as a vector of attribute values. If the rows (polygons) with identical attribute values are merged, then the result (after the addition of a field with polygon area) is known as a "unique conditions" table. The unique conditions table constitutes an invaluable summary of the areas of overlap of the participating maps, and it can be used for many different kinds of analysis. For example, such a table can be exported to multivariate statistical analysis packages, neural network software, and expert system shells.

A discussion of the various types of map models is beyond the scope of this paper; however, it is useful to recognize that in employing combinations of maps in a decision support role, the map model is either subjective (based either on heuristic reasoning or prescribed rules) or objective (based on relationships that may be statistically derived or that come from genetic theory). The subjective category includes simple Boolean or additive weighted models, as well as more sophisticated expert system models that employ knowledge representation methods such as fuzzy membership and Dempster-Shafer belief functions (Chung and Fabbri, 1993). The objective category includes weights-of-evidence models, models based on statistical regression, and many types of neural network models.

These topics are discussed and illustrated in Bonham-Carter (1994) using as examples the selection of a site suitable for a landfill and the selection of areas for gold exploration. A comparison of some of these methods is also given in Wright and Bonham-Carter (1996) in a study of base metal potential of the Snow Lake greenstone belt in Manitoba, Canada. There is considerable potential for applying similar methods to oil and gas exploration as an aid to decision-making based on combinations of spatial evidence for petroleum targets.

DISCUSSION

The experience of many individuals who have been involved in GIS projects can be summed up by the words "integration" and "models." Integration applies not only to the data that are brought together and combined but also to the harmonizing effect of bringing together diverse skills into a team. Models are at the heart of GIS in various ways: data models for defining what the data means and analytical models for defining how data are analyzed and predicted.

At present, many GIS projects are not carried very far into the spatial analysis stages. This is partly because data analysis tools are not well organized and packaged in commercial systems, and partly because many geologists are not trained in quantitative spatial analysis techniques; furthermore, many projects do not justify the additional costs of performing sophisticated data analyses.

The chapters in this volume mainly reflect the successful application of GIS through spatial data management, data integration, and overlay, with the results being evaluated by spatial query and visualization. Nevertheless, several of the authors discuss and illustrate examples of quantitative spatial analyses undertaken with GIS tools. For example, Paulus (this volume) shows how buffering and topological attributes can be used to evaluate the spatial relationship between faults and the composition of formation waters sampled at wells. He also evaluates the relationship beween fault density and fluid migration, and explores the use of cost surfaces for modeling fluid migration. Hood et al. (this volume) describe a study of hydrocarbon resource assessment, involving a sophisticated analysis of exploration plays that is based on historical data from drilling activities. To develop strategies for acquiring or trading lease holdings, they employ a GIS approach to relate risk to the regional geology and to evaluate the combined benefits of various exploration opportunities. Similarly, GIS can be successfully applied to many other aspects of petroleum exploration and development. The objective of this text is to foster greater awareness of GIS concepts and more extensive use of the technology in this broad arena.

REFERENCES CITED

Bisland, R. B., 1989, Database Management: Developing Application Systems Using ORACLE. Englewood Cliffs, NJ: Prentice Hall International, 441 p.

Bonham-Carter, G. F., 1994, Geographic Information Systems for Geoscientists: Modelling with GIS. Oxford: Pergamon, 398 p.

Brodaric, B., 1992, GSC FieldLog, Version 2.83 (Tutorial and Manual). Ottawa, ON: Geological Survey of Canada(ftp://gis.nrcan.gc.ca/pub/fieldlog.283/manual.doc), 83 p.

Brodaric, B., and J. A. Fyon, 1989, OGS FieldLog: A Microcomputer-Based Methodology to Store, Process and Display Map-Related Data. Ottawa, ON: Ontario Geological Survey, Open File Report 5709, 73 p.

Chung, C. F., and A. G. Fabbri, 1993, The representation of geoscience information for data integration. Nonrenewable Resources, vol. 2, p. 122-139.

Denver, L. E., and D. C. Phillips, 1990, Stratigraphic geocellular modeling. Geobyte, vol. 5, p. 45-47.

Hamilton, D. E., and T. A. Jones, 1992, Computer Modeling of Geologic Surfaces and Volumes. Tulsa, OK: American Association of Petroleum Geologists, 297 p.

Houlding, S.W., 1994, 3-D Geoscience Modeling—Computer Techniques for Geological Characterization. Berlin: Springer Verlag, 309 p.

Jones, T. A., D. E. Hamilton, and C. R. Johnson, 1986, Contouring Geologic Surfaces with the Computer. New York: Van Nostrand Reinhold, 314 p.

Mallet, J. L., 1991, GOCAD: a computer aided design program for geological applications, *in* Three-Dimensional Modeling with Geoscientific Information Systems, A. K. Turner, ed., Dordrecht: Kluwer Academic Publishers, p.123-142.

Tipper, J. C., 1979, Surface modeling techniques. Lawrence, KS: Kansas Geological Survey, Series on Spatial Analysis 4, 108 p.

Tomlin, C. D., 1990, Geographic Information Systems and Cartographic Modelling. Englewood Cliffs, NJ: Prentice Hall, 249 p.

Turner, D., 1990, POSC; Petroleum Open Software Corp.; a way forward. Geobyte, vol. 5, pp. 36-37.

US Geological Survey, 1996, National geologic map database. http://wwwflag.wr.usgs.gov/ngmdb/ngmdb_home.html

Watson, D. F., 1992, Contouring: A Guide to the Analysis and Display of Spatial Data. Oxford: Pergamon, 321 p.

Wright, D. F., and G. F. Bonham-Carter, 1996, VHMS favourability mapping with GIS-based integration models, Chisel Lake-Anderson Lake area, *in* EXTECH I: a multidisciplinary approach to massive sulphide research in the Rusty Lake-Snow Lake greenstone belts, Manitoba (G. F. Bonham-Carter, A. G. Galley, and G. E. M. Hall, eds.). Ottawa, ON: Geological Survey of Canada, Bulletin 426, pp. 339-376, 387-401.

Turner, A. K., 2000, Geoscientific modeling: past, present, and future, *in* T. C. Coburn and J. M. Yarus, eds., Geographic information systems in petroleum exploration and development: AAPG Computer Applications in Geology, No. 4, p. 27–36.

Chapter 4

Geoscientific Modeling: Past, Present, and Future

A. Keith Turner
Department of Geology and Geological Engineering, Colorado School of Mines
Golden, Colorado, U.S.A.
and
Faculty of Applied Earth Sciences, Delft University of Technology
Delft, The Netherlands

ABSTRACT

This chapter details the evolution of geologic mapping, from the earliest days of computer-aided drafting and automated cartography to the most recent and advanced applications of geographic information systems (GIS). The technical aspects of GIS are reviewed and discussed, and the developmental progression to *geoscience* information systems (GSIS) and associated modeling approaches is presented. Future trends in GIS technology are also identified, particularly as they impact petroleum exploration and production activities. Enhanced capabilities in mapping and in information access, integration, and management are shown to depend on successful implementation of additional Internet and Web-based functionality.

INTRODUCTION

Geological exploration relies on the collection, analysis, and interpretation of spatial data. One hundred and fifty years ago, geologists debated whether it was better to report their observations primarily in narrative form or graphically, using maps. The use of maps won out, and the familiar geologic map became a standard product of geological exploration.

The first geological maps were produced using computer-aided drafting methods in London in the early 1970s. Early research efforts concentrated on the automation of the cartographic process, but subsequently the capability of the computer to support spatial analysis was the subject of intense research. This research laid the groundwork for the evolution of geographic information systems (GIS). Broad commercial acceptance of GIS technology did not occur in North America until about 1988. This initial acceptance rapidly expanded worldwide over the next few years.

Economic and technological realities caused the economic acceptance of subsurface geological modeling to lag behind the GIS developments. However, by the late 1980s a number of competing three-dimensional geoscientific information systems (defined by the acronym GSIS) were the subject of considerable research and evaluation in both North America and Europe.

Currently available commercial GIS and GSIS products reflect their historical origins. Further evolution and integration of geoscientific modeling within corporate decision-support systems is rapidly occurring, spurred by the evolution of the Internet and Web-browser technologies, hardware and software developments, expanded information transmission capabilities, and modern database concepts.

AUTOMATED CARTOGRAPHY

The geologic map has remarkable data compression and data archival capabilities. It can transform a two-dimensional image into a three-dimensional interpretation of subsurface conditions using simple symbols and colors. Mapping conventions have evolved to the point that the geological interpretations being depicted can be easily understood.

These conventions have not always been entirely universal because over the years, different national geological surveys adopted their own specific standards. In general, however, any geologist could interpret any geological map. Frequently, geologic maps were supported by cross-sections, correlations, facies maps, isometric block diagrams, and contour maps (such as isopach maps). All of these products were manually produced by highly skilled workers.

Thus, it is easy to understand how geoscientists were attracted to the promise of computer graphical automation. The petroleum exploration industry became an important early market for the new graphical plotting devices and mapping software. Initially, the technology could support only relatively simple tasks, such as posting of data values. Early experiments with automated contouring revealed that the interpolation process, at least for most geological applications, was very complex, and the contouring software available at this time was inadequate for the task. This finding led to continuing research in automated contouring capabilities, particularly over the last two decades (Tipper, 1979; Jones et al., 1986; Hamilton and Jones, 1992).

By the early 1970s, computer-controlled graphical hardware and supporting software had sufficiently evolved to allow the first experimental production of geological maps by computers. The Experimental Cartography Unit (ECU), established at the Royal College of Art in London, experimented with the production of topographic, geological, and geochemical maps in cooperation with the Institute of Geological Sciences (Rhind, 1988). These early experiments revealed numerous problems, making rapid implementation of automated cartography impractical. The production by the ECU of the 1:50,000 scale geologic map of Abingdon, probably the first standard series multicolor map produced by computer, revealed the difficulties in matching geologic and topographic features (Rhind, 1971). For this display, the geology had been mapped at a larger scale (1:10,560), and when rescaled and registered to the 1:50,000 scale topographic base, misregistration of small or narrow geologic features made the map logically inconsistent. An extremely expensive and time-consuming redigitizing and editing process resolved these problems. However, this experience revealed the difficulty in coping with multiple representations of geologic features, especially when differing scales lead to different generalizations of the same feature.

Map feature generalization has remained a major research problem of interest to geographers and cartographers. Discussions of cartographic generalization emphasize that the process involves several distinct tools that must be used in combination (Beard, 1987; Brassel and Weibel, 1988; McMaster and Shea, 1992). Chrisman (1997) emphasizes the complexity of the interaction between common GIS operations, such as attribute aggregation, and cartographic generalization. Within most GIS packages, "dissolve" or "drop-line" operations provide for aggregation of contiguous polygons having common characteristics, but such operations do not completely provide full cartographic generalization. For example, selection of a single feature category from a map based on a very detailed classification may not produce a map that meets normal map standards in terms of line sinuosity, feature size, and so on. At first, these problems may appear to be inaccuracies in the data, but in reality they arise from assumptions inherent in diverse data sources.

MANAGING SPATIAL DATA

During the 1970s, many individuals came to realize that computer technology promised much more than automation of the cartography process. The true power of computer technology lay in its ability to store, manipulate, and display spatial data. It was the geographers who first recognized the benefits of these capabilities and who began to espouse the concepts. They were greatly attracted by the promise of the still-crude automated cartographic software to provide displays of their data, thereby facilitating improved description of social, economic, and physical phenomena.

The various disciplines that were required to collaborate in response to the needs of the rapidly evolving environmental movement were more attracted to the ability to compare and combine different maps. Map overlay techniques were applied in an effort to identify areas where phenomena coincided, thereby defining suitability for, or sensitivity to, proposed activities. These techniques were greatly affected by the experiments conducted by landscape architects, most notably described in *Design With Nature* (McHarg, 1969). Initially, the map overlays were manually performed by physically overlaying map transparencies on light tables, but soon computers were used to carry out this task.

These capabilities remained largely experimental. Throughout the 1970s, no readily available commercial products provided the generic tools necessary for handling spatially distributed data (Rhind, 1981). This changed dramatically within a decade. The term "geographic information system" (or sometimes "geographical information system," especially in Europe), along with its acronym "GIS," became widely adopted.

COMMERCIAL GEOGRAPHIC INFORMATION SYSTEMS

GIS analysis became extensively used to support a tremendous variety of applications, including social and economic planning, marketing, facilities management (by utility and transportation interests), and environmental and resource assessment (by government regulators and private-sector investors). As might be expected, such a broad assortment of applications caused GIS to become a growth industry of major proportions in the 1980s. By 1988, the core global market for GIS hardware, software, and services was $529 million, with related spending bringing the total GIS-related expenditure to about $4 billion (Daratech, 1989). Not surprisingly, most major computer hardware and software vendors positioned themselves to take advantage of this opportunity. In 1989, the market review conducted by *GISWorld* magazine identified over 60 GIS or mapping systems having at least some GIS characteristics (GISWorld, 1989). The GIS market continued to grow rapidly throughout the 1990s.

CATEGORIES OF GIS

GIS developed under the auspices of two competing data structures: raster data and vector data. A raster-based GIS represents the area being studied as an array of small, uniform-size cells, usually square or rectangular in shape. Each cell is coded to reflect the dominant values or class of the map feature being represented, and spatial relationships are readily derived from the positions of the cells within the matrix or array. In contrast, a vector-based GIS represents map features as digitized points, lines, and polygonal areas. A hierarchy of features and spatial relationships can thus be defined by carefully expressing the topological relations among the point, line, and areal features forming the mapped region. A raster GIS can represent discrete map features with equal facility using groups of contiguous cells and continuously varying phenomena, such as temperature, pressure, or cost data. Vector GIS polygons, in contrast, cannot represent properties that vary continuously within them. Discretized interpolated surfaces, and hence the raster data structure that supports this approach, are often considered the best way to generate interpolated surfaces extending between widely spaced data points.

Because the capabilities of a raster-based GIS are different from those of a vector-based GIS, the two types are used by different sectors of the GIS marketplace. Petroleum exploration applications often use the vector GIS approach because its capabilities are especially good for displaying geological, land ownership, and facilities maps with high cartographic fidelity, and because they allow for some rescaling of existing data to produce new customized databases for specific applications. Raster-based GIS is also used in petroleum exploration applications. The raster data structure can be linked to many numerical modeling functions, and it provides adequate resolution for the creation of interpretive maps that often involve ambiguous or gradational boundaries between map features. In addition, raster GIS offers a convenient interface to many geophysical and satellite data sources and allows for display of continuously varying phenomena.

LIMITATIONS OF GIS

A GIS is designed to efficiently perform extensive, often tedious, and complex mathematical and spatial-modeling of phenomena occurring on the surface of the earth. Such analyses can be conducted primarily with data that represent two-dimensional spatial relationships. The elevation of the Earth's surface is sometimes unimportant when modeling spatial processes. When required by the analysis, elevation can usually be treated as a dependent variable, or attribute, of spatial location. Accordingly, most systems accept and manipulate two-dimensional spatial data, and they provide only limited three-dimensional data display capabilities, such as contoured maps or isometric views. Most commercial GIS products are restricted in their applicability to geological settings because they are focused on two-dimensional data.

In contrast to the earlier automated cartography development, geological exploration applications formed only a very small part of the global GIS market. Consequently, geological GIS applications had to adapt to a generic set of tools that only partly addressed their specific needs. In some cases, these needs could be readily satisfied, either by adapting the existing GIS tools or by developing customized software modules. Specific software modules were developed to assist in the acceptance of common data formats, and symbol sets were provided to create traditional map displays.

Despite the limitations noted, GIS became a significant tool in petroleum exploration during the 1980s, and these systems continue to play a role in the exploration and production strategies of many companies. Conventional GIS software provides important map creation, manipulation, and archiving capabilities, especially for applications where a three-dimensional view of the subsurface is not required. GIS is frequently an important component of regional studies and resource assessments, as well as the production and maintenance of maps showing leases, options, or infrastructure (the locations and characteristics of wells, pipelines, pumping stations, roads, etc.). Increasingly, GIS is becoming a part of an integrated suite of data handling and analysis tools. This book contains several examples of such applications.

THE INTRODUCTION OF GEOSCIENTIFIC INFORMATION SYSTEMS

By the late 1980s, it was apparent that the conventional GIS products, with their predominantly two-dimensional spatial viewpoint, could not entirely satisfy geological exploration requirements. In particular, detailed definition of subsurface conditions, rather than just the areal extent of geologic, hydrologic, climate, and environmental features, required volumetric representation referenced to three orthogonal axes. Numerical modeling and assessments also required that this volumetric information be further linked to three-dimensional data manipulation procedures. A number of research initiatives began to explore the possible extensions of conventional GIS to better serve the geoscience community.

As far as can be determined, German geoscientists first used the term "geoscientific information system" to describe these new research systems (Vinken, 1986, 1988a, 1992a, b). Initially, they favored the term "digital geoscientific map series" because they believed the terms "information system" did not fully describe the complexities of geological data capture, systemization, and dissemination (Vinken, 1986, 1988b). At about the same time, Dutch geoscientists developed computer-based comprehensive data handling and modeling applications for oil and gas exploration using geoscientific information system and the acronym "GSIS" to describe their own applications (Ritsema et al., 1988). In contrast, Raper (1989) and many British researchers were referring to similar systems as geoscientific mapping and modeling systems and the acronym "GMMS." In the period since about 1989, the terms "geoscientific information system" and "GSIS" have become more frequently used to describe computer-based three-dimensional systems designed to handle the variety of geoscience data and analysis needs (Turner, 1991; Bonham-Carter, 1994). "Geoscientific information system" and "GSIS" appear preferable to the alternatives because they offer some parallelism to the widely defined geographic information systems.

The need for the acronym "GSIS" has been questioned because many consider these geoscientific information systems as providing nothing more than support for standard GIS functions in a three-dimensional context. They prefer to use the term "3-D GIS." The concept of a geoscientific information system includes more than support for a three-dimensional geometry. When attempting to define, model, or visualize subsurface conditions, the geoscientist is faced with extremely sparse data, considerable complexity in subsurface conditions, and various types of discontinuities. Thus, these geoscientific information systems must provide additional capabilities and functions to be able to support subsurface conceptualization. Generally, such systems must include rapid and complex data interpolation and extrapolation procedures, provide for interactive data manipulation, and support rapid visualization. They must also support the range of distinctive data formats commonly encountered within a project, and should provide ways to illustrate the uncertainty inherent in the conceptualization. No entirely satisfactory GSIS products presently exist that meet all these criteria.

Raper (1989, 1991) discusses the key concepts for GSIS. Central to his discussion is the ability of these systems to define and manipulate geo-objects. A geo-object may be a discrete physical entity, such as a rock layer or fault block, or it may represent a spatial volume defined by the distribution of a threshold parameter, such as a pollution plume, an ore body, or a sedimentary facies.

Coincidentally, the late 1980s saw several breakthroughs in computer technology that resulted in the production of a new class of high-performance graphical workstations. These workstations reflected a critical price-performance breakthrough. They provided graphical displays and numerical computational speeds that could support the three-dimensional subsurface geological characterization and modeling requirements defined by the GSIS research teams. They also provided these capabilities at considerably reduced prices compared to the prices of earlier computer hardware. Although the costs of these early graphical workstations were still quite high, petroleum company research laboratories and some government and university research facilities bought limited numbers of them on which to undertake GSIS research.

CLASSES OF GEOSCIENTIFIC INFORMATION SYSTEMS

Three-dimensional GSIS technologies, although less mature than their two-dimensional GIS counterparts, may be divided into two major classes on the basis of their methods for defining spatial volumes (Figure 1). Volume representation methods are the three-dimensional equivalent of two-dimensional raster representation—they divide the volume into very large numbers of discrete volume elements, or "voxels." Surface representation methods, in contrast, are the three-dimensional equivalent of the vector-based GIS approach.

The majority of commercially available GSIS products use voxels. Voxel methods offer a data structure upon which more specialized applications can be rapidly built; consequently, they offer strong advantages. Also, voxel techniques are extensively used in medical imaging, they provide opportunities for displaying continuously varying phenomena, and they are more easily related to the gridded structures employed by many numerical models of physical systems. Voxel representations, however, require very large amounts of data storage capacity; for example, a volume represented by 100 rows by 100 columns by 100 layers requires 1,000,000 voxels, and yet provides only a relatively low-resolution image. Volumes represented by tens or hundreds of millions of cells are often involved.

In medical imaging applications, data compression techniques involving octrees and their derivatives are commonly used (Gargantini, 1991). Such methods have not proved as successful for geological applications, with only a few applications reported (Kavouras, 1987; Bak and Mill, 1989; Samet, 1989, 1990). Because many geological features are relatively thin and widespread, they are more continuous in the horizontal (x- and y-axis) directions, and show much more variability in the vertical (z-axis) direction. Some commercial GSIS products reflect this characteristic by accommodating partially deformable geocellular voxels that have uniform dimensions in the x- and y-directions, and variable dimensions in the z-direction (Denver and Phillips, 1990). Other products can convert voxel data structures into three-dimensional isosurfaces prior to

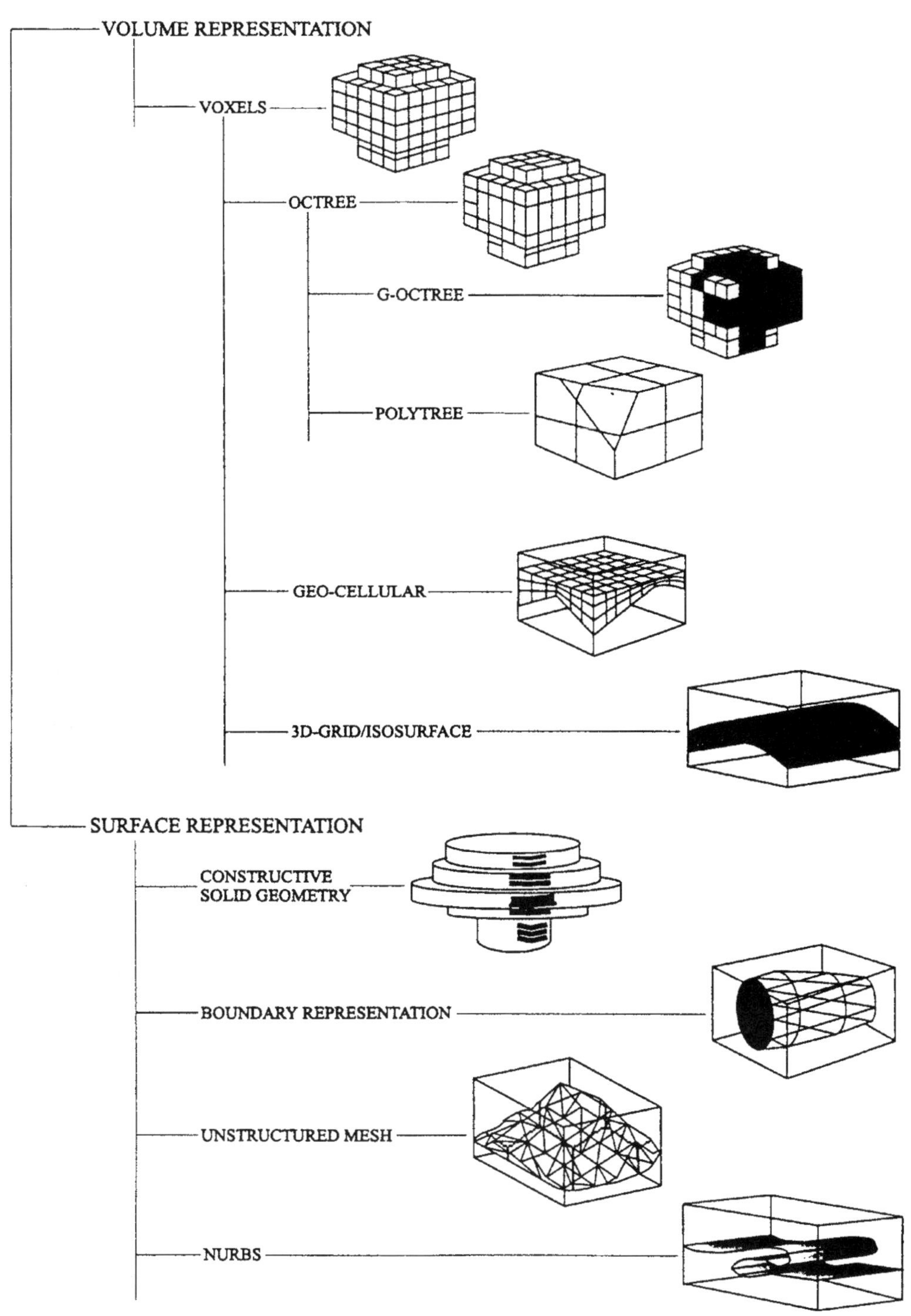

Figure 1—Three-dimensional spatial representations (after D'Agnese et al., 1997).

display (Smith and Paradis, 1989; Belcher and Paradis, 1991).

Surface representation methods have been less widely used in commercial GSIS products. They involve a variety of techniques that are the equivalent of two-dimensional vector GIS methods. Constructive solid geometry facilitates the development of complex shapes by combining simple geometrical primitives, including cylinders, cones, and parallelepipeds (Bak and Mill, 1989; Kavouras, 1991). These methods are often used to describe man-made features, such as excavations, but are not suitable for representing naturally occurring geological features (Houlding, 1994). Boundary representation methods incorporate relatively simple polygon meshes and piecewise linear interpolations to define surfaces and geological features (Mallet, 1988, 1991; Houlding, 1991). More recently, these methods have included the use of unstructured tetrahedral meshes, the three-dimensional equivalents of triangulated irregular network methods (commonly referred to as TINs) used with vector-based GIS. Fisher and Wales (1990, 1992),

Kelk and Challen (1992), and Lasseter (1992) describe experimental applications of curved surfaces and solid modeling based on advanced mathematical splining functions.

TRENDS IN SUBSURFACE CHARACTERIZATION

The process of geological subsurface characterization involves the determination of spatial variation of selected geological parameters. For example, petroleum exploration has emphasized the use of seismic data, almost to the exclusion of alternative data sources. This is beginning to change, however. Other disciplines, including the mining, engineering, hydrogeology, and environmental restoration fields, already routinely use a broader suite of data sources and a more integrated approach to three-dimensional subsurface characterization. In such situations, samples and observations provide isolated direct observations. Spatial prediction and interpolation techniques are used to determine the spatial three-dimensional variation pattern, then various analyses are undertaken to determine the volumes of these measured attributes. Production alternatives can be evaluated, for example, once fluid migration pathways and volumes of oil in place are determined.

The volume and complexity of many datasets make it difficult to accomplish such integrated evaluations; however, three-dimensional visualization makes it easier to decipher trends and correlations. It allows geoscientists to explore multiple hypotheses and to suggest new ideas regarding the nature of their data, often providing new insight and understanding.

To date, three-dimensional visualizations generally have been restricted to one of the three following situations:

- geometric representation of subsurface geo-objects
- visualization of spatial heterogeneity estimated by various simulation methods
- pre- and post-processing of data for numerical models

Unfortunately, these applications are served best by different data representations. Because existing three-dimensional geoscientific visualization systems are evolving, many applications require multiple systems. Data conversion between systems is usually difficult, thereby discouraging all but the most persistent users from fully using the technology.

THE FUTURE FOR GIS IN PETROLEUM EXPLORATION AND PRODUCTION

Current commercially available GIS products have evolved from developments of the late 1970s and early 1980s. Similarly, the current GSIS products are direct descendants of earlier prototypes developed in the late 1980s. Thus, they continue to reflect the data models and other realities imposed by the computing and operating environments available at the time of their development. It appears that an entirely new definition of the functionality and role of GIS within corporate and societal decision-making is now emerging. This new role reflects:

- increasing use and availability of digital data
- cost-performance trends in computer hardware and software
- increasingly seamless integration of software applications
- new database paradigms
- the impact of the Internet and the World Wide Web

Such a change will have considerable impact on the use of GIS in petroleum exploration and production.

Increasingly, governmental agencies are offering baseline environmental and map data in digital form (Hughes, 1998; Turner, 1998a, b). Regulatory agencies are also requiring reports and designs resulting from site investigations to be submitted in digital formats, including maps and other graphics, along with all numerical modeling data. Such trends encourage the adoption of GIS products.

The rapidly expanding power of computers, coupled with declining costs, makes GIS technology increasingly attractive to petroleum exploration, both on technical and economic grounds. Two- and three-dimensional visualizations can materially assist in the determination of trends and correlations exhibited by large and complex multivariate data sets. Such visualizations allow the rapid evaluation of multiple hypotheses and often provide new insights and understanding of the subsurface conditions. In the past, the high cost and complexity of three-dimensional GIS and GSIS software products restricted their use. Now, several lower cost and easier to use PC-based products are being offered that appear extremely attractive in this market.

GSIS techniques have been used as pre- and post-processors for numerical reservoir models, providing valuable data management and graphical display functions. Now GIS tools are increasingly being used to conceptualize the fluid flow systems being modeled, to analyze the uncertainties in values assigned to specific model parameters, and to evaluate risk when using model predictions (D'Agnese et al., 1997). This has led to continued user demands for system and data integration. Current systems and data handling structures make it difficult for users to conduct seamless analyses from beginning to end. However, integrated systems are being introduced that provide data management and analysis capabilities suitable for use in all phases of the exploration process—from conceptualization of the idea, to production and facilities design, to post-production environmental restoration.

A major problem that has constrained GIS developers, and bedeviled GIS users, has been the inability of mainstream relational databases to handle spatial queries. Having controlled the way GIS has developed, the only viable solution up to now was to develop customized approaches to handling geographic data while using database technologies to process the attribute data. The geographic files permitted topological relationships to be developed and preserved, and so allowed for spatial queries to be answered. However, the attribute data were retained in standard, usually relational, database formats. These attribute files required indexing to the geographical files, and so, although they may have been developed and managed using the corporate relational database software, they generally were not part of the corporate database.

Such an approach created two "firewalls;" one between the geographic and attribute files within the GIS, and a second between the GIS applications and the central corporate database. These separations were similar to literal firewalls because they subdivided and isolated the data storage space into compartments. Connections between the compartments were restricted to well-defined and constricted information pipes through the walls. This situation raised significant data management issues because it necessitated replication of data in different databases, which makes keeping the data current a very difficult proposition. Consequently, the need for a specialized GIS database tended to emphasize the uniqueness of GIS, often marginalizing its applications.

In the past year few years, solutions that would break down these firewalls have been proposed and demonstrated by several GIS vendors. In general, these solutions involve additions to mainstream relational database technologies to allow spatial queries and transactions to be conducted within the relational database. Widespread implementation of such a strategy will revolutionize the GIS discipline (Berry, 1998). Not only will it change the internal workings of GIS tools, but it will also change the way GIS is viewed and used within the enterprise. No longer will GIS be considered a specialist's tool, used only for GIS applications. With its new position as part of the corporate information base, GIS will become a "corporate application" (see Mazzo and Burroughs, this volume).

Several vendors have recently provided Web-based mapping products offering at least basic GIS functionality (Limp, 1997; Culpepper, 1998). This rapidly evolving technology uses different approaches to resolve the problems associated with transmission of GIS products across the Internet. The different products are designed to support various applications. Some applications are targeted at novice users, whereas other applications can undertake much more complex GIS analyses; and some easily merge data in many formats, whereas others allow their users to specify customized map displays.

Harder (1998) explains why this diversity in products exists. Many levels of inquiry may be satisfied by serving map information across the Internet. According to Harder, queries may be organized into five levels, ranging from simple to complex. At the simplest level, it is necessary to produce only an electronic copy of a paper map. These maps are static and can be created and displayed as graphics within a standard HTML document. At the second level are maps that show change. The changes may result from display of real-time data, such as weather information. At a third level are maps created in response to user specifications. Such maps may link to extensive databases but display only selected areas or information categories. At more advanced levels, maps may be produced in response to spatial analysis queries (such as buffering or distance calculations) or perform geoprocessing (such as the calculation of slope or shaded relief maps from digital elevation data).

Already, some Internet GIS applications show the way. One of the most geologically oriented Web sites is the Wyoming Internet Map Server (WIMS). Located at <http://wims.sdvc.uwyo.edu/>, it provides spatial data and related information about economic and natural resources in Wyoming. World Wide Web users can access maps of Wyoming, zoom and pan to any area, select an area, and determine what data exist for that area. Closely linked to WIMS is the Natural Resources Data Clearinghouse of the Spatial Data and Visualization Center (SDVC). By using this clearinghouse, Web users can download spatial data sets, as well as browse or query the data documentation (metadata). The clearinghouse is an official node of the National Spatial Data Infrastructure. Several Wyoming state agencies and the National Science Foundation funded WIMS research and development.

WIMS clearly demonstrates how these Internet mapping technologies can influence the distribution and use of geoscience information. In addition to displaying and distributing customized two-dimensional maps, the WIMS site is experimenting with allowing users to display and manipulate geographic and geologic data in three dimensions using Virtual Reality Modeling Language (VRML) (Ames et al., 1997). Currently, WIMS contains two demonstrations designed to show the possibilities of this evolving technology. Further enhancements will allow users to create a three-dimensional display of any portion of Wyoming.

One of the demonstrations available from WIMS provides three-dimensional visualizations of the subsurface geology of the Washakie basin. Depths to formation tops, determined from oil and gas well logs, were gridded to develop surfaces representing the tops of oil- and gas-producing formations within the basin. Users may display selected formations, and then overlay additional reference data, such as major roads, the Public Land Survey grid, and oil and gas fields, or they may include additional scientific information, such as magnetic anomaly data, or hydraulic head data from the Mesaverde Group. Users with a VRML 2.0–enabled Internet browser are able to view, rotate, and explore these three-dimensional models (Ames et al., 1997).

Development of these Web-based applications emphasizes how different users have different requirements and different levels of understanding concerning geographic analysis. Some users—typically the scientists and engineers—require a broad range of extensive and specialized spatial analysis tools. In contrast, a much larger group of users include all those who require broad overviews of the spatial data. This group includes upper management and the public at large. The first group needs software applications, whereas the second group may be satisfied with fairly standardized views of the data that can be produced by widely available software. The two entities really have only one common means of communication, the Web, including various Internet and Intranet components. In many cases, as demonstrated by the existing WIMS site, current graphical Web tools may be sufficient to satisfy the needs of the managers and general public.

The continued evolution of these Web-based GIS capabilities, the increased reliance on digital data, the adoption of integrated information systems that support projects, and the integration of GIS and corporate information systems are mutually supportive. The integration of GIS into the corporate information structure appears to be a crucial step toward the widespread acceptance of GIS as just another tool in the explorationist's toolbox. The chief advantage of this new form of GIS is its configurability—its ability to provide data integrity and security through centralized administration, while providing rapid and economical responses to all users.

The availability of GIS capabilities across the Internet will provide additional flexibility in data access and distribution. Together, they indicate that a rapid expansion of the use of GIS technologies within the petroleum exploration and production fields can be expected. Once this becomes a routine occurrence, geoscientific information, indeed all geospatial information, will become just part of the corporate, or enterprise-wide, information base. GIS and geoscience will then naturally combine to provide appropriate and timely data and information for solving societal problems.

REFERENCES CITED

Ames, A. L., D. R. Nadeau, and J. L. Moreland, 1997, VRML 2.0 Sourcebook, 2nd ed. New York: John Wiley & Sons, 654 p.

Bak, P. R. G., and A. J. B. Mill, 1989, Three dimensional representation in a geoscientific resource management system for the minerals industry, *in* Three-Dimensional Applications in Geographical Information Systems, J. F. Raper, ed., London: Taylor and Francis, p. 155–182.

Beard, M. K., 1987, How to survive on a single detailed database, *in* Proceedings, AUTO-CARTO 8. Falls Church, VA: American Society of Photogrammetry and Remote Sensing (ASPRS), p. 211–220.

Belcher, R. C., and A. R. Paradis, 1991, A mapping approach to three-dimensional modelling, *in* Three-Dimensional Modelling with Geoscientific Information Systems, A. K. Turner, ed., Dordrecht, The Netherlands: Kluwer Academic Publishers, NATO ASI Series C: Mathematical and Physical Sciences, vol. 354, p. 107–122.

Berry, J. K., 1998, GIS software's role is changing. GIS World, vol. 11, no. 9, p. 28–30.

Bonham-Carter, G. F., 1994, Geographic Information Systems for Geoscientists—Modelling with GIS. New York: Pergamon Press, 398 p.

Brassel, K. E., and R. Weibel, 1988, A review and conceptual framework of automated map generalization. International Journal of Geographical Information Systems, vol. 2, p. 229–244.

Chrisman, N., 1997, Exploring Geographic Information Systems. New York: John Wiley & Sons, 298 p.

Culpepper, R. B., 1998, Weave maps across the web, 1998 edition. GEOWorld, vol. 11, no. 11, p. 46–52.

D'Agnese, F. A., C. C. Faunt, A. K. Turner, and M. C. Hill, 1997, Hydrogeologic evaluation and numerical simulation of the Death Valley regional groundwater flow system, Nevada and California. Reston, VA: U.S. Geological Survey, Water Resources Investigations Report 96-4300, 124 p.

Daratech, 1989, Geographical Information Systems—Markets and Opportunities. Cambridge, Massachusetts.

Denver, L. F., and D. C. Phillips, 1990, Stratigraphic geocellular modeling. Geobyte, vol. 5, no. 1, p. 45–47.

Fisher, T. R., and R. Q. Wales, 1990, 3-D solid modeling of sandstone reservoirs using NURBS. Geobyte, vol. 5, no. 1, p. 39–41.

Fisher, T. R., and R. Q. Wales, 1992, Rational splines and multi-dimensional geologic modeling, *in* Computer Graphics in Geology: Three-Dimensional Computer Graphics in Modeling Geologic Structures and Simulating Processes, R. W. Pflug and J. W. Harbaugh, eds., Berlin: Springer-Verlag, Lecture Notes in Earth Sciences, no. 41, p. 17–28.

Gargantini, I., 1991, Modeling natural objects via octree, *in* Three-Dimensional Modeling with Geoscientific Information Systems, A. K. Turner, ed., Dordrecht, The Netherlands: Kluwer Academic Publishers, NATO ASI Series C: Mathematical and Physical Sciences, vol. 354, p. 145–158.

GISWorld, 1989, Results of the 1989 GIS World GIS Survey, *in* GIS Technology '89. Fort Collins, CO: GISWorld, Inc.

Hamilton, D. E., and T. A. Jones, 1992, Computer Modeling of Geologic Surfaces and Volumes. Tulsa, OK: AAPG, 297 p.

Harder, C., 1998, Serving Maps on the Internet. Redlands, CA: Environmental Systems Research Institute (ESRI), 130 p.

Houlding, S. W., 1991, The application of new 3-D computer modelling techniques to mining, *in* Three-Dimensional Modeling with Geoscientific Information Systems, A. K. Turner, ed., Dordrecht, The Netherlands: Kluwer Academic Publishers, NATO ASI Series C: Mathematical and Physical Sciences, vol. 354, p. 303–325.

Houlding, S. W., 1994, 3D Geoscience Modeling—Computer Techniques for Geological Characterization. Berlin: Springer-Verlag, 309 p.

Hughes, J. R., 1998, How about some free data? GIS-World, vol. 11, no. 9, p. 8.

Jones, T. A., D. E. Hamilton, and C. R. Johnson, 1986, Contouring Geologic Surfaces with the Computer.

New York: Van Nostrand Reinhold, 314 p.

Kavouras, M., 1987, A Spatial Information System for the Geosciences. Unpublished Ph.D. Dissertation, University of New Brunswick, Fredericton, New Brunswick, Canada.

Kavouras, M., 1991, A spatial information system with advanced modeling capabilities, *in* Three-Dimensional Modeling with Geoscientific Information Systems, A. K. Turner, ed.), Dordrecht, The Netherlands: Kluwer Academic Publishers, NATO ASI Series C: Mathematical and Physical Sciences, vol. 354, p. 59–67.

Kelk, B., and K. Challen, 1992, Experiments with a CAD system for spatial modelling of geoscientific data, *in* From Geoscientific Map Series to Geo-Information Systems: Proceedings of the Second International Colloquium (R. Vinken, comp.). Bundesanstalt fur Geowissenschaften und Rohstoffe, Hannover, Geologisches Jahrbuch, Reihe A, Heft 122, p. 125–153.

Lasseter, T. J., 1992, An interactive 3-D modelling system for integrated interpretation in hydrocarbon reservoir exploration and production, *in* Proceedings of International Symposium on 3D Computer Graphics in Modeling Geologic Structures and Simulating Geologic Processes, R.W. Pflug and J.W. Harbaugh, eds., Berlin: Springer-Verlag, Lecture Notes in Earth Sciences, No. 41, p. 189–198.

Limp, W. F., 1997, Weave maps across the web: GIS-World, vol. 10, no. 9, p. 46–55.

Mallet, J. L., 1988, Three-dimensional geographic display of disconnected bodies. Mathematical Geology, vol. 20, p. 977–990.

Mallet, J. L., 1991, GOCAD: a computer aided design program for geological applications, *in* Three-Dimensional Modeling with Geoscientific Information Systems (A. K. Turner, ed.). Dordrecht, The Netherlands: Kluwer Academic Publishers, NATO ASI Series C: Mathematical and Physical Sciences, vol. 354, p.123–142.

McHarg, I. L., 1969, Design With Nature. New York: Natural History Press, 197 p.

McMaster, R. B., and K. S. Shea, 1992, Generalization in Digital Cartography.

Washington, DC: Association of American Geographers, 134 p.

Raper, J. F., 1989, The 3D geoscientific mapping and modelling system: a conceptual design, *in* Three-Dimensional Applications in Geographic Information Systems, J. F. Raper, ed., London: Taylor and Francis, p. 11–19.

Raper, J. F., 1991, Key 3D modelling concepts for geoscientific analysis, *in* Three-Dimensional Modeling with Geoscientific Information Systems (A. K. Turner, ed.). Dordrecht, The Netherlands: Kluwer Academic Publishers, NATO ASI Series C: Mathematical and Physical Sciences, vol. 354, p. 215–232.

Rhind, D. W., 1971, The production of a multi-colour geological map by automated means. Nachr. Aus den Karten und Vermessungeswesen, Heft 52, p. 47–52.

Rhind, D. W., 1981, Geographical information systems in Britain, *in* Quantitative Geography (N. Wrigley and R. J. Bennett, eds.). London: Routledge and Kegan, p. 17–39.

Rhind, D. W., 1988, Personality as a factor in the development of a discipline—the example of computer-assisted cartography. American Geographer, vol. 15, p. 277–289.

Ritsema, I. L., M. Riepen, J. Ridder, and S. L. Paige, 1988, Global System Definition for Version 0.x of the Geoscientific Information System. Delft, The Netherlands: SWS Document GSIS 1.1, TNO Institute of Applied Geoscience.

Samet, H., 1989, Applications of Spatial Data Structures—Computer Graphics, Image Processing, and GIS. Reading, MA: Addison-Wesley, 507 p.

Samet, H., 1990, The design and analysis of spatial data structures. Reading, MA: Addison-Wesley, 439 p.

Smith, D. R., and A. R. Paradis, 1989, Three-dimensional GIS for the earth sciences, *in* Three-Dimensional Applications in Geographic Information Systems, J. F. Raper, ed., London: Taylor and Francis, p. 149–154.

Tipper, J. C., 1979, Surface Modeling Techniques. Lawrence, KS: Kansas Geological Survey, Series on Spatial Analysis 4, 108 p.

Turner, A. K. (ed.), 1991, Three-Dimensional Modeling with Geoscientific Information Systems. Dordrecht, The Netherlands: Kluwer Academic Publishers, NATO ASI Series C: Mathematical and Physical Sciences, vol. 354, 443 p.

Turner, A. K., 1998a, U.S. geological survey releases free oil and gas datasets. GISWorld, vol. 11, no. 9, p. 40–41.

Turner, A. K., 1998b, New NOAA datasets describe the U.S. coastal zone. GEOWorld, vol. 11, no. 11, p. 34–36.

Vinken, R., 1986, Digital geoscientific maps: a priority program of the German Society for the Advancement of Scientific Research. Mathematical Geology, vol. 18, p. 237–246.

Vinken, R. (comp.), 1988a, Construction and Display of Geoscientific Maps Derived from Databases: Proceedings of the International Colloquium. Bundesanstalt fur Geowissenschaften und Rohstoffe, Hannover, Geologisches Jahrbuch, Reihe A, Heft 104, 475 p.

Vinken, R., 1988b, Digital geoscientific maps—A research project of the Deutsche Forschungsgemeinschaft (German Research Foundation), *in* Construction and Display of Geoscientific Maps Derived from Databases: Proceedings of the International Colloquium (R. Vinken, comp.). Bundesanstalt fur Geowissenschaften und Rohstoffe, Hannover, Geologisches Jahrbuch, Reihe A, Heft 104, p. 7–20.

Vinken, R. (comp.), 1992a, From Geoscientific Map Series to Geo-Information Systems: Proceedings of the Second International Colloquium. Bundesanstalt fur Geowissenschaften und Rohstoffe, Hannover, Geologisches Jahrbuch, Reihe A, Heft 122, 501p.

Vinken, R., 1992b, From digital map series in the geosciences to a geo-information system, *in* From Geoscientific Map Series to Geo-Information Systems: Proceedings of the Second International Colloquium (R. Vinken, comp.). Bundesanstalt fur Geowissenschaften und Rohstoffe, Hannover, Geologisches Jahrbuch, Reihe A, Heft 122, p. 7–25.

Wade, G., 2000, GIS technology—a vendor perspective of its history and prospects in petroleum exploration and production, *in* T. C. Coburn and J. M. Yarus, eds., Geographic information systems in petroleum exploration and development: AAPG Computer Applications in Geology, No. 4, p. 37–46.

Chapter 5

GIS Technology—A Vendor Perspective of its History and Prospects in Petroleum Exploration and Production

Geoff Wade
Environmental Systems Research Institute
Redlands, California, U.S.A.

ABSTRACT

Many oil and gas companies and related service entities are discovering the spatial nature and true value of their exploration and production data sets as they begin to integrate geographic information systems (GIS) into the management of their everyday business functions. This chapter describes the path that many GIS practitioners have shared in getting to this point of integrated GIS implementation. It also discusses the role of the GIS vendor community, exposes the latest in technology developments, and presents the direction these developments are likely to take in the foreseeable future. Finally, samples of several applications developed by major petroleum industry players are referenced, thereby illustrating the commercial value of GIS and explaining the consequent expansion in take-up that the technology is currently enjoying.

INTRODUCTION

Despite several years of development, an appreciation of the value of integrating the spatial component in everyday datasets with conventional information systems (IS), and the consequent ability to perform combined attribute/spatial analysis, is still in its infancy in many petroleum-oriented organizations. Yet, the early adopters of GIS technology in the mid-1980s have gained significant benefit from its relatively independent use. Over the past few years, however, the pace of GIS acceptance, specifically integration of the spatial data component with corporate IS, has expanded dramatically, such that there are now numerous commercially profitable examples to share. This volume itself contains several such examples from independent sources.

Today, many oil companies, data vendors, and application builders are taking advantage of a new breed of integratable GIS technologies to capitalize on the spatial data component inherent to virtually all petroleum business objects. They are re-engineering the logical petroleum business models to take advantage of integrated attribute and spatial queries. By integrating spatial data into the back-end database, and exposing petroleum business objects through the insulating layer of a logical model, such as provided by PPDM (Public Petroleum Data Model Association) or POSC (Petrotechnical Open Software Corporation), full advantage may be taken of a band of sophisticated commercial application builders that provide integrated, desktop applications that are both focused on exploration and production (E&P) and spatially oriented.

This is not to say that the technology has reached its goal, but that it has just started to realize it in the hands of the more advanced developers and their users. It is now possible to look forward to a very active few years of development as the integration of the spatial component into conventional E&P systems is pursued to the level of its true potential.

Those people integrally involved in this process of development know that there are still many hurdles to overcome, including user acceptance, legacy data conversion, and accessibility issues. Nonetheless, it is an exciting time to be involved as a practitioner, and this is true irrespective of the point along the development chain at which one becomes involved.

HOW WE GOT HERE—A PERSONAL PERSPECTIVE

In 1986, I was working as an Oracle database administrator in the computer department of a UK-based data vendor. As time passed and my role in the further development of the database increased, I became exposed to the most awkward of questions from users: "How do we enter real-world coordinates efficiently? How do we view the data as a map and project it, so it fits with data from another source?" Quite literally, at the time, I didn't have a clue!

Meanwhile, in another building, the cartography department had just invested in something called a GIS at the suggestion of a very bright fellow in my own department. I investigated, both with my friend in computing and these *geography types* in cartography. The cartography guys had purchased something-called ARC/INFO from ESRI in the U.S., and busied themselves re-training from the computer-aided drafting (CAD) packages they were used to.

There was very little opportunity at that time to put a GIS front-end on the Oracle database I was managing. Rather, it would have meant re-engineering the database actually inside the GIS. We didn't want to do that! After all, this was only a small percentage of the user requests (or so we thought at the time). So instead, we learned what we could from reading the GIS manuals and went to work with a homegrown graphical kernel system (GKS) library.

Massive success—the users loved it ! It seemed that when we offered the data as a simple map view, it just came alive. The users could see at once the data coverage, and combined with a simple data form, quickly estimate its value relative to some of their spatially oriented business decisions. That was the beginning of a revelation for me in the value of data visualization, and I've been a firm advocate ever since.

Well, that was the long way around, and a bit of unabashed personal history, but it usefully illustrates the fact that many of us now working in GIS have converged on the technology from quite different angles. This is a good thing because, with such a rapid evolution path, the technology needs, and is better as the result of, diverse user input.

A colleague at ESRI who is currently working on integrating 3D spatial capability uses an analogy to express the value of GIS to the IS community at large. He compares it to the value of color to printing or to the TV. Once you've seen it, he argues, it's almost impossible to imagine going back to monochrome. For me, if the spatial components of leases, seismic lines, wells, reservoirs, and pipelines can be integrated to other systems and analyzed effectively on the desktop for better decision making, then it's hard to believe users would be willing to do without it. Integrated spatial data management is here to stay.

I have, on occasions, met developers who were very GIS-centric in their view of the technology, believing somehow that it provided something of a Holy Grail. I don't think there are many people around now who feel this way. Rather, GIS is being integrated into our existing IS architecture, and for want of a better analogy, many of today's new business-focused applications should come with a sticker that reads "GIS Inside." The fact is, many new application users don't realize they have embraced GIS, and don't care whether they have or not. They just know that the application speaks to them in a context they understand, that it's easy to use, and that it allows them to make business decisions more efficiently.

No doubt many individuals' experiences with the evolution of GIS in their own companies mirror those described. The first GIS use was probably associated with specific islands of development, that is, it was very much decentralized. Through the different ruminations of rightsizing, outsourcing, IS re-engineering, and acceptance of desktop applications, most organizations have arrived at a well-informed IS department, eager to embrace the value of the technology and to expose it to the users through many different application scenarios.

THE GIS VENDOR AND TODAY'S THREE-TIER ARCHITECTURES

Today's IS architectures are probably best described as three tier in structure (Figure 1). This comprises a physical database, which actually contains the information; a logical model, which represents that data in some business context; and an application layer, which delivers the data to the user for a specific business function. Each of these three main layers can be further subdivided by task. Some of the important considerations of each layer are discussed in the following paragraphs.

GIS Integration with the Physical Database

For a long time GIS was not that easy to integrate into the corporate IS structure. Generally, GIS used a suite of proprietary formatted binary "flat-files" to store coordinate geometry, and it held the associated topological information and attributes about the spatial "objects" in simple relational tables. GIS has been able to connect to relational database management

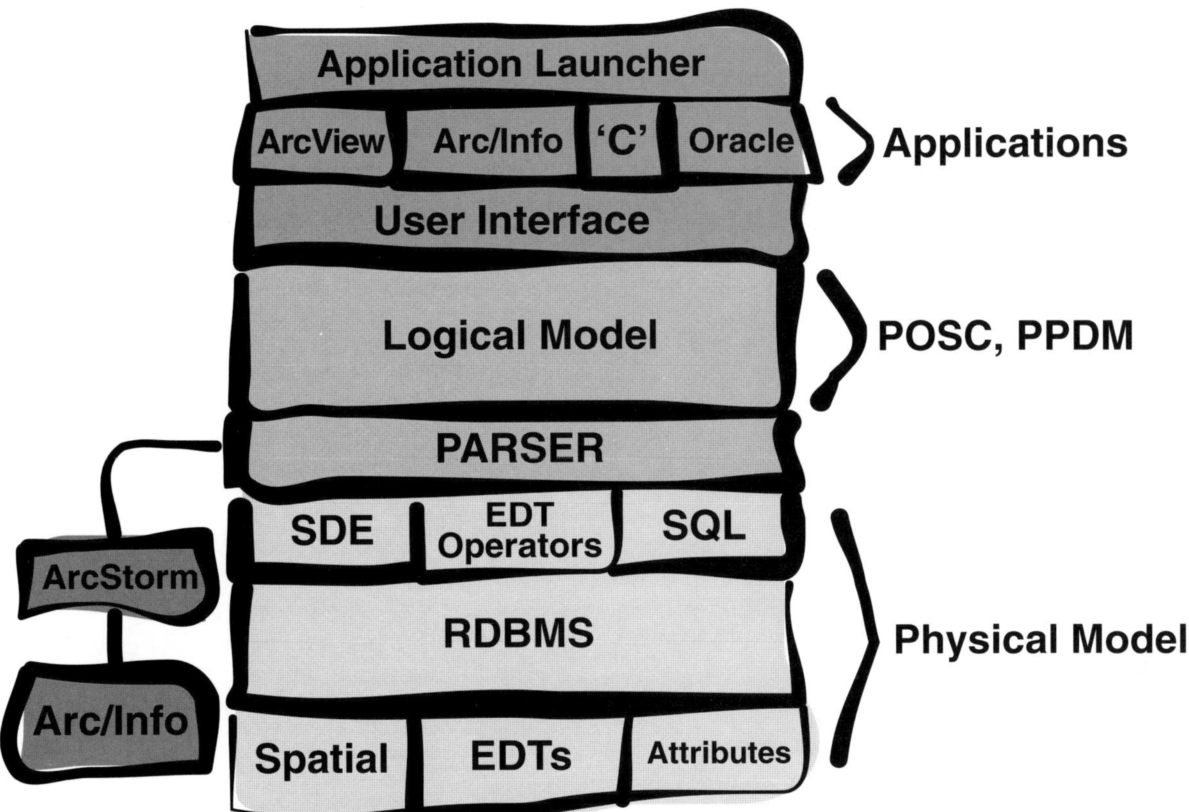

Figure 1—Typical three-tier database/GIS (geographic information system) architecture. ArcStorm, Arc/Info, and ArcView are products of Environmental System Research Institute (ESRI). RDBMS = Relational Database Management System; SQL = Structural Quality Language; EDT = External Object Data Types; SDE = Spatial Data Extensions.

systems (RDBMSs) and a host of other software systems for many years, but on the whole, these connections were not that strong, and there were problems reconciling many-to-many table relationships. The "relate" between the GIS data and the relational table was seen as a potential weak link, requiring a form of batch processing.

Many GIS packages eventually developed and embraced tabular-formatted data tables, CAD integration, some image processing, and the ability to perform quite sophisticated desktop cartography. For many of today's more spatially focused applications, this type of stand-alone system was, and remains, ideal; however, where the principal goal was to integrate the spatial component with the corporate IS network, Oracle database server, and desktop applications, a new approach was required.

The technology is now available to move the spatial data from proprietary "flat-files" so that they actually reside within the RDBMS tables. The most common methodology for this is to make use of the "long-raw" or binary large object (BLOB) data-type to store the spatial geometry, and to store the topological and attribute information in associated RDBMS tables. In effect, there is now no weak link. In fact, no external link is necessary at all. The GIS application can be built directly on top of the RDBMS to take advantage of an IS staff that is very knowledgeable about managing and optimizing RDBMS performance. All the security and referential integrity issues are handed off to the RDBMS application programming interface (API), and the GIS vendor and user alike get to leverage all that existing experience. [In fact, at the time of this writing there were at least three ways to implement the spatial data within a RDBMS: using a BLOB as described, using a set of normalized tables, and using extended object data types (EDTs) as true spatially enabled business objects. This last option is quite new and is able to treat the business objects as classes having inheritance and attached methods. The storage methodology for this last option may be a combination of BLOB and normalized schemas, with the object properties also stored in the database and accessed via specific API code from the vendor.]

The API and Open Standards

Once the spatial data are inside the RDBMS, along with the attribute data that describes it, it is still necessary to have a spatially capable API to operate on it in a true spatial context. Most of the major RDBMS vendors are currently commissioning or cooperating with GIS vendors to integrate a formal spatial API into their existing tool set, which is generally provided by some vendor-applied form of the ANSI 1992, structured query language (SQL). This spatial API, or spatially enhanced SQL2 API, extends the normal RDBMS SQL capability by adding spatial functions to efficiently load, manage, and retrieve/interrogate the spatial geometry data held in the database tables.

Eventually SQL3 will include certain generic spatial functions; but until it does, the GIS vendors are the principal providers of spatial APIs. Several of the major database vendors have formed associations with GIS vendor companies to provide pragmatic solutions until a formal SQL3 standard is in place. Until then, each GIS vendor, and some database vendors, will have their own proprietary API, although most are endeavoring to be open in their attitude and publish their API calls and data formats on the Web.

These open APIs make integration with many programming languages very straightforward and expose the versatility of writing combined attribute and spatial queries to optimize application performance. The API can be integrated with desktop application products and can be made accessible to the host of today's desktop application development languages.

The Open GIS Consortium, Inc.

The Open GIS Consortium, Inc. (OGIS) takes a rather different approach to GIS interoperability. The OGIS mandate is essentially concerned with encouraging an open GIS architecture and API interchangeability. As such, data storage methodologies, and the functions that the various commercial GIS packages should perform, would be somewhat standardized, although the way in which they may be implemented by specific vendors may continue to vary. A certain consensus, however, would at least be erected on the representation of spatial features and perhaps on the major functions and operators. OGIS has already defined a "simple features type" specification and has accepted collaborative proposals on ways to read/write these simple feature types and to address them through geographic functions written from SQL, OLE (object linking and embedding), COM/DCOM (component object model/distributed component object model), and CORBA (common object request broker architecture) environments. The various GIS vendors are now implementing and certifying their products to work with these standards. (Please see the OGIS Web site and individual GIS vendors for the most up-to-date information, because this is changing quickly at the time of this writing.)

E&P Logical Models

For a long time, the petroleum industry has been working toward the publication of an open standard on a consistent logical business model. As previously noted, two principal bodies exist: POSC and PPDM. Essentially, the derivation of a petroleum logical data model is not the concern of any GIS vendor. Rather, it is the business of the oil companies and service companies delivering data or building commercial applications that address particular petroleum business problems. The work of the POSC and PPDM organizations does, however, impact GIS vendors in the sense that whatever logical model a company decides to use, it must be capable of being physically implemented using the vendor's physical data management technologies. Most GIS vendors are continuing to work with both POSC and PPDM to (1) ensure that the logical models being designed can be physically implemented using the appropriate data management tools, (2) use offered portions of the standard, and (3) provide an active contribution to the model itself, where applicable. This collaborative work will become more important as progress is made toward the use of generic, spatially enabled object technologies and the representation of petroleum business objects within this frame.

Functionally Integrated Applications Applied to the E&P Business

Once a company has decided on a logical model that fits its business need and has physically implemented that model, then any number of applications may make use of the information. In time, as SQL3 becomes accepted, OGIS publishes standards, and APIs become more open and integratable, GIS or spatially enabled desktop applications will become "plug and play;" that is, they will largely become interchangeable from different vendors. Of course, there will still be differentiation even where standards exist (such as now exists between the commercial RDBMS vendors, even with SQL2 in place), but the fundamental underpinnings of spatial technology should be more similar.

This integration is the key to taking GIS to the next level of user acceptance. Done correctly, it has the potential to massively increase the inherent use of GIS technology, as it potentially exposes easy-to-use spatial functions to otherwise non–spatially enabled applications without substantial recoding.

Systems Issues

It is the goal of most software vendors to be hardware and system independent to the greatest degree possible. This provides the largest choice to the end user in terms of system selection, and makes the later introduction of software to an existing environment much easier; however, certain styles and methodologies of code development are not possible in some domains; or, they are not practical to implement with present technologies on some hardware platforms. Each vendor essentially attempts to support as many Unix, PC, and Mac platforms as possible by standardizing the bulk of the code, while retaining optimized features for certain platform environments.

In summary, there are some choices to make when considering the way in which to implement GIS technologies in a given environment. The vendors currently distinguish themselves on base functionality and integrability, but as time proceeds, these factors will become less defining as standards force convergence.

The goal for a vendor must be integration/connectability at the database level and embeddability at the application level, coupled with a thorough understanding of, and capability to implement, the specific requirements of a particular set of industry-led business functions.

TODAY'S TECHNOLOGY AND TOMORROW'S DIRECTION

There are many ways to portray a GIS. Figure 2 illustrates a functional description of a GIS lying at the intersection of five potentially disparate tasks: digital cartography, CAD, image processing, database connectivity, and the ability to perform some form of integrated spatial analysis. The spatial functions may be as simple as calculating the area of a lease polygon or as complex as defining an oil migration pathway, modeling a steam injection reservoir, designing a pipeline gathering system, planning for environmental contingencies, or siting a new retail location. The unifying factor is that all these processes embody substantial spatial analysis that is uniquely provided for by GIS (see Figure 3). It is appropriate, then, to look at some of the technological trends that will influence the industry direction over the next few years.

Physical Database

Most GIS data generally are still stored in flat binary files encapsulating their geometry with a set of associated attribute tables holding some elements of the spatial topology, as well as other attribute data belonging to the spatial objects. When databases became large, many GIS vendors devised some form of tiling system so as not to lock very large datasets during update. Eventually, user demand for contiguous data layers and better transactional support led to the design of client-level contiguous spatial layers. These are managed by a form of middle layer or spatial broker that can retrieve spatial data and handle transactional consistency, rollback, and many of the features that are common to RDBMSs. More recent technical development with RDBMS has lead to many of the present storage architectures described and is revolutionizing both the scale of GIS deployments and their integration with mainstream IS programs.

Application Programming Interfaces

As was mentioned, SQL3 is currently being revised to include several spatial operators, as well as many other object manipulation clauses, and OGIS has begun the task of defining generic spatial object types. Gradually, the GIS and relational database vendors are converging on an open standard for the representation of the features in the database and perhaps the spatial meta-data to be held along with them. What then is left for individual vendor definition in the marketplace? Well, just as there is a standard for SQL2 now, but several commercial implementations with vendor additions, there will be plenty of room left to differentiate the market even with the standards because standards only take care of the underlying fundamentals.

Applications

Another important trend in the marketplace is the transition to, and redevelopment of, proprietary desktop application software using Microsoft- or Java-based "component-ware." Some time ago, several GIS vendors started to deliver GIS toolboxes alongside their commercial-off-the-shelf (COTS) desktop GIS products. Product growth has been steady, as desktop developers (particularly in the Microsoft environment) who currently use Visual Basic (VB), Visual C++, or one of the scripting languages like Delphi or Power-Builder, gradually pick up the GIS toolboxes and start to use them. This is not to say that COTS products will go away, but that they will be available as both the ready-to-run compiled application code and as a source-based spatial toolbox or software development kit (SDK). The important point is that they will share the same libraries, be better integrated in the desktop environment, and be easier to upgrade and "talk" to each other seamlessly.

Recently, GIS vendors have been delighted to find that many leading petroleum application developers are also re-engineering based on Microsoft standards, or are intending to build on Java. Such development can only help the further integration of GIS to the desktop application environment for the bulk of users. Also on the horizon are a large number of specialized application extensions to the desktop and professional GIS products. Extensions will allow a host of more specific functions to be executed; for example, raster analysis, network modeling, street geocoding/route planning, image analysis, three-dimensional and temporal modeling, etc. Such functionality will likely be accessible to users on an as-needed basis from perhaps a departmental application server, thereby minimizing the number of concurrent licenses needed by a corporation. Still further specialization of certain business functions will lead to the development of specific application code designed to perform particular applied business functions (for example, routing trucks or siting retail outlets). These have already started hitting the shelves of the computer superstores, and they are likely to increase in number fairly rapidly.

Finally, the World Wide Web is bringing societal standardization of some application environments (for example, the use of Netscape or Microsoft's Internet Explorer). For this reason, several vendors are currently releasing Web-based GIS browsers in the hope of emulating how Netscape became a de-facto standard for awhile by being there at the right time. Several GIS vendors will also bring slim-line upgradeable products to the Web, probably built from a combination of HTML/Java-based client widgets and COM/DCOM-based Web servers. With the Web, it seems that everything is about ease-of-use and user acceptance,

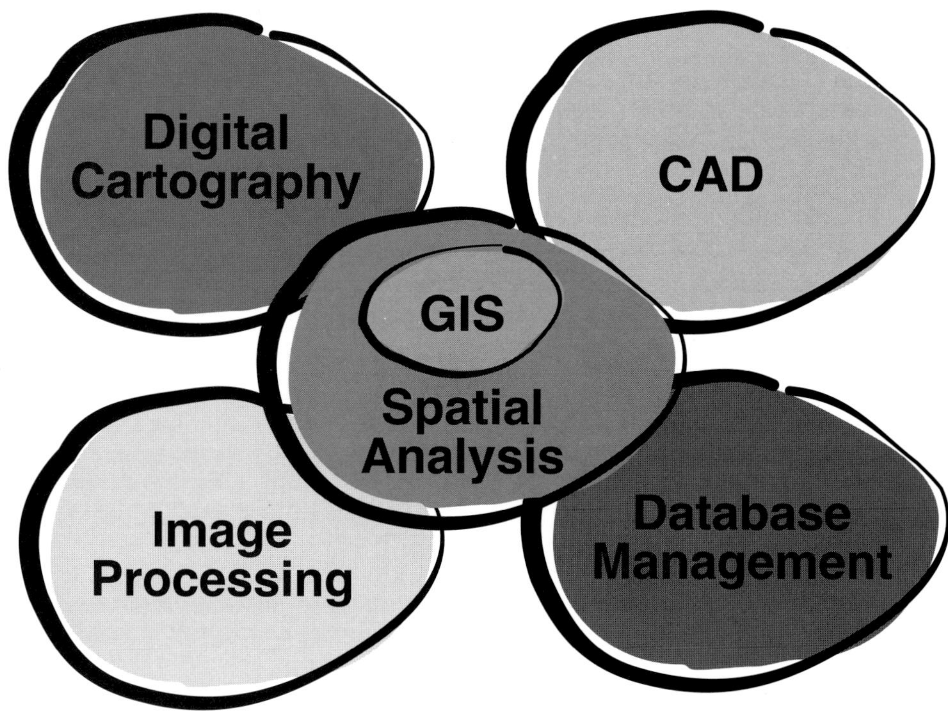

Figure 2—Functional capabilities of a geographic information system.

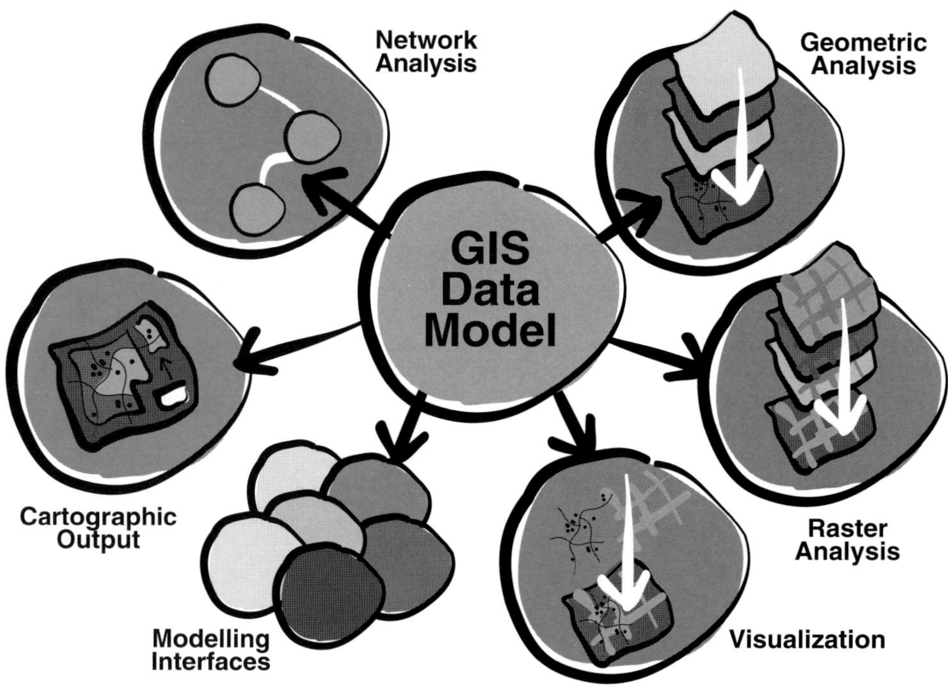

Figure 3—Aspects of spatial analysis enabled by a geographic information system.

which will provide its own drivers back into the next generation of applications developed for the user's desktop environment.

For the Systems Guys

Like it or not, Microsoft has the lion's share of the desktop market, so applications constructed on Microsoft's libraries and object code will likely be around for some time. It looks like the jury is still out with many vendors regarding the use of CORBA, except for confirmed Unix development shops looking for a standard to bear. It seems likely that it will play a role in terms of distributed database connectivity, and then hand-off in some way to COM as the principal application development standard. Java is suggested as a get-around to this whole issue, being cross-platform from its very inception; however, because of Java's commercial immaturity and certain limitations in functionality, it will struggle to be used as anything more than as a Web-based programming language for the foreseeable future.

What Does It All Mean?

For end users, all this probably doesn't mean a lot because on the whole they get what the developers give them. Developers, however, trying to plan a long-term strategy, economize on code re-engineering, and maximize their marketplace have some decisions to make.

WHO IS APPLYING GIS EFFECTIVELY?

This question is easy to answer: lots of people are, and they are making a real impact on the bottom line of their businesses because of it. As described, GIS vendors supply business partners and end-users with somewhat different technologies, depending on their focus. Some examples from the different business partner and end-user perspectives are referenced in the following paragraphs, and some of these are described elsewhere in this volume.

The Data Vendor

The business objective of a petroleum data vendor essentially surrounds the efficient dissemination of its data and not, on the whole, the design of end-user systems to accommodate that data in an E&P context. As such, data vendors tend to use out-of-the-box desktop applications that can quickly be made to encapsulate their data products. This approach makes the data immediately available at the client site upon arrival and minimizes the vendor's requirement to employ applications design engineers, or respond to the requirements of a varied customer base, over and above the data's efficient delivery. Facilities to supply the dataset in an open format, and to place a desktop viewer directly on the data CD for just a few dollars, have significantly increased the use of desktop GIS applications in recent years. The more elaborate task of loading the supplied data to the corporate master archive can then be undertaken by the customer's IS department and need not hold up the data's immediate use by an end user. The recent advances with RDBMS extensions, combined with the latest Web developments described, are hastening the time when data vendors will supply information on demand from a central data warehouse. Some of the major vendors already use this type of technology solution (for example, IHSEnergy Group, QC Data, Tobin International, Veritas, and PGS) and are thus approaching on-demand e-commerce data supply.

The Application Vendor

The aim of the application vendor is to supply the best applications for one's business needs. With this objective, rather different demands are placed on the GIS technologies. The application vendors are concerned with a robust development environment, technical capability, extensibility, run-time pricing, integrability, and desktop ease-of-use. The power of visualization supplied by the GIS is important here, as are the abilities to use standard industry nomenclature and symbols, honor geological rule sets (for example, on contouring), easily construct hard copy, and perhaps visualize the data in three or four dimensions. All these desktop facilities are available, at least in a rudimentary form, within desktop GIS today, and there are now a host of petroleum application developers selectively building products to address everything from E&P, through pipeline and facilities management, to retail site planning and environmental impact analysis. Previously, the desktop and professional development environments have been used to produce GIS-focused applications. Now, with the advent of extended RDBMSs, these applications are being re-addressed to work directly off of a spatially enabled, extended RDBMS to produce dynamic database-generated mapping (for example, Landmark's OpenExplorer or IHSEnergy's Edge products).

The End User

The oil companies themselves are extensive users of GIS technology. This is partly because they find they can use the generic deliverables right out of the box. It is also partly because they either have a desire to construct their own products or they are forced to do so because they cannot find the applications they require in the commercial marketplace.

The oil companies are thus users of a broad range of GIS tools, both the professional high-end GIS toolboxes and the desktop application products that they frequently use to disseminate the collected data. Within oil companies, desktop applications are being installed on intranets for easier application management and faster data dissemination. On the whole, oil companies are very Oracle literate, and they are almost all spatially enabling their RDBMSs using the new extended technologies described earlier.

The first oil company departments to realize the benefits of and to implement GIS were characteristically from the E&P arena, but now, as the corporations reorganize and streamline themselves, and as staff members are gradually introduced to other corporate venues, the technology is rapidly spreading downstream. In many vertically integrated companies such as BP/Amoco, Texaco, Chevron, and Shell, most departmental sectors are using GIS in some form or another. Senior IT management is beginning to leverage advances and technology knowledge in one area of the company for corporate-wide benefit (see Figure 4). Still to be realized on a significant scale are the cross-industry technology references that could be leveraged to their full potential (for example, truck routing, network planning, and retail site positioning/customer profiling) that are used more extensively by other industries. Cross-industry technology

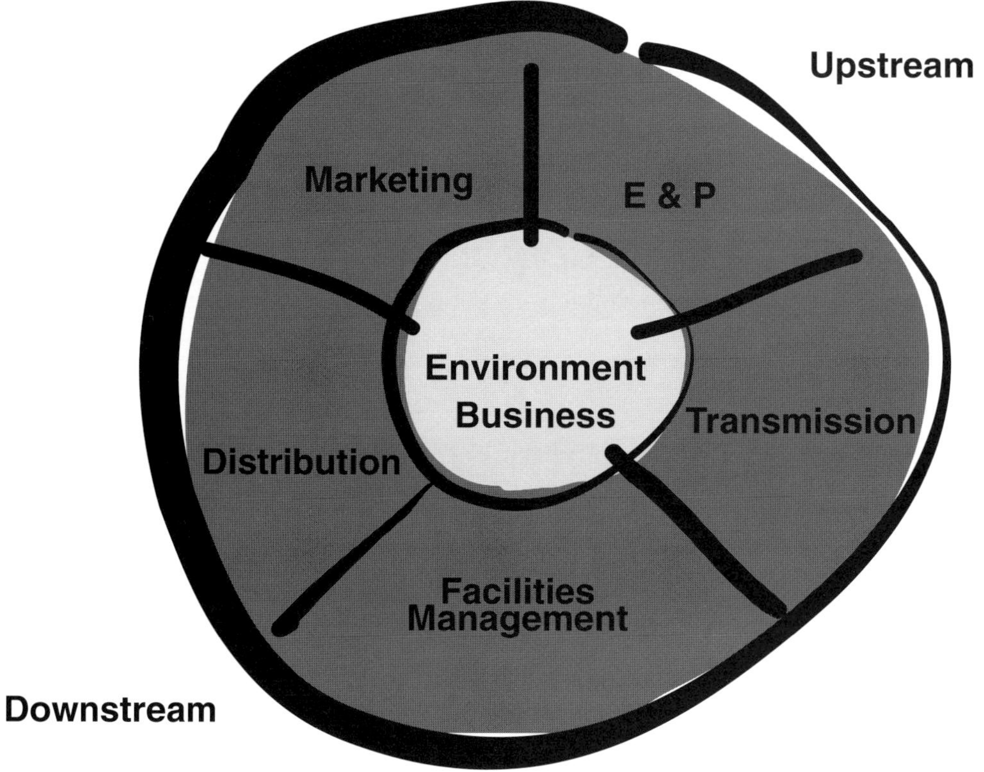

Figure 4—Enterprise model for a large integrated oil company.

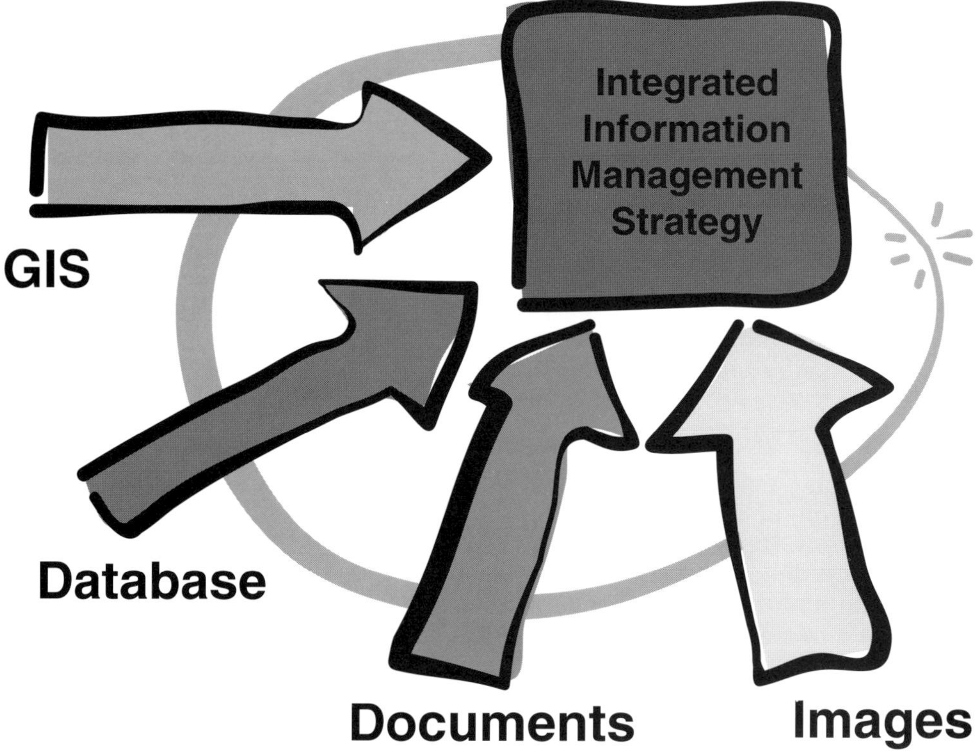

Figure 5—GIS as an integrated technology.

user groups are starting to address these issues, but until the corporations themselves accept wide-scale interdisciplinary technology exchange and open industry communications, there will continue to be islands of development and, consequently, unaddressed islands of opportunity for the budding application builder or consultant.

WHERE IS IT ALL GOING? A CONCLUSION

It is somewhat awkward to prepare a paper of this nature because the pace of technology change is such that the material quickly becomes out of date; consequently, an effort has been made to keep the presentation as generic as possible. The value of spatial data management technologies, their embodiment to date within separate GIS systems, and now as worthwhile functions to integrate with other applications nevertheless appear to be almost universally accepted in the industry. It seems clear that the use of GIS as a separate or stand-alone technology will fade because its very nature is to complement the existing systems rather than to replace them. Now that it can be truly integrated, it is likely to lose its separate technological identity, which is good. Soon, map and spatial interfaces will no doubt be found alongside desktop word processors, spreadsheets, databases, and other desktop utilities; and within just a short time, everyone will wonder how they ever got along without them. In addition, spatial data will be generated directly from the corporate archive or from a chosen vendor data warehouse and streamed on request across the Web straight into the application, such that the most current information set is always used.

Harder to predict are next year's specific winners and losers in the technology race. What is clear is that there will be more modular code convergence, greater database integration, increased use of the Web, and much, much more use of integrated spatial data. GIS, along with electronic document management systems and image analysis, will become a mainstream application, and will be integrated for everyday use within the general corporate information management strategy and desktop work environments of all users (see Figure 5).

GIS is truly approaching the integration goal. Petroleum E&P staff, in particular, will soon have more and better applied, integrated, and business-oriented applications, and they will take spatial data management and analysis for granted. There will probably be no more separatist GIS; just spatially enabled management information systems (MIS). Many of them, hopefully, may come packaged with a label that reads "GIS Inside!".

Practical Realities

It is extremely easy to be enthusiastic about geographic information systems (GIS)—especially applications of the technology to petroleum exploration and development. After all, once constructed, a GIS can yield extremely attractive, colorful maps of all sorts of geoscientific properties and related exploration information, and the data used to produce the maps is easily accessible, updatable, and transportable virtually at the click of a computer mouse.

The capabilities of GIS encompass the data access, data management, and data visualization dreams of almost all exploration and exploitation geologists. Yet, behind the pretty displays and the seamless integration of information lies a history of long hours devoted to editing, processing, matching, and rectifying, as well as extensive efforts devoted to project decision making, strategizing, database population, file conversion and reformatting, software coding, hardware maintenance and various other tasks that all must be accomplished to achieve the final product. It would be a substantial breach of trust not to acknowledge that a good, effective GIS takes a lot of elbow grease to construct; but hardly anyone would disagree that the work is well worth undertaking.

This section contains two papers that address the practical considerations and realities of constructing a GIS. The authors and contributors bring their years of experience to this pragmatic discussion of what it really takes to construct and maintain good GIS products to support petroleum exploration or development activities. The first chapter—**A Conversation with Petroleum Information**—is an annotated transcript of a multi-way conversation between the editors and the staff of the GIS Data Services Group of Petroleum Information/Dwights in Denver, Colorado. In this chapter, compiled by the editors and Pete Stark, a long-time participant in Petroleum Information's efforts to promote and develop GIS technology, the GIS experts from this leading data services company paint a forceful, detailed picture of what it really takes to make GIS work. In the second chapter—**On the Implementation of GIS for Petroleum Exploration and Development: Issues and Perspectives**—Tim Coburn, who, for nearly 10 years was Supervisor of Quantitative Geosciences and of Mapping and Database at Marathon Oil Company, describes the organizational, philosophical, and economic nuances that must be recognized and addressed to effectively incorporate this technology into a typical petroleum exploration and development environment.

Coburn, T. C., et al., 2000, A conversation with Petroleum Information, *in* T. C. Coburn and J. M. Yarus, eds., Geographic information systems in petroleum exploration and development: AAPG Computer Applications in Geology, No. 4, p. 49–60.

Chapter 6

A Conversation with Petroleum Information

Timothy C. Coburn
Department of Mathematics and Computer Science, Abilene Christian University
Abilene, Texas, U.S.A.
and
Mathematical Geology Section, Kansas Geological Survey
Lawrence, Kansas, U.S.A.

Jeffrey M. Yarus
Roxar, Inc.
Houston, Texas, U.S.A.

Phillip H. Stark
IHS Energy Group
Denver, Colorado, U.S.A.

ABSTRACT

Recently, the GIS Data Services group at PI/Dwights was interviewed about its experiences with GIS in the commercial arena. This insightful discussion provides the reader with an informative view of the inner workings of a GIS organization and some of the issues and problems to be considered when implementing GIS projects.

INTRODUCTION

Over the years, Petroleum Information (PI) has arguably been the leader in collecting, managing, and reporting data about the oil and gas industry in the United States. PI has continually maintained a large share of the market in sales of commercial data and services to the petroleum industry. In that regard, PI has become recognized for its focus on the technology necessary to provide its products in the most timely and accurate way possible. As the result of its 1996 merger with Dwights Energydata, the combined entity of PI/Dwights (now, a part of IHS Energy Group) represents the principal source of oil and gas information in the United States.

Geographic information systems (GIS) represent one of the technological advances that have kept PI, and now PI/Dwights, at the forefront of the commercial data market. PI was one of the earliest players in this technology, recognizing the power that GIS brought to the visualization and mapping of data. In fact, it was PI's work in this area—its early efforts to store data as layers that could be superimposed on one another—that is largely responsible for the industry's more recent moves toward data integration, computer-based data management, and more powerful computer mapping systems. Although the staff of PI would likely be the first to acknowledge that they do not have all the answers, nor all the expertise in this arena, it simply would not be acceptable to publish a book about GIS applications in the petroleum industry without acknowledging PI's contributions, discussing some of its experiences, and including its reflections about the past and future growth of this technology.

Recently, the editors of this volume and the staff of the GIS Data Services group in the PI/Dwights Denver, Colorado, office participated in a taped, three-way conference call, which was subsequently transcribed for presentation here. Participating in the call from PI/Dwights were Paul Forbess, supervisor, GIS Data Services, Dave Evinsin, Doug Mean, and Jim Lindauer. Tim Coburn, Abilene Christian University, and

Jeffrey Yarus, Roxar, Inc., served as questioners and moderators. A set of wide-ranging questions designed to stimulate the flow of the conversation was provided to Paul Forbess in advance. A copy of these questions, which generally pertain to the applications of GIS technology in petroleum exploration and production (E&P) settings, is included here as an appendix.

What follows is a somewhat abridged transcript of the audiotape of that conference call. Although every attempt has been made to maintain the flavor and integrity of the questions and comments of all the participants, the quality of the audio transmission made it impossible to create an exact reproduction of all conversations. Except for some changes to correct punctuation and to facilitate sentence flow, the discussion is presented largely in unedited, conversational style. All comments about specific commercially available products are the personal opinions and assessments of the participants, and in no way do they reflect the position of PI/Dwights, IHS Energy Group, the editors, or the American Association of Petroleum Geologists (AAPG), nor do they in any way constitute positive or negative endorsements by these parties.

In addition to the transcript of this conference call, Pete Stark, representing PI/Dwights, has provided the following brief historical account of the introduction and growth of GIS ideas and concepts within the company. We hope you enjoy reading all of the remarks of these gentlemen who have spent so much of their professional lives working in this exciting arena.

PERSPECTIVE ON THE USE OF GIS AT PETROLEUM INFORMATION

During the early 1980s, PI investigated new tools to improve both the construction processes and accuracy of its petroleum lease maps and base maps. ArcInfo, a new GIS product from Environmental Systems Research Institute (ESRI), was included in the evaluation. A pilot project was launched to compare the time, cost, and benefits of using GIS technology in the map construction process. Although initial results favored the use of computer-aided drafting (CAD) systems for map construction, PI soon learned that they did not meet all of its emerging map construction and product requirements. PI began to build digital lease databases during the late 1980s and realized that GIS technologies supporting the linkage of map features and databases would better meet the company's business objectives of reducing costs while improving map accuracy and product flexibility. ArcInfo finally was installed at PI's Denver facilities in 1988 to initiate the company's first GIS mapping system.

Substantial development effort was required to implement needed functionality. Format converters and data loaders were developed for PI well, production, and lease data, as well as for commercial and government land grid sources. A routine was developed to convert land legal descriptions to points, lines, and polygons. Specialized data edits for both graphic and text accuracy and placement were written to support map database maintenance. Processes to verify and correct well locations and coordinates were a high-priority item. A user interface providing point-and-click access to tools such as projection conversions, polygon processing, and edge matching across geographic boundaries enhanced the map construction process. Finally, map database management functions, an enhanced Oracle interface, and menu-driven map generation capabilities, such as automatic title block and legend generation, as well as calls to predefined map options, were created. These tools still provide the foundation for PI/Dwight's custom map services.

From initial emphasis on construction and delivery of standard land/lease and base maps, PI has shifted its GIS focus to the realm of graphic database browsers and generation of custom maps using its large proprietary E&P databases. Although this commercial focus has limited the scope of PI's GIS experience, it provides an interesting perspective. Once the GIS map construction and generation tools were in place, it was expected that market forces would push utilization into spatial analysis and enhanced visualization, involving both remotely sensed images and other kinds of images. In fact, international projects have tended to involve new map layers, such as surface geology and digital elevation models, interpreted subsurface geology, and interfaces to new databases, well logs, and seismic information. On the other hand, most U.S. customer mapping requests continue to focus on summary and display of basic data and information from PI's master E&P databases.

This type of "bread-and-butter" petroleum mapping is data-centric rather than GIS-technology-centric. The heart of most custom mapping projects is devoted to identifying what data are in the databases, how to retrieve and condition these data, and how to best display them on a map to meet the user's needs.

The know-how and data management process involved in transforming the content of E&P databases to map-ready files are the most critical components of the PI/Dwights GIS service. A project begins with an initial customer interview, which is targeted at defining the objectives and the types of data, and map displays required to address the objectives. The GIS specialist must have a thorough understanding of the database formats, contents, data relationships, and possible implications of spatial data distributions to assist the customer in this critical project specification phase. In the next phase, the specifications are transformed into database queries, and the necessary data summaries, computations, and codification are performed. Such steps are required before the data is ready to "load" into the GIS. At the map generation level, additional rigorous specifications and commands are required to display attributes to highlight the desired patterns and trends. The PI/Dwights petroleum mapping processes and workflow are shown in Figure 1.

This perspective is intended to provide background and understanding of the dialogue that resulted in the taped conversation with PI/Dwight's custom mapping team in Denver. Their responses to key questions

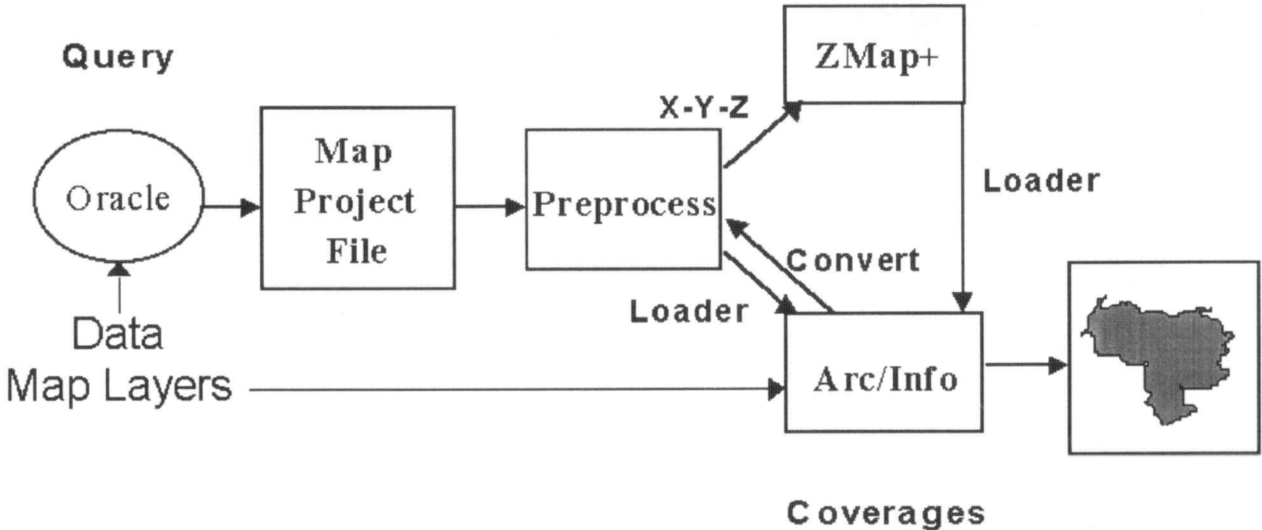

Figure 1—PI petroleum mapping process. Arc/Info is a product of Environmental Systems Research Institute (ESRI); ZMap is a computer mapping package developed by Zycor, Inc., now a part of Landmark Graphics, Inc.

illuminate the importance of data knowledge in creating effective petroleum GIS maps.

TRANSCRIPT OF THE TAPED CONVERSATION

PF: Let's go around the table. There are four of us. I'm Paul Forbess.

DE: And I'm Dave Evinsin.

DM: I'm Doug Mead.

JL: And I'm James Lindauer.

TC: OK. Thank you. Jeff Yarus is in Houston. Let me remind you that we are recording this. For those of you who may not know what all of this is about, Jeff and I are editing a new AAPG text entitled *Geographic Information Systems in Petroleum Exploration and Development.* We don't think we can publish a book of this nature without some information from PI because we think that you all have a great deal of knowledge to impart to us. We would like to have you tell us from your perspective, not only in a commercial environment, but just as professionals working in this area, what GIS is about, particularly as it relates to the geosciences. Paul, we sent you an e-mail with a list of the kinds of things we are interested in. There may be items on that list which you don't think are important, or there may be other subjects which you'd rather talk about.

PF: Right. I do have the list. You're right. There are some things on there that really touch the things that we do. However, we don't really build GIS systems for people. We're actually users of GIS—mainly ArcInfo—we actually produce products. We produce maps and data products. Some of the products we provide are in GIS format—like shapefile format.

We have a lot of databases that PI and Dwights have built over the many years. Big oil and gas databases. Pipeline databases. Things like that. We have pretty much carte blanche permission to go against any of those databases to pull data together, reformat it into a GIS mapping format, and then to use that to create a suite of products for custom mapping customers.

JY: It might be a good idea to start with a particular vantage point here, and reference Tim's second question. Given someone comes in the door and they want a project done, what are you going to ask that person for, how well are his/her questions defined, etc.?

DE: The requirements have to be very well defined from the client. It's my contention that a map is designed to answer a question or set of questions, and we have to get the client to formulate what his question or questions are to make the map useful. It's more than just the basic skeleton—what part of the world are you in, state, county, this kind of stuff. We need to know the objective. Several of us have geological backgrounds, and we can help the client refine his/her question better.

JY: You may, for example, elucidate in order to further define that question, so that, at the end, the natural questions are adequate enough.

DE: Well, I don't know. It's kind of hard to state what would be an initial question, or a final question, that we have worked on. We go back and forth with discussions so much. The client could come in and say, I'm working in the Morrow Trend in thus-and-such county, and I need a map to see all the Morrow producers. He might be interested in, as well as all producers, dry holes that penetrated the formation—age range of wells, when they were drilled. The different technologies they might have been using. Who knows? It's hard to say.

PF: You may be looking at two or three formations. You may be looking at a certain mapping scale that will determine how the map is constructed. We may want to use colors to denote ranges of production. We may want to use bubbles—sizes of bubbles—to show ranges of production. May want to use the colors to show the different formations of producers and non-producers. There's all sorts of scenarios like that. So we kind of help the customer go through a planning process and just work to get him to define exactly what he wants to do his job.

JY: Would it be fair to say, perhaps, that the common types of questions might come from geoscientists who are assigned to work an area that have no particular knowledge of it at all, and who are interested in gathering ideas or obtaining a knowledge about the type of data that is available, and then searching, perhaps, for the best type of data that might help them along their way.

PF: Yes, I think what you've said is very true. We get a lot of geologists who are working in a certain area, and they are using us as the technical tools to further their knowledge about that area. Interestingly enough, a lot of the people we work with are from major oil companies that already have this data—that have their own mapping outfits—and sometimes they just can't get things done fast enough in-house. They can't get the time and scheduling to work out, so they need someone like us to do it in a quick manner.

JY: Do you often get questions from those that are interested in trying to figure out whether or not there's an economic project in a particular area? In other words, do you gather information that might be used in risk analysis. For example, we've been considering entering basin XYZ. We need to know that field sizes—what we're looking for has some data indicating one million barrels of oil—the average well production is such and such, and the depth of the objective is something else. Does that make sense?

PF: Absolutely. Yes. And we've actually designed several standard products for that particular case. One of the things we've done is to build field outlines of the whole United States in a GIS format that can be used in ArcView, where a person can throw the field outlines up on the screen and make queries, just like you mentioned there—how many fields here produced over a certain amount, and are still actively producing, etc., and, then color-code those things on the screen—red or blue or orange. This is actually a data product that we've put together instead of a mapping product—an actual hardcopy map that would be delivered to people—this is actually a product that is delivering the data files. It consists of a graphical field outlines database that has region topology in it, and that allows for polygons to be overlapping—lain on top of each other, etc. The GIS still has the intelligence to know which field is which. It allows people to actually go and click on the field with their cursor and upload part of the table with the information about that field showing the discovery date, the date of last production, the last production of oil, last gas, last month's gas, number of wells in the field—things like that; so it gives them a real good reconnaissance or regional view in that sense.

JY: What is the client's need in terms of hardware and software to be able to benefit from that; in other words, to be able to call up such and such file from the database that you deliver in electronic files? What do they need to be running to be able to do those queries adequately?

PF: ArcView. It's the only thing they need. And that will run on a PC or a workstation environment. There are several other systems now that will use ArcInfo shape files, which is what ArcView likes to use. ArcInfo, the main GIS program from ESRI, will read shape files; and actually Geographix—the PC-based geological mapping software—is coming out with a version that will read shapefiles which will allow this product to be used. So, either ArcView, ArcInfo, or the new Geographix that is coming out. There are also a lot of applications out there right now that are using ESRI's MapObjects product, which allows people to look at ArcInfo coverages and shapefiles—actually running them from C programs and visual basic—all kinds of applications like that now can include what they call map objects, which would allow this data set to be used.

TC: Do you all get involved at all in changing the data in any way to customize it for the customer.

PF: All the time. Really, a lot of times we'll get the data into a GIS by manipulating them in many different ways. So, there have certainly been a lot of cases where we have even delivered data to customers that have been either tabulated or statistically summarized, or manipulated in all those different kinds of ways.

JY: The various databases that you have access to are obviously those from PI/Dwights, but do you also know of other industry databases—public domain or otherwise—that are relevant to your projects; or are you restricted only to those which you have?

PF: No, we'll use just about anything we can get our hands on. We've even got some government data sets that we're taking in from the Minerals Management Service and the USGS. We've even cleaned them up a little bit and made them a little more tidy and GIS-ready. We have all those to pull together; and really, anything that the customer might have—any sort of image data, photography—anything that he might have—seismic lines, dxf AutoCAD files—we have taken in all kinds of things like that and merged it with our data and made maps from it.

Let me back up a bit, too. You were asking me earlier about regional reconnaissance views of data. Like you said, a customer might come in and say, "I'm moving into this basin for the first time, and I've never worked here before. I'd kind of like to know just general things about it." We also have a hardcopy mapping product off the shelf now that's part of the basin display series. It's a set of six maps for every producing basin in the United States. These six maps will show you production, the activity that's going on, well locations and permits, well starts, active drilling, etc. They will also show you a regional structure map of the basin, as well as the producing formations and how they lay out across the geology.

TC: Suppose I, the geoscience manager, walk in off the street, and I've heard that GIS can work wonders for me. If I walk into your shop, and ask you—tell you I want to build a GIS—what are you going to tell me.

PF: What were the things you were going to ask us?

TC: I'm just going to walk into the door and say I've

heard GIS is a great tool. I don't know anything about it, but I want you to build me one, for whatever trend I'm talking about. What are you going to tell me?

PF: Well, the term "build a GIS"—we probably need to talk about that for a minute here. What I've just told you about are products that we have made from the GIS, but when you use the term that "you want us to build a GIS," that's a different animal.

JL: Instead of going in and buying a car off the showroom floor, that's like asking, "Well, I'm interested in building a factory that manufactures cars, trucks, and such—it's an order of magnitude different than what we normally do—our charge within the organization really isn't to build a GIS; but certainly, if someone wants to get into that, it's something we have some expertise with.

JY: Let me just modify Tim's question to say that we're interested in what it takes in terms of hardware, software, and substantive work to generate products to sell. We're also interested in how one goes about building a GIS. Tim, can you elaborate a little bit?

DE: The project definition of a GIS says that it is a system of hardware, software, data, and staff that all work together to answer questions of a geographical nature. From the data side, we've gathered well and production data since the 1920s, and—

JL: Petroleum Information, in the mid-1980s, established a department that began the process of taking information and building a GIS out of it. Initially, the department basically took information from a USGS quad, digitized it, and then using an industry-agreed—a corporate—data dictionary standard, went ahead and separated out the information into separate data layers. These data layers were then put into an ArcInfo library—and library is synonymous with database—which has grown to approximately 40 gigabytes of disk space. The library includes culture data, political boundaries, hydrology, transportation, key well and production data; and it also has legal descriptions and land surveys. About that same time, Petroleum Information moved initially from a VAX environment to now where it's on a Unix platform. A lot of the information we use for large projects gets ported off and put on tape for future analysis. Consequently, in the future, if a customer requests that the project would be updated, we basically restore the data as such and update it accordingly.

Jumping back to—we might want to bring this up—Paul talked about field outlines. The field outlines constitute a nationwide file; and basically, Paul wrote a lot of mainframe programs that merged production data with well information. The result of that work was basically this product, which has summary information at the field level. It's very valuable because it enables the customer to do relational queries. Jumping back to the databases for a bit, PI started in the mid-1980s, and we went through a couple different mapping types of software before finally deciding on ArcInfo in the late 1980s. One of the differences between ArcInfo and ArcView is that ArcInfo is a full professional GIS, whereas ArcView—when in first came out—was basically just a viewing software where you could use to look at data and print out reports. Since that time, ESRI, which is the company that makes ArcInfo and ArcView, has continued to go ahead and add more functionality every year.

We've also built databases, along with software. What the software enables us to do is to create some of the framework around maps, expedite data retrieval, expedite data conversion—for example, we take USGS CDs (we basically take them the way they come out—in zipped DLG format) and we convert the national USGS DLG file, on the fly, into a file format that is very compatible with ArcInfo and ArcView.

DE: Jim has been primarily discussing the base map type of information. For about 15 years we have invested in extracting information from the USGS quad sheets. We now have the capability to use the digital line graph product on CD from the USGS, that will let us, on the fly, improve on the maps.

JL: Other types of databases that we have. Our department in Houston has a product called GRID—and basically what they do is accumulate pipeline information on a flat file. We take that flat file up here in Denver, and then we convert it into a GIS file. Now, this pipeline file is nationwide, and one of the services we offer to an external client is reformatting of the file into any type of format that the customer wants. We also use these pipelines as a supplementary data set in constructing some of our hardcopy maps.

TC: When you say "any type of format that the customer wants," what does that mean?

DM: Well, I work on the conversions of the pipeline data. Essentially, Petroleum Information has rights to an ArcInfo workshop for customers that will take these flat files and build ArcInfo coverages, either as .E00 files or as Arc shapefiles that can be imported into ArcView. If a customer has a different type of GIS software other than one of ESRI's products, we can reformat the data as needed. For example, we might make a .cdf file, which is what a Geographix user would require to bring the data into his system. We can also take data and make AutoCAD files that someone who doesn't have a full-blown GIS could use. What we're saying is, we can—even though we are an ArcInfo shop—we can manipulate the data and put them in any format that other users will be able to take advantage of.

DE: ArcInfo has probably fifteen or so conversion routines to use in going back and forth between its products and other products. Sometimes the way the data are handled on either side may or may not be compatible, but there is always some functionality that can be used to go to other systems—for example, to ERDAS to MOSS to GRASS, or similar systems.

DM: Let me just interject. Some of these conversion processes we do outside of ArcInfo, we actually do on the mainframe—we feed in the flat files and do the conversions there, putting the data out in a format that, for example, Geographix users are able to take advantage of.

JY: To what extent does having a GIS system inside PI help you in responding to data conversion.

PF: Like I said—and like Dave said—there's maybe ten or fifteen conversion routines that are included with the ArcInfo software. This allows us to do a lot of the conversions in a real slick and easy way because, in Denver, we've basically written the software. Yet, there's a lot of behind-the-scenes processing we have to do on our own. There's a lot of computer programming that we do to make the ends fit, not to mention, just sometimes gathering our own data out of our databases. There's no routine set up in ArcInfo to do that. We have to do a lot of our own programming to get the data from point A into GIS format.

JY: But once there . . .

PF: Absolutely. Once there—you know, a picture is worth a thousand words! And GIS has moved us tremendously forward as a company in being able to show how our data can be used. Yes.

JL: I would like to complete the "build a GIS" discussion and interject some more about the hardware and software requirements for a GIS.

JY: Please.

JL: OK. Based on our experience with ArcInfo, the first requirement for a PC installation would be having Windows NT 4.0 as an operating system. The thing with a lot of GIS software is that it's very system intensive—it takes up a lot of CPU processing. So, what we recommend is 64 megabytes of RAM. If you're developing a really basic type of GIS, which would be a one-machine type of environment, you could do it in ARCInfo with 64 megabytes of RAM. You want to have a hard drive of at least two gigabytes. The processor itself needs to be a Pentium II 266 or above. You want to have a backup system, whether that be a tape drive, a CD rewrite, or something else.

For ArcView, the requirements are the same; however, ArcView can also be used on Windows 95, and it doesn't require as much RAM as ArcInfo. Thirty-two megabytes should be enough, but ESRI experts recommend as much RAM as possible. On the Unix side of things, it just depends on your scope and your GIS—basically how much data you want to use.

Like I mentioned earlier, corporately we have several terabytes of data; and even our group by itself uses approximately 40 gigabytes of data. It really depends on the scope of your operation. To get this data into its existing format, it took thousands and thousands of hours to do the necessary editing, massaging, and processing of the data. One of the reasons why companies come to us is that when people begin to count the cost of a GIS, they begin to recognize that most of the budget will be consumed by data conversion—it's approximately 75 % to 80 % of the initial cost.

And now to answer one of your other questions. When you get a GIS project—it's like any software development project—you need to have corporate support. When I say corporate support, this means people who are also outside of the GIS chain of command. This would be, let's say, your vice president and your chief financial officer. This is true because, initially, if you want to start your own GIS, a lot of the costs upfront take lots of years to be recouped.

Now, where the GIS industry is trending is—and this is the same for the most computer-based technology—is that we're starting to take real-time information, convert it on the fly, and analyze it immediately. There's a lot of such applications being developed right now across the country, and that's a whole different ball of wax.

JY: For example?

JL: Highway systems, for example. Forest fires. Natural disasters. What we do, for example, with images—and Doug Mead has done a lot of this recently with geo-rectification. Sometimes we'll have to incorporate images with our project data. We basically take those images, and we register them with our real-world coordinates. If you want, this would be a good time for Doug to go ahead and talk about that. Would that be OK with you guys?

TC: Yes.

JL: Go ahead, Doug.

DM: OK. Besides our huge in-house corporate databases—our current well databases, our production databases, our GRID database—a lot of times the data that a customer needs for his mapping product isn't in digital form. It might be like a hardcopy map, an areal photograph, a satellite image, or something similar. We can merge these types of data with ours, but first we need to rectify it—which means that we need to find points that are on the image that match up with similar points in our ArcInfo coverages, and then register all the points. And after this is done, we do the geo-rectification, which is a transformation program built in ArcInfo that hooks up the points on the image to the points on the coverage. The software mathematically transforms all of the points in the image to register with the coverage.

TC: Is that a time-consuming process?

DM: It can be. It depends on a lot of factors. It depends on what we know about the raster image that we're rectifying with your coverage. If you know that it's in some sort of real-world projection, then we're able to convert our coverages to the same projection that the image is in. Sometimes we'll need to scan a hardcopy map; and knowing what projection the original hardcopy map is in, we can convert our coverages into the same projection. This facilitates the actual geo-rectification process. When we have to rectify an unprojected image, it's a lot more time-consuming. The ArcInfo software identifies as many points as possible on the image that match up with points in our coverage. It also identifies points within the transformation that seem to be moving too far, and alerts the user to the fact that some of them may have been registered erroneously—that they've been matched up improperly—and they need to be taken out of the registration before the whole image rectification process is completed.

TC: What percent of the time of a typical project would be involved in the rectification process?

DM: I'm sorry. I didn't quite understand the question.

TC: What percent of a "typical project" would be devoted to the rectification process?

JY: How big a problem is rectification?

DM: It's not something that we do all the time; but, it *is* an option for data that aren't yet in digital form. It's a means for those data to be brought in and merged with data that are already in your GIS. So, it is a small percentage.

PF: Yes; and once again, it kind of depends on what shape the image is in. Someone might give you a photograph, or something like that, and they don't know where in the world it came from. Or it might be a hand drawing—maybe some draftsman had drawn some contour lines, or something like that—gosh, you know, if you don't know anything about it, or where it came from, or the mapping projection it was originally drafted in, then it's going to take a little more time. So then, the quote might be a little higher because then you've got to worry about stretching and shaping and kind of rubber sheeting the image into place. On the other hand, if someone gives you an image where they know exactly the projection, the scale, etc., then it's a lot faster because you already have tools set up in ARCInfo to allow you to do the transformation.

JY: In your own experience, does the latter or the former seem to happen more frequently?

PF: About 50-50.

JY: About 50 % of the time you're dealing with those types of situations.

PF: Right.

JY: And about 50 % of the time with stuff that they got—

PF: Right.

TC: How do you handle requests to integrate data that's not lateral—or horizontal—in nature—well log data, seismic data?

JY: Vertical data, huh?

TC: Yeah.

DE: OK. Well, that's the point at which we differentiate between a geographic information system and a geoscience information system. Vertical data are probably most often displayed as contour lines or as color-shadings on a map to indicate what's happening in that third dimension.

JY: I don't think that's what Tim had in mind.

PF: Well, usually, when people come to us, they're looking for a certain interval on the log, for example. They want us to be able to display for them everything from the top of the hole to the bottom. They're usually after a certain play—they're after a certain formation—something that they suspect may have oil or gas. There may be two or three of those. If so, we have to use different symbols, different colors to show things like that.

DE: Most of the work that we do is not really involved with the well log. The ESRI software isn't built to address that sort of thing. There are software products that are built to store the well log traces to be able to pull them up and look at them in cross-section. There's one product I know of that is designed to store the vertical information. It will draw plat maps of where the wells are, identify which wells to call up, and do on-screen vertical correlations—that sort of thing; however, its GIS, or lateral capabilities, are not as extensive as something like ArcInfo.

TC: So we wouldn't be able to go to the map, at a well spot, click on the well spot, and have an image of a well log—say, 30 feet of sonic log—come up on the screen?

DE: That could be built into ArcView, I believe. I think it has the capabilities to store images and display them on pop-up windows.

DM: Our group has recently been involved in some foreign mapping in South America where we have been doing things similar to this. Paul's been in charge of some projects in Venezuela that could shed some light on this. Basically, through the hardcopy map, we've been doing structure contouring on the surface map, and then showing the well logs on the map margin as a scanned image; and then Paul has incorporated these using ArcView into mapping projects. This would be similar to what you're talking about; but I'll let Paul tell you more about that.

PF: That's pretty much the substance of it right there; but "yes" is the answer to your question. That's certainly something that GIS can do—to be able to click on a spot and pull up any kind of a picture or image or report or table or what-have-you.

JY: So, in a sense, one could store vertical information digitally at a location . . . and be able to retrieve it.

DE: Sure.

PF: You bet. In fact, you know, the GIS has a database underneath it. So, you could load anything that you wanted into that database; and then it becomes an object to query—anything that you want to query on—as long as it is inside of the database, you can pull it up at will, and use it to do all kinds of spatial queries.

JY: So, one could pull up a scanned image of a well log, which is normal, or one could actually analyze the digital form.

JL: Yes. One of the things that ArcView offers is basically a hot link, which the user can spatially, within bounds, go and select a feature. This feature could be a well, or it could be, let's say, a city, or anything of interest. If it's structured right, the user could select that feature, and up would pop your image, text file, etc.

JY: OK.

JL: Are you interested in an overview of our group, per se, or do you have that already?

TC: I have that already, pretty much.

JL: So, you're aware that we do data products, standard products, custom products, like litigation mapping?

TC: Yes. I think we would be more interested in having you say the kinds of things that GIS would be useful for, in terms of the geosciences.

DE: Jim earlier alluded to our vast corporate resources. I guess a few additional words ought to be said about the major corporate databases. PI and Dwights both had somewhat similar databases that are currently in the process of being brought together. In general, Dwights has a well history database—something on the order of one to two million wells in the U.S.—that tracks all the information on the location of a well, the drilling that went on, the types of liners, the liner sizes that were used, the bit sizes, the formation tops, the tests conducted and test results—essentially everything that has happened to a well except the actual production figures. The production information is stored in a separate database that tracks the volumes of oil, gas, water, and condensate on a month-by-month basis for all producing wells in the U.S. There's a reservoir information database that looks at petrographic things like porosity, permeability, and engineering aspects of the rocks in the reservoirs. These are all databases that either reside on tape or on large mass storage, and they are not necessarily the GIS format. So, in any given project we would retrieve data from those databases and convert them by various methods.

TC: No doubt you guys—in fact, I know you do—you see all the technology around, you go to conferences, you see things—there are some concepts about GIS—there are some "GIS products" out there—that probably don't follow your definition. How would you distinguish between products that are GIS products—or GIS systems—and those that aren't?

JL: One quick characteristic to distinguish between non-GIS systems and GIS systems is being able to take a file and match its information to real-world coordinates. I've worked with a lot of different people in my lifetime; and, basically, information in real-world coordinates is accurate. If you're trying to play a specific area to scale that's not in a real-world coordinate system, then it comes down to a matter of how accurate you want to be.

TC: OK.

JL: That's one difference right there.

PF: A major difference that will affect the GIS is GIS, right? You have a database attached to it. For two-dimensional or three-dimensional maps—even *n*-dimensional maps—even *n*-dimensional data—you've got data and a database underneath. I think that's what really sets it apart. Instead of just having a drawing—something where you can draw lines, shapes, or stuff like that—you've got something that has intelligence up underneath.

DE: Is the question in the terms of ArcInfo versus AutoCAD versus CPS-1?

TC: No.

DM: GIS versus non-GIS?

TC: Yes.

DM: One of the basic definitions of a GIS system is one that is capable of relating spatial data with flat files, or a tabular database.

TC: OK.

PF: Connection.

TC: Now, I have heard the terminology used that GIS is like an onion. You can peel off layers and get down to the core. How would you characterize that?

PF: It makes a lot of sense. Like you said, there's the initial graphics, which would be on the screen. You can click on something on the screen, and you can pull up tables and charts. There's a whole bunch of information loaded up underneath it. In that case, the layer concept makes a lot of sense. We use it all the time, where each component of a map is a separate layer—like the land grid is one layer, well spots is another layer, field outlines is another layer, and maybe colors that we use to denote production is another layer. So you can take these layers all one-by-one, or you can build them on top of each other—stack them on top of each other one-by-one. One of the beauties of GIS is that you take one layer and query others. For example, you can have a pipeline; and you can say, OK, now find the well that is closest to the pipeline. All that sort of stuff can be done, from one layer to another.

JY: Now, how hard and fast is the difference between GIS and CAD?

JL: In a roundabout way to answer your question, five years ago there was a big difference between CAD types of applications; and at that point in time you could take Geographix and associate it more as a CAD application than what it is now. What's happened in the industry in the last five years, with technology getting more powerful, is that more and more of these CAD applications are becoming more and more true GIS-oriented. They're getting more full-fledged functionality. Intergraph is another example, with its Microstation product. As years have gone on, they've added more and more functionality. There's actually a convergence of GIS and CAD applications to where they're all going to become GIS.

TC: OK.

PF: Everybody is having to keep up with the Jones. You know, when the big, dedicated GIS packages came out, all the systems that worked with the GIS had to keep up—they had to do whatever they could to advance their products to try to join in and compete with the powerful GIS. It's just a sign of the times. Whenever you've got a piece of software out there that has the world market share, and it is powerful, why not use it to build applications. We've built a mapping system here that is really something built around ArcInfo. When ArcInfo comes to you, it's basically a collection of tools—a giant tool bag—of different commands, of programs—and it's up to you to string those together, and to write programs that will hold it all together and do the kinds of things you want it to. We've spent years on that.

DM: I just want to interject something about GIS versus non-GIS. One of the main differences is that GIS has topology. GIS systems can basically be divided into two types—(1) a vector GIS, which stores things as points, lines, polygons, and recently, regions, which are collections of polygons; or (2) a raster GIS, which stores information in a series of cells or grid paths. Topology consists of points, lines, and polygons, or cells in a raster database. A non-GIS is a system whose tabular databases aren't able to be connected to this topology. A non-GIS, such as a strict CAD system, does have a set of layers of different lines that have attributes attached to the lines, but they're not capable of getting intelligence where you can do spatial queries, as you can in a GIS, between the different types of topology.

TC: OK. Can you guys speak to all the ramifications of the data models and data standards? Do those affect the quality of the GIS, or the efficacy of a GIS?

JL: In the process of developing a GIS, let's say you're using Oracle files to create it. Without that data model, subsequent changes can cost a lot of time, money, and effort, particularly if you want to make any changes to your GIS. So, your data model is very important in the upfront process of defining and building a GIS. It's basically part of the structure, and you really need it.

TC: And is there one model that you all consider more acceptable?

JL: Excuse me. Is there a model that is more successful?

JY: More acceptable, huh? Like PPDM, POSC, and others.

JL: Yes. PI/Dwights has adopted the PPDM data model for its corporate standard. PPDM accommodates all of our commercial data files. To date, however, most of our applications have been extracted from the legacy M204 database, which has been the de facto standard for U.S. well and production data. Since all of us have the expertise to customize layers and coverages to whatever specification the client requires, having a data model where all the data are standardized in layers allows us to rapidly query and update our database.

TC: What happens when they come to you and their data are not in a model that you need?

JL: We reformat it.

TC: OK.

JL: It is very easy to reformat data.

PF: Yes. About 30 percent of our time is spent outside the GIS functions—you know, outside of ArcInfo programming—actually manipulating data and changing its format.

JL: Lots of times we'll get a flat file—an ASCII flat file—from a client or customer, and they'll want to do something with it. It could even be a Microsoft Excel file. One of the things we have to be able to do is link such files to a GIS coverage; so, there has to be a primary key. The key is something you have to match up with in the flat file, say, with something in the attributes file of a coverage. To do that, we have to basically be able to reformat the data.

Let's say, for example, that a client's doing a project in a couple of counties, and she has some well information she wants us to map out. She's got a list or file of APIs (well numbers), but also some other information she wants us to have mapped. What we have to do is basically merge that file with our ArcInfo well coverage file—which comprises, you know, approximately 2.67 million wells. Then we have to take the subset of wells and their information that results from this data merger and merge that with other appropriate coverages to give her the information that she wants. It's a multi-step process. Lots of times, the client may not even have the APIs.

JY: What data sets are you using? One specific data model?

JL: We have our standard corporate data model for our GIS data layers, and we've used it for a lot of years. Having that data model enables us to retrieve data out of the statewide libraries and our nationwide fields really quickly. This type of capability is not readily available industry-wide. It's one of our industry initiatives; and it's why people come to us. Having a data model where all the data are standardized in layers, we have no problems pulling the data out or even updating our database.

JY: So, in essence, you've got your own proprietary data model.

JL: Yes.

TC: I know you've got some international projects going on? What do you do in those cases? What's different?

JL: We've done some. We've done a China project. We've done a Venezuela project.

Domestically, we have various services. Another thing we have, which also isn't available industry-wide, is an ability to adjust digital base maps in time. We've got this plot operator who uses a menu interface that's been customized in ArcInfo. Our customers and clients—which range from small independents to large companies—can request a digital base map from any area in the United States where we have coverage. We can do over 30 states, and we offer wells, culture, lots and tracts, land grids, and also pipelines to be put on it. This system is automated, so a customer can order—let's say, ten Texas maps—and we can ship that order out the same day, even at a couple of different scales.

Another type of product that we have is the offshore field records guide in Houston. We basically generate maps up here, which they use for a five-volume publication for the offshore field records guide. One of our newer products is the federal offshore lease data product.

TC: What I was getting at is this—what are the differences between a domestic project and an international project? Or are there any?

JL: Besides the language barrier?

TC: Yes, what I'm really getting at here is this—are there technical differences in the data that you are required to accommodate when you do an international project that you're not required to accommodate in a domestic project?

JL: I would say no. The difference in data files is just a projection; and you can use projections in the U.S., and use certain different projections for international. One of the data products out there for the international market is the Digital Chart of the World. The information contained there is standard across the world. It's basically a database that's broken up by five degrees.

TC: OK. Does PI/Dwights cooperate with Petroconsultants?

PF: Yes, we do.

TC: So, you have access to their data?

PF: Yes.

TC: OK.

PF: And they ours.

TC: OK. So, if you are doing a Venezuela project, what do you use for well spots and well locations?

PF: In many cases, we've actually contracted with the government to obtain information.

TC: OK.

PF: So, that's probably one of the big differences for international mapping. The luxury of U.S. maps—as far as oil and gas go, and our versions of them—is that we have our databases. For domestic projects, we're used to things being put into fields, you know, the way that we like to see them. For international mapping, we end up probably dealing with a lot more strange formats—things that are not really in a GIS-ready format per se—things we have to manipulate a little bit more. Those kinds of projects happen in the United States, too, when you have a company that has been keeping its own files on its own computers, and no one really thought about putting the township in one column and the range in another column. They may have just typed out the information, or something like that. You run into that situation in the United States a lot; and of course, you run into that in international data a tremendous amount. The real luxury we have with the PI/Dwights databases is that somebody had the foresight many, many years ago—in the 1960s—to begin putting together fixed-format fields—layering the data into fields where they could be retrieved easily.

As far as the actual mapping goes—like everybody is saying here—a mapping system, especially one with ArcInfo's power, doesn't matter where you are on the globe. Is the data collection different? Yeah. The language barrier is a real problem. We may laugh, but it is a real problem—in spelling, phonetics, and things like that; the way the data are used; marks over the tops of letters and things like that. Sometimes the database of a text editor that you're using won't carry some of that stuff with it. There's all kinds of little things to watch out for.

TC: So, is it fair to say that, domestically, there would be no difference in building a GIS for the Rocky Mountains, the Offshore, California, or the East Coast? There's nothing unusual in those areas that would make a difference?

DE: There would be the type of report that is required by the regulatory agencies in the different states—the data that have to be reported and the way in which they are kept. That's from the petroleum data—the downhole information. As far as the underlying basemap information goes, a rectangle on the earth is a rectangle on the earth.

TC: OK.

JL: The hardware and software should work everywhere.

TC: OK. Is there anything peculiar to the Gulf of Mexico? Offshore?

PF: Anything peculiar in what way?

TC: Are the data—the availability of data different? Is the data handled any differently?

PF: Well, we certainly don't have a very good handle on things like formations. It's an industry-wide problem. Sure, they're harder to identify. You're dealing with a different kind of graphic representation of the well spot because you're drilling from platforms. You'll have a platform location, and then you'll have all these little spikes coming off, going under water and then under ground. So yes, there's a difference that way.

TC: OK. I just have one more question to ask. Maybe Jeff has one more. But my last question has to do with maintenance and support. What's it take to keep a GIS going?

JL: To keep a GIS going, sometimes you have to update the software that's been developed by a corporation in years past because the GIS software, itself, has changed through the years. Right now, software is largely being developed which is made using more object-oriented types of computer languages. So, part of the maintenance would be basically modifying the in-house applications, which are affected by the changes in GIS software. Hardware also has to be maintained and enhanced. Both fortunately and unfortunately, the software for a specific application grows and grows and grows. For example, ArcInfo on the Unix platform now takes up maybe 150 megabytes, so you have to accommodate for necessary disk space. This is really true whether it's on a Unix workstation or on a PC. Further, as images become more prevalent in GIS—and they are doing so quite rapidly—you also have to accommodate for that because images take up a lot of space. Finally, you have to get as much RAM as possible; because, when you work with larger images, that takes a lot more horsepower to do the processing.

DE: There is also a maintenance issue on the data themselves. You know, the organization has to have the political muster, if you will, to go in and correct data errors as they're found. Also, in a dynamic industry like the petroleum industry, wells are being started and completed every day; and you have to have a method to gather that new information. That might mean buying it from us because we have the people on the ground who are gathering it; or you simply might not choose to obtain it or to maintain it internally.

JL: We update our well file and pipeline file pretty regularly. It's bimonthly for the wells, and approximately quarterly for the pipelines. That requires running programs to get that information converted into a GIS coverage. We usually run our well conversion at night because we're processing over two-and-a-half million vectors. So, it does take some time.

PF: You have to keep watch over your hardware systems to be sure that, as the industry moves on, and things get faster, you continue to update your hardware so you have the fastest possible equipment and that you can move and grow with the rest of the industry.

JY: Tell us a little bit about the future. For PI/Dwights. What are you guys planning? One possible idea that we've heard about is being able to get things off the Web.

DE: Yes. As an aside right now, let me mention that we've got a Web site that you might want to look at—it's <www.gisds.com>. It shows who we are and the kinds of things we do. Yes, there are Web map servers. We've been talking here about the type of application that the client could walk through on the Web, query the data he wants, see the map drawn on the screen, and download it so that he could plot it elsewhere. One issue is how to protect the corporate assets. We don't have an answer yet, but it's certainly something that's going to be the future. Mapping that's on the Web right now seems mostly to be street maps and things like that that are relatively straightforward. Ours would be a lot more complicated, just because of the complexity of the data.

TC: So, it's safe to say that we're not going back to the old times of basemapping?

PF: Probably not. No—probably not. Though people still like those.

DE: No; however, basemaps are a basic tool for the geologists. As Jim mentioned, we have the runtime generation of our basemaps, and then the wells are updated every two weeks. So, instead of having a draftsman keep up with all the wells in your area and repeatedly plotting on basemaps, the basemap that we pull down will be current every two weeks.

TC: Do you guys have anything else you want to add?

PF: I just remembered something else on—you mentioned—how to keep a GIS updated, and what kinds of problems are involved in keeping it going. One thing we left out was backups. That's a real important thing to know. People have to be prepared with all GIS applications to be able to back up their work daily. That takes a lot of storage space involved in that.

TC: Do you guys see more interest by the majors or by the independents, or is it pretty much the same?

DE: We do projects for the majors, and for large and small independents as well. We also do projects for individual consulting geologists, and for individual deal sellers who are putting together a deal, and are looking to us for the technical support in their hypotheses. So we deal across the spectrum.

DM: Acquisitions.

PF: Acquiring. And then even bankers. And like we said before, lawyers. These days, there's a lot of requests from folks like these.

JL: Banks, lawyers, and utilities are all good customers.

TC: OK. Jeff, do you have any more questions?

JY: No, I think that we're about done here. I've got most of my questions answered.

TC: Guys, we really appreciate the time.

JL: Please stop by and visit us any time. We'd greatly appreciate it.

APPENDIX: DISCUSSION QUESTIONS

The following questions were prepared and submitted to the staff of the PI/Dwights GIS Data Services group for their consideration and review prior to a conference call with the editors early in 1998. The conversations, which encompassed discussion of these and other issues, form the basis of the material and information presented in this paper.

(1) What are the basics involved in initiating a GIS project, from the following perspectives?
  (a) Planning
  (b) Organization of data and other resources
  (c) Execution
  (d) Follow-up/maintenance/support

(2) What response is given when a client walks in the door and says she/he wants to do a GIS project? More precisely:
  (a) What questions must be asked to understand exactly what clients want, what they think they want, and why they want it?
  (b) What must be done just to get started?
  (c) Is it possible to outline the basic steps necessary to construct a GIS?

(3) What are the typical features that clients/companies want in a GIS? What dynamic aspects do they request? How is the interface customized to produce accurate representations of map features, such as conventional surface seismic lines? How is accurate, updatable labeling produced?

(4) What kinds of data are most often requested to be contained in a GIS? What other kinds of data can be incorporated?

(5) Other questions about data include:
  (a) How does one go about obtaining all the necessary data?
  (b) Where is the data physically located?
  (c) What steps have to be taken to get the data "in shape?"
  (d) What are the costs of obtaining data? Of editing/massaging data? What kinds of costs are involved?
  (e) How does PI/Dwights, in particular, interact with other data providers to obtain the necessary data resources? Who are some of those other data providers?
  (f) Of the data typically used, what formats does it reside in? How does one go about reformatting data? Under what conditions does data need to be reformatted? What costs (time, money, effort, etc.) are involved in reformatting data?
  (g) What is involved in data coding? What about data that is not currently in digital form (for example, photos, images, etc.)?
  (h) What is meant by rasterization, what is involved in the process, and what does the integration of raster images entail?

(6) What does the concept of geo-rectification involve? What does it mean?

(7) What is the big deal about data modeling? What is it and what is involved? What are "industry data standards" and why are they important?

(8) How are data that are horizontal in nature (e.g., roads, county boundaries, and other information that can be drawn on a conventional map) integrated with data that is vertical in nature (e.g., well logs, seismic data, and other information that has a depth component)?

(9) Within the context of a GIS, is there a way for clients/companies to share data? Or does each client/company need to have its own GIS and its own copy or copies of data (certainly they would want to keep their own proprietary data)?

(10) What is the difference between the map interface and the larger user interface? How is it constructed?

(11) How does the map interface relate to other data sets, some of which are possibly relational (like Oracle files), and some of which may be physically located in other locales (that is, on different computers in different office buildings, etc.)? How successful is a GIS in a diverse, cross-platform, cross-site environment?

(12) In defining a GIS, what are the basic parameters that govern its structure? From the context of these parameters, are there some systems that are called GIS, but really aren't? Why? Do they or do they not meet certain criteria?

(13) What kind of computer resources are needed to construct a good GIS? Are there any minimum requirements?

(14) What would be the difference between a small and large GIS? Does "small" versus "large" pertain more to amount of computer equipment and resources involved, the amount of data involved, the geographic extent of the project, or some other factor(s)?

(15) What is the difference between a "geographic information system" and a "geoscience information system?" Is a "geoscience information system" simply more all-encompassing than a "geographic information system" that might be largely map based?

(16) What applications are GIS products most often used for? Is GIS strictly a more contemporary method of making base maps, or are other applications gaining in popularity (for example, environmental assessment)? Is GIS used more for conventional basemapping, or is it playing a more specific exploration role (for example, well siting, reservoir characterization, etc.)?

Coburn, T. C., 2000, On the implementation of GIS for petroleum exploration and development: issues and perspectives, *in* T. C. Coburn and J. M. Yarus, eds., Geographic information systems in petroleum exploration and development: AAPG Computer Applications in Geology, No. 4, p 61–68.

Chapter 7

On the Implementation of GIS for Petroleum Exploration and Development: Issues and Perspectives

Timothy C. Coburn
Department of Mathematics and Computer Science, Abilene Christian University
Abilene, Texas, U.S.A.
and
Mathematical Geology Section, Kansas Geological Survey
Lawrence, Kansas, U.S.A.

ABSTRACT

Data and information are clearly crucial to the success of any petroleum exploration and development effort. People who have been around the oil business for any length of time recognize that most companies are virtually swimming in data that they have paid large sums of money to obtain. Yet, the industry has been plagued with the enigma of how to get at these data, and how to package and deliver them in the forms that are most easily and efficiently used by geoscientists. Over the past decade, geographic information systems (GIS) have come to the forefront of industry-wide attempts to transform the way that petroleum exploration and development information is obtained, processed, managed, and delivered. This is exciting and captivating technology of which everyone seems to want a part, and yet it has not experienced the rapid assimilation into day-to-day operations that might have been anticipated. This chapter explores the many reasons for the lag in implementation, and it details the unique management issues that are involved. Topics include the definition of what a GIS constitutes, product and project costs, staffing requirements, equipment needs, corporate/departmental philosophies about data, and many others. Also discussed are the rewards that can be achieved through a patient, but deliberate, pursuit of the GIS approach to upstream data issues.

INTRODUCTION

Geographic information systems, or GIS as it is commonly called, is a captivating and exciting technology. It's one of those ideas that everyone wishes they'd thought of first. The technology works and performs the way lots of people work and perform, and so, for individuals whose job tasks are spatial or geographic in nature (which is lots of folks), it's one of those rare technologies that's a natural fit, contributing to a highly synergistic and productive work environment. The output of its functionality is colorful and visual, having lots of eye-catching appeal, and so audiences (like prospective partners in business ventures, colleagues in seemingly unrelated disciplines, and "the bosses" up the chain of command) are readily attracted to it and turned on by its potential. The bottom-line results have been a financial boon—over the last ten years or so, this seemingly simplistic idea about the way people instinctively process information, effectively transferred to the computer, has mushroomed into a multi-million-dollar (perhaps multi-billion-dollar) enterprise.

For the petroleum industry, particularly the upstream sector, GIS is just a natural fit. The search for oil and gas is a highly visual, data-rich, and spatially oriented endeavor. It is no wonder this technology (or the essence of the technology, its ideas and concepts) has taken the industry by storm. After all, from every appearance (and every commercial advertisement), it is a failproof means of finally pulling together the storerooms of geoscientific data, reports, and interpretations that have piled up in oil company warehouses for decades, and of dynamically producing multi-dimensional maps (the lifeblood of geoscience!) that can lead to more rapid and effective decision making.

From all accounts, there should have been an immediate incorporation of GIS into the technological toolbox of every earth scientist. After all, spatial relationships are ubiquitous to geoscientific investigation. Unfortunately, this has not been the case. In oil company offices around the world, the implementation of GIS has presented many operational and organizational dilemmas. In fact, the proliferation of GIS into these companies, particularly its application to exploration and development activities, has had a somewhat controversial history.

There are lots of reasons why GIS hasn't been more fully embraced in the oil patch; some are real, but some are also contrived. The overriding reason is that implementation of GIS requires a change in operating paradigm, a change that some individuals, groups, and companies have simply been unwilling to make. Shifts in paradigms are always threatening to managers, workers, and vendors alike; and so, the proliferation of a technology like GIS, even one that is so clearly beneficial (an efficiency raiser, a money maker, etc.), is often stifled due to some personal or corporate fear of losing control. Organizations that have been able to get beyond such fear and forge a role of technological leadership have realized some very significant returns on their investments.

The following questions, then, might be asked: "Why can't every company achieve similar results?" or "Just what does it take to make GIS work, particularly in the petroleum exploration and development arena?" The answers are found in the essence of the technology itself, for effective application of GIS demands a certain working philosophy that is more open, flexible, and consensus-building than is found in most traditional, hierarchical business ventures.

First and foremost, companies and organizations will have to move from the command-and-control philosophy of centralized project and data management to one of decentralization. One of the beauties of GIS is that it is capable of accessing and linking various types of data that physically reside in diverse locations or belong to different groups. Consequently, continued tolerance of internal disputes about who's going to be in control of "the data" (and even "the GIS")—for example, the individual geoscientists, the exploration or production organizations, the corporate information services (IS) department, etc.—is clear evidence that the technology is probably not going to be a good fit. Second, companies and business groups have to evolve into organizations in which machines perform most of the tasks that people used to do, in which employees have new and different skills (for example, computer-generated maps versus hand-contouring, which requires less time, a different set of personnel skills, and possibly fewer people), and in which formal groups with specialized functions might be dispersed or even eliminated (for example, the computer mapping function might be relegated to individual geoscientists rather than centralized in a formal department). Third, there must be a general recognition and acknowledgment that technology is ever-changing, that keeping up requires constant upgrading, which, in turn, demands a reliable funding stream. Fourth, workers and managers alike must learn to think in 3D. Spatial relationships are not just two-dimensional, and earth science is absolutely spatial in nature. Finally, GIS is team technology; to be effectively employed, it requires the synergistic interaction of individuals from all sorts of disciplines. Geoscientists, for example, who prefer an individualistic workstyle will not be able to successfully incorporate GIS into their projects because they will not have the skills necessary to manage all the facets of the technology by themselves. Each of these features encompasses structural changes that are necessary for GIS to thrive, regardless of the discipline (Dangermond, 1989).

To those who are familiar with the operating environment of many petroleum exploration and development organizations, the foregoing discussion may suggest that this segment of the oil industry might be somewhat unfriendly and unreceptive to new technology, particularly technology like GIS. However, nothing could be further from the truth. The continued success of oil and gas exploration as a business venture depends in large part on the availability of new ideas and new technology, and much of the industry's past prosperity can be directly attributed to its willingness to embrace innovation. Nonetheless, because of their constant association with risk, explorationists are, by their very nature, a cautious lot, and the adoption of new technology is never instantaneous. In fact, many of these individuals openly prefer to be followers rather than leaders, just to make sure that new ideas are adequately "proved up" before sinking hard-won dollars into them. Over time, the GIS-approach to petroleum exploration and development will no doubt become more universally accepted. It's just that it may take longer than some enthusiasts would like.

Against this backdrop it is necessary to view the proliferation of GIS as an exploration and development tool. It is unlikely that anyone will debate the viability of the technology. As already noted, it's a cool, sexy invention that everyone seemingly wants a part of. Yet, there are some real issues to be considered and tackled, and some hard questions to be answered, before GIS can be expected to fully penetrate the petroleum exploration and development milieu (Judd, 1995; Bernhardsen, 1999). To put the situation more in perspective, it is important to first understand what this technology actually is and does in terms that the layman can understand.

Although the real GIS gurus might cringe at such a definition, a GIS is simply the combination of a multifaceted software package that performs some unique functions and a database that is organized in a particular way. Clearly, GIS can manifest itself in many forms and expressions (Antenucci, 1988). Although the acronym "GIS" is often used in the vernacular to mean either the software package or the database, it is really the specific combination of the two into an overall system. The basic goal of constructing a GIS is to produce highly informative pictures of data (like maps), and to be able to do so "on the fly." Such maps might encompass some or all of the following data elements: well locations, seismic lines, transportation infrastructure, elevation contours, wetlands boundaries, and the like. To the geoscientist, GIS is ultimately a better, more efficient way to make maps that are more reliable, informative, and detailed.

Next, it is important to consider the individual implications of the software, data, and system components of a GIS. *As a software package,* a GIS is generally purchased "off the shelf," having some basic mathematical and numerical capabilities already written into it. However, it mainly consists of a collection of functional routines that an individual must piece together in a computer programming sort of way. These functions, like "go out and find all producing wells within Weld County," are mostly geographic or spatial in nature, and have already been pre-programmed for use, but the numerical boundaries within which they are to operate must be specified. The stitching together of these functions in such a way as to produce a logical and coherent flow of analysis is a major component of the "system construction" aspect of GIS, one which, depending on project requirements, is usually not a trivial exercise. To end up with the desired end product (a map, for example) requires much thought, planning, and doing (sometimes through trial and error), as well as continuous interdisciplinary interaction between the individual doing the "programming" and the scientist/engineer who has the need.

Two implications are embedded in this characterization. First, in most (but certainly not all) cases, the individual doing the programming will not be a scientist or engineer (most engineers and scientists, certainly geoscientists, do not have the time, skill, and inclination to do this kind of work). Second, constructing a GIS requires a full-time "programmer," at least for the life of the particular project of interest, and not just any programmer will do. Such an individual must be fluent in the language and functionality of GIS, have some knowledge and understanding of the subject matter involved, and have broad computational and computer system literacy, being highly skilled in the interaction among machines, code, and data.

As a database, a GIS is a facility in which many different types of information can be simultaneously organized and housed. The unique aspect of a GIS database is that different data elements about the same physical construct (for example, a well, field, or basin) can be linked together with a common thread, or identifier, so that recalling a particular piece of information about that construct is no different, no more challenging, and no more time-consuming than recalling them all. Now, a GIS database is not necessarily just one huge computer file. The linking of data elements implies that any number of files can exist, possibly in different physical locations or on different computers; but this, in turn, also implies the existence of a reliable, high-speed computer network through which files and data elements can be accessed. In addition to equipment, software has to be written, purchased, or otherwise installed as part of the GIS to ensure such ready data access.

While in simplest terms a GIS is a place or facility to store data, the manner in which it is stored, or the organization of the data elements, usually has a particular scheme to it. This scheme, which is often referred to as a data model, is usually established so that everyone who adds data to the GIS, or constructs a new file to be accessed by the GIS, does so in the same way (e.g., porosity follows permeability) and uses the same measurement units. The objective is to minimize, and ideally eliminate, the possibility of confusion. Standard data models exist for various industries, including the petroleum industry [e. g., the Petrotechnical Open Software Corporation (POSC) model and the Public Petroleum Data Model (PPDM)].

Two additional notions about GIS data are very important. First, from the GIS standpoint, not all data have to be digital. This is one of the most unique and powerful aspects of this technology. Standard digital computer files can be directly linked to satellite images, photographs, hand-drawn maps, copies of written reports, and various other types of information (seismic sections, well logs, etc.) that previously could only be juxtaposed in massive, manually compiled portfolios or reproduced in part in large wall displays or montages. Within the context of GIS, all such data can be stored, compacted, and immediately recalled, the benefits of which are enormous. Development of the related technologies of scanning, digitization, and rasterization, and the invention of low-cost, high-speed mass storage devices (i.e., disks, tapes, and other computer-related media) have made this capability possible.

The second notion has to do with the idea of spatial registration, or geo-rectification, of data. In the context of GIS, all the data elements about a particular construct (e.g., a well, field, or basin) are stored or presented (mapped) in such a way that they properly line up with each other relative to their common geographic location. These locations are accurately positioned relative to the earth. For example, using the geo-rectification functionality of GIS, two different vintage maps of the same region, perhaps constructed using different projection systems, can be redrawn or reprojected in such a way that all their features properly overlay. Well spots from a different computer file can then be accessed and properly positioned on the new rendition of the map.

The metaphor of an onion is often used to characterize the nature of data as it exists within the confines of a GIS. Different layers of data can be superimposed on one another, or extracted from each other, in much the way that layers of an onion can be peeled off, all the while maintaining accurate positional relationship, like the layers of an onion maintain their spherical, geometrical shape as it becomes smaller and smaller with the removal of each successive layer. This is again a powerful and unique capability of GIS, the results of which could not otherwise be produced without extensive amounts of time and manual labor. The ability to establish and maintain correct spatial positioning does not, however, come easily or cheaply. It requires careful and thorough data entry and editing during the initial stages of project definition, and it necessitates accurate locations of data elements to be attached and stored in the database, all of which are expensive and time-consuming operations.

As a system, a GIS is more than a software package and a database. It's an interwoven, interconnected collection of data, information about data, data entry and storage conventions, processing routines, analytical procedures, connecting protocols, and machines that spew forth facts, answers, and pictures on command via a sophisticated user interface. The user interface, itself, is a well-organized arrangement of radio buttons, fill-in boxes, pull-down menus, pictures, maps, and similar mechanisms that appear as graphical items on the landscape of a computer screen through which users interact with the GIS by entering/selecting information or specifying commands for it to perform. The user interface is what people physically see when they first sit down to work with the GIS, and it is what they commonly refer to as being "the GIS." Consequently, designing the architecture of the user interface and actually constructing it are some of the most important, and time-consuming, parts of establishing a GIS. In this sense, the GIS for every project or application is unique.

The hallmarks of a GIS, then, are these: the interconnectedness of all kinds of data and information; immediacy of access to data, including links to remote sites through the wonders of electronic communications; visualization of data with high-resolution graphics, including dynamic or three-dimensional displays; and rapidity and repeatability of operations on data, which is the ability to do things fast, and to reproduce/update them with ease. The final characteristic is perhaps the most important one. A GIS is uniquely spatial, reflecting the way people think and the way the physical world exists.

As noted, given such highly desirable features, it is difficult to understand why this technology would not be immediately embraced; and yet, there are barriers and reluctance, and yes, even legitimate concerns (Table 1, for example, provides a list of potential barriers). When the implementation of GIS is debated in oil company offices around the world, some of the first questions out of managers' mouths are typically "Who's going to use it?," "What's it going to cost?," and "Is this just another toy for the computer whizzes?" At first blush such questions might sound a bit raw and insensitive, but they represent truly valid concerns. They are questions to be expected from responsible managers, and they are ones that every proponent of the technology must be prepared to answer.

A careful, thorough analysis of staff requirements, cost, and product efficacy is required before the implementation of any technological innovation can be brought to fruition; and even then, success is not absolutely guaranteed. In the specific case of GIS applied to petroleum exploration and development, it's all well and good to take the approach of: "If we build it, they will come," but many good intentions have become sidetracked along the way and many excellent projects have failed to be completed due to inadequate planning, insufficient resources with which to follow through, lack of broad-based organizational support, or expectations that were too high or too unrealistic (Aronoff, 1991; Obermeyer and Pinto, 1994; Huxhold and Levinsohn, 1995).

Among these three issues, staffing is perhaps the most important because it is so tightly interwoven with the other two. There are two different aspects of staffing to consider. First, effective implementation of GIS requires a strong and sufficient cadre of software engineers and geo-information specialists to support the overall development effort. Second, there has to be a core of users who are capable of bridging the geoscience and quantitative disciplines, and who can bring reality and perspective to the results that are produced.

In most cases it is logical to initially assume that the geoscientists, themselves, will be the users of GIS because they ostensibly have the most to gain from the technology. On second thought, however, the fallacy of this assumption begins to play out. Geoscientists are not typically well trained in quantitative and computational techniques; and as a group, they don't have broad-based familiarity with hardware, software, and other facets of computer technology. In most instances, when geoscientists see what GIS can do, they will overwhelmingly embrace, even actively lobby for, its implementation within their companies, but when asked to become personally involved to the point of "running the GIS" themselves, their enthusiasm quickly wanes. That's because this technology requires a shift in the personal paradigms under which most geoscientists operate. It requires them to become trained to work in a way that emphasizes computer literacy as much as, or more so than, earth science literacy. Geologists, for example, must learn to make maps using the landscape of the computer monitor rather than paper montages and displays, they

Table 1. Some Barriers to Implementation of GIS in Petroleum Exploration and Development.

- Paying lip service only to data integration
- Lack of understanding about the power of data integration
- Unwillingness or inability to fund the costs of the necessary data, equipment, and manpower
- Absence of a personal knowledge of the merits of the technology on the part of key management personnel
- Too much familiarity and comfort with the "old" way of doing things

must learn to explore within the confines of the computer as much so, perhaps, as by "going to the field," and they must learn to look for data on the network rather than going to the map room or the log library. In fact, to be facile GIS users, they must become "computer geologists," which connotes a different mentality, skill set, aura, peer group, and reward system. In many cases, geologists would really just rather someone else do this work for them. Consequently, before the decision is made from on high that geoscientists are going to be the primary users of any GIS, it is necessary to ask and answer the following questions: "Is it actually possible to get them to use it (can it be made simple enough, attractive enough, etc.)?," "Can adequate training be provided?," and "Will it really improve their individual decision-making processes and capabilities—more so than if someone else is the user?"

The other alternative, of course, is to relegate both the development and the functional use of GIS to master analysts who are more technology literate (these are not the same individuals as the software engineers and geo-information specialists). Clearly, having access to such individuals is a boon to most exploration and development organizations that set out to implement GIS technology, but it represents an entirely different organizational approach. Typically, GIS experts, individuals specially and formally trained in GIS theory and methodology, spatial data handling, geographic referencing, computer systems, and the like, are in very high demand, particularly those with geoscience backgrounds. Often, they are synergy seekers, discipline integrators, and consistency builders. Although perhaps not the team leaders because of the very nature of their skills, they are the kinds of individuals that project teams are constructed around. These are often the single individuals who know and understand all the data interrelationships. The question then becomes whether or not companies have the wherewithal to actively recruit such individuals, pay them enough to make them stay, give them the resources and freedom they need to produce results, and listen to—and act on—their recommendations. While such individuals may not be a luxury every company can afford, good geoscience-oriented GIS specialists can cut years off the technology learning curve, and can instill a degree of efficiency that generally cannot be achieved without them (Medyckyj-Scott and Hearnshaw, 1993).

The point is sometimes made that GIS experts can and should be "grown" from existing staff. Although there is a certain attractiveness to this idea (such individuals would presumably already know a lot about the company's business), experience has shown that it doesn't usually work out very well, particularly if there is a need to rapidly develop a skill set for a pressing and high-profile new venture. In an exploration and development organization, there are always too many other project distractions and high-priority tasks, resulting in insufficient time, money, and intensity being devoted to internal training.

The issue of staffing, then, when it comes to implementing a GIS, is a multi-faceted and challenging one. Companies engaged in petroleum exploration and development that want to develop a GIS capability must ask themselves whether or not they have the commitment to recruit and pay for the right kind of staff, or whether they have the time, patience, and financial resources to train existing personnel and the commitment to effectively use them after they are trained. Clearly, without the right kinds of people, and enough of them, all attempts to implement GIS technology are likely to ultimately fail.

Cost, both the initial expense and the ultimate price tag, is clearly an issue that requires thorough analysis before embarking on any project involving GIS, much less a corporate decision to fully embrace and integrate the technology. The discussion by Shah (1997) underscores this position. Part of the cost has already been identified and addressed; that is, the cost of recruiting and retaining adequate and appropriate staff (mostly in the form of salaries) and the cost of training current employees to become specialists and users (mostly in the form of training expenses and staff development time). There are more cost aspects to think about, however, in the broader context of GIS implementation, and many of them are technology driven.

A decision that needs to be made early on in the planning and implementation process is whether or not to "buy or build." Purchasing a GIS, perhaps in the form of a turnkey operation, is an entirely different approach than creating one internally essentially from scratch; and it involves different funding obligations. At the outset, it is important to acknowledge and underscore this fact: it is impossible to purchase a fully operational GIS *system* "off the shelf," even for moderately sized individual projects. The very nature of the technology demands interaction between the vendor of the GIS product and the customer, if only to the extent of "populating the GIS database" with information and facts to which only the customer has access.

Clearly, because all projects, particularly petroleum exploration and development projects, involve unique and different data, no GIS vendor has the capability and resources to sell products that are totally customized for all situations.

Consequently, purchasing a GIS generally means acquiring the collection of routines and commands, graphical functionality, data storage files, and the like that comprise any computer-based software package, and no more than that. The purchaser is often left on its own to install the software, make it work, add the data, and link to other systems or software (the added value of customization). In some instances, a GIS can be purchased with certain data preloaded (for example, well information from Petroleum Information/Dwights). In other cases, the vendor is retained to load information or to "hard wire" certain functionality on behalf of its customer. Certainly, lots of GIS companies are willing and anxious to provide the services necessary to develop project-level, and even corporate-level, GIS products. Yet, for the GIS to be truly useful and effective, someone other than the vendor eventually has to keypunch data, move, copy, or download files, and push the buttons necessary to make the GIS work! The point is this: obtaining and implementing a GIS is a much more involved exercise than buying one in a box at the local computer software store, and this is true no matter how extensive its ultimate application is envisioned to be.

There is a second way to acquire a GIS. Sometimes GIS functionality comes imbedded within other commercially available products, and this is particularly true in the case of petroleum-related GIS. For example, a GIS mapping feature might be included in a software package that is primarily intended for seismic or well log interpretation, or it might be part of a more specialized information management system, such as might be used for records management. In the strictest sense, these are not geographic information systems. They don't usually encompass the full-blown capabilities of a software package that exists as a stand-alone GIS, nor are they usually priced as such.

The third way to acquire a GIS is to construct it from the ground up. Depending on the size of the project, this may or may not be a more costly route to take, but without a doubt it requires a different kind of commitment on the part of the sponsoring organization. Before a single line of code is written or a single image is scanned, lots of questions have to be answered and lots of decisions have to be made. Table 2 provides a sampling of the issues to be addressed.

Without a doubt, the "build-it-yourself" approach is a direct and effective way to incorporate GIS technology into a project or organization. Major benefits can be realized because ownership is established in the resulting product. However, this approach is definitely the least timely one. Companies, managers, and project teams simply must realize that GIS products and capabilities don't just materialize over night. Software and database development is a tedious and sometimes drawn-out process that requires a long-term commitment for success. Failure to recognize the time requirements is the single most frequent reason why such ventures fall by the wayside and even fail.

In addition, if the decision is made to pursue the "build-it-yourself" approach, it cannot be done so in a vacuum. There has to be "buy in," participation, cooperation, and promotion from all affected parties (not the least of which is the senior management representative). On the other hand, it will not be possible to acquiesce to the whims and wishes of every potential player. It's a simple reality—geoscientists will ultimately have to change the way they work because the new *system*, no matter how bedazzling or how computer-intelligent, won't be able to mimic exactly what each geoscientist can or might do by hand.

Regardless of which approach to acquiring a GIS is taken, there are some other common, yet very important, decisions to be made that will affect cost. These have to do with hardware, and with the computer/information network in general. First and foremost, effective use of a GIS requires a fast computer, generally a Unix-based workstation equipped with lots of memory and disk storage space (in most cases, more than one such computer is required). GIS users and operators need to be prepared to handle data sets that are at least on the order of gigabytes in size, and this generally means mass storage devices are required. Second, effective use of a GIS demands access to a robust, high-speed network. A GIS works best in an environment of nearly instantaneous access to data, and without a high-speed network, the movement and transfer of large data files is time-consuming at best. Without any kind of network, about the best that can be hoped for is managed duplication of existing files, which is certainly not cost and resource effective. Efficient use of a GIS also requires high-quality data display devices to be available, including large-screen computer monitors that provide the user with the broadest possible information landscape on which to work. It goes without saying that plotters, printers, scanners, and other such peripheral devices must be part of the combined hardware suite. The cost of all this equipment, as well as the expense to maintain it on an ongoing basis, rapidly adds up, and it is a significant part of the overall cost of implementing GIS technology.

Finally, organizations that seek to implement GIS technology must consciously and continuously address their rationale for doing so. They must pointedly ask these questions: "Why are we doing this?," "How are we going to use it?," and "Is a significant return on productivity going to be realized from the cost and effort that is committed?" Developing a GIS for a special project, say, a regional exploration study in Argentina, involves an entirely different level of implementation than developing a GIS-based corporate data management system. Consequently, a decision to embrace this exciting technology requires considerable discipline and a willingness to constantly focus on what's absolutely important and necessary to achieve.

For petroleum exploration and development, GIS can, without a doubt, yield substantial dividends in terms of efficiency, productivity, problem conceptualization and understanding, and decision-making. Yet, as for all innovations, there are barriers to overcome,

Table 2. Some Issues to be Addressed when Constructing a GIS

- the data engine to be used
- the software engine to be used
- which internal databases and which legacy software systems are to be linked
- whether or not data will be purchased, developed internally, or both; and the formats that will be accepted/used
- the way in which databases will be populated, integrated, and accessed
- how data will be processed, presented, and displayed
- the type of user interface to be employed
- how special data will be handled (e.g., seismic lines and well logs)
- the projection/coordinate systems that are needed

and there is a cost to pay before the rewards can be realized. As described here, the costs of GIS technology are both economic (in terms of dollars and cents) and structural (in terms of changed paradigms and ways of conducting business); successful implementation requires good planning and a commitment to follow through. Despite the upfront expense, groups and organizations that are diligent in their pursuit of a technological advantage are likely to reap handsome rewards down the road using GIS, not only in terms of the organizational benefits listed, but in terms of financial strength as well.

REFERENCES CITED

Antenucci, J. C., 1988, Defining requirements: GIS, what is it to you? (abs.), *in* Abstracts, GIS Symposium: Integrating Technology and Geoscience Applications (Hugo F. Thomas, ed). Washington, D. C.: National Academy of Sciences, p. 32.

Aronoff, S., 1991, Geographic Information Systems: A Management Perspective. Ottawa, ON: WDL Publications, 294 p.

Bernhardsen, T., 1999, Geographic Information Systems: An Introduction, 2nd ed. New York: John Wiley and Sons, 372 p.

Dangermond, J., 1989, The organizational impact of GIS technology. ARC News, Environmental Systems Research Institute, vol. 11 (Summer), p. 25-26.

Judd, D., 1995, Thoughts on implementing a GIS, Earth Observation Magazine, vol. 4, no. 3, pp. 50-51.

Huxhold, W. E., and A. G. Levinsohn, 1995, Managing Geographic Information System Projects. Oxford: Oxford University Press, 247 p.

Medyckyj-Scott, D., and H. M. Hearnshaw, 1993, Human Factors in Geographical Information Systems. London: Belhaven Press, 266 p.

Obermeyer, N. J., and J. K. Pinto, 1994, Managing Geographic Information Systems. New York: The Guilford Press, 226 p.

Shah, N., 1997, Unmasking nine hidden budget busters in your GIS project, Earth Observation Magazine, vol. 6, no. 11, p. 41.

Case Studies

At the heart of any good technology lie its many successful applications to real, live physical situations. This is especially true of geographic information systems (GIS). This third section—the real focus of the book—contains eight chapters involving GIS case studies from around the world that encompass a wide range of petroleum exploration and development activities.

The first five chapters specifically pertain to early aspects of the petroleum exploration process. The first of these is by Carl Mazzo and Bill Burroughs at Exxon Exploration Company (now ExxonMobil). **Building a Regional Study in GIS** is a superb presentation of the techniques, methodologies, and requirements necessary to apply GIS technology to this important early phase in the search for oil and gas. The specific application involves Exxon's Circum Arctic Regional Project. In **GIS-Based Analysis of Fluid Migration in Sedimentary Basins**, Gernot Paulus, from the University of Salzburg, Austria, discusses an application of a GIS to another important aspect of the initial exploration process. In this case, techniques are used to investigate the relationship between subsurface faults and well locations, to calculate fault density, and to model fault-related fluid flow in an area of the Molasse Basin at the northern margin of the Alps. The third chapter is a joint collaboration among David Jefferis and Skip Pack at Dynamic Graphics and Michael Connelly of Connelly Exploration. In **The Gap Mountain Prospect: A New Approach to Conceptual Geological Modeling,** these authors combine the spatial analysis tools of GIS and three-dimensional modeling to develop a better understanding of the geologic domain of the White River Valley area of Nevada, to evaluate its oil-in-place potential, and to assess the suitability and accessibility of a proposed drilling site. Al Haertlein and Dave Rich describe the use of a GIS and its mapping products in evaluating properties and concessions on a competitive bid or negotiated sale basis. Examples are presented from South America, the West Coast of Africa, and the North Sea. Their chapter is entitled **Evaluating International Acquisitions Using a Spatial Database System—A Low-Cost Approach**. Finally, Ken Hood, Bernie South, Dennis Walton, Don Baldwin, and Bill Burroughs of Exxon Exploration Company (now Exxon Mobil) discuss a similar example of using GIS techniques to evaluate the petroleum potential of future acquisitions in **Use of Geographic Information Systems in Hydrocarbon Resource Assessment and Opportunity Analysis**.

The last three chapters pertain to the exploitation and development of existing fields. Kirk Barrell of Geodynamic Solutions describes the processes, benefits, and results of using GIS technology as the basis of a full-fledged field study. His chapter is entitled **Conducting a Field Study with GIS: Port Hudson Field, The Tuscaloosa Trend, East Baton Rouge Parish, Louisiana**. Gene Safley and Mike Fowler representing BDM Petroleum Technologies describe the effective use of a GIS approach to develop and support a reservoir management strategy for an aging giant oil field. Their chapter is entitled **Waterflood Optimization Using GIS Techniques in Citronelle Field, Mobile County, Alabama, USA**. The last chapter is entitled **Reservoir Characterization of a Stratigraphically and Structurally Complex Gulf of Suez Oil Field Using Three-Dimensional Modeling and GIS Techniques** by Ron Day of Amoco Trinidad Oil Company (now part of BP/Amoco), Khaled Talaat of GUPCO, and Karen Hoffman of Dynamic Graphics. These authors collaborate to illustrate how GIS processes and functionality can be combined with three-dimensional modeling to construct more detailed realizations of subsurface architecture that can be used to rehabilitate reservoirs once thought to be depleted or uneconomic.

Mazzo, C. R., and W. A. Burroughs, 2000, Building a regional study in GIS, *in* T. C. Coburn and J. M. Yarus, eds., Geographic information systems in petroleum exploration and development: AAPG Computer Applications in Geology, No. 4, p. 71–120.

Chapter 8

Building a Regional Study in GIS

Carl R. Mazzo
William A. Burroughs
Exxon Exploration Company (now part of Exxon Mobil)
Houston, Texas, U.S.A.

INTRODUCTION

Geographic information systems (GIS) have been applied by the petroleum industry to manage, display, and analyze map-based data since the mid-1980s (Calkins, 1990; Feineman, 1994; Foley and Ghazi, 1995). Although a GIS is broadly used, it is often considered distinctly different from conventional exploration computing tools, such as gridding, contouring, or well-log analysis. Exxon's historical GIS usage has focused on either the play or prospect level, where GIS archival and spatial data-access capabilities were not fully demonstrable. In 1996, Exxon applied a GIS to the capture of data for the Circum Arctic Regional Study (CARP) as an integral part of the ongoing work process. CARP became, in a sense, a laboratory where the process of real-time data capture using GIS would be tested, modified, and ultimately applied. The goal was to create a suite of regional study products that could be used in a shared work environment, updated on an as-needed basis, and integrated for required analysis within budgetary and time constraints. The CARP project largely met or exceeded these goals and objectives while validating a work process that is being implemented on a company-wide basis.

GIS OVERVIEW

In the digital mapping industry, a GIS generally refers to a software system that has at its core a spatial database for managing suites of map layers. The database allows data to be attached to individual maps and map features (Figure 1). This characteristic provides, for example, a means to review detailed permitting information about a target land parcel polygon on a graphics screen. The other unique feature of a GIS allows information from different types of map features to be combined to answer topical questions, such as "where is the best place to site a new landfill such that various environmental considerations are minimized?"

In the geosciences, the definition seems much less clear-cut and has generally come to embrace the broad topic of computer data management, analysis, and map-based data display. For instance, one of the most comprehensive overviews of GIS in the geosciences (Bonham-Carter, 1994) discusses everything from the statistical analysis of multiple map layers to algorithms for shaded color relief displays of grids. Furthermore, a GIS in the geosciences is closely bound up with the field of image processing, although it is generally considered distinct from the conventional gridding, contouring, well-log analysis, and seismic interpretation systems that have come to occupy important niches in the exploration industry. It is still common to find GIS software used in some geoscience organizations largely as elaborate, and not very efficient, computer-aided drafting (CAD) systems.

The wide range and disparity of data types found in regional geological studies are ideal for GIS capture. The GIS's inherent organizational structure provides a mechanism to handle a wide range of data types, including text, data tables, grids, scanned maps, satellite images, drafted geological maps, and detailed information about the features on such maps. Ultimately, a GIS can serve as the repository for virtually all study information.

Specifically, seven main features are required of a GIS for regional studies. Such a GIS must:

(1) Have the ability to deal with many different formats of mappable and non-mappable data. This includes complex polygons common in small-scale geological maps typical of regional studies, and simpler exploration contour maps common at a larger scale;
(2) Encompass coordinate systems that can be easily rescaled and transformed;
(3) Be able to readily organize large volumes of information. A given regional study may be composed of 1000 or more products. A GIS provides maximum flexibility in setting up a hierarchical data structure with appropriate slots in which to organize the products. This feature, in turn, allows distinct geological elements and interpretations to be instantly accessible for use in many

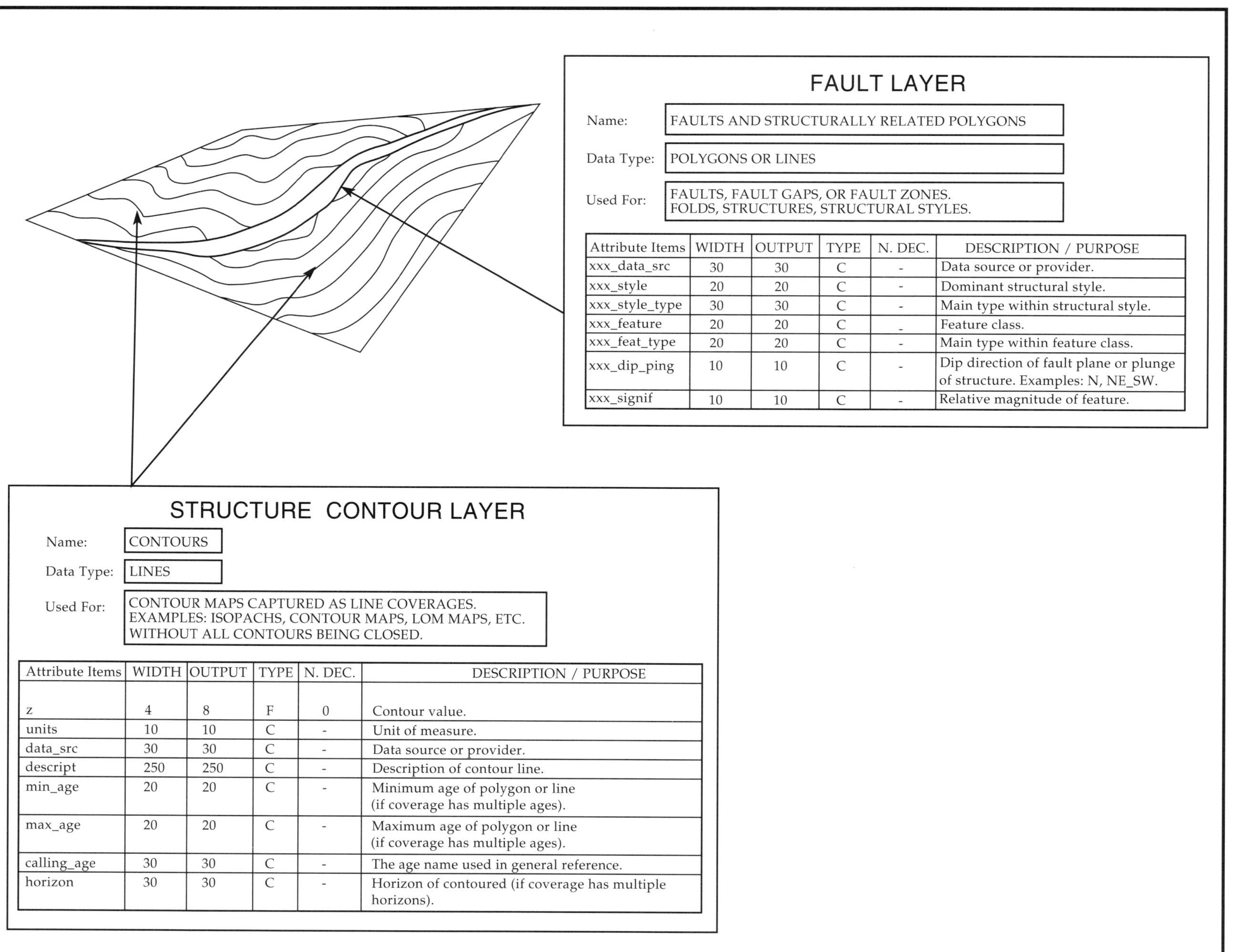

Figure 1—Conceptual diagram of relational attribute database structures attached to spatial map features.

different types of maps. The data structure allows data to be captured and stored only once in the system without having to copy it into every drawing that uses the information;

(4) Provide a database that stores meaningful geological information (attributes) for map features, such as faults, field closures, or geological polygons. This property eliminates redundant data and facilitates the creation of a variety of products from a few sets of input layers. The database must be flexible because many of the map features found on regional study maps (e.g., basement-type polygons) are not conventionally found in industry-standard data models;

(5) Have the ability to digitally overlay and suture layers of data together into new graphic and tabular relationships. Overlay requires a series of "proximity" operators, which act to join multiple tables of data together, based on the fact that the geographic features they are attached to are "close" to each other in space. This capability is fundamental to quantitative exploration resource appraisal, which seeks to identify zones of high oil and gas potential in regional work (Hood et al., this volume);

(6) Have the ability to store metadata (data about the data). That is, such a system must be able to produce a kind of a card catalog that references all the maps in the system;

(7) Have the ability to provide high-quality graphic output.

Although no off-the-shelf systems are preprogrammed to provide all of these features, several of the larger, commercial systems can be easily configured to meet these criteria.

REGIONAL STUDY OVERVIEW

Exxon's regional studies are the technical foundation for exploration, generating a suite of maps that provide the geological context for both strategic and prospect-level decisions. Regional studies are characterized by broad aerial extent, often covering multiple basins and a wide variety of tectonic settings. A regional study is typically staffed by a core team of geoscientists and technical support staff, with additional specialists providing support on an as-needed basis. The team focuses on a number of products, including plate tectonic reconstructions for key time intervals, a major structural features map, sediment accumulation maps, potential fields data, lithofacies for key time slices, known hydrocarbon occurrences, and hydrocarbon-systems elements maps. Additional map suites are often created to address local business needs.

The work required to build a regional study can be subdivided into five process steps: project planning, data capture, interpretation and map construction, basin and play analysis, and final review and documentation. Planning depends on the project goals and the project manager. Traditionally, many plans often focus on the list of final products without providing due consideration to the process of actually making those products. Data capture is often the most time-consuming effort and entails building a database from a myriad of sources and an array of formats. Map construction is often divided among the team geographically or thematically, with team members working in their media or software of choice and in concert only as required. Interpretations usually follow the geographic or thematic team division, and regional compilations are not undertaken until late in the project. Analysis depends on the capabilities of two-dimensional mapping software or a light table to combine play elements. Final products either remain in the software of origin or are captured by drafting. Product documentation and archival is usually undertaken at the end of the project, often after key team members have been reassigned and are unable to assist.

GIS APPROACH

Requirements and Process Overview

The GIS approach does not alter the fundamental steps used to build a regional study. In fact, GIS-based projects are often described in terms of traditional processes (Miller, 1992; Vitek et al., 1996). The GIS approach does provide a means for improving the process, however, by requiring better planning, standardization, and consistency (Figure 2). It also saves time by allowing documentation to take place during, rather than at the end, of a project. In essence, the GIS approach improves the efficiency of the conventional regional-study process.

Planning is the critical step in the GIS process and includes developing a final product list, generating product component (or layer) critical paths and interdependencies, and scheduling the correct mix of team skills at the appropriate time. Once the product list is established, individual layers used to create the product are identified and incorporated into a master project plan. For example, a GIS-based final sediment-fill map can be divided into a series of layers that must be available before the final product is completed. These layers may include, for example, a sediment-fill isopach, basement outlines, faults, continent-ocean boundaries, coastlines, and graticule. In addition, gravity-magnetic compilations and structural elements must be completed to support interpretations while constructing portions of the sediment-fill isopach. To build the sediment-fill map using the GIS approach, each layer is scheduled on a plan for the final product, the interdependency of layers and components is determined, team members are assigned to coordinate building the layers, and critical deadlines are established (Figure 3). As the project progresses, layers are incorporated or used as required to build products. Deviation from the plan impacts not only the immediate product but any dependent products as well.

Data capture makes use of the GIS organizational and spatial database capabilities. The spatial database can be fully used only if data are captured in a consistent

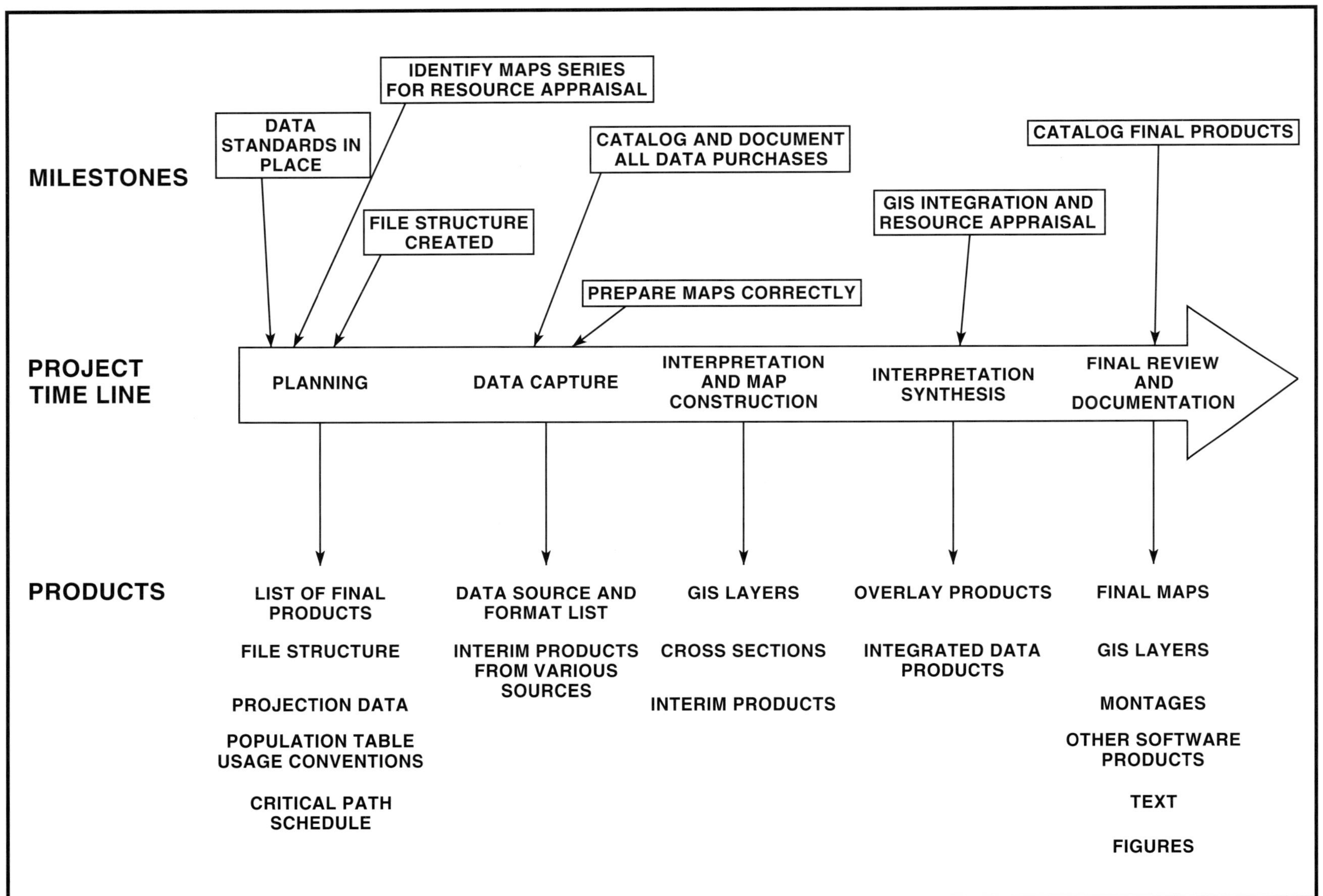

Figure 2—Overview of geographic information system (GIS) work process with flow chart illustrating major processes and issues in a regional geological study.

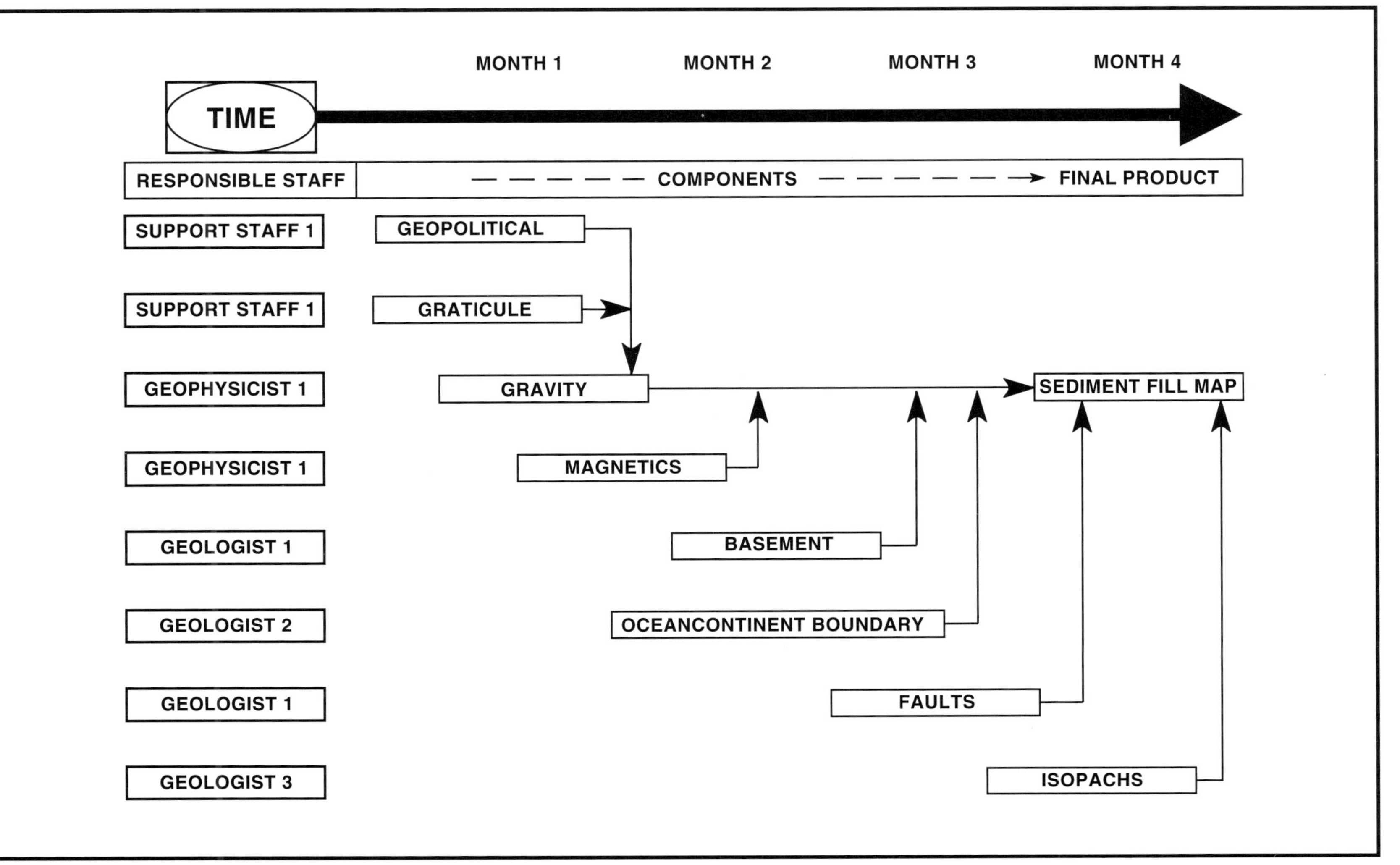

Figure 3—Scheduling final products using a critical path chart, sediment fill example.

format by using some type of standard database structure and code tables. GIS organizational functionality allows data to be captured one time only and made available in a shared environment. In practice, this means that each line, be it a basin boundary, coastline feature, fault, or facies limit is digitized *once* as a thematic layer and incorporated into whatever products require it. Data purchases have become an increasingly important facet of the data capture process, and one that may be undermanaged. When data are ordered, it is also critical to understand their form and format, content, and data-transport media. A lack of this information can cause interminable delays in making the data available to the work group. Even when data meet these criteria, difficulties can occur when, for example, a vendor supplies separate layers for every unique type of structural element where the project standard stipulates a single layer for all types of structural elements. Merging multiple layers into a single layer may require significant extra work prior to use. Managing data purchases must be an integral part of the GIS work process and must be given appropriate attention.

Map construction also relies on GIS organizational, database, and editing capabilities as raw data are converted to actual component layers and products. While the GIS approach is based on GIS software, it is not all-inclusive as a software application. Spatial representations of seismic interpretations are more readily handled in two-dimensional mapping software. Some workers may still prefer to work on Mylar or paper. The key to map construction is identifying and using the best method to complete the immediate task, with the understanding that products are moved to the GIS for analysis and archival. Populating a database, in addition to drawing features on a map, is extra work for geoscientists. Experience at Exxon has shown that the database-population effort can be minimized effectively when geoscientists use standard procedures when preparing maps for digitizing. Once a map's projection has been determined, the geoscientist identifies a standard database structure that is appropriate for the map theme (Appendix A), color-codes features on the map according to this structure, and attaches a sheet to the map, which links what is drawn on the map to a list of the attribute codes for each class of feature. The technical support staff can then build the database as they digitize the maps. The GIS software can be used for any map editing, suturing, or reprojecting required to build component layers. Corrections resulting from new interpretations or data are made to individual layers, which, in turn, can be used to update final products. Once layers are constructed, they are copied into slots within a standard file structure and are available to the rest of the team.

Analysis takes place digitally using GIS overlay capabilities. Overlay acts as an electronic light table, allowing the combination of any number of polygons from any number of component layers. One of the most powerful aspects of using the GIS is the ability to experiment with combinations of maps never considered in traditional regional studies. For example, combining gravity, structural features, and hydrocarbon occurrence maps may provide hints identifying new play fairways.

Effective use of the GIS approach includes documentation of individual layers on an ongoing basis (assuming a standard database structure is used) and, when finalized, filing layers according to a standard archive configuration. Documentation becomes a part of the day-to-day work process rather than an addendum or afterthought to the project. Final montages, cross sections, and figures can also be stored as created in a number of compressed file formats. Essentially, the project is complete when the last figure is created. For a traditional regional study, the possibility exists for several CAD personnel to be employed full time for up to one year capturing a series of maps, which have no spatial registration, which cannot be reprojected or easily scaled, and which contain much redundant data.

The successful implementation of the GIS work process is contingent on detailed forward planning, the availability of a standard database and archival file structures, the correct mix of team skills, and a team commitment to the process. Planning is the keystone for the entire process. Without a detailed plan, any efficiency gained using GIS would be negated by lack of organization. Database standards ensure that data are consistent and can be shared within the team. An archival structure provides a common, well-understood data location and format for products. Not every team member needs to be a GIS expert. Regardless of the size or scope of the team, however, people having three critical skills are required: GIS specialists to coordinate data capture and composition, geoscientists with a broad understanding of GIS to bridge between other team geoscientists and the GIS specialists, and a data administrator to ensure that standards are maintained. If the process is not accepted universally by all team members, it will put undue strain on the rest of the team and will ultimately reduce project efficiency.

Advantages

Although a rigorous and structured effort is required by the GIS approach to regional studies, experience at Exxon has shown that the long-term advantages of this effort outweigh any perceived short-term disadvantages. Advantages to the GIS approach include the following:

(1) *All spatial information is available for interpretation in essentially one system.* The GIS database acts as a digital light table, allowing different layers of information to be combined to help constrain new interpretations. For instance, to define basin outlines, color relief grids of gravity and magnetics can be combined with sedimentary-fill contours, structural elements, and basement type and zonation maps. The quality and consistency of interpretation are improved by such an approach.

(2) *New maps can be created from old maps.* All new interpretive maps loaded into the GIS have an

attached database (a suite of information about each fault, igneous body, contour, lithofacies polygon, etc. for each feature on the map). It is therefore a simple process to subset, isolate, highlight, and combine any layers from many maps to create a new collection of products.

(3) *Multiple products can be easily "zippered."* Regional studies covering multiple basins are often compiled on a geographic or thematic basis. A regional team, in collaboration with explorationists working on discrete plays or prospects, may create a whole suite of separate products for an area. At some point, the individual projects are compiled into the final regional interpretation. Merging or zippering maps in a GIS saves extensive hand fitting, rotating, cutting, and redrafting to resolve boundary interpretation differences. In addition, maps can be generated directly from the GIS in the proper scale and projection for use.

(4) *The capability to develop early correlations between hydrocarbon occurrence and mapped geology is provided.* This understanding is fundamental to any regional study undertaken in an exploration setting. This correlation helps to better define hydrocarbon plays, petroleum systems, and the ultimate quality or potential of basins for further exploration. The early focus guides later efforts in the study, and using a GIS simplifies the ultimate reporting at the end of the study. Typical products include field size distributions and estimates of oil and gas reserves per basin and horizon, all of which can be derived using GIS overlay capabilities on different layers of data (Hood, et al., this volume).

(5) *The system is self-documenting.* Methodically loading the data into a standard database and cataloging layers and maps in a standard archival structure results in a largely documented study by the end of a project. Using the GIS approach, a project becomes a continual stream of work products, which are compiled into finalized review products. Customary drafting and documentation can be virtually eliminated. In addition, products are easily located and are retrievable.

(6) *Projects are completed in a timely manner.* The planning and discipline required for the GIS approach pay significant dividends throughout the life of the project. Defining a project in terms of a predefined map series creates efficiencies in process flow, documentation, and drafting, as well as elimination of duplicate effort.

(7) *Projects are completed at a reduced cost compared to more traditional methods.* Based on the CARP experience, it is estimated that, on average, the cost of a regional study completed using GIS methodology will be 80% of the cost of completing the same study by conventional methods. A portion of the cost savings stems from the near elimination of a separate documentation phase at the end of the project. However, it is equally clear that "front loading" a large study with a carefully defined product list and production plan, although initially more time-consuming, allows a much more measured and structured pace of work throughout the course of the project. Such planning also results in a generally more efficient work process on both the geological and database sides.

USING THE GIS APPROACH

Exxon Experience

On average, CARP employed five geoscientists and two GIS support personnel over a span of three years. The study area itself is immense, covering all of the earth's surface north of 60°N latitude, and going as far south as 45°N in the Atlantic Ocean (Figure 4). CARP was designed as a number of independent subprojects, each in its own projection system to accommodate the conventional usage of Exxon local-office contributors or international scientific and institutional partners. The subprojects were to be zippered together into a series of regional compilations during the last year of the project. Coordinate transformation of data was a major requirement in the final compilation process.

Originally, the project was planned as a traditional regional study, with products developed essentially on paper or with two-dimensional mapping software and subsequently drafted. Midway through the project, the availability of enhanced GIS systems and geologic database standards, the necessity of dealing with the ever-increasing volume of diverse data in the project, and a change in project management resulted in a mandate to evaluate the GIS approach with CARP. After an initial six months of experimentation with staffing requirements, training project geologists, and standards definition, the project methodology described in this paper was developed and used in the project. Following are some of the particular difficulties and successes that stand out from this experience.

Difficulties

Perhaps the greatest hurdle for the CARP geoscience team was the buy-in to a new approach to the work process. This was especially apparent when timing pressures caused the geoscientists to backslide into more traditional work habits. This problem ultimately manifested itself in regards to three major issues: consensus on standards, timing of individual contributions, and resolution of geological differences among interpreters. These issues are discussed in more detail in the following paragraphs. Ultimately, the project manager refocused the team on the final deliverables, which proved to be key in keeping the project on track.

Consensus on Standards

Exxon began developing a set of standards for GIS-based data management in 1994. Key issues addressed by the standards were the ultimate location of final products (archival file structure), the structure and format of the database for various map themes, and creating the

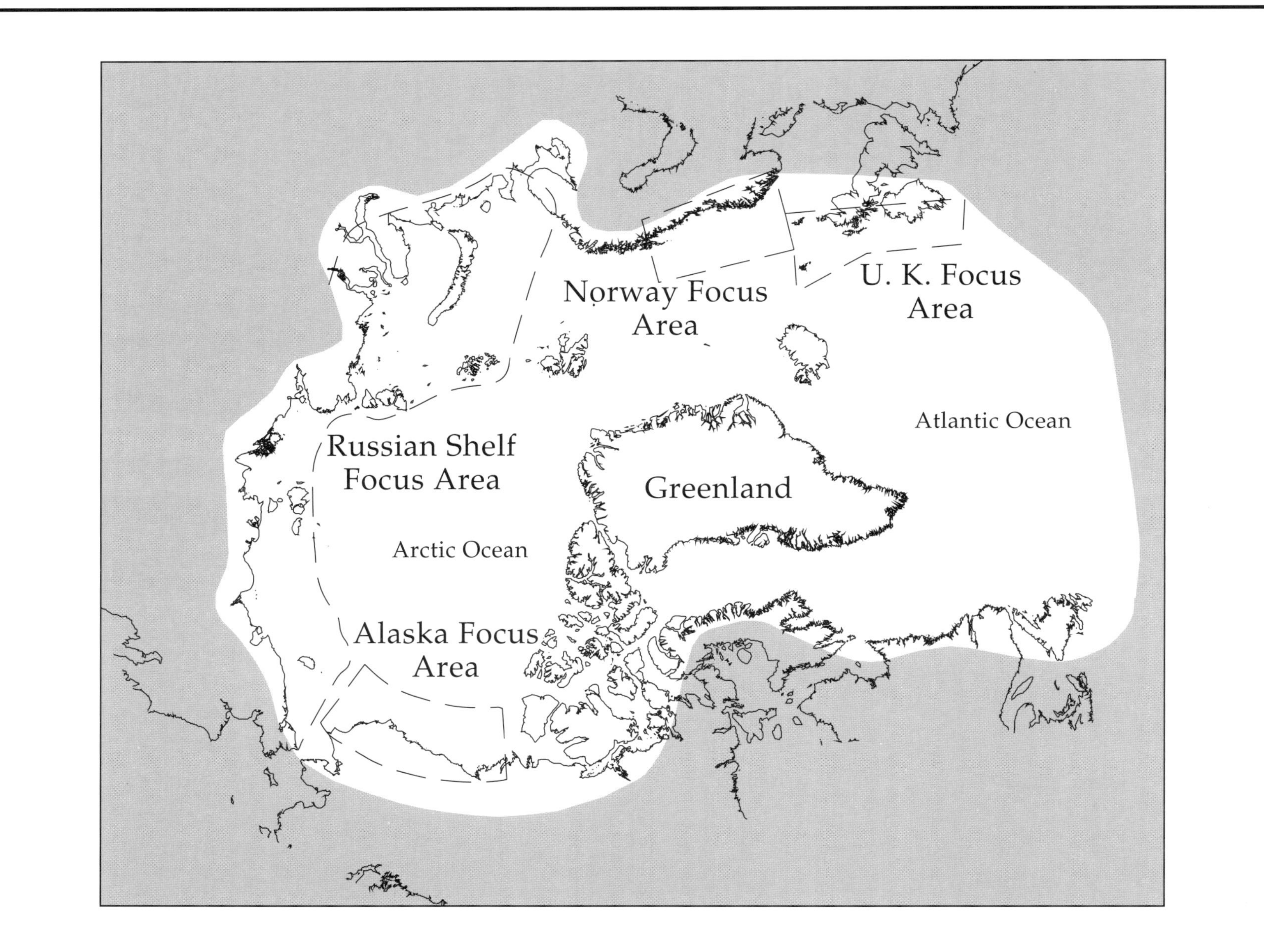

Figure 4—Overall study area for the Circum Arctic Regional Project (CARP) showing subregional focus areas.

most effective method to systematically populate the data structures. The standards were completed during CARP and were adopted in time for project use.

The archival file structure mirrors the hierarchical Unix directory structure and is presented using the analogy of a file cabinet with *project, drawer,* and *folder* levels. The project level generally includes all products developed in a given study. The project is subdivided into a series of drawers, each based on a scale, geographic, or geologic subdivision of the project. Within each project is a standard set of folders, into which maps can be filed thematically in accordance with the standards (Figure 5). Using this system allows products to be filed in standard locations, regardless of the specific project theme.

Separate databases were created for different categories or themes of map layers. For example, the database for structural features (faults and folds) is distinctly different from the database used for paleogeography. All paleogeography maps, however, have the same database structure. The databases are created with macros and populated through a series of data-entry templates at the time the coverage is created (Appendix A).

Each database is composed of a related set of data tables containing standardized codes, which provide a systematic means for describing individual features found on geological maps. For example, the MAR_WD (marine water depth) variable in the *sedimentary lithofacies* database structure can only be coded as ONSHORE, LITTORAL, NERITIC, or BATHYAL as described in the code lists (see Appendix A). Although the standards cause data to be somewhat regimented, this approach provides a level of consistency that simplifies planning, data capture, searchability, and ultimate reuse of the data.

As might be expected, the added effort of populating a database constrained by standard tables and codes caused the work group some concern. Much of the discussion revolved around determining which portions of each database should be populated and which codes should be used to describe given map layers and layer features.

Although Exxon databases provide slots for virtually all meaningful data that could be assigned to a mapped geologic feature, *it is not required* that all such slots be populated. Data availability in regional or frontier areas is too sparse and inconsistent for such a requirement. Consequently, it was left to the CARP work group to determine the best subset of each thematic database to meet its needs. Each database was subsequently reviewed, and the work group determined the appropriate fields to populate.

Populating the respective databases also required some attention. The flexibility of the data population tables allowed a map feature to be described in the database in a number of ways. The team had to agree to the combinations that would be used to describe certain map features. For example, shale with sandstone stringers could be described in the LITHOLOGY and LITH_TYPE variables of the sedimentary lithofacies database structure as [SHAL, NONE] or [MXD, SANDSHAL] (see Appendix A). Consistency in coding was especially important where several contributors were working portions of a map that would be merged at a later date. Reviewing database issues ahead of time minimized reworking during map compilation, although there were several instances of miscommunication resulting in different codes being used to populate the databases.

The discussion over regional structural elements was a case in point. Portions of the maps were to be constructed by different individuals and then zippered together into a regional compilation. Two schools of thought developed regarding information capture. One school of thought advocated capturing the relative importance and timing of activity for each feature, whereas the other group felt that there was insufficient time or data to acquire and code this information. This particular instance was not adequately resolved, and both methods were applied to the individual map contributions. The ramifications of this lack of agreement are described in the next section.

Timing of Individual Contributions

Because a fundamental consideration of the GIS approach is forward planning, the availability of each layer is critical to the completion of final products. Each map product is essentially an amalgam of earlier developed products. In the ideal case, a diagram or flowchart can be developed for each product, and this diagram, in fact, has the same significance as a Gantt chart in critical-path planning methodology. Critical-path methodology requires a project plan to identify the dependency of each task on earlier completed tasks. The critical path is the dependent chain of tasks that ultimately delay the whole project if the scheduled completion of any one task in the chain is delayed. In the context of a regional study, failure to deliver a designated component or subcomponent on time affects the timing and integrity of the final deliverable (Figure 3).

An example in CARP involved the final basin outline map, which was derived from the regional sediment-fill compilation, gravity/magnetic interpretations, bedrock distribution maps, and the major structural features map. As noted, the structural features compilation was zippered from individual contributions. The plan, however, was developed using the base case, which emphasized capture of features without timing data. As a result, portions of the structural features map were not ready as planned, and the ultimate delivery occurred in several phases rather than as a single compilation. This approach required multiple revisions to the basin outline map and, subsequently, to a whole family of maps derived from the basic basin map. In all, the basins map was modified twice with requisite changes to over 30 dependent products.

Agreement on Data Elements

The GIS approach seeks to eliminate redundant storage of geological features. A map element that is used on several different products or by more than one interpreter should exist once and only once in the

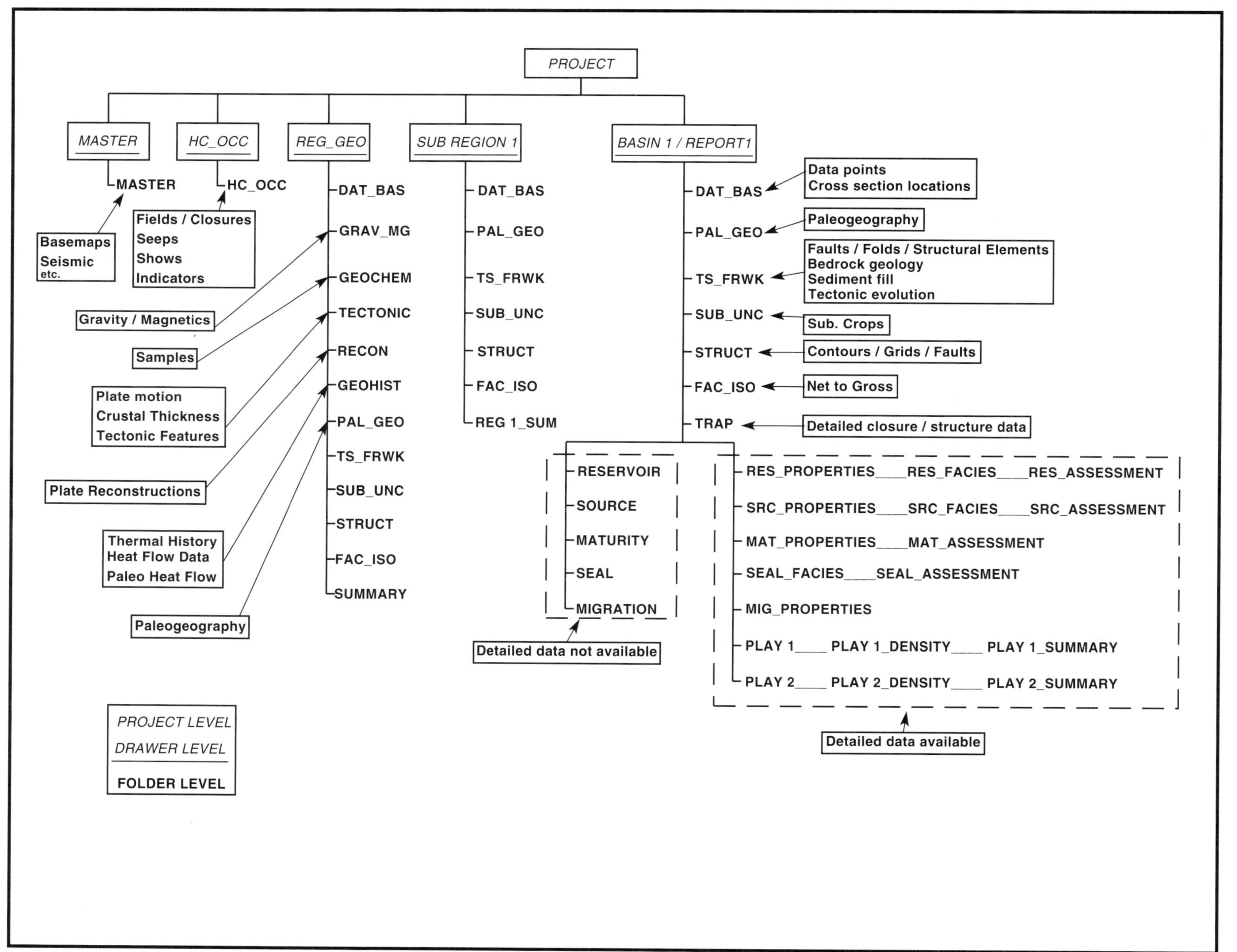

Figure 5—Archive computer file structure for regional geological studies.

system. With this approach, discrepancies that were simply overlooked in stand-alone, individually drafted products in previous studies stand out in GIS products created by overlaying multiple thematic elements. With the GIS approach, different interpreters must work actively to maintain interpretation consistency.

This objective proved particularly difficult to maintain with features produced on paleo-reconstructed bases (i.e., maps illustrating continents and ocean basins rotated back to a past configuration). It was not uncommon for inconsistencies to appear between maps drawn by the paleogeographer and the geochemist early in the project. For example, a lithofacies limit for shale on a reconstructed base as drawn by the paleogeographer might not match the same feature on the geochemist's map. This effect was partly the result of working on reconstructed bases, but it was also a function of the different sources of data traditionally used by each discipline (primarily tectonic and stratigraphic summary literature by the paleogeographer versus well and core sample databases by the geochemist). The magnitude of the problem gains perspective considering that there were 21 maps in each map set, one on a paleo-reconstructed base and the other on a present-day base. The potential for disagreement was compounded in that over 40 basins were being evaluated. The problem was discovered early, and close coordination was required to ensure consistency. Even with a heightened awareness, however, several iterative steps were initially required to maintain map consistency.

Successes

In addition to the advantages of high-quality databases and work efficiencies, the main success of the GIS approach to CARP was the high degree of geological integration built into the final interpretations and regional syntheses. Agreement on common data elements in itself is a significant contribution to a more consistent set of regional interpretations.

In a more mechanical sense, consensus interpretations within a common database made compilation and interpretation easier than would otherwise have been possible. Three products stand out in this regard: the sediment-fill map, the tectonic product suite, and the contribution of a Russian scientific organization.

Sediment-Fill Map

As noted, the sediment-fill map was constructed by compositing several layers within GIS. The key layer for this map composition was the sediment-fill isopach. The isopach was built using a variety of sources, including hand-contoured maps from earlier studies, detailed local compilations based on seismic interpretations, published data, and grids derived from gravity, magnetic, and bathymetric data. Although any two-dimensional modeling software could have been used to build the sediment-fill isopach, the editing, reprojection, and spatial integration capabilities of the GIS allowed a high level of synthesis on a regional basis. The isopach was constructed in several phases, using an older, hand-contoured map as the basic map and adding new interpretations as they were obtained or generated. The update process included the following steps: the geoscientist created isopachs (often on Mylar); in-house geotechnical or vendor personnel digitized the map draft; a GIS specialist merged the data into the master sediment-fill isopach and plotted a proof copy; and, finally, the geoscientist reviewed, and often further refined, the map to ensure seamless zippering. This last step often resulted in a complete process recycle. In CARP, efficiencies were gained when one of the geoscientists took the initiative to learn the editing tools available in the GIS software and make on-screen changes to the isopach. By the end of the project, this geoscientist had taken responsibility for the sediment-fill isopach, completing on-screen interpretations rather than using more traditional paper-based methods. This approach proved especially effective, so much so that the time spent learning the software was more than offset by the efficiency gained through on-screen editing. In the case of the sediment-fill isopach, the product was actually completed ahead of schedule. In addition, the geoscientist was able to play a role as a facilitator between the GIS specialists and other geoscientists on the team.

Basin Delineation and Tectonics

Considerable effort was expended to work out the major tectonic interrelationships characterizing and delineating the Arctic basins. The work was largely based on a several-year effort integrating high-resolution gravity and magnetic data sets, migrating them into the GIS, and merging them with topography, bathymetry, sediment fill, structural features, and interpreted seismic. Historically, merging potential field data, bathymetry and topography, and interpreted maps was relegated to the realm of the light table. Using the GIS overlay and edit capabilities, mapsets that had been mutually exclusive were now used to refine interpretations for the entire suite of products. The analysis of the tectonic elements was critical to the construction of the basin outline map.

Basin outlines were derived from this suite of maps using a rigorous set of definitions. Basin outlines are one of the last map layers to be built because they require the use of a range of products, including sediment fill, structural elements, potential fields grids, a mid-ocean ridge, a continent-ocean boundary, and basement terrane maps. The outlines were built by the work group, using various bases consisting of different combinations of data layers built in the GIS and plotted out. On-screen digitizing was not feasible due to the logistical considerations (primarily the size of the work group). A total of 41 basin outlines was drawn and digitized. Once captured, interim outlines were plotted for team review, discussion, and refinement. In areas where critical supporting layers (for example, structural elements) were not complete, outlines were drawn with the understanding that they would have to be revised. The resulting basin outlines were far superior to any previous versions due to the ability of the GIS to integrate many geotechnical constraints on a relatively small number of maps. In all, most basin outlines within the project area were modified, 14 basin or subbasin outlines were fundamentally changed, and several new subbasins were identified where none had been previously recognized.

Institutional Partners

During the project, a large number of data products was acquired from outside digital data vendors and research institute partners who, under contract, developed basic interpretations for portions of the study area. The CARP team spent considerable effort ensuring that the geological products addressed significant problems required to unravel the geology of the Arctic. The team also ensured that the contributing organizations and vendors employed the Exxon standards for archiving files, attaching thematic databases, and populating the databases from the code lists. For the research institutes, this included significant training and deployment of equipment. Initial skepticism regarding the GIS work process was fairly common, but as the partners used the process to meet Exxon standards, they recognized the inherent advantages and became accepting and enthusiastic about its potential. From Exxon's standpoint, an immediate benefit of this effort was the fact that final products were readily integrated into the regional compilations with minimal difficulty. This benefit was significant because most of the products were scheduled to arrive close to the end of the CARP project to be incorporated into a number of end products. Without the standards in place, the research institute's products could not have been assimilated into the final compilation within the time constraints of the project.

CORPORATE LEARNINGS

It is clear that three factors are essential to the GIS work process: product-based planning, application of GIS data standards, and commitment to the process by management and project geologists.

Product-based planning maximizes the benefits of using a GIS. It requires the work group to think in terms of products and their component layers, schedules, and individual contributions. As a result, the group concentrates on essential tasks and has a clear understanding of the interdependencies of component products, as well as project deadlines. The CARP experience clearly illustrated that deviation from, or a lack of consensus on, the plan could negatively impact the final compilation phase of a project.

Standards for the GIS database and file structure provided a level of consistency that proved critical in zippering smaller compilations into a regional synthesis. The standards also aided in systematic data analysis because data were coded in such a way as to leave no doubts about their meaning. Final products were produced in a consistent manner, which, in itself, aided in documentation and archiving. CARP adopted a uniform and geologically intuitive way of organizing its computer data to make it easy for current and future users to find the study's 1200 computer products. Not only were products filed in a standard structure, but the maps were also catalogued in a spatial map index to ensure easy search and retrieval efforts in the future.

Management commitment was a critical driving force early in the implementation phase of the process, when the work group was introduced essentially to a new way to work. Resistance to changing work habits, while varying among individuals, nonetheless required effort to maintain group alignment until the benefits of the process became apparent. In addition, it was noted that, until the process was firmly established, the work group would backslide to old work habits that were injurious to the process. At these times, management commitment was key in holding the team to common goals.

CONCLUSION

Traditional regional studies typically devote much less effort to the planning and production of the final product suite than they do in meeting short-term analytic and review display goals. Throughout such studies, data capture and map construction are undertaken in a variety of ways, including manual construction on paper, two-dimensional computer modeling, and nongeographically oriented drafting packages, all of which conspire to produce a myriad of different kinds of maps, both on a regional and subregional scale. These sundry components have to be integrated into a consistent set of regional products. Because a typical regional study generates hundreds of documents, assembling final versions is a task that can take up to a full year. Because it is often impossible to keep a team assembled for that period, the task often falls to one or two individuals, along with the drafting department.

A GIS-based work process, as used in CARP, offers considerable benefits to an organization. Foremost is an estimated 20% cost savings over more traditional regional study processes. Documents are completed and updated on a real-time basis. Standardized maps facilitate a level of integration and analysis unobtainable in a traditional study. Final documentation in a standardized archival structure is ready within weeks of completing the project. Most important, final products are updatable and retrievable for ongoing studies.

REFERENCES CITED

Bonham-Carter, G. F., 1994, Geographic information systems for geoscientist: modeling with GIS. Ottawa: Pergamon, 398 p.

Calkins, J. F., 1990, Tenneco's GIS: a rough start: Geobyte, vol. 5, no. 3, p. 31–35.

Feineman, D. R., 1994, GIS and petroleum geology in retrospect: 1994 International Association for Mathematical Geology Annual Conference, Papers and Extended Abstracts for Technical Programs, p. 122–127.

Foley, D. C., and T. Y. Ghazi, 1995, How Conoco uses GIS technology to map geology, geography through time: Oil and Gas Journal, vol. 93, no. 19, p. 74–77.

Miller, B. M., 1992, Guide to the development and application of geographic information systems for sedimentary basin analysis—case study for the San Juan basin, New Mexico and Colorado: U.S. Geological Survey, Bulletin 2025, 33 p.

Vitek, J. D., J. R. Giardino, and J. W. Fitzgerald, 1996, Mapping geomorphology: a journey from paper maps, through computer mapping to GIS and virtual reality: Geomorphology, vol. 16, no. 3, p. 233–249.

ACKNOWLEDGMENTS

The authors wish to thank the management of Exxon Exploration Company for permission to publish this manuscript and for ongoing support in developing and fostering this technology. With regard to this last point in particular, we would like to acknowledge D. M. Maughan and B. A. Vining. We also wish to thank the AAPG reviewers of this paper for many thoughtful suggestions and improvements.

Evolution of the methodology itself has indeed been a team effort involving many groups and individuals in Exxon Exploration Company too numerous to enumerate, much less to thank adequately. In particular, however, we wish to credit the following individuals: C. D. Norman, D. A. Leary, and J. E. Hughes for their work in developing the map attribute standards (Appendix A); B. C. South for the archival file structure; and the members of the core CARP team, C. A. Johnson, D. Gilbert, and A. Kaplan, for aggressively translating theory into practice. A special debt is also due E. B. Neslon, CARP's last project leader, who was pivotal in implementing the technology at home and with outside partners.

APPENDIX A

Exxon Templates and Data Population Tables

Template	Pages	General Usage
Template index	84	Selecting the correct database structure for a map
Bedrock	85–93	Surface geology, subcrops, crustal genesis, paleotectonics
Contours	94–95	All isolines
Faults and structures	96–97	Faults and related information, folds, structures, structural types
Miscellaneous features	98–99	Polygon or line maps that do not fit into specific categories (e.g., erosional limits)
Point data	100–102	Various types of point data
Polygonized values	103–104	Maps containing polygons with quantitative attributes such as LOM or chance of adequacy or risk
Sedimentary lithofacies	105–119	Sedimentary facies, depositional environment, paleography, or lithofacies
Closures	120	Fields, prospects, leads, or shows

SELECTING THE CORRECT ATTRIBUTE STRUCTURE FOR A MAP

MAP SUBJECT	COVERAGE TYPE
anticline axes	Faults and Structures
assessment - BAU outlines	Assessment BAU's
assessment - GPLAY input/output	PlaysumZ
basement grains	Miscellaneous Polygons & Lines
basement warps	Faults and Structures
chance of oil field/play	Play Element Adequacy
clastic grain lithology	Bedrock and Tectonics
contour maps (all contours closed)	Polygonized Values
contour maps (all contours not closed)	Contours
contour maps (control points)	Oracle or Non-Oracle Point Data
cross section index lines	Miscellaneous Polygons & Lines
depositional axis	Miscellaneous Polygons & Lines
depositional environments	Sedimentary Lithofacies
dip & strikes	Oracle or Non-Oracle Point Data
dry tests	Closures/Closure Related Data Table
enhanced permeability zones	Miscellaneous Polygons & Lines
extent of reservoir characteristic	Miscellaneous Polygons & Lines
extent of seal	Play Element Adequacy
facies control points	Oracle or Non-Oracle Point Data
fault block (structural contour polygon)	Polygonized Values
fault block outline	Faults and Structures
fault gap control points	Oracle or Non-Oracle Data Points
fault traces	Faults and Structures
field hydrocarbon volumes and details	ORACLE/PCDB
field outlines	Closures/Closure Related Data Table
flexures	Faults and Structures
fold axes	Faults and Structures
fracture zones	Miscellaneous Polygons & Lines
igneous geochemistry	Bedrock and Tectonics
igneous or metamorphic facies	Bedrock and Tectonics
isopachs (all contours closed)	Polygonized Values
isopachs (all contours not closed)	Contours
lead outlines	Closures/Closure Related Data Table
lineaments	Miscellaneous Polygons & Lines
litho-tectonic map control points	Oracle or Non-Oracle Point Data

MAP SUBJECT	COVERAGE TYPE
litho-tectonic maps	Bedrock and Tectonics
lithologies	Bedrock and Tectonics
lom map (all contours closed)	Polygonized Values
lom map (all contours not closed)	Contours
magnetic anomalies	Miscellaneous Polygons & Lines
magnetic reversal data points	Non-Oracle Point Data
migration chance of adequacy	Play Element Adequacy
migration limit	Miscellaneous Polygons & Lines
migration timing chance of adequacy	Play Element Adequacy
ocean floor age polygon	Bedrock and Tectonics
oil family polygons	Miscellaneous Polygons & Lines
outcrop maps	Bedrock and Tectonics
paleo-depositional environments	Sedimentary Lithofacies
paleotectonic maps	Bedrock and Tectonics
plate boundaries	Bedrock and Tectonics
play element boundaries	Play Element Boundaries
plunge data	Oracle or Non-Oracle Point Data
prospect outlines	Closures/Closure Related Data Table
reservoir chance of adequacy	Play Element Adequacy
seal/trap limits	Play Element Adequacy
sedimentary lithofacies	Sedimentary Lithofacies
source chance of adequacy	Play Element Adequacy
source limits	Play Element Adequacy
spreading centers	Miscellaneous Polygons & Lines
structural style	Faults and Structures
structural terranes	Faults and Structures
subcrop maps	Bedrock and Tectonics
subduction zones	Miscellaneous Polygons & Lines
surface geology	Bedrock and Tectonics
sutures	Miscellaneous Polygons & Lines
syncline axes	Faults and Structures
tectonic terranes	Bedrock and Tectonics
transforms	Faults and Structures
trap type polygons	Miscellaneous Polygons & Lines
trap/seal chance of adequacy	Play Element Adequacy

Name: BEDROCK AND TECTONICS

REFER TO CODE LISTS FOR SHADED ATTRIBUTE ITEMS

Data Type: POLYGONS OR LINES

Used For:

Paleotectonic coverages
Surface geologic coverages
Detailed subcrop or outcrop coverages
Igneous and Metamorphic compositions
Ocean floor age coverages
Crustal genesis coverages
Paleogeographic coverages

Attribute items:

	WIDTH	OUTPUT	TYPE	N. DEC.	ALT. NAME	DESCRIPTION / PURPOSE OF THE ITEM
xxx_bu_id	8	8	C		bu_id	User defined. Relate key to related data table.
xxx_class	20	20	C	-	class	Code lists. Major geologic class of polygon or class of closure.
xxx_data_src	30	30	C	-	data_src	User defined. Data source or provider. Example: EPR.44EX.87
xxx_origin	4	4	C	-	origin	Code lists. Major genetic subdivision within geologic class.
xxx_lithology	8	8	C	-	lithology	Code lists. Major lithology type.
xxx_depth	20	20	C	-	depth	Code lists. Depth of origination of rock.
xxx_lith_type	16	16	C	-	lith_type	Code lists. Type or subdivision of lithology.
xxx_sublith	50	50	C	-	sublith	User defined. Sub-lithology descriptor.
xxx_txtr_strat	30	30	C	-	txtr_strat	Code lists. Textural or stratal pattern descriptor.
xxx_lith_mod	30	30	C	-	lith_mod	Code lists. Lithologic composition modifier.
xxx_group	30	30	C	-	group	Code lists. Major formative group.
xxx_subgroup	30	30	C	-	subgroup	Code lists. Secondary formative group.
xxx_tect_regime	16	16	C	-	tect_regime	Code lists. Tectonic regime during formation.
xxx_margin	16	16	C	-	margin	Code lists. Type of continental margin during formation.
xxx_tect_assem	16	16	C	-	tect_assem	Code lists. Tectonic assemblage of which polygon is a part.
xxx_crust_resp	20	20	C	-	crust_resp	Code lists. Crustal response.
xxx_descript	250	250	C	-	descript	User defined. Description of polygon or line. Example: MIXED SANDSTONE/SILTSTONE/SHALE.
xxx_min_age	20	20	C	-	min_age	Code lists. Minimum age of polygon or line (if coverage has multiple ages).
xxx_max_age	20	20	C	-	max_age	Code lists. Maximum age of polygon or line (if coverage has multiple ages).
xxx_calling_age	30	30	C	-	calling_age	User defined. The age name used in general reference.
xxx_horizon	30	30	C	-	horizon	User defined. Horizon of polygon or line (if coverage has multiple horizons). Use BU horizon names
xxx_con_lith	10	10	C	-	con_lith	Code lists. Confidence of lithology interpretation.
xxx_con_env	10	10	C	-	con_env	Code lists. Confidence of environment interpretation.
xxx_age_mng	20	20	C	-	age_mng	Code lists. Type geologic event to which age of magnetism applies.
xxx_name	32	32	C	-	name	User defined. Business Unit name for polygon or line. Example: LOST RIVER CHERT.
xxx_text	250	250	C	-	text	User defined. Annotation to appear on coverage. May be the same as another item.
xxx_subject	30	30	C	-	subject	User defined. Overall coverage subject or entity. Example: UPPER CRETACEOUS.
xxx_title	100	100	C	-	title	User defined. Information mapped about the subject. Example: SURFACE GEOLOGY.
xxx_file	50	50	C	-	file	User defined. Name of original, pre-Arc file. Example: CRFCTAR4.DWG

xxx = ITEM PREFIX REQUIRED FOR POLYGON AND POINT COVERAGES ONLY. EXAMPLE: IF COVERAGE IS BAU_CRET_LOW THEN XXX COULD BE BCL..
BOLDFACE = REQUIRED ITEM

BEDROCK AND TECTONICS ATTRIBUTE STRUCTURE - DATA OPTIONS
CHOOSE ONE ITEM FROM EACH CATEGORY AS APPROPRIATE

SEDIMENTARY

xxx_rock_class	xxx_origin	xxx_lithology	xxx_lith_type	xxx_depth
SEDIMENTARY	CARB *(Carbonate)*	LMST *(Limestone)*	MUDSTN *(Mudstone)*	ABYSSAL
			MICRITE	BATHYL
			MCRT_MUD *(Micrite to Mudstone)*	HADAL
			BNDSTN *(Boundstone)*	INNER_NERITIC
			CHALK	LITTORAL
			XLN *(Crystalline)*	LOWER_BATHYL
			FLOATSTN *(Floatstone)*	MIDDLE_BATHYL
			GRNSTN *(Grainstone)*	MIDDLE_NERITIC
			NONE *(Limestone Undiff.)*	NERITIC
			MARL	ONSHORE
			PACKSTN *(Packstone)*	OUTER_NERITIC
			RUDSTN *(Rudstone)*	UPPER_BATHYL
			WACKSTON *(Wackestone)*	
		DLMT *(Dolomite)*	MCRT_MUD *(Micrite to Mudstone)*	
			BNDSTN *(Boundstone)*	
			XLN *(Crystalline)*	
			DOLOMITE	
			NONE *(Dolomite Undiff.)*	
			FLOATSTN *(Floatstone)*	
			GRNSTN *(Grainstone)*	
			PACKSTN *(Packstone)*	
			RUDSTN *(Rudstone)*	
			WACKSTON *(Wackestone)*	
		CARB *(Carbonate Undiff)*	NONE *(Carbonate Undiff)*	
	CLST *(Clastic)*	SAND *(Sandstone)*	QTZ_AREN *(Quartz Arenite)*	
			ARKOSE	
			LTH_AREN *(Lithic Arenite)*	
			NONE *(Sandstone Undiff.)*	
			CLC_AREN *(Calcarenite)*	
			SUBARKSE *(Subarkose)*	
		SHAL *(Shale)*	MUD *(Mudstone)*	
			CLY_MUD*(Claystone to Mudstone)*	
			CLY_SHAL *(Claystone to Shale)*	
			CLAY *(Claystone)*	
			MUD_SHAL *(mudstone to Shale)*	
			NONE (Shale Undiff.)	
		SILT *(Siltstone)*	NONE *(Siltstone Undiff.)*	
		CNGL *(Conglomerate)*	OLIGMICT *(Oligomictite)*	
			NONE *(Conglomerate Undiff.)*	
			PTROMICT *(Petromictite to Polymictite)*	
			TILLITE	
		COAL	PEAT	
			ANTHRC *(Anthracite)*	
			ANTHRX *(Anthraxolite)*	
			BIT *(Bituminous)*	
			BOGHEAD *(Boghead-Borganite)*	
			CANNEL *(Cannel Coal)*	
			NONE *(Coal Undiff.)*	
			LIGNITE	
			SEMI_BIT *(Semi-Bituminous)*	
CONTINUED NEXT PAGE	*CONTINUED NEXT PAGE*		SUB_BIT *(Sub-Bituminous)*	

BEDROCK AND TECTONICS ATTRIBUTE STRUCTURE - DATA OPTIONS
CHOOSE ONE ITEM FROM EACH CATEGORY AS APPROPRIATE

SEDIMENTARY

xxx_rock_class	xxx_origin	xxx_lithology	xxx_lith_type
SEDIMENTARY *CONTINUED FROM PREVIOUS PAGE*	CLST *(Clastic)* *CONTINUED FROM PREVIOUS PAGE*	BRCC *(Breccia)*	SCREE *(Scree Talus)* AGGLOM *(Agglomerate)* BONE *(Bone Breccia)* NONE *(Breccia Undiff.)* COLLAPSE *(Collapse Breccia)* FAULT *Fault Breccia)* FISSURE *Fissure Breccia)* INTRFORM *(Intraformational Breccia)*
		CLST *(Clastic)*	NONE *(Clastic Undiff.)*
	CRYP *(Cryptocrystalline)*	CHRT *(Chert)*	FLINT NONE *(Chert Undiff.)* *DIATOM* *(Diatomite)* JASPER JASPERD *(Jasperoid)* NOVACLT *(Novaculite)* PORCNLT *(Porcellanite)* RADIOLR *(Radiolarite)* SPICULT *(Spiculite)* TRIPT *(Tripolite)*
		CRYP *(Cryptocrystalline)*	NONE *(Cryptocrystalline Undiff.)*
	EVAP *(Evaporite)*	EVAP *(Evaporite)*	GYPSUM
	EVAP *(Evaporite)*	EVAP *(Evaporite)*	ANHYD *(Anhydrite)* NONE *(Evaporite Undiff.)* PHOSPHT *(Phosphatite)* MAGNST *(Magnesite)* SALT *(Salt-Halite)*
	MIX *(Mixed Sediment)*	MXD *(Mixed Sediments)*	*SANDSHAL (Sandstones And Shales)* SANDLMST *(Sanstones And Limestones)* LMSTSILT *(Limestone and Silts)* LMSTEVAP *(limestones and Evaporites)* DLMTSHAL *(Dolomites and Shales)* DLMTEVAP *(dolomites and Evaporites)* SANDSILT *(Sandstones and Siltstones)* SANDUDCB *(Sandstones and Carbonates)* CARBEVAP *(Carbonates and Evaporites)* SHALEVAP *(Shales and Evaporites)* SILTEVAP *(Silstones and Evaporites)* SILTCOAL *(Siltstones and Coals)* SHALSILT *(Shales and Sitlstones)* SANDEVAP *(Sandstones and Evaporites)* CNGLSAND *(Conglomerates and Sandstones)* SANDCOAL *(Sandstones and Coals)* DLMTSILT *(Dolomites and Silts)* SANDDLMT *(Sanstones and Dolomites)* UDCBUDCS *(Carbonates and Clastics)* LMSTSHAL *(Mixed Limestones and Shales)* LMSTDLMT *(Mixed Limestones and Dolomites)*
	SED *(Sedimentary)*	SEDS *(Sedimentary Undiff.)*	NONE *(Undifferenciated Sediments)*
IGNEOUS	IGNS *(Igneous)*	IGNS *(Igneous)*	INTRA *(Intrusive)*
			EXTR *(Extrusive)*
			NONE *(Ingeous Undiff.)*
METAMORPHIC	META *(Metamorphic)*	META *(Metamorphic)*	NONE *(Metamorphic Undiff.)*
UNKNOWN	UNK *(Unknown)*	UNK *(Unknown)*	NONE
UNDIFFERENTIATED	CLST *(Clastic)*	UNK *(Unknown)*	NONE

BEDROCK AND TECTONICS ATTRIBUTE STRUCTURE - DATA OPTIONS

CHOOSE ONE ITEM FROM EACH CATEGORY AS APPROPRIATE

SEDIMENTARY

xxx_txtr_strat	xxx_lith_mod	xxx_group	xxx_subgroup
BIOTURBATED	ACTINOLITE	MARINE	xxx_subgroup
COARSENING_UPWARD	CALCAREOUS	MARGINAL_MARINE	BASIN (or BASINAL)
CONCORDANT	DIOPSIDE	NON_MARINE	COASTAL
CROSS_BEDDED	DOLOMITIC		COASTAL_PLAIN
CYCLIC	EPIDOTE		EOLIAN
DISCORDANT	FOSSILIFEROUS		ESTUARINE
DISSEMINATED	HEMATITIC		EVAPORITIC
FINELY_PARTED	HORNBLENDE		FLUVIAL
FINING_UPWARD	ILLITE		FLUVIAL_ALLUVIAL
INTERBEDDED	KAOLINITE		FLUVIAL_DELTAIC
INTERCALATED	LIMEY		FLUVIAL_ESTUARINE
LAMINATED	MIXED		GLACIAL
MASSIVE	OOLITIC		HEMIPELAGIC
NODULAR	SHALY		INNER_SHELF
POORLY_SORTED	SILTY		INTERTIDAL
THIN_BEDDED	SKELETAL		LACUSTRINE
WELL_SORTED	SMECTITE		LAGOONAL_BAY
	TREMOLITE		MIDDLE_SHELF
			NEARSHORE
			OFFSHORE
			ONSHORE
			OPEN_INTERIOR
			OUTER_SHELF
			PELAGIC
			PERITIDAL
			PLATFORM
			RAMP
			RESTRICTED_INTERIOR
			SABKHA
			SHELF (or SHELFAL)
			SHELF_MARGIN
			SLOPE
			SUBTIDAL
			SUPRATIDAL
			TIDAL_FLAT

Additional values for all items in this column:

UNK *(Unknown)*
UNDIFF *(Undifferenciated)*

BEDROCK AND TECTONICS ATTRIBUTE STRUCTURE - DATA OPTIONS

CHOOSE ONE ITEM FROM EACH CATEGORY AS APPROPRIATE

METAMORPHIC

xxx_class	xxx_origin	xxx_lithology	xxx_lith_type	xxx_lith_mod	xxx_depth
METAMORPHIC	CNTC *(Contact Metamorphism)*	ALBT_EPI *(Albite-Epidote-Hornfels Facies*	HORNFLES MARBLE NONE QUARTZITE SERPENTINITE SKARN SPOT_HORNFELS	*User-defined. Examples:* ACTINOLITE ALBITE BIOTITE CALCITE CHLORITE DIOPSIDE DOLOMITE EPIDOTE GARNET HORNBLENDE MUSCOVITE PLAGIOCLASE QUARTZ SILLIMANITE STAUROLITE TREMOLITE ↑ *Additional items for all items in this column:* UNK *(Unknown)* UNDIFF *(Undifferenciated)*	ARCHIZONAL CATAZONAL EPIZONAL MESOZONAL
		HRNBLND *(Hornblende-Hornfels Facies)*	GRANOFELS HORNFELS MARBLE NONE QUARTZITE SERPENTINITE SKARN SPOT_HORNFELS		
		NONE *(Metamorphic Undiff.)*	GRANOFELS HORNFELS MARBLE NONE QUARTZITE CATACLASITE MYLONITE PHYLLONITE SERPENTINITE SKARN SPOT_HORNFELS		
		PYROXENE *(Pyroxene-Hornfels Facies)*	GRANOFELS HORNFELS NONE QUARTZITE		
		SANIDITE *(Sanidite Facies)*	GRANOFELS HORNFELS MARBLE NONE QUARTZITE		
	RGNL *(Regional Metamorphism)*	AMPHIBOL *(Amphibolite Facies)*	AMPHIBOLITE GNEISS SCHIST		
		ANATEXIS	MIGMATITE		
		BLUSCHST *(Blueschist Facies)*	ARGILLITE BLUESHCST MARBLE NONE PHYLLITE QTZ_ARGILLITE QTZ_SLATE QUARTZITE SCHIST SERPENTINITE SLATE SPOT_PHYLLITE SPOT_SLATE		

BEDROCK AND TECTONICS ATTRIBUTE STRUCTURE - DATA OPTIONS
CHOOSE ONE ITEM FROM EACH CATEGORY AS APPROPRIATE

METAMORPHIC

xxx_class	xxx_origin	xxx_lithology	xxx_lith_type
METAMORPHIC	*(Regional Metamorphism) (Continued)*	COESITE *(Coesite-Diamond-grade Facies*	GNEISS
			GRANULITE
			MARBLE
			MIGMATITE
			NONE
			QUARTZITE
		EGLOGITE *(Eclogite Facies)*	ARGILLITE
			ECLOGITE
			MARBLE
			NONE
			PHYLLITE
			QTZ_ARGILLITE
			QTZ_SLATE
			QUARTZITE
			SCHIST
			SERPENTINITE
			SLATE
			SPOT_PHYLLITE
			SPOT_SLATE
		GRNSCHST *(Greenschist Facies)*	ARGILLITE
			MARBLE
			NONE
			PHYLLITE
			QTZ_ARGILLITE
			QUARTZITE
			SCHIST
			SLATE
			SPOT_PHYLLITE
			SPOT_SLATE
		GRNULITE *(Granulite Facies)*	GNEISS
			GRANULITE
			MARBLE
			MIGMATITE
			NONE
			QUARTZITE
		NONE *(Undifferenciated Metamorhics Fac*	AMPHIBOLITE
			ARGILLITE
			BLUESCHIST
			GNEISS
			GRANULITE
			GREENSCHIST
			MARBLE
			MIGMATITE
			PHYLLITE
			QTZ_ARGILLITE
			QTZ_SLATE
			QUARTZITE
			SCHIST
			SERPENTINITE
			SLATE
			SPOT_PHYLLITE
			SPOT_SLATE

BEDROCK AND TECTONICS ATTRIBUTE STRUCTURE - DATA OPTIONS
CHOOSE ONE ITEM FROM EACH CATEGORY AS APPROPRIATE

METAMORPHIC

xxx_class	xxx_origin	xxx_lithology	xxx_lith_type
METAMORPHIC	*(Regional Metamorphism)* *(Continued)*	PREHNITE *(Prehnite-Pumpellyite Facies)*	ARGILLITE MARBLE NONE PHYLLITE QTZ_ARGILLITE QTZ_SLATE QUARTZITE SCHIST SERPENTINITE SLATE SPOT_PHYLLITE SPOT_SLATE
		UNDIFF *(Regional Undifferenciated)*	UNDIFF
		ZEOLITE *(Zeolite Facies)*	ARGILLITE MARBLE NONE PHYLLITE QTZ_ARGILLITE QTZ_SLATE QUARTZITE SCHIST SERPENTINITE SLATE SPOT_PHYLLITE SPOT_SLATE

xxx_txtr_strat	xxx_group	xxx_subgroup
User-defined. Examples:	METAIGNEOUS	HIGH_GRADE
GNEISSIC	METASEDIMENTARY	LOW_GRADE
MASSIVE		MEDIUM_GRADE
POICHIOBLASTIC		VERY_LOW_GRADE
SCHISTOSE		

BEDROCK AND TECTONICS ATTRIBUTE STRUCTURE - DATA OPTIONS
CHOOSE ONE ITEM FROM EACH CATEGORY AS APPROPRIATE

IGNEOUS

xxx_class	xxx_depth	xxx_origin	xxx_lithology	xxx_lith_type
IGNEOUS	*(Plutonic Rocks)*	EXTR	FELSIC *(Felsic-Acidic-Silicic)*	ALK_RHYOILTE
	CATAZONAL	*(Volcanic Extrusive)*		COMMENDITE
	EPIZONAL			PANTELLERITE
	HYPABYSSAL			PHOLLITE
	MESOZONAL			RHYOLITE
				UNDIFF
			INTERMED *(Intermediate)*	ANDESITE
				DACITE
				LATITE
				QTZ_LATITE
				QTZ_TRACHYTE
				TRACHYANDESITE
				UNDIFF
			MFC_BASC *(Mafic Basic)*	ALK_BASALT
				ALK_TRACHYTE
				BASALT
				FELDS_BASALT
				FLOOD_BASALT
				HI_AL_BASALT
				OLIVN_THOLEITE
				QTZ_THOLEITE
				SPLTC_BASALT
				TARCHYBASALT
				THOLEITE_BASALT
				UNDIFF
			ULT_MFC *(Ultramafic-Ultrabasic)*	EXTR_CRBNATITE
				KOMATITE
				UNDIFF
			UNDIFF *(Undiff. Volcanic Extrusive)*	UNDIFF *(Undiff. Volcanic Extrusive)*
		IGNS	FELSIC (Felsic-Acidic-Silicic)	UNDIFF
		(Igeous Undifferenciated)	MFC_BASC (Mafic Basic)	UNDIFF
			UNDIFF (Undiff. Volcanic Extrusive)	UNDIFF (Undiff. Volcanic Extrusive)
		INTR	FELSIC *(Felsic-Acidic-Silicic)*	2MICA_GRANITE
		(Plutonic Intrusive)		ANTCT_GRANITE
				A_GRANITE
				GRANITE
				GRANODIORITE
				I_GRANITE
				QTZ_MONZONITE
				S_GRANITE
				UNDIFF
			INTERMED *(Intermediate)*	DIORITE
				MONZODIORITE
				MONZOGABBRO
				MONZONITE
				SYENITE
				TONALITE
				UNDIFF
			MFC_BASC *(Mafic Basic)*	ANORTHOSITE
				GABBRO
				LAMPROPHYRE
				NORITE
Continued Next Page)	Continued Next Page)	Continued Next Page)		UNDIFF

BEDROCK AND TECTONICS ATTRIBUTE STRUCTURE - DATA OPTIONS
CHOOSE ONE ITEM FROM EACH CATEGORY AS APPROPRIATE

IGNEOUS

xxx_class	xxx_depth	xxx_origin	xxx_lithology	xxx_lith_type
IGNEOUS	*(Plutonic Rocks)*	INTR	ULT_MFC *(Ultramafic-Ultrabasic)*	DUNITE
	CATAZONAL	*(Plutonic Intrusive)*		HARZBURGITE
	EPIZONAL			HORNBLENDITE
	HYPABYSSAL			INTR_CRBNATITE
	MESOZONAL			KIMBERLITE
				LAYERED_INTR
				LHERZOLITE
				PERIDOTITE
				PYROXENITE
				UNDIFF
				WEHRLITE
			UNDIFF *Undiff. Plutonics)*	UNDIFF *Undiff. Plutonics)*

xxx_lith_mod	xxx_txtr_strat	xxx_group	xxx_subgroup
User-defined. Examples:	*User-defined.Examples:*	*(Magma Series)*	*(Chemistry)*
ALPINE	BOMBS	METALUMINOUS	ALKALI_CALCIC
FERROUS_RICH	COARSE_GRAINED	PERALKALINE	ALKALIC
GARNET_BEARING	CRYPTIC_LAYERING	PERALUMINOUS	CALC_ALKALIC
SPILITIC	FINE_GRAINED		CALCIC
TIN_BEARING	FLOW		
EPIDOTE_BEARING	FOLIATED		
	GRADED		
	GRANOPHYRIC		
	IGNEOUS_LAYERING		
	IGNIMBRITE		
	INCH_SCALE_LAYERING		
	INTERCALATED		
	MINERAL_LAYERING		
	PALAGONITE_TUFF		
	PILLOW_LAVA		
	PIPES		
	POIKILITIC		
	PORPHYRITIC		
	PUMICE		
	PYROCLASTIC		
	RHYTHMIC_LAYERING		
	SCORIAE		
	TRACHYTIC		
	TUFFACEOUS		
	WELDED_TUFF		

Additional values for all items in this column:
UNK *(Unknown)*
UNDIFF *(Undifferenciated)*

BEDROCK AND TECTONICS ATTRIBUTE STRUCTURE - DATA OPTIONS
CHOOSE ONE ITEM FROM EACH CATEGORY AS APPROPRIATE

ALL

xxx_class	xxx_tect_regime	xxx_margin	xxx_tect_assem	xxx_crust_resp
IGNEOUS	COMPRESSIONAL	ACTIVE	ACCRETIONARY	THICKENED
METAMORPHIC	EXTENSIONAL	CONT_INT	ARC_ASSEMBLAGE	THINNED
SEDIMENTARY	STABLE	CONVERGENT	BACK_ARC	
	TRANSVERSE	CRATONIC	COLLISION_CMPLX	
		DIVERGENT	CONT_CRUST	
		INTRACRATONIC	FORE_ARC	
		OBLIQUE	FOREDEEP	
		OCEANIC	HOT_SPOT	
		OROGENIC	INTRA_ARC	
		PASSIVE	MANTLE_PLUME	
		TRANSFORM	OCEANIC	
			OPHIOLITE	
			OROGENIC_BELT	
			PLATFORM	
			PR_PLATFORM	
			PR_RISE	
			PR_SLOPE	
			PRISM	
			PSV_MRGN_WEDGE	
			RIFT	
			RISE	
			SHELF	
			SLOPE	
			SUBDCT_ZONE	
			TRENCH	

xxx_tect_regime — *Additional values for all items:*
UNK *(Unknown)*
UNDIFF *(Undifferenciated)*

xxx_margin — *Additional values for all items:*
UNK
UNK *(Unknown)*
UNDIFF *(Undifferenciated)*

xxx_tect_assem — *Additional values for all items:*
UNK *(Unknown)*
UNDIFF *(Undifferenciated)*

xxx_con_lith
INFERRED
SUPPORTED

xxx_con_env
INFERRED
SUPPORTED

xxx_age_mng
DEFORMATION
DEPOSITION
EMPLACEMENT
METAMORPHISM

Additional values for all items:
UNK *(Unknown)*
UNDIFF *(Undifferenciated)*

Name: CONTOURS

Data Type: LINES

REFER TO CODE LISTS FOR SHADED ATTRIBUTE ITEMS

Used For: Contour maps captured in Arc as line coverages.
Examples: Isopachs, contour maps, bathymetry contours, LOM maps, etc. without all contours being closed

Attribute items:

	WIDTH	OUTPUT	TYPE	N. DEC.	ALT. NAME	DESCRIPTION / PURPOSE OF THE ITEM
bu_id	8	8	C	-	-	User defined. Relate key to related data table.
z	4	8	F	0*	-	User defined. Contour value.
units	10	10	C	-	-	Code lists. EEC units of measure codes.
data_src	30	30	C	-	-	User defined. Data source or provider. Example: EPR.44EX.87
descript	250	250	C	-	-	User defined. Description of contour or line.
min_age	20	20	C	-	-	Code lists. Minimum age of polygon or line (if coverage has multiple ages).
max_age	20	20	C	-	-	Code lists. Maximum age of polygon or line (if coverage has multiple ages).
calling_age	30	30	C	-	-	User defined. The age name used in general reference.
horizon	30	30	C	-	-	User defined. Horizon of contour (if coverage has multiple horizons). Use BU horizon names
con_int	10	10	C	-	-	Code lists. Confidence in interpretation of contour or line.
name	32	32	C	-	-	User defined. BU name for line or polygon to which line is related. Example: SMITH UNIT BOUNDARY.
text	250	250	C	-	-	User defined. Annotation to appear on coverage. May be the same as another item.
subject	30	30	C	-	-	User defined. Overall coverage subject or entity. Example: UPPER CRETACEOUS SOURCE.
title	100	100	C	-	-	User defined. Information mapped about the subject. Example: PRESENT DAY LOM.
file	50	50	C	-	-	User defined. Name of original, pre-Arc file. Example: CRFCTAR4.DWG

BOLDFACE = REQUIRED ITEM
* ADD DECIMALS AS NEEDED

CONTOUR ATTRIBUTE STRUCTURE - DATA OPTIONS

SELECT ONE ITEM FOR EACH CATEGORY AS APPROPRIATE

xxx_units
Use Units Population Table

xxx_min_age
Use Age Population Table

xxx_max_age
Use Age Population Table

xxx_con_int
INFERRED
SUPPORTED

Name: FAULTS AND STRUCTURALLY RELATED POLYGONS

Data Type: POLYGONS OR LINES

REFER TO CODE LISTS FOR SHADED ATTRIBUTE ITEMS

Used For: Faults, fault gaps, or fault zones.
Folds, structures, structural styles.

Attribute items:

	WIDTH	OUTPUT	TYPE	N. DEC.	ALT. NAME	DESCRIPTION / PURPOSE OF THE ITEM
xxx_bu_id	8	8	C	-	bu_id	User defined. Relate key to related data table.
xxx_data_src	30	30	C	-	data_src	User defined. Data source or provider. Example: EPR.44EX.87
xxx_style	20	20	C	-	style	Code lists. Dominant structural style.
xxx_style_type	30	30	C	-	style_type	Code lists. Main type within structural style.
xxx_feature	20	20	C	-	feature	Code lists. Feature class.
xxx_feat_type	20	20	C	-	feat_type	Code lists. Main type within feature class.
xxx_feat_subtype	30	30	C	-	feat_subtype	Code lists. Sub-type within feature class.
xxx_feat_mod	20	20	C	-	feat_mod	Code lists. Feature modifier.
xxx_dip_plng	10	10	C	-	dip_plng	User defined. Dip direction of fault plane or plunge of structure. Examples: N, NE_SW.
xxx_signif	10	10	C	-	signif	Code lists. Significance or relative magnitude of feature.
xxx_detect	20	20	C	-	detect	Code lists. Method used to detect, discover or map feature or closure.
xxx_descript	250	250	C	-	descript	User defined. Description of polygon. Example: CRETACEOUS FAULT ZONE.
xxx_min_age	20	20	C	-	min_age	Code lists. Minimum age of polygon or line (if coverage has multiple ages).
xxx_max_age	20	20	C	-	max_age	Code lists. Maximum age of polygon or line (if coverage has multiple ages).
xxx_calling_age	30	30	C	-	calling_age	User defined. The age name used in general reference.
xxx_horizon	30	30	C	-	horizon	User defined. Horizon of polygon or line (if coverage has multiple horizons). Use BU horizon names.
xxx_con_feat	10	10	C	-	con_feat	Code lists. Confidence of interpretation of feature, feat_type or feat_subtype.
xxx_con_style	10	10	C	-	con_style	Code lists. Confidence in interpretation of style and assemblage.
xxx_name	32	32	C	-	name	User defined. Business Unit name for polygon or line. Example: TABEI FAULT ZONE
xxx_text	250	250	C	-	text	User defined. Annotation to appear on coverage. May be the same as another item.
xxx_subject	30	30	C	-	subject	User defined. Overall coverage subject or entity. Example: UPPER CRETACEOUS
xxx_title	100	100	C	-	title	User defined. Information mapped about the subject. Example: FAULTS AND STRUCTURES.
xxx_file	50	50	C	-	file	User defined. Name of original, pre-Arc file. Example: CRFCTAR4.DWG

xxx = ITEM PREFIX REQUIRED FOR POLYGON AND POINT COVERAGES ONLY. EXAMPLE: IF COVERAGE IS BAU_CRET_LOW THEN XXX COULD BE BCL..
BOLDFACE = REQUIRED ITEM

FAULTS AND STRUCTURES ATTRIBUTE STRUCTURE - DATA OPTIONS
SELECT ONE ITEM FOR EACH CATEGORY AS AS APPROPRIATE

xxx_style	xxx_style_type
BASEMENT_INVOLVED	BASEMENT_WARP
	CONTRACTIONAL
	EXTENSIONAL
	IGNEOUS_CORED
	TRANSVERSE
	TRANSVERSE_CONVERGENT
	TRANSVERSE_DIVERGENT
DETACHED	FOLD_THRUST
	NORMAL_FAULT
	SALT_NON_PIERCEMENT
	SALT_PEIRCEMENT
	SALT_WITHDRAWAL
	SHALE_NON_PIERCEMENT
	SHALE_PIERCEMENT
COMBINATION	BASEMENT_CONCORDANT
	DETACHED_CONCORDANT
	DISCONCORDANT
POST_TECTONIC	DEPOSITIONAL_DRAPE
	DIFFERENTIAL_COMPACTION
UNKNOWN	CONTRACTIONAL
	EXTENSIONAL
	TRANSVERSE
	UNKNOWN

xxx_con_style
INFERRED
SUPPORTED

xxx_feature	xxx_feat_type	xxx_feat_subtype
FAULT	INVERTED_NORMAL	HIGH_ANGLE
	NORMAL	LEFT_LATERAL
	REVERSE_THRUST	LISTRIC
	STRIKE_SLIP_WRENCH	LOW_ANGLE
		RIGHT_LATERAL
FOLD	ANTICLINE	ANTICLINE
	MONOCLINE	ANTITHETIC_FLT_BLOCK
	SYNCLINE	ARCH_PLATFORM
		CROSS_FAULTED
		CROSS_FOLDED
		DOME
		DOME
		FLOWER
		FLT_BEND_FOLD
		FLT_PROPAGATION_FOLD
		INTERFERENCE
		INVERTED_CON_FLT_BLK
		INVERTED_EXT_FLT_BLK
		INVERTED_NORM_FLT_BLK
		INVERTED_THR_FLT_BLK
		NON_PIERCEMENT
		NORMAL_FLT_BLOCK
		OTHER_FOLD
		OVER_DEPOSITION
		OVER_EROSION_BAS_INV
		OVER_EROSION_DETACHED
		OVER_STRUC_BAS_INV
		OVER_STRUC_DETACHED
		PIERCEMENT
		PILLOW
		REGIONAL_DIP_SHELF
		RESIDUAL_HIGH
		SAG
		SILL
		SOLUTION_FRONT_HIGH
		SOLUTION_INVERSION_HIGH
		SYNTHETIC_FLT_BLOCK
		THRUST_FLT_BLOCK
		TOE_STRUCTURE
		TURTLE
		WALL
		WRENCH_FLT_BLOCK
UNKNOWN	UNKNOWN	UNKNOWN
	If xxx_feat_type is not known, repeat value used in xxx_feature.	*If xxx_feat_type is not known, repeat value used in xxx_feat_type.*

xxx_feat_mod
User-defined. Examples:
ACTIVE
ANTITHETIC
BEDDING_DIP
DIAPIRIC
DISHARMONIC
GENERATIVE
GRAVITY_SLIDE
GROWTH
HINGED
INACTIVE
INVERTED
LONGITUDINAL
NON_GROWTH
OBLIQUE_SLIP
OVERTURNED
SUPRATENUOUS
SYNTHETIC
TRANSVERSE_TEAR

xxx_signif
MAJOR
MINOR

xxx_detect
GRAVITY
MAGNETIC
MAPPED
OUTSIDE_SRC
SEISMIC
SURFACE
WELL

xxx_min_age
Use Data Population Tables See Appendix A

xxx_max_age
Use Data Population Tables See Appendix A

xxx_con_feat
INFERRED
SUPPORTED

xxx_con_style
INFERRED
SUPPORTED

Name: MISCELLANEOUS POLYGONS AND LINES

REFER TO CODE LISTS FOR SHADED ATTRIBUTE ITEMS

Data Type: POLYGONS OR LINES

Used For:

Outlines of structural, tectonic, or sedimentary features or provinces.
Outlines of metamorphic belts, igneous or sedimentary complexes.
Generally used for major, basin or sub-basin scale features.
Non-geologic / non-cartographic features such as areas of interest or proposed areas of investigation.
Seismic or cross section index lines.
Any line not representing a play element boundary or contour.
Not to be used for limits of seal, source, migration.
Not used for outlining fields, prospects, or leads.
Not used for mapping lithologies, facies, crustal genesis, depositional environments, detailed surface mapping.
Not to be used for assessment related polygons or points.

Attribute items:

	WIDTH	OUTPUT	TYPE	N. DEC.	ALT. NAME	DESCRIPTION / PURPOSE OF THE ITEM
xxx_bu_id	8	8	C	-	bu_id	User defined. Relate key to related data table.
xxx_data_src	30	30	C	-	data_src	User defined. Data source or provider. Example: EPR.44EX.87
xxx_class	20	20	C	-	class	Code lists. Major geologic class of polygon or line data.
xxx_feature	20	20	C	-	feature	Code lists. Feature class.
xxx_feat_type	20	20	C	-	feat_type	Code lists. Main type within feature class.
xxx_feat_subtype	30	30	C	-	feat_subtype	Code lists. Sub-type within feature class.
xxx_feat_mod	20	20	C	-	feat_mod	Code lists. Feature modifier.
xxx_signif	10	10	C	-	signif	Code lists. Significance or relative magnitude of feature.
xxx_descript	250	250	C	-	descript	User defined. Description of polygon or line. Example: CRETACEOUS MAGNETIC ANOMALY.
xxx_min_age	20	20	C	-	min_age	Code lists. Minimum age of polygon or line (if coverage has multiple ages).
xxx_max_age	20	20	C	-	max_age	Code lists. Maximum age of polygon or line (if coverage has multiple ages).
xxx_calling_age	30	30	C	-	calling_age	User defined. The age name used in general reference.
xxx_horizon	30	30	C	-	horizon	User defined. Horizon of polygon or line (if coverage has multiple horizons). Use BU horizon names
xxx_con_int	10	10	C	-	con_int	Code lists. Confidence in interpretation of polygon or line.
xxx_name	32	32	C	-	name	User defined. Business Unit name for polygon or line. Example: TABEI UPLIFT.
xxx_text	250	250	C	-	text	User defined. Annotation to appear on coverage. May be the same as another item.
xxx_subject	30	30	C	-	subject	User defined. Overall coverage subject or entity. Example: UPPER CRETACEOUS.
xxx_title	100	100	C	-	title	User defined. Information mapped about the subject. Example: STRUCTURAL PROVINCES.
xxx_file	50	50	C	-	file	User defined. Name of original pre-Arc file. Example: CRFCTAR4.DWG

xxx = ITEM PREFIX REQUIRED FOR POLYGON AND POINT COVERAGES ONLY. EXAMPLE: IF COVERAGE IS BAU_CRET_LOW THEN XXX COULD BE BCL..
BOLDFACE = REQUIRED ITEM
* ADD DECIMALS AS NEEDED

MISCELLANEOUS POLYGONS AND LINES ATTRIBUTE STRUCTURE - DATA OPTIONS
SELECT ONE ITEM FOR EACH CATEGORY AS APPROPRIATE

xxx_class
User-defined. Examples:
DEPOSITIONAL
DIAGENETIC
NA
PALEOGEOGRAPHIC
SOURCE_MIG
STRUCTURAL
SURFACE
TECTONIC
UNKNOWN

xxx_feature
User-defined. Examples:
BELT
AREA
AXIS
BASIN
BREAK
CENTER
COMPLEX
CONTACT
DEPRESSION
DIAPIR
DIRECTION
DOWNLAP
DOWNWARP
EDGE
EXTENT
GRABEN
GRAIN
INDEX
LIMIT
MARGIN
ONLAP
OUTCROP
PROVINCE
SAG
SUBCROP
SWELL
SYSTEM
TERRANE
TRANSITION
TROUGH
TRUNCATION
UNCONFORMITY
UPLIFT
UPWARP
WEDGE
ZONE

xxx_feat_type
User-defined. Examples:
3D_SURVEY
BASIN
CALIBRATION
CRESTAL
CROSS_SECTION
DEPOSITIONAL
DETACHED
EROSIONAL
FACIES
FAULT
FLEXURE
FOLD
FOREARC
FRACTURE
HYDROCARBON
IGNEOUS
INTEREST
LITHOLOGY
MAGNETIC
MATURATION
METAMORPHIC
MIGRATION
NA
OCEANIC
PRESERVATIONAL
RESERVOIR
SALT
SEAL
SEDIMENTARY
SHALE
SHEAR
SHELF
SLOPE
SOURCE
SPREADING
SUBDUCTION
TRANSPORT
UNKNOWN

xxx_feat_subtype
User-defined. Examples:
ACTIVE
BASE_INVOLVED
BUSINESS_UNIT
DETACHED
GPLAY
INACTIVE
NA
PALEO
PROPOSED
UNKNOWN

xxx_feat_mod
User-defined.

xxx_signif
MAJOR
MINOR

xxx_min_age
Use Age Population Table
See Appendix A

xxx_max_age
Use Age Population Table
See Appendix A

xxx_con_int
INFERRED
SUPPORTED

Name: NON ORACLE POINT DATA

REFER TO CODE LISTS FOR SHADED ATTRIBUTE ITEMS

Data Type: POINTS

Used For:

Control points and other point data not currently in ORACLE.
Should NOT duplicate data or data models currently in ORACLE.
Can be used to link to ORACLE and/or a Business Unit database.
Examples include: outcrop locations.
Not to be used for closure point data.
NOTE: BEFORE ENTERING POINT DATA, CHECK WITH ORACLE DATA MODELER TO ENSURE THAT DATA ARE NOT CURRENTLY IN ORACLE OR CANNOT BE ADDED TO ORACLE.

Attribute items:

	WIDTH	OUTPUT	TYPE	N. DEC.	ALT. NAME	DESCRIPTION / PURPOSE OF THE ITEM
xxx_bu_id	8	8	C	-	bu_id	User defined. Relate key to related data table.
xxx_class	20	20	C	-	class	Code lists. Major class of point data.
xxx_data_src	30	30	C	-	data_src	User defined. Data source or provider. Example: EPR.44EX.87
xxx_type	30	30	C	-	type	Code lists. Type of point data within major class.
xxx_subtype	30	30	C	-	subtype	Code lists. If point data, secondary or sub-type of data.
xxx_feature	30	30	C	-	feature	Code lists. Purpose of data.
xxx_value	4	8	F	0*	value	User defined. Numerical value assigned to point.
xxx_units	10	10	C	-	units	Code lists. EEC units of measure codes.
xxx_method	30	30	C	-	method	User defined. Method or analytical technique used to obtain value.
xxx_descript	250	250	C	-	descript	User defined. Description of point. Example: PLATE LOCATION 500 MYBP.
xxx_min_age	20	20	C	-	min_age	Code lists. Minimum age of polygon or line (if coverage has multiple ages).
xxx_max_age	20	20	C	-	max_age	Code lists. Maximum age of polygon or line (if coverage has multiple ages).
xxx_calling_age	30	30	C	-	calling_age	User defined. The age name used in general reference.
xxx_horizon	30	30	C	-	horizon	User defined. Horizon of point (if coverage has multiple horizons). Use BU horizon names
xxx_con_loc	16	16	C	-	con_loc	Code lists. Confidence in location of point.
xxx_con_int	10	10	C	-	con_int	Code lists. Confidence in interpretation of point.
xxx_name	32	32	C	-	name	User defined. Business Unit name for point. Example: E&S INPUT 32-20.
xxx_text	250	250	C	-	text	User defined. Annotation to appear on coverage. May be the same as another item.
xxx_subject	30	30	C	-	subject	User defined. Overall coverage subject or entity. Example: UPPER CRETACEOUS.
xxx_title	100	100	C	-	title	User defined. Information mapped about the subject. Example: CROSS SECTION INDEX POINT.
xxx_file	50	50	C	-	file	User defined. Name of original pre-Arc file.

xxx = ITEM PREFIX REQUIRED FOR POLYGON AND POINT COVERAGES ONLY. EXAMPLE: IF COVERAGE IS BAU_CRET_LOW THEN XXX COULD BE BCL..
BOLDFACE = REQUIRED ITEM
* ADD DECIMALS AS NEEDED

NON-ORACLE POINT DATA ATTRIBUTE STRUCTURE - DATA POINTS

SELECT ONE ITEM FOR EACH CATEGORY AS APPROPRIATE

xxx_class
User-defined. Examples:
IMAGINARY_LOCATION
PLATE_RECONSTRUCTION
SHOTPOINT
SURFACE_LOCATION
WELL

xxx_type
User-defined. Examples:
CONVENTIONAL_CORE
DROP_CORE
HAND_SAMPLE
MAGNETIC_STRIPE
MUDLOG
WIRELINE_LOG

xxx_subtype
User-defined. Examples:
CONTROL_POINT
PLATE_MOTION
SEQUENCE_STRAT

xxx_feature
User-defined. Examples:
DATABASE_SUBSET
DOWNLAP_POINT
ONLAP_POINT
TRUNCATION_POINT

xxx_min_age
Use Age Population Table

xxx_max_age
Use Age Population Table

xxx_con_loc
APPROXIMATE
KNOWN

xxx_con_int
INFERRED
SUPPORTED

Name: **ORACLE POINT DATA**

REFER TO CODE LISTS FOR SHADED ATTRIBUTE ITEMS

Data Type: **POINTS**

Used For: Point data stored in ORACLE.

Attribute items:

	WIDTH	OUTPUT	TYPE	N. DEC.	ALT. NAME	DESCRIPTION / PURPOSE OF THE ITEM
wlbr_id	2	6	B	-	wlbr_id	User defined. Key field to link via relate to BU database. Choose exact name of field.
xxx_bu_id	8	8	C	-	bu_id	User defined. Relate key to related data table.
xxx_data_src	30	30	C	-	data_src	User defined. Data source or provider. Example: EPR.44EX.87
xxx_name	32	32	C	-	name	User defined. Business Unit name for point. Example: MA 564 #1.
xxx_subject	30	30	C	-	subject	User defined. Overall coverage subject or entity. Example: TARIM BASIN.
xxx_title	100	100	C	-	title	User defined. Information mapped about the subject. Example: BASE MAP.
xxx_file	50	50	C	-	file	User defined. Name of original pre-Arc file. Example: CRFCTAR4.DWG.

xxx = ITEM PREFIX REQUIRED FOR POLYGON AND POINT COVERAGES ONLY. EXAMPLE: IF COVERAGE IS BAU_CRET_LOW THEN XXX COULD BE BCL..
BOLDFACE = REQUIRED ITEM

Name: POLYGONIZED VALUES

REFER TO CODE LISTS FOR SHADED ATTRIBUTE ITEMS

Data Type: POLYGONS

Used For:
Polygons that have numeric values or ranges of values.
Polygons that have descriptive values or ranges that don't fit under miscellaneous polygons and lines.
Isopachs or other closed contours mapped as polygons.
Not to be used for assessment related polygons, features, play element coverages, assessment BAU's, play element boundaries, chance of oil

Attribute items:

	WIDTH	OUTPUT	TYPE	N. DEC.	ALT. NAME	DESCRIPTION / PURPOSE OF THE ITEM
xxx_bu_id	8	8	C	-	bu_id	User defined. Relate key to related data table.
xxx_val	4	8	F	0*	val	User defined. Single assigned value to a polygon.
xxx_descript	250	250	C	-	descript	User defined. Description of polygon. Example: WATER DEPTHS GREATER THAT 500 METERS
xxx_data_src	30	30	C	-	data_src	User defined. Data source or provider. Example: EPR.44EX.87
xxx_val_min	4	8	F	0*	val_min	User defined. Minimum value. Usually the value of one line of the polygon.
xxx_val_max	4	8	F	0*	val_max	User defined. Maximum value. Usually the value of one line of the polygon.
xxx_val_rng	20	20	C	-	val_rng	User defined. The range of values within a polygon. Examples: 10-20, >12, <8.
xxx_units	10	10	C	-	units	Code lists. EEC units of measure codes.
xxx_min_age	20	20	C	-	min_age	Code lists. Minimum age of polygon or line (if coverage has multiple ages).
xxx_max_age	20	20	C	-	max_age	Code lists. Maximum age of polygon or line (if coverage has multiple ages).
xxx_calling_age	30	30	C	-	calling_age	User defined. The age name used in general reference.
xxx_horizon	30	30	C	-	horizon	User defined. Horizon of polygon (if coverage has multiple horizons). Use BU horizon names.
xxx_con_int	10	10	C	-	con_int	Code lists. Confidence in interpretation of polygon.
xxx_name	32	32	C	-	name	User defined. Business Unit name for polygon. Example: BACHU UPLIFT.
xxx_text	250	250	C	-	text	User defined. Annotation to appear on coverage. May be the same as another item.
xxx_subject	30	30	C	-	subject	User defined. Overall coverage subject or entity. Example: UPPER CRETACEOUS SOURCE.
xxx_title	100	100	C	-	title	User defined. Information mapped about the subject. Example: Source TOC Polygon.
xxx_file	50	50	C	-	file	User defined. Name of original pre-Arc file. Example: CRFCTAR4.DWG.

xxx = ITEM PREFIX REQUIRED FOR POLYGON AND POINT COVERAGES ONLY. EXAMPLE: IF COVERAGE IS BAU_CRET_LOW THEN XXX COULD BE BCL..
BOLDFACE = xxx_data_src AND ONE OF THE TWO OTHER ITEMS ARE REQUIRED
* ADD DECIMALS AS NEEDED

POLYGONIZED VALUES ATTRIBUTE STRUCTURE - DATA OPTIONS

SELECT ONE ITEM FOR EACH CATEGORY AS APPROPRIATE

xxx_units
Use Units Population Table

xxx_min_age
Use Age Population Table

xxx_max_age
Use Age Population Table

xxx_con_int
INFERRED
SUPPORTED

Name: **SEDIMENTARY LITHOFACIES**

Data Type: **POLYGONS OR LINES**

[ ] — REFER TO CODE LISTS FOR SHADED ATTRIBUTE ITEMS

Used For:

Sedimentary facies.
Sedimentary depositional environments.
Sedimentary lithologies.
Capture Igneous or metamorphic assemblages as a separate coverage with "Bedrock and Tectonics" attribute structure.

Attribute items:

	WIDTH	OUTPUT	TYPE	N. DEC.	ALT. NAME	DESCRIPTION / PURPOSE OF THE ITEM
xxx_bu_id	8	8	C	-	bu_id	User defined. Relate key to related data table.
xxx_rock_class	20	20	C	-	rock_class	Code lists. Major geologic classification of rocks.
xxx_data_src	30	30	C	-	data_src	User defined. Data source or provider. Example: EPR.44EX.87
xxx_origin	4	4	C	-	origin	Code lists. Major genetic subdivision within geologic class.
xxx_lithology	8	8	C	-	lithology	Code lists. Major lithology type.
xxx_lith_type	16	16	C	-	lith_type	Code lists. Type or subdivision of lithology.
xxx_lith_mod	30	30	C	-	lith_mod	Code lists. Lithologic composition modifier.
xxx_sublith	50	50	C	-	sublith	User defined. Sub-lithology descriptor.
xxx_txtr_strat	30	30	C	-	txtr_strat	Code lists. Textural or stratal pattern descriptor.
xxx_grn_size	30	30	C	-	grn_size	Code lists. Sedimentary grain size.
xxx_dep_class	8	8	C	-	dep_class	Code lists. Major depositional class.
xxx_environ	16	16	C	-	environ	Code lists. Sedimentary depositional environment.
xxx_env_type	16	16	C	-	env_type	Code lists. More detailed breakdown of the environment.
xxx_env_mod	30	30	C	-	env_mod	Code lists. Modifier of environment type.
xxx_subenv	50	50	C	-	subenv	Code lists. General descriptor of gross geometry of depositional unit.
xxx_subenv_mod	50	50	C	-	subenv_mod	Code lists. Descriptor of degree of amalgamation of depositional unit.
xxx_mar_wd	20	20	C	-	mar_wd	Code lists. Water depth of marine deposition.
xxx_paleo_geo	20	20	C	-	paleo_geo	Code lists. Paleogeographic setting.
xxx_paleo_geo_md	20	20	C	-	paleo_geo_md	Code lists. Modifier for paleogeographic setting.
xxx_stack_pat	20	20	C	-	stack_pat	Code lists. Sequence stratigraphic stacking pattern.
xxx_descript	250	250	C	-	descript	User defined. Description of polygon or line. Example: MIXED SANDSTONE/SILTSTONE/SHALE.
xxx_min_age	20	20	C	-	min_age	Code lists. Minimum age of polygon or line (if coverage has multiple ages).
xxx_max_age	20	20	C	-	max_age	Code lists. Maximum age of polygon or line (if coverage has multiple ages).
xxx_calling_age	30	30	C	-	calling_age	User defined. The age name used in general reference.
xxx_horizon	30	30	C	-	horizon	User defined. Horizon of polygon or line (if coverage has multiple horizons). Use BU horizon names
xxx_con_lith	10	10	C	-	con_lith	Code lists. Confidence of lithology interpretation.
xxx_con_env	10	10	C	-	con_env	Code lists. Confidence of environment interpretation.
xxx_name	32	32	C	-	name	User defined. Business Unit name for polygon or line. Example: LOST RIVER CHERT.
xxx_text	250	250	C	-	text	User defined. Annotation to appear on coverage. May be the same as another item.
xxx_subject	30	30	C	-	subject	User defined. Overall coverage subject or entity. Example: UPPER CRETACEOUS.
xxx_title	100	100	C	-	title	User defined. Information mapped about the subject. Example: DEPOSITIONAL FACIES.
xxx_file	50	50	C	-	file	User defined. Name of original, pre-Arc file. Example: CRFCTAR4.DWG

xxx = ITEM PREFIX REQUIRED FOR POLYGON AND POINT COVERAGES ONLY. EXAMPLE: IF COVERAGE IS BAU_CRET_LOW THEN XXX COULD BE BCL..
BOLDFACE = REQUIRED ITEM

SEDIMENTARY LITHOFACIES ATTRIBUTE STRUCTURE - DATA OPTIONS
SELECT ONE ITEM FOR EACH CATEGORY AS APPROPRIATE

xxx_rock_class	xxx_origin	xxx_lithology	xxx_lith_type
SEDIMENTARY	CARB *(Carbonate)*	LMST *(Limestone)*	MUDSTN *(Mudstone)*
			MICRITE
			MCRT_MUD *(Micrite to Mudstone)*
			BNDSTN *(Boundstone)*
			CHALK
			XLN *(Crystalline)*
			FLOATSTN *(Floatstone)*
			GRNSTN *(Grainstone)*
			NONE *(Limestone Undiff.)*
			MARL
			PACKSTN *(Packstone)*
			RUDSTN *(Rudstone)*
			WACKSTON *(Wakestone)*
		DLMT *(Dolomite)*	MCRT_MUD *(Micrite to Mudstone)*
			BNDSTN *(Boundstone)*
			XLN *(Crystalline)*
			DOLOMITE
			NONE *(Dolomite Undiff.)*
			FLOATSTN *(Floatstone)*
			GRNSTN *(Grainstone)*
			PACKSTN *(Packstone)*
			RUDSTN *(Rudstone)*
			WACKSTON *(Wakestone)*
		CARB *(Carbonate Undiff)*	NONE *(Carbonate Undiff)*
	CLST *(Clastic)*	SAND *(Sandstone)*	QTZ_AREN *(Quartz Arenite)*
			ARKOSE
			LTH_AREN *(Lithic Arenite)*
			NONE *(Sandstone Undiff.)*
			CLC_AREN *(Calcarenite)*
			SUBARKSE *(Subarkose)*
		SHAL *(Shale)*	MUD *(Mudstone)*
			CLY_MUD *(Clatstone to Mudstone)*
			CLY_SHAL *(Claystone to Shale)*
			CLAY *(Claystone)*
			MUD_SHAL *(mudstone to Shale)*
			NONE (Shale Undiff.)
		SILT *(Siltstone)*	NONE *(Silstone Undiff.)*
		CNGL *(Conglomerate)*	OLIGMICT *(Oligomictite)*
			NONE *(Conglomerate Undiff.)*
			PTROMICT *(Petromictite to Polymictite)*
			TILLITE
		COAL	PEAT
			ANTHRC *(Anthracite)*
			ANTHRX *(Anthraxolite)*
			BIT *(Bituminous)*
			BOGHEAD *(Boghead-Borganite)*
			CANNEL *(Cannel Coal)*
			NONE *(Coal Undiff.)*
			LIGNITE
			SEMI_BIT *(Semi-Bituminous)*
CONTINUED NEXT PAGE	*CONTINUED NEXT PAGE*		SUB_BIT *(Sub-Bituminous)*

xxx_lith_mod
User-defined. Examples:
ACTINOLITE
CALCAREOUS
DIOPSIDE
DOLOMITIC
EPIDOTE
FELSPATHIC
FOSSILIFEROUS
HEMATITIC
HORNBLENDE
ILLITE
KAOLINITE
LIMEY
MICRITIC
MIXED
NONE
OOLITIC
SHALY
SILTY
SKELETAL
SMECTITE
TREMOLITE

xxx_sublith
Free-form text.
User defined.

xxx_txtr_strat
User-defined. Examples:
BIOTURBATED
COARSENING_UPWARD
CONCORDANT
CROSS_BEDDED
CYCLIC
DISCORDANT
DISSEMINATED
FINELY_PARTED
FINING_UPWARD
INTERBEDDED
INTERCALATED
LAMINATED
MASSIVE
NODULAR
POORLY_SORTED
THIN_BEDDED
WELL_SORTED

xxx_con_lith
SUPPORTED
INFERRED

Additional values for all items in this column:

UNK *(Unkown)*
UNDIFF *(Undifferenciated)*

SEDIMENTARY LITHOFACIES ATTRIBUTE STRUCTURE - DATA OPTIONS
SELECT ONE ITEM FOR EACH CATEGORY AS APPROPRIATE

xxx_rock_class	xxx_origin	xxx_lithology	xxx_lith_type
SEDIMENTARY *CONTINUED FROM PREVIOUS PAGE*	CLST *(Clastic)* *CONTINUED FROM PREVIOUS PAGE*	BRCC *(Breccia)*	SCREE *(Scree Talus)*
			AGGLOM *(Agglomerate)*
			BONE *(Bone Breccia)*
			NONE *(Breccia Undiff.)*
			COLLAPSE *(Collapse Breccia)*
			FAULT *Fault Breccia)*
			FISSURE *Fissure Breccia)*
			INTRFORM *(Intraformational Breccia)*
		CLST *(Clastic)*	NONE *(Clastic Undiff.)*
	CRYP *(Cryptocrystalline)*	CHRT *(Chert)*	FLINT
			NONE *(Chert Undiff.)*
			DIATOM (Diatomite)
			JASPER
			JASPERD *(Jasperoid)*
			NOVACLT *(Novaculite)*
			PORCNLT *(Porcellanite)*
			RADIOLR *(Radiolarite)*
			SPICULT *(Spiculite)*
			TRIPT *(Tripolite)*
		CRYP *(Cryptocrystalline)*	NONE *(Cryptocrystalline Undiff.)*
	EVAP *(Evaporite)*	EVAP *(Evaporite)*	GYPSUM
	EVAP *(Evaporite)*	EVAP *(Evaporite)*	ANHYD *(Anhydrite)*
			NONE *(Evaporite Undiff.)*
			PHOSPHT *(Phosphatite)*
			MAGNST *(Magnesite)*
			SALT *(Salt-Halite)*
	MIX *(Mixed Sediment)*	MXD *(Mixed Sediments)*	*SANDSHAL (Sandstones And Shales)*
			SANDLMST *(Sanstones And Limestones)*
			LMSTSILT *(Limestone and Silts)*
			LMSTEVAP *(limestones and Evaporites)*
			DLMTSHAL *(Dolomites and Shales)*
			DLMTEVAP *(dolomites and Evaporites)*
			SANDSILT *(Sandstones and Siltstones)*
			SANDUDCB *(Sandstones and Carbonates)*
			CARBEVAP *(Carbonates and Evaporites)*
			SHALEVAP *(Shales and Evaporites)*
			SILTEVAP *(Silstones and Evaporites)*
			SILTCOAL *(Siltstones and Coals)*
			SHALSILT *(Shales and Sitlstones)*
			SANDEVAP *(Sandstones and Evaporites)*
			CNGLSAND *(Conglomerates and Sandstones)*
			SANDCOAL *(Sandstones and Coals)*
			DLMTSILT *(Dolomites and Silts)*
			SANDDLMT *(Sanstones and Dolomites)*
			UDCBUDCS *(Carbonates and Clastics)*
			LMSTSHAL *(Mixed Limestones and Shales)*
			LMSTDLMT *(Mixed Limestones and Dolomites)*
	SED *(Sedimentary)*	SEDS *(Sedimentary Undiff.)*	NONE *(Undifferenciated Sediments)*
IGNEOUS	IGNS *(Igneous)*	IGNS *(Igneous)*	INTRA *(Intrusive)*
			EXTR *(Extrusive)*
			NONE *(Ingeous Undiff.)*
METAMORPHIC	META *(Metamorphic)*	META *(Metamorphic)*	NONE *(Metamorphic Undiff.)*
UNKNOWN	UNK *(Unknown)*	UNK *(Unknown)*	NONE
UNDIFFERENTIATED	CLST (Clastic)	UNK *(Unknown)*	NONE

xxx_min_age
Use Data Population Tables
See Appendix A

xxx_max_age
Use Data Population Tables
See Appendix A

SEDIMENTARY LITHOFACIES ATTRIBUTE STRUCTURE - DATA OPTIONS
SELECT ONE ITEM FOR EACH CATEGORY AS APPROPRIATE

xxx_dep_class
MRG_MRN *(Marginal Marine)*
MRN *(Marine)*
NON_MAR *(Non-Marine)*
UNDIFF *(Undifferenciated)*
UNK *(Unknown)*

xxx_environ	xxx_env_type
ALLUVIAL	ABSENT
	ALLUVIAL_PLAIN
	BAR
	BRAIDED_STREAM
	CHANNEL
	CHANNEL_COMPLEX
	CHANNEL_FILL
	DEBRISFLOW_FAN
	DUNE
	DUNE_COMPLEX
	EPHEMERAL_LAKE
	FAN
	FAN_FRINGE
	FAN_LOBE
	FLAT
	FLOOD_PLAIN
	GRAINFLOW_FAN
	INTER_DUNE
	MASS_TRANS_CMPLX
	MIXED_FLAT
	MUD_FLAT
	NONE
	OUTWASH_FAN
	PLAIN
	PLAYA
	PLAYA_LAKE
	SABKHA
	SAND_BAR
	SAND_FLAT
	SHEET_COMPLEX
	SOIL_HORIZON
BASIN	ABSENT
	ABYSSAL_PLAIN
	ALGAL_MOUND
	BIOHERM
	BUILDUP
	CHANNEL_COMPLEX
	DEBRISFLOW_FAN
	DELTA_FRONT
	FAN
	FAN_FRINGE
	FAN_LOBE
	FOREREEF
	GRAINFLOW_FAN
	LVEE_LVEE_CMPLX
	MASS_TRANS_CMPLX
	MOUND
	MUD_BANK
	NONE
	PINNACLE_REEF
	PLAIN
	PRODELTA
	REEF
	SHEET_COMPLEX
	SHELF_MARGIN
	SINGLE_CHANNEL
	SINGLE_SHEET
	SLOPE_OVERBANK
	SUBMARINE_FAN
	TURBIDITE
	TURBIDITE_FAN

(Continued on Next Page)

xxx_env_mod
User-defined. Examples:
ABANDONED
ABSENT
ALGAL
ALGAL_LAMINATED
ALLOCHTHONOUS
AMALGAMATED
ANNUAL
ARID
AUTOCHTHONOUS
AXIS
COMPENSATING
DEEP
DISTAL
DISTAL_OVERBANK
DISTALLY_STEEPENED
EPHEMERAL
EVAPORITIC
FLOOD
FRESH
HIGH_ANGLE
HIGH_ENERGY
HUMID
HYPERSALINE
INNER
ISOLATED
LAYERED
LOW_ANGLE
LOW_ENERGY
LOWER
MARGIN
MESOSALINE
MIDDLE
MIXED
NA
NON_AMALGAMATED
NONE
OFF_AXIS
OPEN
OUTER
PERENNIAL
PERMANENT
PROXIMAL
PROXIMAL_OVERBANK
RESTRICTED
SEMI_AMALGAMATED
SHALLOW
SHINGLED
STARVED
SUBAQUEOUS
SUBAREAL
UNCONFINED
UPPER

xxx_subenv
Free-form text or
CHANNEL SAND
LEVEE SAND
MOUNDED THICK
SHEET SAND

xxx_mar_wd
ABYSSAL
BATHYL
HADAL
INNER_NERITIC
LITTORAL
LOWER_BATHYL
MIDDLE_BATHYL
MIDDLE_NERITIC
NERITIC
ONSHORE
OUTER_NERITIC
UPPER_BATHYL

xxx_paleo_geo
BASIN_FLOOR
CONTINENTAL
HIGHLAND_SOURCE
INTRASLOPE_BASIN
LOWLAND_BYPASS
SHELF
SHELF_MARGIN_BASIN
SHELF_SLOPE_BREAK
SLOPE
SLOPE_UPPER
SLOPE_MIDDLE
SLOPE_LOWER
TOE_OF_SLOPE

xxx_stack_pat
AGGRADATIONAL
CONSTRUCTIVE
HIGHSTAND
LOWSTAND
PROGRADATIONAL
REGRESSIVE
RETROGRADATIONAL
TRANSGRESSIVE

xxx_con_env
SUPPORTED
INFERRED

Additional values for all items in this column:

UNK *(Unkown)*
UNDIFF *(Undifferenciated)*

xxx_subenv_mod
AMALGAMATED
SEMI_AMALGAMATED
NON_AMALGAMATED

xxx_paleo_geo_md
PROXIMAL
MEDIAL
DISTAL

SEDIMENTARY LITHOFACIES ATTRIBUTE STRUCTURE - DATA OPTIONS
SELECT ONE ITEM FOR EACH CATEGORY AS APPROPRIATE

Back to Home Page

xxx_environ	xxx_env_type
CANYON	ABSENT
	BRAIDED_STREAM
	CHANNEL_COMPLEX
	CHANNEL_FILL
	DEBRISFLOW_FAN
	DELTA_CHANNEL
	DELTA_FRONT
	FAN
	FAN_LOBE
	GRAINFLOW_FAN
	LEVEE
	LVEE_LVEE_CMPLX
	MASS_TRANS_CMPLX
	NONE
	OUTWASH_FAN
	PRODELTA
	SHEET_COMPLEX
	SINGLE_CHANNEL
	SINGLE_SHEET
	SLOPE_OVERBANK
	SUBMARINE_FAN
	TURBIDITE
	TURBIDITE_FAN
	UNDIF_SLP_BN_FLR (Undifferenciated Slope/Basin Floor)
CANYON_COMPLEX	CHANNEL_COMPLEX
	LEVEE
	MASS_TRANS_CMPLX (Mass Transit Complex)
	SHEET_COMPLEX
	SINGLE_CHANNEL
	SINGLE_SHEET
	UNDIF_SLP_BN_FLR (Undifferenciated Slope/Basin Floor)
CHNLZ_SLOP (Channelized Slope)	CHANNEL_COMPLEX
	LEVEE
	MASS_TRANS_CMPLX (Mass Transit Complex)
	SHEET_COMPLEX
	SINGLE_CHANNEL
	SINGLE_SHEET
	UNDIF_SLP_BN_FLR (Undifferenciated Slope/Basin Floor)
CHNLZ_SLOP_CMPLX (Channelized Slope Complex)	CHANNEL_COMPLEX
	LEVEE
	MASS_TRANS_CMPLX (Mass Transit Complex)
	SHEET_COMPLEX
	SINGLE_CHANNEL
	SINGLE_SHEET
	UNDIF_SLP_BN_FLR (Undifferenciated Slope/Basin Floor)
COASTAL	ABSENT
	BAR
	BARRIER_BAR

(Continued on top right of this Page)

xxx_environ	xxx_env_type
COASTAL *(Continued)*	BEACH
	BEACH_RIDGE
	BRAIDED_STREAM
	CHANNEL
	CHANNEL_FILL
	CHENIER_PLAIN
	DUNE
	DUNE_COMPLEX
	EPHEMERAL_LAKE
	FLAT
	FLOOD_PLAIN
	FORESHORE
	INLET
	INLET_CHANNEL
	INTER_DUNE
	LAGOON
	LAKE
	LAKE_MARGIN
	LONGSHORE_BAR
	LOWER_SHOREFACE
	MARSH
	MEANDER_STREAM
	MIXED_FLAT
	MUD_BANK
	MUD_FLAT
	NONE
	PERMANENT_LAKE
	PLAIN
	PLAYA
	PLAYA_LAKE
	POND
	SABKHA
	SALINA
	SALT_PAN
	SAND_BAR
	SAND_FLAT
	SHOREFACE
	SOIL_HORIZON
	STREAM
	SWAMP
	TIDAL_CHANNEL
	TIDAL_CREEK
	TIDAL_INLET
	TIDAL_POND
	UPPER_SHOREFACE
COASTAL_PLAIN	ABSENT
	BAR
	BEACH

(Continued at top left of next page)

SEDIMENTARY LITHOFACIES ATTRIBUTE STRUCTURE - DATA OPTIONS
SELECT ONE ITEM FOR EACH CATEGORY AS APPROPRIATE

Back to Home Page

xxx_environ	xxx_env_type
COASTAL_PLAIN	BEACH_RIDGE
(Continued)	BRAIDED_STREAM
	CHANNEL
	CHANNEL_COMPLEX
	CHANNEL_FILL
	CHANNEL_LEVEE
	CHENEIR_PLAIN
	CREVASSE_SPLAY
	DELTA_CHANNEL
	DISTRIB_CHANNEL
	DUNE
	DUNE_COMPLEX
	EPHEMERAL_LAKE
	FLAT
	FLOOD_PLAIN
	FORESHORE
	INLET
	INLET_CHANNEL
	INTER_DUNE
	LAGOON
	LAKE
	LAKE_MARGIN
	LOWER_SHOREFACE
	LVEE_LVEE_CMPLX
	MARSH
	MEANDER_STREAM
	MIXED_FLAT
	MUD_FLAT
	NONE
	OVERBANK
	OVERBANK_LEVEE
	PERMANENT_LAKE
	PLAIN
	PLAYA
	PLAYA_LAKE
	POND
	SABKHA
	SALINA
	SALT_PAN
	SAND_FLAT
	SHOREFACE
	SINGLE_CHANNEL
	SLOPE_OVERBANK
	SOIL_HORIZON
	SPLAY
	STREAM
	SWAMP
	TIDAL_CHANNEL

(Continued on top right of this Page)

xxx_environ	xxx_env_type
COASTAL_PLAIN	TIDAL_CREEK
(Continued)	TIDAL_FLAT
	TIDAL_INLET
	TIDAL_POND
	UPPER_SHOREFACE
DELTAIC	ABSENT
	BAR
	BARRIER_BAR
	CHANNEL
	CHANNEL_COMPLEX
	CHANNEL_LEVEE
	CHENIER_PLAIN
	CREVASSE_SPLAY
	DELTA
	DELTA_CHANNEL
	DELTA_FRONT
	DISTRIB_CHANNEL
	FAN
	FAN_FRINGE
	FAN_LOBE
	FLAT
	FLOOD_PLAIN
	INLET
	INLET_CHANNEL
	LAGOON
	LAKE
	LVEE_LVEE_CMPLX
	MARSH
	MASS_TRANS_CMPLX
	MIXED_FLAT
	MUD_FLAT
	NONE
	OVERBANK
	OVERBANK_LEVEE
	PLAIN
	POINT_BAR
	POND
	PRODELTA
	SAND_BAR
	SAND_FLAT
	SINGLE_CHANNEL
	SOIL_HORIZON
	SPLAY
	STREAM
	STR_MTH_BAR
	SWAMP
	TIDAL_CHANNEL
	TIDAL_CREEK

(Continued at top left of next page)

SEDIMENTARY LITHOFACIES ATTRIBUTE STRUCTURE - DATA OPTIONS
SELECT ONE ITEM FOR EACH CATEGORY AS APPROPRIATE

Back to Home Page

xxx_environ	xxx_env_type
DELTAIC	TIDAL_INLET
(Continued)	TIDAL_POND
	TURBIDITE
EOLIAN	ABSENT
	BEACH
	BEACH_RIDGE
	CHENIER_PLAIN
	DEBRISFLOW_FAN
	DUNE
	DUNE_COMPLEX
	EPHEMERAL_LAKE
	FAN
	FAN_FRINGE
	FAN_LOBE
	FLAT
	FLOOD_PLAIN
	INTER_DUNE
	LAKE
	LAKE_MARGIN
	MASS_TRANS_CMPLX
	MIXED_FLAT
	MUD_FLAT
	NONE
	OUTWASH_FAN
	PERMANENT_LAKE
	PLAIN
	PLAYA
	PLAYA_LAKE
	SABKHA
	SALINA
	SALT_PAN
	SAND_BAR
	SAND_FLAT
	SOIL_HORIZON
ESTUARINE	ABSENT
	BAR
	BARRIER_BAR
	CHANNEL
	CHANNEL_COMPLEX
	CHANNEL_FILL
	FLAT
	FLOOD_PLAIN
	INLET
	INLET_CHANNEL
	LAGOON
	MARSH
	MEANDER_STREAM
	MIXED_FLAT

(Continued on top right of this Page)

xxx_environ	xxx_env_type
ESTUARINE	MUD_FLAT
(Continued)	NONE
	POINT_BAR
	POND
	SAND_BAR
	SAND_FLAT
	SOIL_HORIZON
	STREAM
	STR_MTH_BAR
	SWAMP
	TIDAL_CHANNEL
	TIDAL_CREEK
	TIDAL_INLET
	TIDAL_POND
FAN	ABSENT
	ALLUVIAL_PLAIN
	BRAIDED_STREAM
	CHANNEL
	CHANNEL_COMPLEX
	CHANNEL_FILL
	DEBRISFLOW_FAN
	FAN
	FAN_FRINGE
	FAN_LOBE
	FLAT
	FLOOD_PLAIN
	GRAINFLOW_FAN
	LEVEE
	LVEE_LVEE_CMPLX
	MASS_TRANS_CMPLX
	MIXED_FLAT
	MUD_FLAT
	NONE
	OUTWASH_FAN
	SAND_FLAT
	SHEET_COMPLEX
	SINGLE_CHANNEL
	SINGLE_SHEET
	SLOPE_OVERBANK
	SUBMARINE_FAN
	TURBIDITE
	TURBIDITE_FAN
	UNDIF_SLP_BN_FLR (Undifferenciated Slope/Basin Floor)
FAN_COMPLEX	CHANNEL_COMPLEX
	LEVEE
	MASS_TRANS_CMPLX (Mass Transit Complex)
	SHEET_COMPLEX
	SINGLE_CHANNEL

(Continued at top left of next page)

SEDIMENTARY LITHOFACIES ATTRIBUTE STRUCTURE - DATA OPTIONS
SELECT ONE ITEM FOR EACH CATEGORY AS APPROPRIATE

Back to Home Page

xxx_environ	xxx_env_type
FAN_COMPLEX (Continued)	SINGLE_SHEET
	UNDIF_SLP_BN_FLR (Undifferenciated Slope/Basin Floor)
FLUVIAL	ABSENT
	BAR
	BRAIDED_STREAM
	CHANNEL
	CHANNEL_COMPLEX
	CHANNEL_FILL
	CHANNEL_LEVEE
	CREVASSE_SPLAY
	FLAT
	FLOOD_PLAIN
	LVEE_LVEE_CMPLX
	MEANDER_STREAM
	MIXED_FLAT
	MUD_FLAT
	NONE
	OVERBANK
	OVERBANK_LEVEE
	POINT_BAR
	SAND_BAR
	SAND_FLAT
	SINGLE_CHANNEL
	SOIL_HORIZON
	SPLAY
	STREAM
	SWAMP
FLUVIAL_ALLUVIAL	ABSENT
	ALLUVIAL_PLAIN
	BAR
	BRAIDED_STREAM
	CHANNEL
	CHANNEL_COMPLEX
	CHANNEL_FILL
	CHANNEL_LEVEE
	CHENIER_PLAIN
	CREVASSE_SPLAY
	DEBRISFLOW_FAN
	DUNE
	DUNE_COMPLEX
	EPHEMERAL_LAKE
	FAN_FRINGE
	FAN_LOBE
	FLAT
	FLOOD_PLAIN
	GRAINFLOW_FAN
	INTER_DUNE
	LAKE_MARGIN

(Continued on top right of this Page)

xxx_environ	xxx_env_type
FLUVIAL_ALLUVIAL *(Continued)*	LVEE_LVEE_CMPLX
	MASS_TRANS_CMPLX
	MEANDER_STREAM
	MIXED_FLAT
	MUD_FLAT
	NONE
	OUTWASH_FAN
	OVERBANK
	OVERBANK_LEVEE
	PERMANENT_LAKE
	PLAIN
	PLAYA
	PLAYA_LAKE
	POINT_BAR
	SABKHA
	SAND_BAR
	SAND_FLAT
	SHEET_COMPLEX
	SINGLE_CHANNEL
	SOIL_HORIZON
	SPLAY
	STREAM
	SWAMP
FLUVIAL_DELTAIC	ABSENT
	BAR
	BRAIDED_STREAM
	CHANNEL
	CHANNEL_COMPLEX
	CHANNEL_FILL
	CHANNEL_LEVEE
	CREVASSE_SPLAY
	DELTA
	DELTA_CHANNEL
	DISTRIB_CHANNEL
	FLOOD_PLAIN
	INLET
	INLET_CHANNEL
	LAKE
	LVEE_LVEE_CMPLX
	MARSH
	MASS_TRANS_CMPLX
	MEANDER_STREAM
	NONE
	OVERBANK
	OVERBANK_LEVEE
	PLAIN
	POINT_BAR
	POND

(Continued at top left of next page)

SEDIMENTARY LITHOFACIES ATTRIBUTE STRUCTURE - DATA OPTIONS
SELECT ONE ITEM FOR EACH CATEGORY AS APPROPRIATE

Back to Home Page

xxx_environ	xxx_env_type
FLUVIAL_DELTAIC	SAND_BAR
(Continued)	SINGLE_CHANNEL
	SOIL_HORIZON
	SPLAY
	STREAM
	SWAMP
FLUVIAL_ESTUARN	ABSENT
	BANK
	BAR
	BARRIER_BAR
	BRAIDED_STREAM
	CHANNEL
	CHANNEL_COMPLEX
	CHANNEL_FILL
	CHANNEL_LEVEE
	CREVASSE_SPLAY
	FLAT
	FLOOD_PLAIN
	INLET
	INLET_CHANNEL
	LVEE_LVEE_CMPLX
	MARSH
	MEANDER_STREAM
	MIXED_FLAT
	MUD_FLAT
	NONE
	OVERBANK
	OVERBANK_LEVEE
	POINT_BAR
	POND
	SAND_BAR
	SAND_FLAT
	SINGLE_CHANNEL
	SOIL_HORIZON
	SPLAY
	STREAM
	STR_MTH_BAR
	SWAMP
	TIDAL_CHANNEL
	TIDAL_CREEK
	TIDAL_INLET
	TIDAL_POND
GLACIAL	ABSENT
	ABYSSAL_PLAIN
	BAR
	BRAIDED_STREAM
	CHANNEL_COMPLEX
	DEBRISFLOW_FAN

(Continued on top right of this Page)

xxx_environ	xxx_env_type
GLACIAL	FAN_FRINGE
(Continued)	FAN_LOBE
	GLACIAL_LAKE
	GRAINFLOW_FAN
	MASS_TRANS_CMPLX
	MEANDER_STREAM
	NONE
	OUTWASH_FAN
	PLAIN
	SOIL_HORIZON
	STREAM
HEMIPELAGIC	ABSENT
	ABYSSAL_PLAIN
	DEBRISFLOW_FAN
	FAN
	FAN_FRINGE
	FAN_LOBE
	GRAINFLOW_FAN
	MASS_TRANS_CMPLX
	NONE
	SUBMARINE_FAN
	TURBIDITE
	TURBIDITE_FAN
INNER_SHELF	ABSENT
	ALGAL_MOUND
	BACKREEF
	BANK
	BAR
	BARRIER_BAR
	BEACH
	BIOHERM
	BUILDUP
	CHANNEL
	CHANNEL_FILL
	DELTA
	DELTA_CHANNEL
	DELTA_FRONT
	DISTRIB_CHANNEL
	FOREREEF
	FORESHORE
	GRAIN_FLAT
	GRAIN_SHOAL
	INLET
	INLET_CHANNEL
	LONGSHORE_BAR
	LOWER_SHOREFACE
	LVEE_LVEE_CMPLX
	MOUND

(Continued at top left of next page)

SEDIMENTARY LITHOFACIES ATTRIBUTE STRUCTURE - DATA OPTIONS
SELECT ONE ITEM FOR EACH CATEGORY AS APPROPRIATE

Back to Home Page

xxx_environ	xxx_env_type
OFFSHORE *(Continued)*	LOWER_SHOREFACE
	LVEE_LVEE_CMPLX *(Levee Complex)*
	MASS_TRANS_CMPLX *(Mass Transist Complex)*
	MOUND
	MUD_BANK
	NONE
	PATCH_REEF
	PINNACLE_REEF
	PRODELTA
	REEF
	SAND_BAR
	SHEET_COMPLEX
	SHELF_MARGIN
	SINGLE_CHANNEL
	SINGLE_SHEET
	SLOPE_OVERBANK
	STR_MTH_BAR
	SUBMARINE_FAN
	TURBIDITE
	TURBIDITE_FAN
ONSHORE	ABSENT
	ALGAL_MOUND
	BAR
	BEACH
	BEACH_RIDGE
	BIOHERM
	BRAIDED_STREAM
	BUILDUP
	CHANNEL
	CHANNEL_COMPLEX
	CHANNEL_FILL
	CHANNEL_LEVEE
	CREVASSE_SPLAY
	DEBRISFLOW_FAN
	DELTA_CHANNEL
	DISTRIB_CHANNEL
	DUNE
	DUNE_COMPLEX
	EPHEMERAL_LAKE
	FAN
	FAN_FRINGE
	FAN_LOBE
	FLAT
	FLOOD_PLAIN
	FORESHORE
	GRAINFLOW_FAN
	INTER_DUNE
	LAKE

(Continued on top right of this Page)

xxx_environ	xxx_env_type
ONSHORE *(Continued)*	LAKE_MARGIN
	LOWER_SHOREFACE
	LVEE_LVEE_CMPLX
	MARSH
	MASS_TRANS_CMPLX
	MEANDER_STREAM
	MIXED_FLAT
	MOUND
	MUD_FLAT
	NONE
	OUTWASH_FAN
	OVERBANK
	OVERBANK_LEVEE
	PERMANENT_LAKE
	PLAIN
	PLAYA
	PLAYA_LAKE
	POINT_BAR
	POND
	SABKHA
	SALINA
	SALT_PAN
	SAND_BAR
	SAND_FLAT
	SHOREFACE
	SINGLE_CHANNEL
	SOIL_HORIZON
	SPLAY
	STREAM
	SWAMP
	UPPER_SHOREFACE
OPEN_INTERIOR	ABSENT
	ALGAL_MOUND
	BACKREEF
	BANK
	BAR
	BARRIER_BAR
	BEACH
	BEACH_RIDGE
	BIOHERM
	BUILDUP
	CHANNEL
	FOREREEF
	FORESHORE
	GRAIN_FLAT
	GRAIN_SHOAL
	INLET
	INLET_CHANNEL

(Continued at top left of next page)

SEDIMENTARY LITHOFACIES ATTRIBUTE STRUCTURE - DATA OPTIONS
SELECT ONE ITEM FOR EACH CATEGORY AS APPROPRIATE

Back to Home Page

xxx_environ	xxx_env_type
OPEN_INTERIOR	LONGSHORE_BAR
(Continued)	LOWER_SHOREFACE
	MIXED_FLAT
	MOUND
	MUD_BANK
	MUD_FLAT
	NONE
	PATCH_REEF
	REEF
	SAND_BAR
	SAND_FLAT
	SHELF_MARGIN
	SHOAL
	SHOAL_COMPLEX
	SHOREFACE
	TIDAL_CHANNEL
	TIDAL_CREEK
	TIDAL_INLET
	TURBIDITE
	UPPER_SHOREFACE
OUTER_SHELF	ABSENT
	ALGAL_MOUND
	BACKREEF
	BAR
	BIOHERM
	BUILDUP
	CHANNEL
	CHANNEL_COMPLEX
	CHANNEL_FILL
	DEBRISFLOW_FAN
	DELTA_FRONT
	FAN
	FAN_FRINGE
	FAN_LOBE
	FOREREEF
	GRAINFLOW_FAN
	GRAIN_SHOAL
	LONGSHORE_BAR
	LVEE_LVEE_CMPLX
	MASS_TRANS_CMPLX
	MOUND
	MUD_BANK
	NONE
	PINNACLE_REEF
	PRODELTA
	REEF
	SAND_BAR
	SHEET_COMPLEX

(Continued on top right of this Page)

xxx_environ	xxx_env_type
OUTER_SHELF	SHELF_MARGIN
(Continued)	SHOAL
	SHOAL_COMPLEX
	SINGLE_CHANNEL
	SINGLE_SHEET
	SLOPE_OVERBANK
	SUBMARINE_FAN
	TURBIDITE
	TURBIDITE_FAN
PARALIC	ABSENT
	ALLUVIAL_PLAIN
	BARRIER_BAR
	BEACH_RIDGE
	CHANNEL
	CHANNEL_LEVEE
	CHENIER_PLAIN
	CREVASSE_SPLAY
	DELTA
	DELTA_CHANNEL
	DELTA_FRONT
	DISTRIB_CHANNEL
	DUNE_COMPLEX
	EPHEMERAL_LAKE
	FLAT
	FLOOD_PLAIN
	FORESHORE
	GRAIN_FLAT
	GRAIN_SHOAL
	INLET
	INLET_CHANNEL
	INTER_DUNE
	LAGOON
	LONGSHORE_BAR
	LOWER_SHOREFACE
	LVEE_LVEE_CMPLX
	MARSH
	MEANDER_STREAM
	MIXED_FLAT
	MUD_BANK
	NONE
	OUTWASH_FAN
	OVERBANK
	OVERBANK_LEVEE
	PATCH_REEF
	PLAIN
	POINT_BAR
	POND
	SABKHA

(Continued at top left of next page)

SEDIMENTARY LITHOFACIES ATTRIBUTE STRUCTURE - DATA OPTIONS
SELECT ONE ITEM FOR EACH CATEGORY AS APPROPRIATE

Back to Home Page

xxx_environ	xxx_env_type
PARALIC *(Continued)*	SAND_BAR
	SAND_FLAT
	SHEET_COMPLEX
	SHOAL
	SHOAL_COMPLEX
	SHOREFACE
	SINGLE_CHANNEL
	SPLAY
	STREAM
	STR_MTH_BAR
	SWAMP
	TIDAL_CHANNEL
	TIDAL_CREEK
	TIDAL_INLET
	TIDAL_POND
	UPPER_SHOREFACE
PELAGIC	ABSENT
	ABYSSAL_PLAIN
	DEBRISFLOW_FAN
	FAN
	FAN_FRINGE
	FAN_LOBE
	GRAINFLOW_FAN
	MASS_TRANS_CMPLX
	NONE
	SUBMARINE_FAN
	TURBIDITE
	TURBIDITE_FAN
PERITIDAL	ABSENT
	ALGAL_MOUND
	BANK
	BAR
	BARRIER_BAR
	BEACH
	BEACH_RIDGE
	CHANNEL
	CHANNEL_COMPLEX
	CHANNEL_FILL
	CHENIER_PLAIN
	DUNE
	DUNE_COMPLEX
	FLAT
	FLOOD_PLAIN
	FORESHORE
	GRAIN_FLAT
	GRAIN_SHOAL
	INLET
	INLET_CHANNEL

(Continued on top right of this Page)

xxx_environ	xxx_env_type
PERITIDAL *(Continued)*	INTER_DUNE
	LAGOON
	LONGSHORE_BAR
	LOWER_SHOREFACE
	MARSH
	MIXED_FLAT
	MOUND
	MUD_FLAT
	NONE
	PATCH_REEF
	PLAIN
	SABKHA
	SAND_BAR
	SAND_FLAT
	SHOAL
	SHOAL_COMPLEX
	SHOREFACE
	SINGLE_CHANNEL
	SOIL_HORIZON
	STREAM
	SWAMP
	TIDAL_CREEK
	TIDAL_INLET
	TIDAL_POND
	UPPER_SHOREFACE
PLATFORM	ABSENT
	ALGAL_MOUND
	BACKREEF
	BANK
	BAR
	BARRIER_BAR
	BEACH
	BEACH_RIDGE
	BIOHERM
	BUILDUP
	CHANNEL
	CHANNEL_FILL
	FLAT
	FOREREEF
	GRAIN_FLAT
	GRAIN_SHOAL
	LAGOON
	LONGSHORE_BAR
	MIXED_FLAT
	MOUND
	MUD_BANK
	MUD_FLAT
	NONE

(Continued at top left of next page)

SEDIMENTARY LITHOFACIES ATTRIBUTE STRUCTURE - DATA OPTIONS
SELECT ONE ITEM FOR EACH CATEGORY AS APPROPRIATE

Back to Home Page

xxx_environ	xxx_env_type
SABKHA	INLET_CHANNEL
(Continued)	INTER_DUNE
	LOWER_SHOREFACE
	MASS_TRANS_CMPLX
	MIXED_FLAT
	MUD_FLAT
	NONE
	OUTWASH_FAN
	PLAIN
	PLAYA
	PLAYA_LAKE
	SABKHA
	SALINA
	SALT_PAN
	SAND_BAR
	SAND_FLAT
	SHEET_COMPLEX
	SHOAL
	SHOAL_COMPLEX
	SHOREFACE
	SINGLE_SHEET
	SOIL_HORIZON
	TIDAL_CHANNEL
	TIDAL_CREEK
	TIDAL_INLET
	UPPER_SHOREFACE
SHELF	ABSENT
	ALGAL_MOUND
	BACKREEF
	BANK
	BAR
	BARRIER_BAR
	BEACH
	BEACH_RIDGE
	BIOHERM
	BUILDUP
	CHANNEL
	CHANNEL_FILL
	DELTA
	DELTA_FRONT
	FOREREEF
	FORESHORE
	GRAIN_FLAT
	GRAIN_SHOAL
	LAGOON
	LONGSHORE_BAR
	LOWER_SHOREFACE
	MOUND

(Continued on top right of this Page)

xxx_environ	xxx_env_type
SHELF	MUD_BANK
(Continued)	NONE
	PATCH_REEF
	PINNACLE_REEF
	PRODELTA
	REEF
	SAND_BAR
	SHELF_MARGIN
	SHOAL
	SHOAL_COMPLEX
	SHOREFACE
	TURBIDITE
	UPPER_SHOREFACE
SHELF_MARGIN	ABSENT
	ALGAL_MOUND
	BACKREEF
	BANK
	BAR
	BIOHERM
	BUILDUP
	CHANNEL
	CHANNEL_FILL
	DELTA
	DELTA_FROMT
	FOREREEF
	GRAINFLOW_FAN
	GRAIN_FLAT
	GRAIN_SHOAL
	INLET
	INLET_CHANNEL
	LONGSHORE_BAR
	MOUND
	MUD_BANK
	NONE
	PATCH_REEF
	PINNACLE_REEF
	PRODELTA
	REEF
	SAND_BAR
	SHELF_MARGIN
	SHOAL
	SHOAL_COMPLEX
	TIDAL_CHANNEL
	TIDAL_INLET
	TURBIDITE
SLOPE	ABSENT
	ALGAL_MOUND
	BIOHERM

(Continued at top left of next page)

SEDIMENTARY LITHOFACIES ATTRIBUTE STRUCTURE - DATA OPTIONS
SELECT ONE ITEM FOR EACH CATEGORY AS APPROPRIATE

Back to Home Page

xxx_environ	xxx_env_type
SLOPE	BUILDUP
(Continued)	CHANNEL
	CHANNEL_COMPLEX
	CHANNEL_FILL
	DEBRISFLOW_FAN
	FAN
	FAN_FRINGE
	FAN_LOBE
	FOREREEF
	GRAINFLOW_FAN
	LVEE_LVEE_CMPLX
	MASS_TRANS_CMPLX
	MOUND
	MUD_BANK
	NONE
	PINNACLE_REEF
	PRODELTA
	REEF
	SHEET_COMPLEX
	SHELF_MARGIN
	SINGLE_CHANNEL
	SINGLE_SHEET
	SLOPE_OVERBANK
	SUBMARINE_FAN
	TURBIDITE
	TURBIDITE_FAN
SUBTIDAL	ABSENT
	ALGAL_MOUND
	BACKREEF
	BANK
	BAR
	BARRIER_BAR
	BIOHERM
	BUILDUP
	CHANNEL
	CHANNEL_COMPLEX
	DEBRISFLOW_FAN
	FAN
	FAN_FRINGE
	FAN_LOBE
	FOREREEF
	GRAINFLOW_FAN
	GRAIN_FLAT
	GRAIN_SHOAL
	INLET
	INLET_CHANNEL
	LAGOON
	LOWER_SHOREFACE

(Continued on top right of this Page)

xxx_environ	xxx_env_type
SUBTIDAL	MASS_TRANS_CMPLX
(Continued)	MOUND
	MUD_BANK
	NONE
	PATCH_REEF
	PINNACLE_REEF
	REEF
	SAND_BAR
	SHELF_MARGIN
	SHOAL
	SHOAL_COMPLEX
	SINGLE_CHANNEL
	SUBMARINE_FAN
	TIDAL_CREEK
	TIDAL_INLET
	TURBIDITE
	TURBIDITE_FAN
SUPRATIDAL	ABSENT
	BAR
	BEACH
	BEACH_RIDGE
	CHANNEL
	CHANNEL_COMPLEX
	CHANNEL_FILL
	CHENIER_PLAIN
	DUNE
	DUNE_COMPLEX
	EPHEMERAL_LAKE
	FAN_FRINGE
	FAN_LOBE
	FLAT
	FLOOD_PLAIN
	GRAIN_FLAT
	INLET
	INLET_CHANNEL
	INTER_DUNE
	LAKE_MARGIN
	MARSH
	MASS_TRANS_CMPLX
	MIXED_FLAT
	MUD_FLAT
	NONE
	OUTWASH_FAN
	PLAIN
	PLAYA
	PLAYA_LAKE
	POND
	SABKHA

(Continued at top left of next page)

SEDIMENTARY LITHOFACIES ATTRIBUTE STRUCTURE - DATA OPTIONS
SELECT ONE ITEM FOR EACH CATEGORY AS APPROPRIATE

Back to Home Page

xxx_environ	xxx_env_type
SUPRATIDAL	SALINA
(Continued)	SALT_PAN
	SAND_BAR
	SAND_FLAT
	SHEET_COMPLEX
	SINGLE_CHANNEL
	SINGLE_SHEET
	SOIL_HORIZON
	STREAM
	SWAMP
	TIDAL_CHANNEL
	TIDAL_CREEK
	TIDAL_INLET
	TIDAL_POND
	UPPER_SHOREFACE
SWAMP	ABSENT
	BARRIER_BAR
	BEACH
	BEACH_RIDGE
	CHANNEL
	CHANNEL_COMPLEX
	CHANNEL_FILL
	CHANNEL_LEVEE
	CHENIER_PLAIN
	CREVASSE_SPLAY
	DISTRIB_CHANNEL
	EPHEMERAL_LAKE
	FAN_FRINGE
	FAN_LOBE
	FLAT
	FLOOD_PLAIN
	INLET
	INLET_CHANNEL
	LAGOON
	LAKE
	LAKE_MARGIN
	LONGSHORE_BAR
	LVEE_LVEE_CMPLX
	MARSH
	MEANDER_STREAM
	MIXED_FLAT
	MUD_BANK
	MUD_FLAT
	NONE
	OVERBANK
	OVERBANK_LEVEE
	PERMANENT_LAKE
	PLAIN

(Continued on top right of this Page)

xxx_environ	xxx_env_type
SWAMP	POINT_BAR
(Continued)	POND
	SAND_BAR
	SAND_FLAT
	SHEET_COMPLEX
	SINGLE_CHANNEL
	SINGLE_SHEET
	SOIL_HORIZON
	SPLAY
	STREAM
	STR_MTH_BAR
	TIDAL_CHANNEL
	TIDAL_CREEK
	TIDAL_INLET
	TIDAL_POND
UNK *(Unknown)*	UNK *(Unknown)*

Name:	CLOSURES (FIELD, DRY, PROSPECT)	[shaded box]	REFER TO CODE LISTS FOR SHADED ATTRIBUTE ITEMS
Data Type:	POLYGONS OR POINTS		

Used for:

Field polygons or point locations of field centroids
Composite field outlines
Linkage to PCDB and assessment related data tables
For assessment maps.
Contains one point for each structure regardless of status of feature (field, prospect, dry) or the number of assessment intervals that it contains
Feature coverage is an interim solution until the general closure model is fully implemented
Information specific to various stratigraphic or assessment intervals is organized in a related data table

Attribute items:

	WIDTH	OUTPUT	TYPE	N. DEC.	ALT. NAME	DESCRIPTION / PURPOSE OF THE ITEM
xxx_closure_no	8	8	I	-	closure_no	User defined. Number assigned to closure in study of area.
xxx_name	32	32	C	-	name	User defined. Business Unit name for closure.
xxx_data_src	30	30	C	-	data_src	User defined. Data source or provider. Example: EPR.44EX.87
xxx_arc_fld_id	16	16	C	-	arc_fld_id	User defined. Key field (oil & gas field ID) that links to all related data tables.
xxx_class	20	20	C	-	class	Code lists. Major geologic class of closure.
xxx_detect	20	20	C	-	detect	Code lists. Method used to detect, discover or map feature or closure.
xxx_signif	10	10	C	-	signif	Code lists. Significance or relative magnitude of feature.
xxx_occ	20	20	C	-	occ	Code lists. Field type or largest show.
xxx_type	30	30	C	-	type	Code lists. Method of discovery determination or show detection.
xxx_eur_moeb	4	8	F	3	eur_moeb	User defined. Estimated ultimate recovery of MOEB (use gcf to mbc factor of 6.0).
xxx_eur_mbo	4	8	F	3	eur_mbo	User defined. Estimated ultimate recovery in millions of barrels of oil.
xxx_eur_gcf	4	8	F	1	eur_gcf	User defined. Estimated ultimate recovery in billions of cubic feet of gas.
xxx_eur_mbc	4	8	F	3	eur_mbc	User defined. Estimated ultimate recovery in millions of barrels of condensate.
xxx_closure_area	4	12	F	0	closure_area	User defined. Maximum area of closure used for volumetric calculations in square meters.
xxx_edit_date	8	8	I	-	edit_date	User defined. Number to indicate the last edit date. Example: 19980101 = January 1, 1998.
xxx_exxon_def	1	1	C	-	exxon_def	User defined. Identifies if field outline is defined by Exxon. Y= Exxon defined; blank = non-Exxon defined
xxx_well_api	12	12	C	-	well_api	User defined. Well API number.
xxx_fld_stat	4	4	C	-	fld_stat	Code lists. Describes field status.
xxx_hc_acml_type	8	8	C	-	hc_acml_type	Code lists. Hydrocarbon accumulation type.
xxx_con_loc	16	16	C	-	con_loc	Code lists. Confidence in location of closure.
xxx_subject	30	30	C	-	subject	User defined. Overall coverage subject or entity. Example: UPPER CRETACEOUS.
xxx_title	100	100	C	-	title	User defined. Information mapped about the subject. Example: LEADS AND DRY TESTS.
xxx_file	50	50	C	-	file	User defined. Name of original pre-Arc file. Example: CRFCTAR4.DWG

xxx = ITEM PREFIX REQUIRED FOR POLYGON AND POINT COVERAGES ONLY. EXAMPLE: IF COVERAGE IS BAU_CRET_LOW THEN XXX COULD BE BCL..
BOLDFACE = REQUIRED ITEM

CLOSURES ATTRIBUTE STRUCTURE - DATA OPTIONS
SELECT ONE ITEM FOR EACH CATEGORY AS APPROPRIATE

xxx_class
DRY_TEST
FIELD
LEAD
PROSPECT
TESTED_NO_DETAILS

xxx_detect
GRAVITY
MAPPED
OUTSIDE_SRC
SEISMIC
SURFACE
WELL

xxx_signif
MAJOR
MINOR

xxx_occ
GAS_CONDENSATE
NO_HC_OCC
OIL
OIL_GAS
OTHER

xxx_type
CONVENTIONAL_CORE
DRILLING
DST
LOG_ANALYSIS
MUDLOG_CUTTINGS
PERF
RFT
SIDEWALL_CORE
(Oil on pit, blowout, etc.)

xxx_fld_stat	Description
	Data not available
UNK	Unknown
UNDV	Undeveloped
UNDS	Undeveloped - static
UNDC	Undeveloped - commercial
DEV	Developed
PROD	Producing
DPLT	Depleted
SHUT	Shut-In
ABND	Abandoned
.	Not Available
DCVR	Discovery

xxx_hc_acml_type	Description
GASCND	Gas Condensate
GASDRY	Dry Gas
OIL	Oil without associated gas cap
OILGC	Oil with associated gas cap
OILGAS	Oil and gas
OILNGC	Oil without gas cap and gas reservoirs
OILGCG	Oil with gas cap and oil reservoirs
OILWOG	Oil and gas
PROSPT	Prospect: hydrocarbons not yet discovered
UNK	Unknown
	Data not available
.	Not available
DRY	Dry
OGC	Oil, Gas condensate
GAS	Gas reservoir w/without condensate
COND	condensate
OC	oil and condensate

xxx_con_loc
APPROXIMATE
KNOWN

NOTE

Additional codes for all items in these columns:

UNK *(Unknown)*
UNDIFF *(Undifferenciated)*

Paulus, G., 2000, GIS-based analysis of fluid migration in sedimentary basins, *in* T. C. Coburn and J. M. Yarus, eds., Geographic information systems in petroleum exploration and development: AAPG Computer Applications in Geology, No. 4, p. 121–136.

Chapter 9

GIS-Based Analysis of Fluid Migration in Sedimentary Basins

Gernot Paulus
Department of Geology and Paleontology, University of Salzburg
Salzburg, Austria

ABSTRACT

Subsurface fluid migration is a multi-parameter process depending on many different components (e.g., tectonics, facies, lithology, hydrochemistry, and hydraulics). Knowledge of subsurface fluid migration is critical to the success of petroleum and geothermal exploration, and to petroleum production in sedimentary basins. A geographical information system (GIS) has been used for the present study to integrate different data sets related to subsurface fluid migration and to perform spatial analysis on the resulting comprehensive database. The discussion illustrates the reliance on geographic location as the common frame of reference during the integration phase. The area of investigation is the Molasse Basin in Upper Austria, which is a typical asymmetric, compressional foreland basin at the northern margin of the Alps. Due to petroleum exploration and production, an excellent data set exists with more than 500 wells ranging in depth from 1000 to 4000 m. Three different fault systems have been active at different times from pre-Tertiary until the present and have a large impact on hydrocarbon migration and fluid flow. There are two components to the fluid migration modeled in this study: (1) a north-directed migration of hydrocarbons and (2) an east-directed fluid flow of deep groundwater at basin scale in a Jurassic carbonate aquifer. The aquifer is the main target for geothermal exploration in this area. The datasets used in the spatial analysis contain well data, subsurface fault data, and hydrocarbon reservoir data. Applications and examples of various spatial GIS analysis techniques are presented here, including (1) investigation of the spatial relationship between subsurface faults and well locations using buffer zones and topological overlay, (2) calculation of fault density and assessing its implication on subsurface fluid flow, (3) modeling of fault-related fluid flow using cost surfaces, and (4) spatial overlay of hydrocarbon reservoir information and fault-related permeability distributions. These examples provide a comprehensive overview of how to use the unique capabilities of a GIS in subsurface geologic analysis.

INTRODUCTION

Subsurface fluid migration is a multi-parameter process and a key component of petroleum and geothermal exploration and production in sedimentary basins. Primary parameters, such as stratigraphy, lithology, facies, and tectonics, describe the geological setting within a study area. Secondary parameters represent derived information by qualitative and quantitative measurements on these primary parameters, such as well-logging, seismic, petrophysical, hydraulic, geochemical, and isotopic studies (Figure 1). Furthermore, the degree of contribution of the different parameters to the process of fluid migration in a sedimentary basin may vary in space and time. For example, faults may act as conduits and barriers for fluids at different times during the evolution of a sedimentary basin. They may also act as conduits and barriers for fluids in different areas within a study area during one evolutionary stage of a basin (Nachtmann, 1995; Wagner, 1996). It is exactly this spatial variation of distinctive parameters within a study area that is of interest. One approach to study spatial and non-spatial parameters in sedimentary basins is to use geostatistics (Yarus and Chambers, 1994).

For a comprehensive understanding of fluid migration, it is necessary to consider and analyze *all* the primary and secondary parameters that are available. This subsurface information is usually generated by specialists on a single field from many institutions, departments, and divisions with different levels of quality in the sense of digital representation and metadata description. A broad range of information is available from written reports, scientific papers, and analytical results in table-form to simple analog maps and fully digital data representations encompassing metadata descriptions meeting the spatial data transfer standard (SDTS).

In summary, a tremendous amount of special subsurface data exists. This information, in the worst case, is stored compartmentalized and isolated or, in the best case, it is comprehensively organized in a relational database management system (RDBMS). The main providers of this gigantic subsurface spatial dataset on sedimentary basins throughout the world are mostly the major oil companies, simply because they are the only ones who have the financial resources to produce such expensive data. One major goal of this paper is to present a different look at subsurface data by using them as "fuel" to launch a geographical information system (GIS) "rocket." The mission of this "rocket" is to explore new insights on fault-related fluid migration in sedimentary basins.

Why GIS ?

The main reason for using a GIS for this fluid migration project is quite simple. It is the only comprehensive tool available that supports the integrated digital analysis of multi-component processes by considering any needed attribute combination of the various components. Two definitions out of many for GIS are offered.

- A GIS is a system that uses a spatial database to provide answers to queries of a geographical nature. Because putting spatial data into a computer for the sole purpose of getting it out again would be expensive and pointless, a GIS must allow a variety of manipulations to be carried out, such as sorting, selective retrieval, calculation, and spatial analysis and modeling; furthermore, a full range of functions to allow input of data (in map form) and cartographic output is expected (Goodchild, 1985).
- A GIS is a computer system designed to allow users to collect, manage, and analyze large volumes of spatially referenced and associated attribute data (Hemenway, 1989).

Benefits and Uncertainty

A lot of benefits may arise from applying and using the tools of GIS. Consider the following:

- GIS is an appropriate tool to perform continuous digital workflow, beginning with data capture over data management and map generation. Once a GIS is organized properly, it will definitely rationalize traditional workflow, improve efficiency to a significant degree, and, therefore, have a high potential of reducing costs.
- Another benefit from an organizational point of view derives from the integrated use of GIS and digital communication systems. Such an integration might be referred to as a management information system, and might be characterized as an umbrella technology (Parker, 1993). All activities of an oil company, regardless of scale, size, or location, are organized and processed within one system. Such activities may include management of production parameters of reservoirs, pipeline network management, management of human resources (e.g., by optimization of travel time and travel costs for experts or drilling crews), processing of exploration data, and reevaluation of existing plays.
- The most important power of a GIS is its capability to spatially analyze large datasets. That is what really makes it unique and different. A GIS allows all of the individual thematic information sets of the fluid migration puzzle to be brought together using geographic location as the common denominator, and there it facilitates their analysis by spatial techniques and functions. New information is delineated for further investigation by overlaying existing information, represented as digital geoscience maps, to create one multi-layer map stack. The traditional procedure used to do this is "sandwiching" of map sheets on a light table. Once such a multi-layer map stack is organized, it is possible to query it for potential dependencies, patterns, and spatial coincidence between the single map layers. Such a query may comprise point-to-point

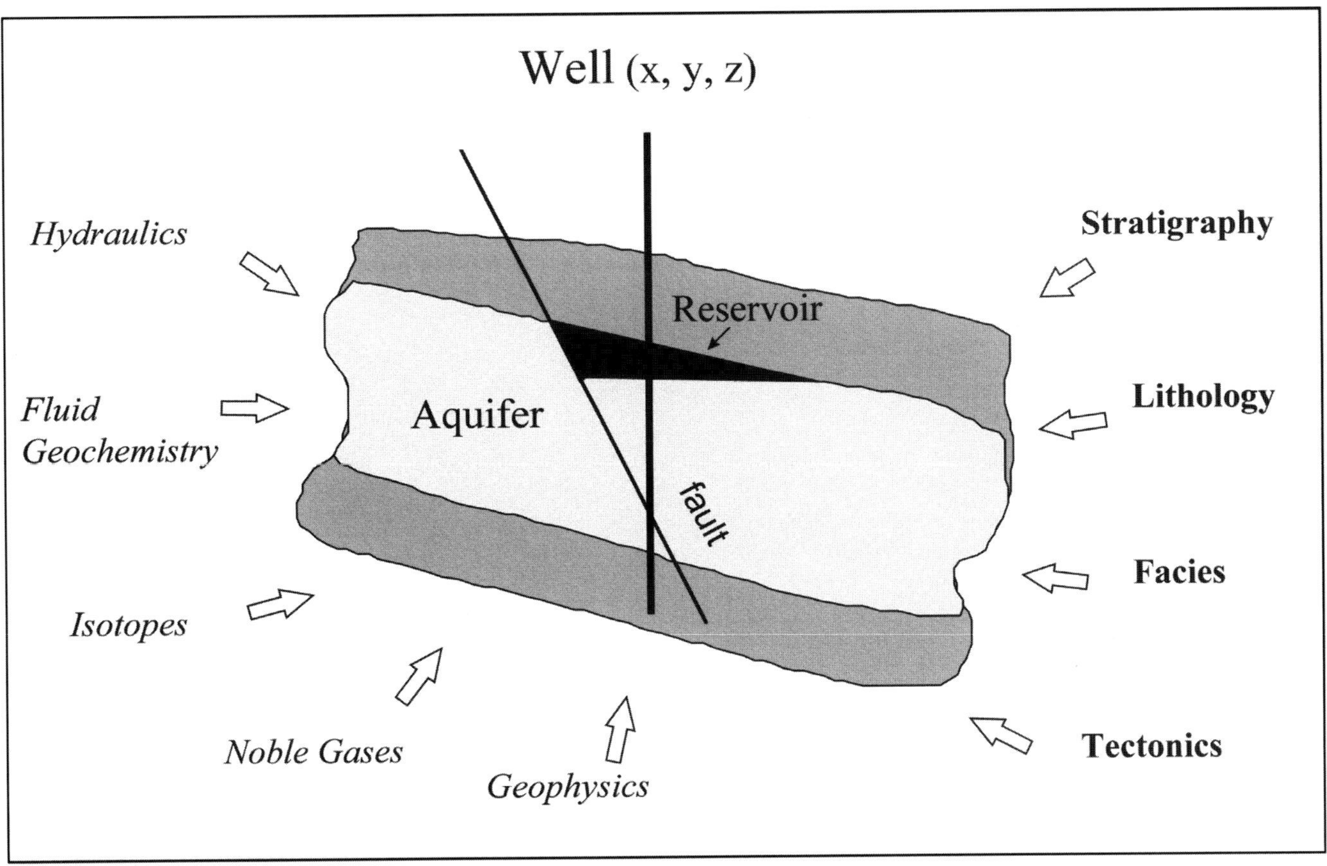

Figure 1—Fluid migration in sedimentary basins as a multi-component process (Bold: primary field parameters influencing fluid flow; italics: secondary parameters that are physical measurements for indirect analysis of fluid migration).

overlay or region-wide and map-wide overlay, which is similar to asking the computer to identify a certain pattern (or more generally spoken, a condition) from each map layer for every location in a projected area (Figure 2). The result of such a query will be combinatorial, computational, and statistical summaries of map coincidences (Berry, 1993).

- Another important function of a GIS is decision support. Together with capabilities to incorporate errors and uncertainty into a process, multiple-objective and multiple-criteria analysis provide powerful tools for resource allocation decision makers (Eastman, 1997). One example may be the problem of choosing the optimal site for exploration or production wells.

The crucial point concerning geoscientific thematic maps as input for spatial analysis procedures is the transformation of observations about the real world into a computational domain. This process requires various levels of abstraction and is therefore largely subjective and a potential source of error.

Map data may represent different types of observations, such as continuous measurements (e.g., 3-D seismic survey data, satellite images), discrete samples (e.g., core-sample measurements), and distinct geometric objects (polygons, lines, and points, e.g., hydrocarbon reservoirs, faults, and hydrochemical sample points referred to as points, lines, and polygons). In addition, geoscience maps may often also contain interpretations derived by various interpolation techniques based on a limited number of observations due to incomplete sampling (e.g., seismic lines and well-logs).

Again, experts are responsible for such classifications, which are undoubtedly needed to produce digital data. These classifications depend on the establishment of definition criteria for observations (e.g., maturity distribution of source rocks or delimitation of formation water and meteoric water). Observations are often translated by Boolean logic into attribute classes assuming that the attributes can be described and measured exactly. In the real world, however, it is often impossible to exactly record attributes of environmental data because of measurement errors, spatial variation, and the existence of transition zones. From this point of view it is quite obvious that maps can differ in quality and in the level of uncertainty assigned to each occurrence of a map unit (Chung and Fabbri, 1993).

Extensive studies have been undertaken to find better and more realistic techniques to consider the uncertainty of spatially distributed data and to minimize the error propagation of uncertain data. Such techniques

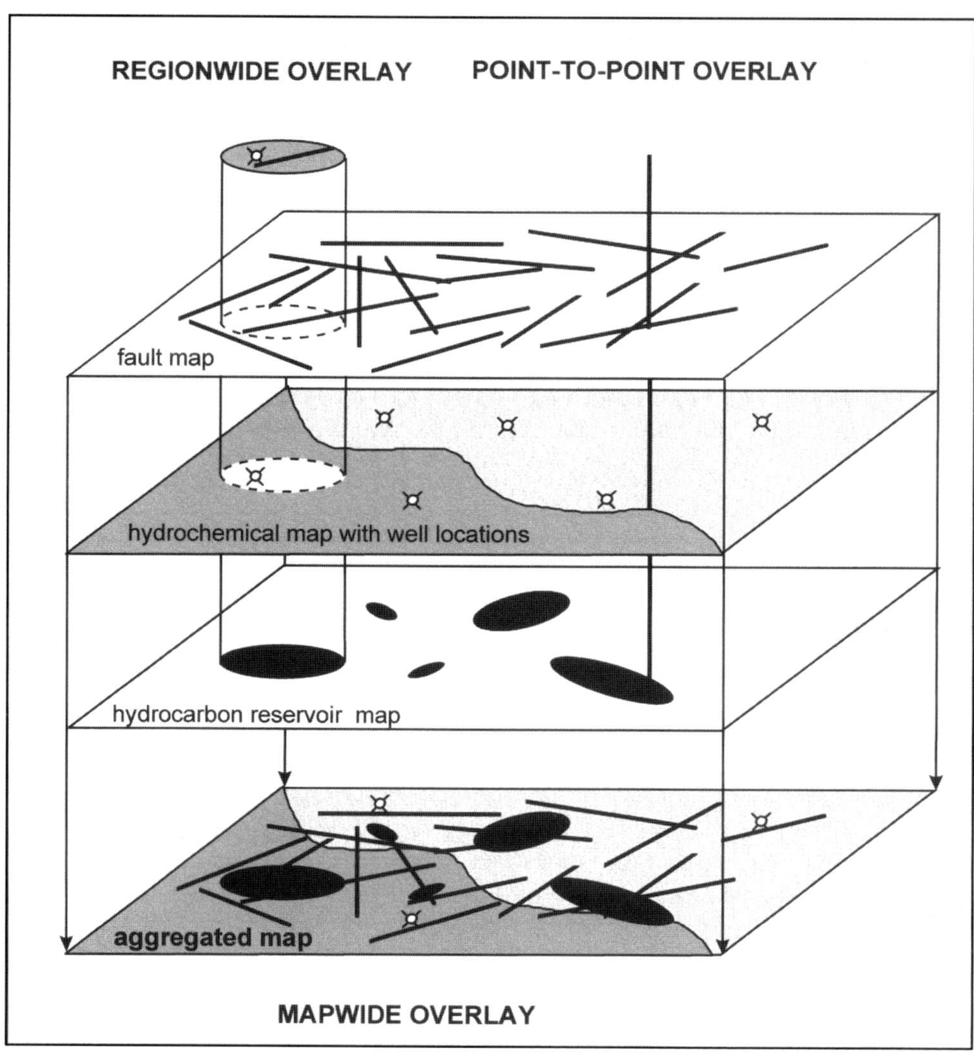

Figure 2—Spatial analysis techniques for multi-layer map stacks.

encompass the application of probability functions, Dempster-Shafer belief functions, membership functions in fuzzy sets, and epsilon- or Perkal-bands, and others (Chung and Fabbri, 1993; Heuvelink and Burrough, 1993; Blakemore, 1984).

STUDY AREA

The Molasse Basin of Austria is a typical asymmetric, compressional foreland basin at the northern margin of the Alps. Due to petroleum exploration and production in that area (Upper Austria), there exists an excellent dataset of more than 500 wells ranging from 1000 to 4000 m in depth. The stratigraphic record consists of metamorphic and magmatic basement rocks, covered with a transgressive Mesozoic shelf sequence, which constitute the base for the Tertiary clastic basin fill (Figure 3). The area of investigation has an extent of approximately 10,000 km^2 and a volume of 20 million km^3, assuming an average basin depth of 2000 m.

Three different fault systems can be distinguished:

- In the Early Cretaceous a basin inversion took place along this pre-Tertiary, northwest-trending fault system with erosion and karstification of the Jurassic carbonates. In the Late Cretaceous to Early Tertiary, a transpressive reactivation of this fault system occurred. This transpressive tectonic regime is due to the beginning collision of the European foreland and the Adriatic plate. Different large-scale fault blocks were tilted to the east with vertical displacements of up to 1000 m.
- An Eocene–Oligocene, west-trending compressional fault system is attributable to tectonic loading of the Alpine nappe stack on the European foreland (Nachtmann and Wagner, 1987). This second fault system exhibits vertical displacements of up to 300 m, which act as major structural traps for hydrocarbon reservoirs in the Upper Cretaceous and Eocene strata.
- Reactivation of the preexisting fault systems by Miocene time conjugate strike-slip faults, having lateral displacement ranging from several hundred meters up to approximately 1000 m (Nachtmann, 1995). This Miocene modification of the structural

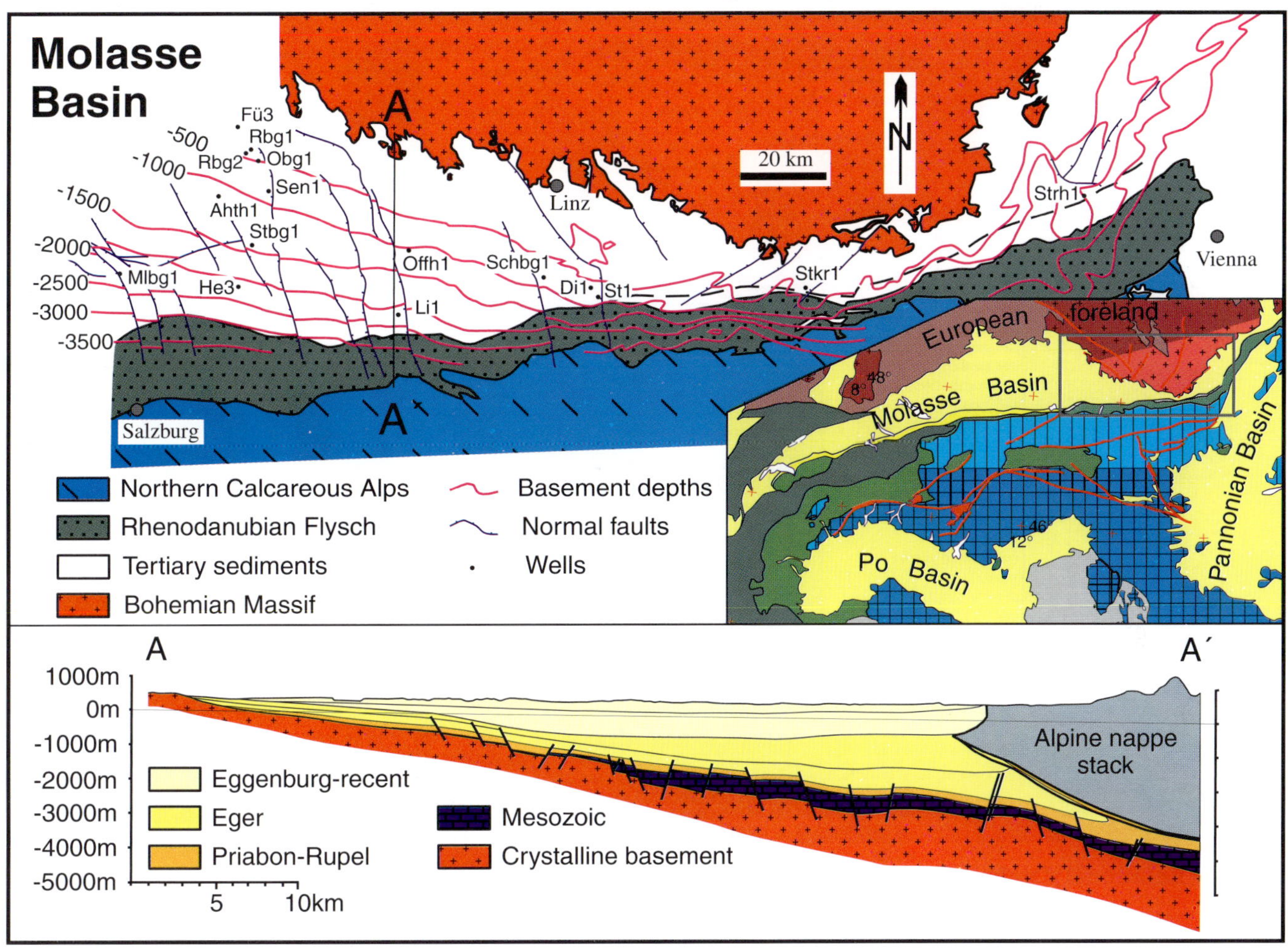

Figure 3—Tectonic map and north-south profile through the Molasse Basin, Upper Austria (modified after Genser et al., 1997).

pattern is considered to represent the reason for final sealing of fault-bounded hydrocarbon traps.

The fluid flow in the Molasse Basin consists of two components, which are both strongly influenced by the subsurface fault pattern:

- North-directed migration of hydrocarbons from the "oil kitchen" below the Alpine thrusts to the present-day traps in the Molasse Basin, having migration paths of up to 70 km. Generation started in the Miocene and is still continuing (Wagner, 1996; Schmidt and Erdogan, 1996). Major targets for hydrocarbon exploration have been seismic-mapped structures at base Tertiary and in the Upper Cretaceous.
- An east-directed fluid flow of deep groundwater at basin scale, postulated from hydrochemistry (Andrews et al., 1985; Goldbrunner, 1988; Schubert, 1996), suggesting the existence of a dynamic system at approximately 2000 m depth in the Jurassic carbonate aquifer. This deepest system has likely been active since the Pleistocene; furthermore, there exists a vertical and lateral hydrochemical layering in terms of mineralization for the deep groundwater systems in the Molasse Basin. Three aquifer systems can be distinguished with increasing depth: (1) a meteoric aquifer system at shallow depth, (2) a connate aquifer system at intermediate depth (i.e., high-mineralized formation waters without any connection to a present-day recharge area at the surface), and (3) a deep meteoric aquifer system (Upper Cretaceous sandstones, Jurassic carbonates) exhibiting a lateral hydrochemical variation. The meteoric, low-mineralized system is restricted to the northern, shallower part of the basin, whereas high-saline, high-mineralized formation waters occur in the southern, deeper part. This indicates different hydrodynamic conditions in the northern and southern part of the Molasse Basin. With exploitable aquifer temperatures of up to 100°C, this aquifer is also the main target for geothermal exploration in the Molasse Basin (Goldbrunner, 1988; Schubert, 1996).

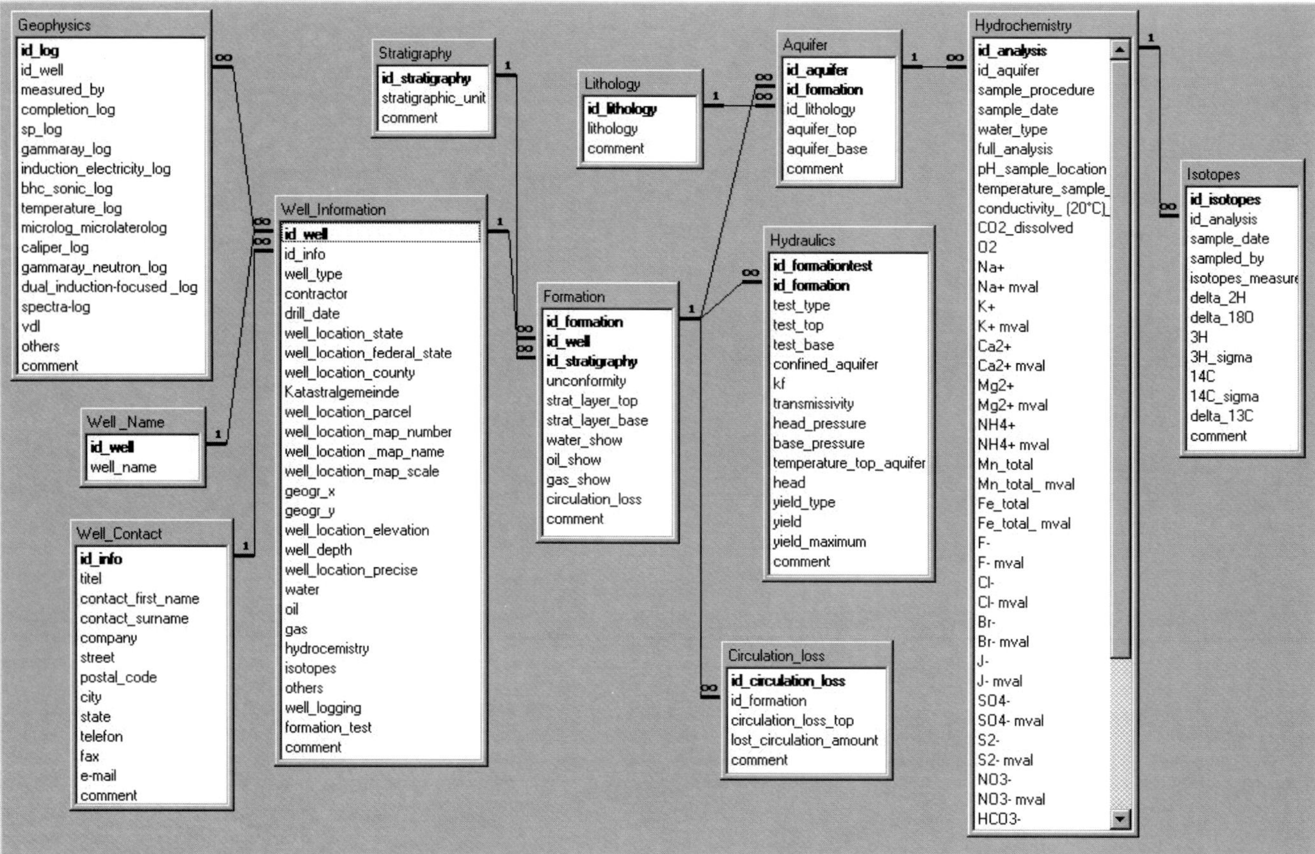

Figure 4—Relational database design for well attribute data.

INPUT DATA SETS

The dataset for this project is a heterogeneous one consisting of a mixture of analog maps, paper tables of well reports and analyses (e.g., well logs, hydraulic tests, hydrochemical analyses), and, to a minor extent, digital data (well locations). For GIS analysis, it is necessary to provide the data sets in digital form. Therefore, the analog information must be transformed or modeled using either a vector model or a raster model as the appropriate digital representation. In both models a graphic element has a link to an attribute record that contains information about that particular graphic element. In the case of the vector model, points (e.g., wells), line segments (e.g., faults), and polygons (e.g., reservoirs) are used as graphic elements to model spatially distributed data. In the raster model, an object is organized into regular grid cells where each grid cell is assigned one number (e.g., geochemical anomalies, certain map units). If a high degree of precision is required for capturing data for point or line features, then a very fine grid resolution may be required, but that also necessitates having very large amounts of data (Laurini and Thompson, 1992). Choosing the model for representation depends on the characteristics of the objects to be modeled and on the type of spatial analysis to be performed. Also, transformations between vector and raster models are possible. Vector models are very efficient for representing objects with low spatial variability, such as map units with constant values over large areas, whereas raster models are used to represent data with high spatial variability, such as sensed images or digital elevation models (Chung and Fabbri, 1993).

The first step in GIS analysis is to create a comprehensive, integrated digital GIS data set where graphic elements are linked to attribute records. These attribute records are organized in an RDBMS to provide very efficient data storage, data security, and data query capabilities. This build-up of a GIS database and the design of the appropriate RDBMS structure is a very sensitive, cost effective, and time-consuming process, and its effectiveness largely depends on subjective expert decisions. Important questions may concern each of the following: the data model, data quality, selection of attributes, RDBMS design, resolution, and additional capture and storage of metadata. It should also be strongly emphasized that every decision in this phase has large, subsequent implications on the entire GIS project, and any one of them has the potential to determine its eventual success or failure.

The real-world objects that are analyzed in the present fluid migration project are well data, fault data, and reservoir data. They are modeled using both the vector and raster approaches.

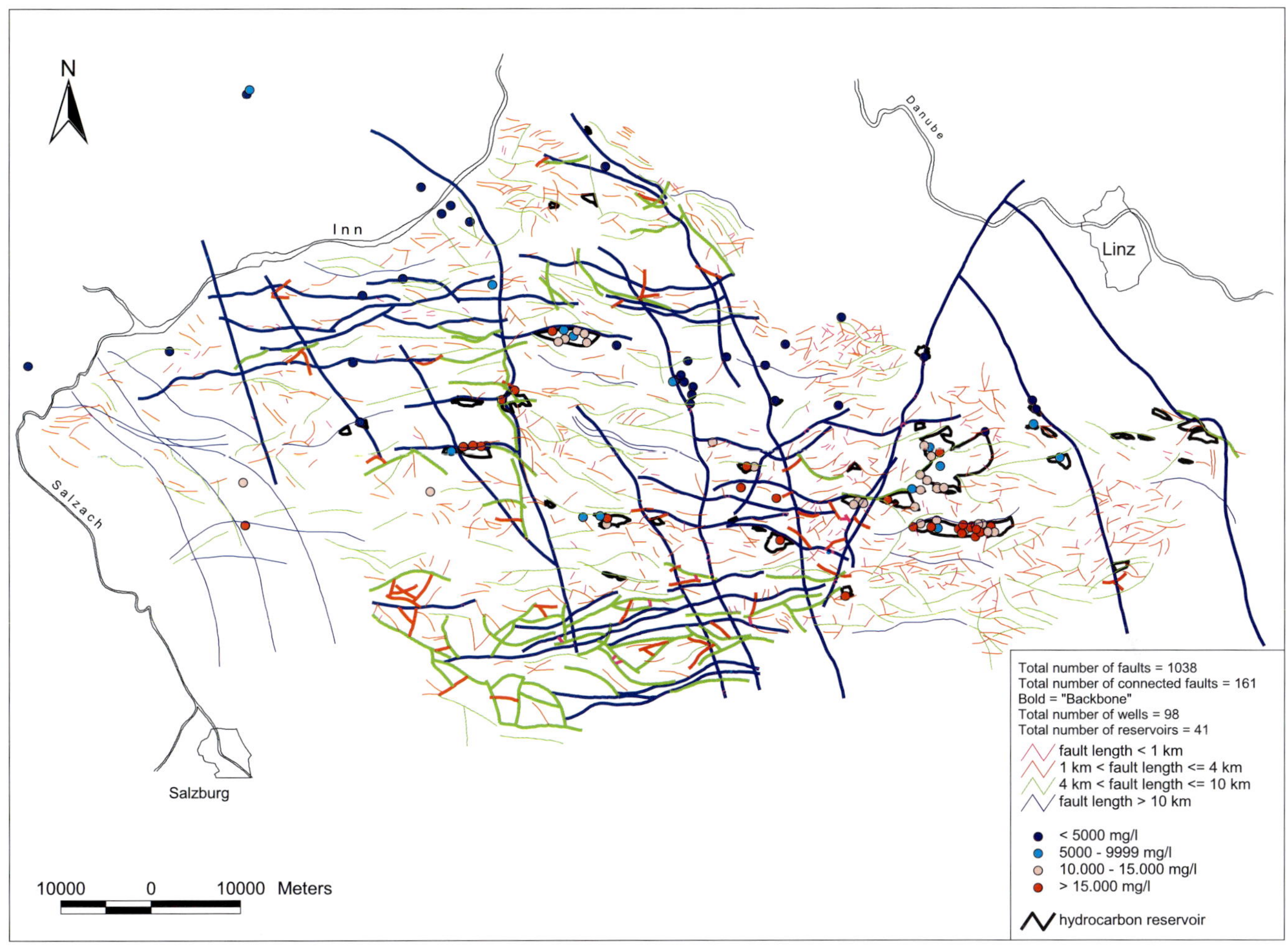

Figure 5—Distribution of input data for spatial analysis in the study area (i.e., wells, faults, and hydrocarbon reservoirs).

Well Data

Figure 4 shows the RDBMS database design of the well data and its corresponding attributes. Here, the different well attributes are organized using one-to-many relationships between single tables with the same identifier in each table. For example, one well may contain many formations, or a single formation within a well may contain many aquifers or hydrocarbon-bearing layers at different depths. All wells for which hydrochemical data from Jurassic to base Tertiary formations are available have been selected and used as input for spatial analysis (Figure 5).

Subsurface Fault Data

More than 500 wells, along with compiled and interpreted maps based on seismic profiles, were used for the regional analysis of the subsurface fault data and reservoir distribution (Nachtmann and Wagner, 1987). The fault maps were scanned and georeferenced by using the exact coordinates of selected wells as pass points for an affine transformation. The faults were digitized as line strings and linked to a database table containing various attributes, such as strike, dip-direction, fault length, and whether a fault is part of the backbone of the fault network. Mean strike directions were calculated by using the coordinates of the first and last vertex of a line-string as input, and then calculating the arctangent. Fault lengths and their database records were determined by a standard GIS function after digitizing. Faults were selected according to their length using structured query language (SQL) database calls against the fault attributes stored in an RDBMS. The results can then be graphically displayed on a digital map (for example, see Figure 5). (For example, a query for selection of all faults from the attribute table "fault" with a fault length > 1 km and < 4 km has to be translated into the following SQL statement: SELECT * FROM FAULT WHERE FAULT_LENGTH IS BETWEEN 1 AND 4.)

Table 1 shows descriptive statistics on the pre-Tertiary and Tertiary subsurface faults. The connected part of the fault network ("backbone") has been detected only by visual analysis, but it was not used for any weighting procedure for the calculation of fault density distributions (de Marsily, 1985).

Table 1. Descriptive Statistics on Subsurface Faults, Molasse Basin

Subsurface Faults, Molasse Basin, Upper Austria	Strike (°)	Dip (°)	Fault length (km)
Observations	1038	1038	1038
Mean vector (m)	91.34	215.13	
Length of mean vector (r)	0.78	0.12	
Circular variance	0.22	0.88	
Circular standard deviation	40.16	116.93	
Mean			4
Standard deviation			5.8

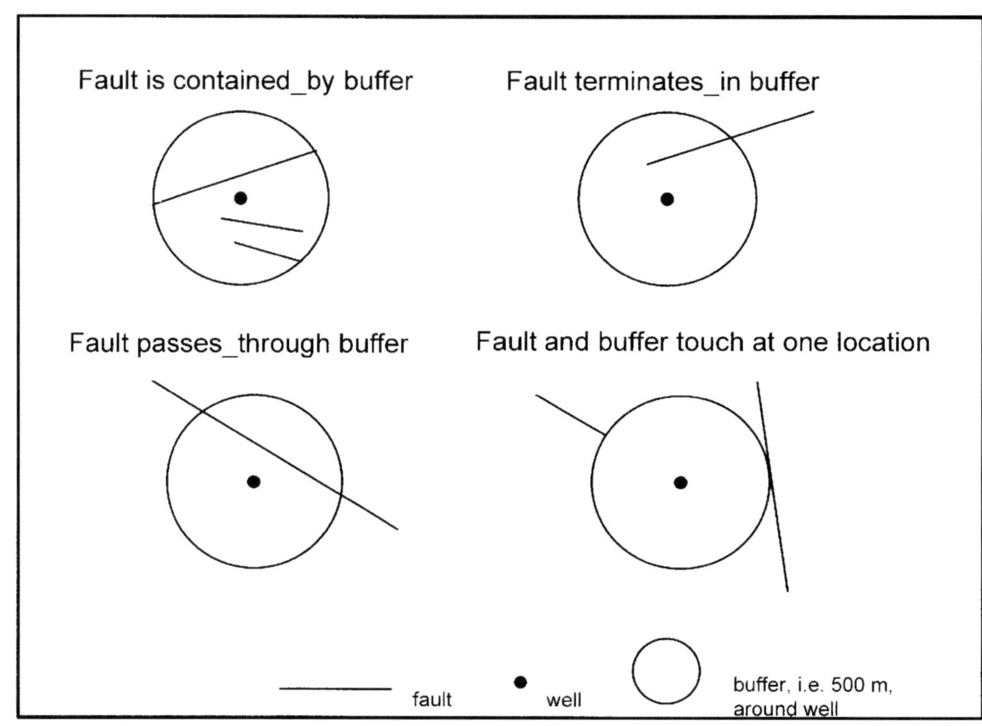

Figure 6—Selected proximity functions between faults (line segments) and a buffer (circle = special representation of a polygon)—This combined spatial operator has been used in the topological analysis.

Hydrocarbon Reservoir Data

In this project, the reservoir attributes are simple ones. They are the name of the reservoir and its area, which is also calculated by a standard GIS function. Figure 5 shows a detailed overview of the study area with the input data wells (classified in terms of mineralization of formation waters), faults (classified in terms of fault length), and base Tertiary reservoirs, which will be used in the following sections to demonstrate various examples of spatial analysis.

APPLICATIONS OF GIS ANALYSIS

Spatial Relationship Between Faults and Selected Well Locations

For many geoscientific applications, it is important to know the spatial relationship between different data objects of interest, such as wells, faults, and reservoirs. Such a relationship between these objects and their relative positions can be easily analyzed by a GIS. The concept behind this kind of spatial analysis is called topology, and it refers to those characteristics of a data object, such as adjacency, connectivity, and containment, that are not affected by spatial coordinate transformations (Bonham-Carter, 1996). In such an analysis, relative spatial relationships between the three principal map features—area (e.g., a reservoir), lines (e.g., faults), and points (e.g., wells)—are translated in a mathematical way to obtain the so-called topological structure. This structure uses faces (or polygons), edges (or chains and arcs), and vertices (or nodes) as topological entities to represent the real world (Laurini and Thompson, 1992). In the context of topological vector overlay, it is important to consider that more than one realization for the spatial relations between a line and a polygon will exist. Following Egenhofer et al. (1993) and Mark and Egenhofer (1992), 19 realistic topological relations between a simple line and a polygon exist.

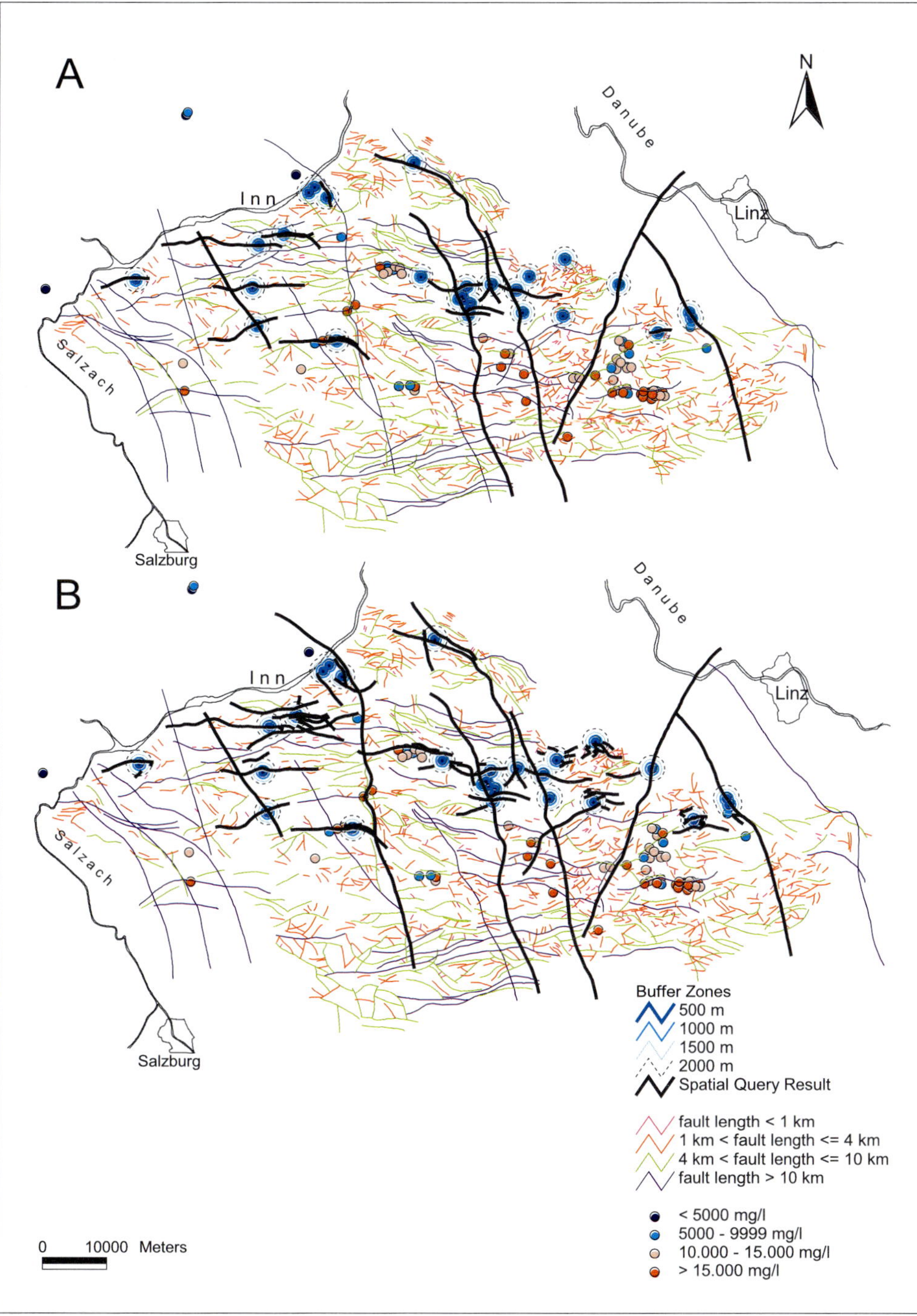

Figure 7—Selection of faults in the proximity of wells with low mineralized formation waters using a combined spatial operator (see Figure 6)—(A) Buffer zone radius = 500 m. (B) Buffer zone radius = 2000 m. The following number of faults have been selected for different buffer zones radii: 500 m, 22 faults; 1000 m, 33 faults; 1500 m, 56 faults; 2000 m, 76 faults.

Figure 6 shows the four proximity functions between faults (line segments) and a buffer zone (circle = special representation of a polygon) that have been considered in the present analysis. These four cases were combined into one single spatial operator, which was the selection criteria for the topological analysis.

SQL queries are used first to select those wells where a low total mineralization of formation waters [total dissolved solids (TDS) < 5000 mg/l] is observed. In a second step, four buffer zones ranging from 500 to 2000 m in diameter are constructed. In a third step, the relationship between the various buffer areas and the faults in their vicinity is analyzed by vector overlay. The main question is whether low TDS content in wells and the proximity of faults is spatially correlated and how such a fault pattern will look.

Figure 7 depicts the topological vector overlay between buffer zones of various radius (500, 1000, 1500, 2000 m) and subsurface faults. The number of selected faults increases with increasing buffer zone radius. The result of this analysis provides qualitative information as to which wells have similar hydrochemical features (e.g., low mineralization in formation waters) that may possibly be hydraulically

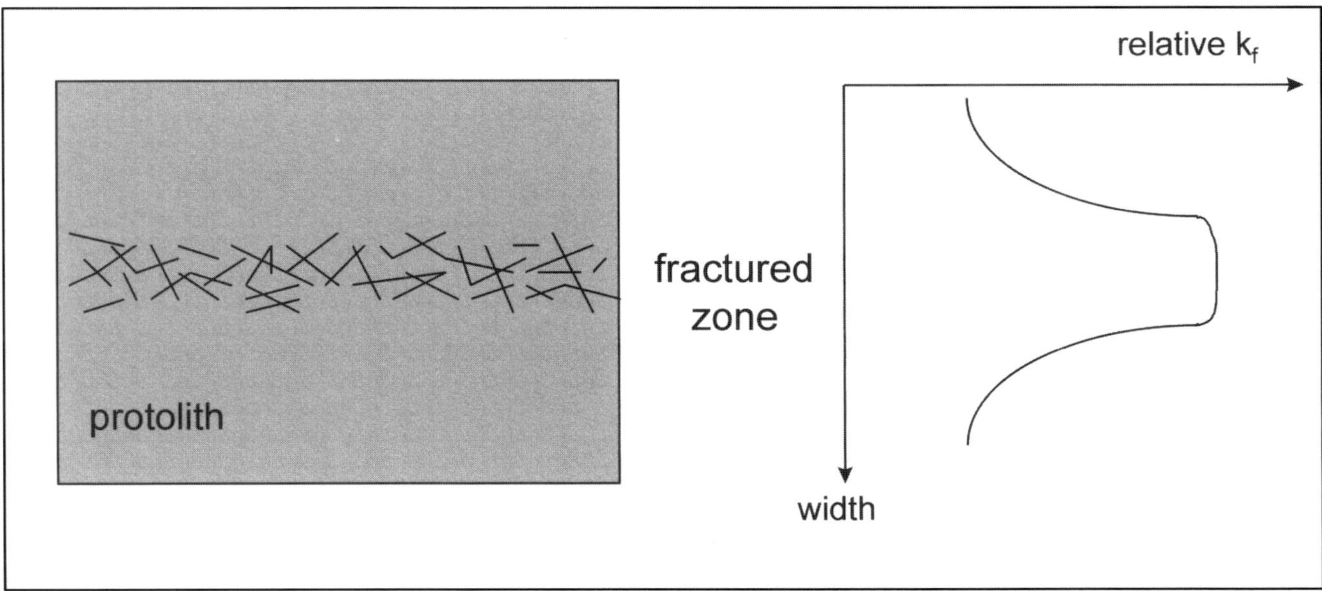

Figure 8—Theoretical mean variation of permeability across a fault zone (modified after Smith et al., 1990).

connected by faults and, therefore, may act as potential conduits within a fault network. This is important information for efficient development of adjacent hydrocarbon reservoirs, and it may help regulatory agencies establish better pumping rate limits for geothermal or irrigation use of deep groundwater to prevent over-exploitation of such systems.

A shortcoming of this method is that the dimensioning of buffer size is arbitrary, based on statistical assumptions rather than on physical properties, and it clearly depends on the scale of investigation. In addition, there are no reliable rules of thumb for combining spatial operators in ways that will best accomplish such an analysis.

Fault-Related Permeability Distribution and Fluid Flow

Fault structures are defined as discrete zones within rock formations where cohesion has been diminished or lost by brittle deformation. Such fault zones, therefore, often provide preferred pathways for fluid migration. Faults are represented on geological maps as discrete line segments with sharp boundaries; but in nature, fault zones show a high degree of deformation at their centers, which decreases with increasing distance from the fault center. As illustrated in Figure 8, a simple approximated permeability profile across a fault zone will follow the same distribution with a maximum of permeability in the fault center and a decrease of permeability with increasing distance from the fault center (Smith et al., 1990). Larger scales and more complex permeability profiles can be expected due to fault gauge or lithologic variation.

The following parameters are important, regardless of whether faults act as conduits for fluids at the basin scale:

(1) *Connectivity of a fault network* (de Marsily, 1985): The connectivity of a fault network is described by percolation theory. A critical percolation threshold, p_c (a measure for the connectivity of a fault system), may exist that depends on the density of faults and a sufficient number of intersections between them. Fluid flow in fault networks is often concentrated only in some parts of the system because many faults in a fault network are either isolated, form unconnected single-fault clusters, or contribute as "dead-end road-faults" to the storage capacity of the system without conducting any fluid. In the present example, the number of potentially connected faults at basin scale derived from simple visual analysis is 161 out of 1038.

(2) *Fault intersections* are another important reason for enhanced fluid migration (Harding and Tuminas, 1989).

(3) *Fault length*: The fault length has a positive qualitative correlation with the width of rock deformation and the existence of smaller faults in the vicinity of a major fault (Gauthier and Lake, 1993).

(4) *Sealing potential and juxtaposition* (e.g., Knipe, 1997).

A method for modeling a normally distributed deformation (or a permeability distribution) across a fault zone is the so-called "tunnel method." This method has been used so far to model and calculate the positional accuracy and density distribution of satellite image lineaments (Buchroithner et al., 1990; Heiss, 1992). The principle of this method is that a "tunnel," or a normal distribution, is calculated for every single fault. The height and width of the tunnel are variables and are defined as pixels. The tunnel vertex represents the maximum and corresponds with the fault center

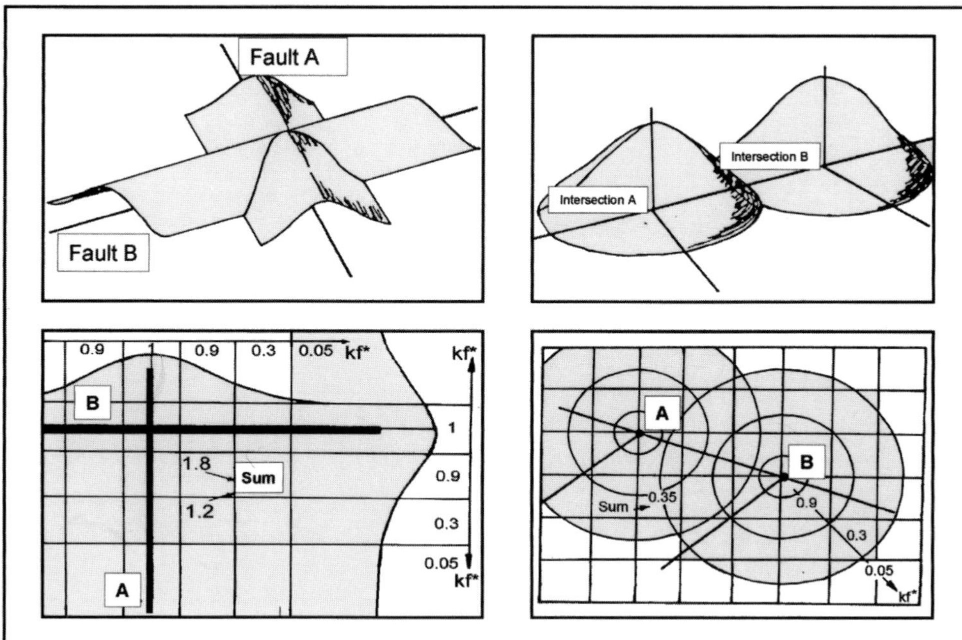

Figure 9—Principle of the "tunnel method" used to calculate spatial density distributions of faults and fault intersections (modified from Heiss, 1992). Note: kf* = relative permeability.

from where the influence will decrease to zero following a normal distribution. This procedure is performed for every single fault, and the results of the individual calculations are summed up. The fault intersection density is calculated in a similar way, but this time only one-half of a normal distribution is rotated around a vertical axis through a fault intersection (Figure 9).

This method has been used to describe the subsurface fault network affecting the strata from crystalline basement to base Tertiary in the Molasse Basin to detect areas and zones with high, fault-related fluid flow potential. The four fault-length classes are weighted with higher tunnel factors (i.e., wider tunnels) for the longer faults and lower tunnel factors for the shorter faults. The weights in width of the tunnel range from 500 m on each side of faults less than 1 km in length to 3000 m on each side of faults greater than 10 km in length (Figure 10).

The fault intersections were not weighted for two reasons: (1) the number of detected fault intersections in the study area is much less than the number of faults (any weighting procedure and refinement in the classification will not provide a remarkable increase in information compared to the non-weighted analysis) and (2) it is very difficult to define meaningful weights for fault intersections unless the hydraulic properties of the intersecting faults themselves are known well enough (therefore the same radius of 5000 m has been assigned around every fault intersection) (Figure 11).

Additional information about whether a loss of circulation in the Jurassic carbonates was observed in a well was added by simple overlay. The loss of circulation in the carbonates shows both positive and negative correlation with the calculated fault densities. Four end members for the qualitative description of relative hydraulic fault properties can be established using this kind of spatial analysis:

- *Loss of circulation + high fault density:* fault may act as potential conduit.
- *Loss of circulation + low fault density:* loss of circulation is not bounded on a fault, but, for example, on karstification; or loss of circulation is bounded on a fault, but the fault has not been detected.
- *No loss of circulation + high fault density:* faults may act as potential barriers and may be sealed.
- *No loss of circulation + low fault density:* these areas may represent potential non-structural hydraulic barriers because of lithology or facies differences.

Modeling Fluid Migration Using Cost Surfaces

In this section the application of so-called *cost surfaces* will be discussed and demonstrated. Cost surface modeling deals with the cost of moving through space, where costs are a function of both the standard costs associated with movement, and of frictions and forces that impede or facilitate that movement (Eastman, 1997). Costs are calculated as distance equivalents originating from a source considering certain amounts of friction. They are radially determined from one or several sources. In a nine-cell pattern, for example, movement is possible from one grid cell to each of the eight surrounding cells, but the costs for diagonal movement are higher by 1.41 times the friction value. Similarly, a distance equivalent to 100 corresponds to the movement over 100 cells with the relative friction 1, or 20 cells with a friction of 5 or 1 cell with the friction 100. Simply spoken, a cost surface can be seen as a hilly landscape where movement straight up a hill will require more energy (= costs = distance equivalents) than following a path along the plain.

The connection between cost, surface analysis and fault-related fluid migration at basin scale has to do

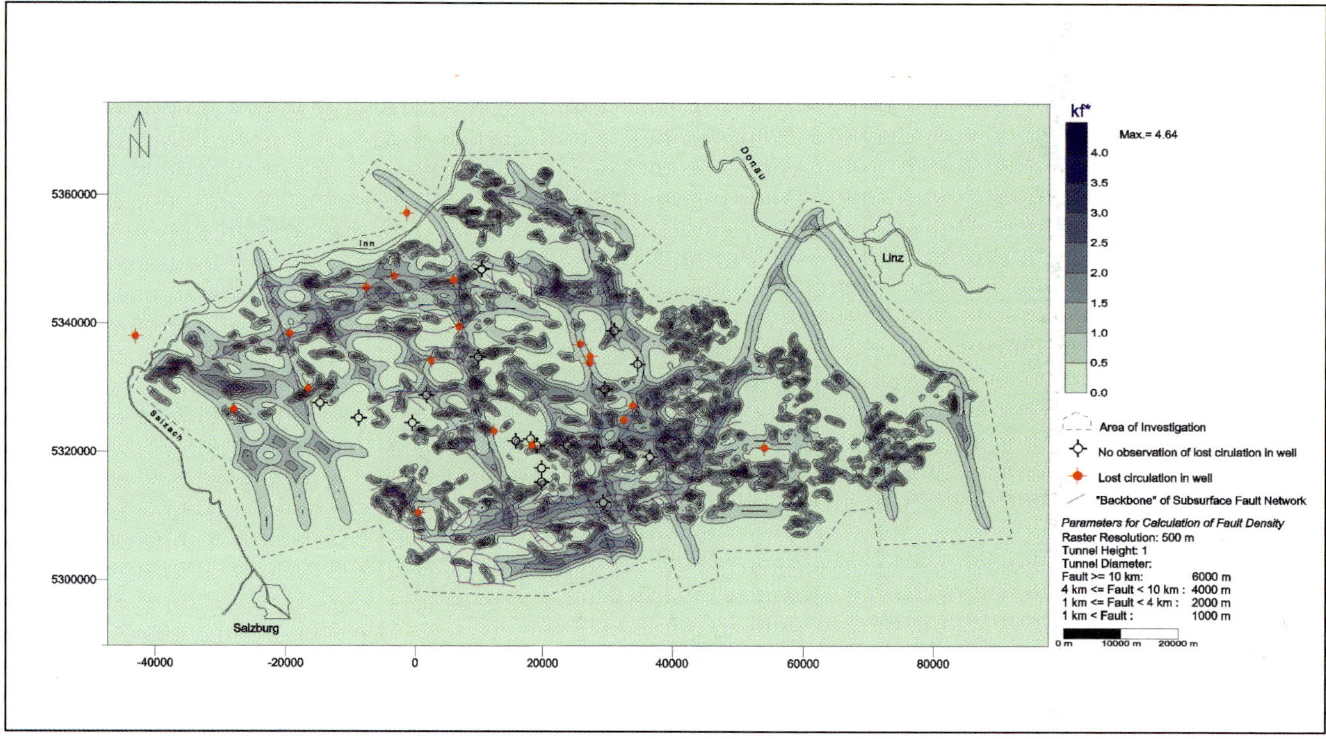

Figure 10—Distribution of fault density (kf* = relative permeability).

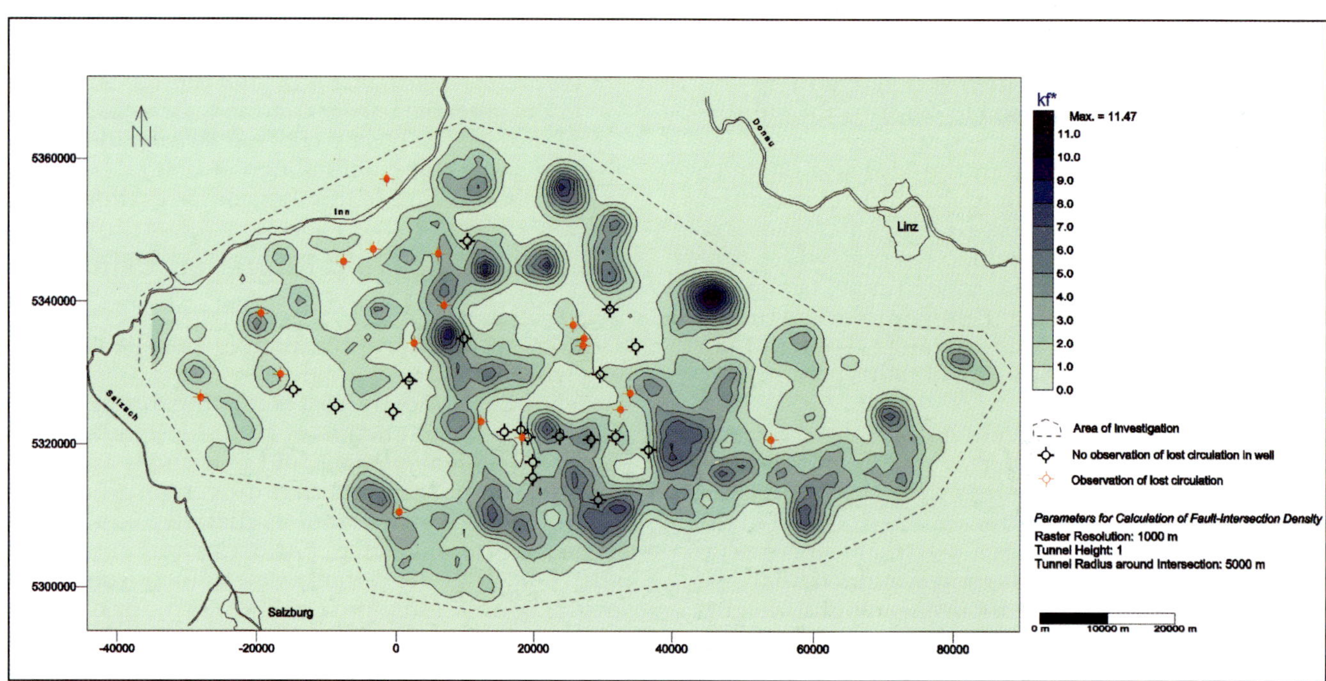

Figure 11—Distribution of fault-intersection density (kf* = relative permeability).

with the hypothesis that fluids will more likely use zones with high fault density than zones with low fault density as preferred flow paths (i.e., flow corridors). For calculating a fluid flow cost surface, the inverse fault density distribution is used as a friction surface (high fault density equals low friction and vice versa).

For the present example, areas of very low fault density were modeled as absolute barriers, whereas all other cells were assigned a very low friction. Furthermore, a shortest path analysis was performed to visualize the shortest path for fluid flow from east to west to simulate the postu-

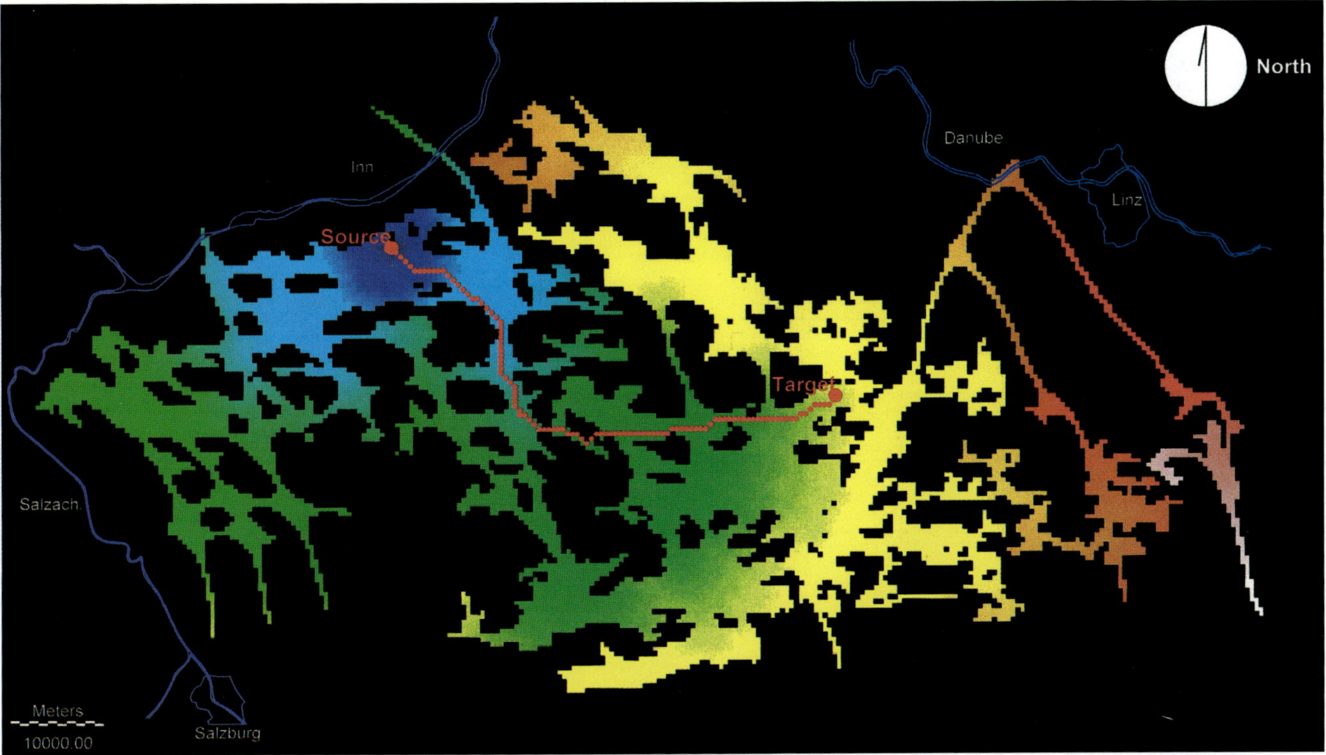

Figure 12—Cost surface modeling and shortest path of potential fault-related fluid flow from west to east. Fluids will require only low costs in the bluish colored areas to migrate from a specified source through the subsurface fault network to a specified target, whereas the reddish colors indicate high costs.

lated east-directed flow system in the Jurassic carbonate aquifer (Figure 12).

Spatial Overlay Between Hydrocarbon Reservoirs and Fault Density Distribution

As previously discussed, one of the most widely used analytical functions of GIS is spatial overlay. For the present study, one of the most important questions concerns the spatial distribution of lower Tertiary hydrocarbon reservoirs in the Molasse Basin and its relationship to the distribution of calculated fault density. Spatial overlay was used to address this question. The following tasks were performed:

(1) Raster overlays between reservoir images and fault-density images (resolution 100 m) were constructed using simple addition of both source images, thereby resulting in a single aggregated target image.
(2) Fault densities for the reservoirs were determined by applying the reclassification and histogram functions (Table 2) to the target image (Figure 13). As a first result from this analysis at basin scale, most of the reservoirs (86%) showed low fault density values ranging from 0.1 to 1.9. This result was actually expected because a reservoir requires sealing structures and low fault-densities to be effective; otherwise, hydrocarbons could not be trapped or retained. The relationship between hydrocarbon reservoirs and low fault-related permeability was further used to interpret hydrodynamic conditions in certain parts of the basin. Areas with reservoirs and low fault density were tagged as regions of low hydraulic potential, providing valuable additional qualitative information for developing an exploration strategy for the play.

CONCLUDING DISCUSSION

The main aim of this paper is to present various GIS applications that may provide new insights into existing data and the ability to extract new information from them. This may be of great advantage, especially in cases where considerable data already exist (e.g., the reevaluation of existing and known plays).

A GIS combined with an RDBMS is a modern and up-to-date tool to manage, store, and analyze these spatial data, regardless of kind, amount, and scale. Just *one* system is necessary for complex processing of these large comprehensive data sets. Special attention has to be paid to the critical GIS project stages of data capture and data transformation because the results of a GIS analysis are always only as good as the input data. Special attention should also be paid to creating the appropriate data classifications and classes. In the best case, such decisions on data limits and thresholds should be made by an interdisciplinary team using GIS as the connecting link. Every assumption and

Table 2. Frequency Distribution of Fault Density in Hydrocarbon Reservoirs at Base Tertiary, Molasse Basin

Fault Density	Frequency for Reservoirs in pixels (100 m)	%
0 (background)	1,126,284	
0.1–0.9	5,092	66
1.0–1.9	1,756	22.8
2.0–2.9	657	8.5
3–3.71 (reservoir maximum)	211	2.7
Minimum study area	0	
Maximum study area	5.81	

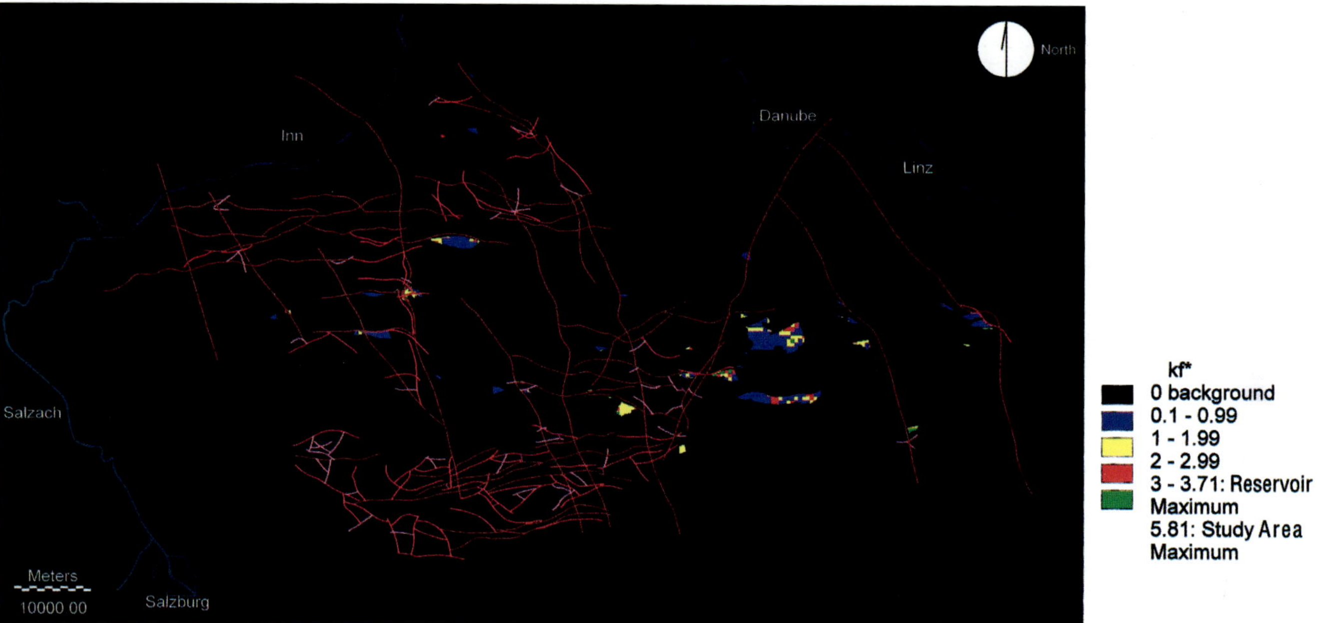

Figure 13—Fault densities in hydrocarbon reservoirs at base Tertiary, Molasse Basin, upper Austria (kf* = relative permeability).

every mistake made at this stage will be promulgated throughout the entire spatial analysis process, which can have a large impact on the results. For example, for the fluid migration study presented here, two important pieces of information have not been considered due to lack of availability of the data. The parameters are the Miocene reactivation of pre-Tertiary and lower Tertiary fault structures and the sealing potential of the subsurface faults.

Sufficient time should be devoted to project documentation and metadata description. To return to the analogy of the "GIS rocket," where the data is the "fuel," an efficient metadata description is the radar system for better orientation. Continuous documentation is further desirable for reasons of transparency, and it will lead to enhanced objectivity and comprehension.

The whole field of GIS is a very dynamic and fast-growing one. Recent developments, such as the foundation of the Open GIS Consortium, are the driving power of this dynamic evolution. All the main information technology (IT) players, e.g., Intergraph Corporation, Oracle Corporation, Sun Microsystems, Environmental Systems Research Institute (ESRI), IBM Corporation, and others, are members of this consortium. Their representatives come together to discuss and develop common standards for further IT developments, so that the exchange of data between different platforms will no longer be a point of serious contention.

Another interesting element in geoscience database technology is the so-called Spatial Cartridge available from Oracle Corporation. It allows the management of multidimensional or spatiotemporal data (i.e., geographic location + time stamp + attribute value) to be undertaken within *one* RDBMS table structure with the consequence that spatial analysis will possibly be directed at the RDBMS level (Oracle Corporation, Inc., 1995). Nevertheless, it should always be kept in mind that such complex natural subsurface systems such as sedimentary basins can never be described and simulated in sufficient detail because the input parameters needed for any kind of spatial modeling are never totally complete or accurate. The GIS analytical methods presented here thus can provide valuable tools to help one understand the processes in such a system. It must still be kept in mind that verification and validation of spatial models on naturally occurring systems is impossible because natural systems are never closed (Oreskes et al., 1994; Mazor, 1995).

ACKNOWLEDGMENTS

I would like to express my appreciation to Erich Riegler, Institute of Digital Image Processing, Joanneum Research, Graz, Austria, for his great help in calculating the fault-related permeability distributions. I also thank Josef Strobl, Eric Lorup, Franz Neubauer, Johann Goldbrunner, Achim Schubert, and an anonymous reviewer for their valuable comments and discussions on this paper. The GIS team at the Department of Geography, University of Salzburg, have provided an excellent and productive working environment. This research project was generously supported by a grant from Intergraph Corporation Austria, which is gratefully acknowledged.

REFERENCES CITED

Andrews, J. N., J. E. Goldbrunner, W. G. Darling, G. B. Hooker, G. B. Wilson, M. J. Youngman, L. Eichinger, W. Rauert, and W. Stichler, 1985, A radiochemical, hydrochemical and dissolved gas study of groundwater in the Molasse Basin of Upper Austria. Earth and Planetary Science Letters, vol. 73, p. 317–332.

Berry, J., 1993, Beyond Mapping: Concepts, Algorithms and Issues in GIS. Ft. Collins, CO: GIS World Books, 246 p.

Blakemore, M., 1984, Generalization and error in spatial databases. Cartographica, vol. 21, p. 131–139.

Bonham-Carter, G. F., 1996, Geographic Information Systems for Geoscientists: Modelling with GIS. Oxford: Pergamon Press, 398 p.

Buchroithner, M., E. Riegler, and D. Strobl, 1990, Computergestützte graphische analyse von fernerkundeten geologischen elementen. 3. Symposium Tektonik-Strukturgeologie-Kristallingeologie, April 19–21, 1990, Graz, Austria, p. 31–35.

Chung, C. F., and A. G. Fabbri, 1993, The representation of geoscience information for data integration. Nonrenewable Resources, vol. 2, p. 122–139.

de Marsily, G., 1985, Flow and transport in fractured rocks: connectivity and scale effect, *in* Hydrogeology of Rocks of Low Permeability (S. P. Neuman and E. S. Simpson, comps.). Montpellier: International Association of Hydrogeologists, p. 267–277.

Eastman, J. R., 1997, IDRISI for Windows User´s Guide, Version 2.0. Worcester, MA: Clark Labs, Clark University, 223 p.

Egenhofer, M., J. Sharma, and D. Mark, 1993, A critical comparison of the 4-intersection and 9-intersection models for spatial relations: formal analysis. International Journal of Geographical Information Systems, vol. 5, p. 161–174.

Gauthier, B. D. M., and S. D. Lake, 1993, Probabilistic modeling of faults below the limit of seismic resolution in Pelican field, North Sea, offshore United Kingdom. AAPG Bulletin, vol. 77, p. 761–777.

Genser, J., S. Cloething, and F. Neubauer, 1997, Slab detachment, delamination and oblique Alpine convergence: New constraints from subsidence analysis of the Austrian Molasse Basin. Unpublished manuscript.

Goldbrunner, J. E., 1988, Tiefengrundwässer im oberösterreichischen Molassebecken und im Steirischen Becken. Steirische Beiträge zur Hydrogeologie, vol. 39, p. 5–94.

Goodchild, M. F., 1985, Geographic information systems in undergraduate geography: a contemporary dilemma. Operational Geographer, vol. 8, pp. 34–38.

Harding, T. P., and A. C. Tuminas, 1989, Structural interpretation of hydrocarbon traps sealed by basement normal block faults at stable flanks of foredeep basins and at rift basins. AAPG Bulletin, v. 73, p. 812–840.

Heiss, H. P., 1992, Geologische Fernerkundung von Ausgewählten Gebieten der Suleiman Range und des Hindukusch – Pakistan. Unpublished Masters Thesis, Department of Geology and Palaeontology, University of Graz, Austria, 109 p.

Hemenway, D., 1989, GIS Observer. Photogrammetic Engineering and Remote Sensing, vol. LV, p. 402.

Heuvelink, G. B. M., and P. A. Burrough, 1993, Error propagation in cartographic modeling using Boolean logic and continuous classification. International Journal of Geographical Information Systems, vol. 7, p. 231–246.

Knipe, R. J., 1997, Juxtaposition and seal diagrams to help analyze fault seals in hydrocarbon reservoirs. AAPG Bulletin, vol. 81, p. 187–195.

Laurini, R., and D. Thompson, 1992, Fundamentals of Spatial Information Systems. New York: Academic Press, The A.P.I.C. Series, no. 37, 680 p.

Mark, D., and M. Egenhofer, 1992, An evaluation of the 9-intersection for region-line relations, *in* Proceedings, GIS/LIS '92. Bethesda, MD: American Society of Photogrammetry and Remote Sensing (ASPRS), p. 513–521.

Mazor, E., 1995, Stagnant aquifer concept: Part 1. Large-scale artesian systems—Great Artesian Basin, Australia. Journal of Hydrology, vol. 173, p. 219–240.

Nachtmann, W., and L. Wagner, 1987, Mesozoic and early Tertiary evolution of the alpine foreland in Upper Austria and Salzburg. Tectonophysics, vol. 137, p. 61–76.

Nachtmann, W., 1995, Fault-bounded structures as hydrocarbon traps in the Upper Austrian Molasse Basin, Austria. Geologische Paläontologische Mitteilungen Innsbruck, vol. 20, p. 221–230.

Oracle Corporation, Inc., 1995, Oracle7 Multi-Dimension: Advances in Relational Database Technology for Spatial Data Management. Redwood Shores, CA: Oracle Corporation, Oracle White Paper, 31 p.

Oreskes, N., K. Shrader-Frechette, and K. Belitz, 1994, Verification, validation, and confirmation of numerical models in the earth sciences. Science, vol. 263, p. 641–646.

Parker, D., 1993, GIS, the umbrella technology, *in* International GIS Sourcebook. Ft. Collins, CO: GIS World, Inc., p. 1–2.

Schmidt, F., and L. T. Erdogan, 1996, Paleohydrodynamics in exploration, *in* Oil and Gas in the Alpidic Thrustbelts and Basins of Central and Eastern Europe (G. Wessely, and W. Liebel, eds.). Houten, The Netherlands: European Association of Geoscientists and Engineers, EAGE Special Publication no. 5, p. 255–265.

Schubert, A., 1996, Grundwasserhydrogeologische Untersuchungen im Tiefengrundwasserentlastungsgebiet des Molassebeckens westlich von Linz. Unpublished Ph.D. Thesis, Department of Applied Geosciences II, Technical University of Berlin, Germany, 127 p.

Smith, L., C. Forster, and J. Evans, 1990, Interaction of fault zones, fluid flow, and heat transfer at the basin scale, *in* Hydrogeology of Low Permeability Environments (S. P. Neuman and I. Netetnieks, eds.). Hannover: Verlag Heinz Heise, p. 41–67.

Wagner, L., 1996, Stratigraphy and hydrocarbons in the Upper Austrian Molasse Foredeep (active margin), *in* Oil and Gas in the Alpidic Thrustbelts and Basins of Central and Eastern Europe (G. Wessely and W. Liebel, eds.). Houton, The Netherlands: European Association of Geoscientists and Engineers, EAGE Special Publication No. 5, p. 214–236.

Yarus, M. Y., and R. L. Chambers, 1994, Stochastic Modeling and Geostatistics. Tulsa, OK: American Association of Petroleum Geologists, 379 p.

Jefferis, D. R., et al., 2000, The Gap Mountain prospect: a new approach to conceptual geological modeling, *in* T. C. Coburn and J. M. Yarus, eds., Geographic information systems in petroleum exploration and development: AAPG Computer Applications in Geology, No. 4, p. 137–150.

Chapter 10

The Gap Mountain Prospect: A New Approach to Conceptual Geological Modeling

David R. Jefferis[1]
Reservoir Visualization Applications
Plains Resources, Inc.

Michael S. Connelly
Connelly Exploration, Inc.
Denver, Colorado, U.S.A.

Skip Pack
Dynamic Graphics, Inc.
Alameda, California, U.S.A.

ABSTRACT

The Gap Mountain prospect is located in the southern end of the White River Valley, Nevada. Approximately 60 km to the northwest, in Railroad Valley, lies the Grant Canyon field, which has produced nearly half of the oil extracted from Nevada fields to date. The Gap Mountain area is considered prospective for exploration of hydrocarbons, in that several structures have been identified from geologic mapping and through the analysis of gravity and magnetic data. The geology of this play is defined by Mesozoic compressional thrusting that has juxtaposed Mississippian source rock and Devonian reservoir rock. Structure is further complicated by Basin and Range extension and the generation of listric normal faults. Tertiary volcanic rocks in the area are also considered to be possible reservoirs.

To better understand the geologic domain and to evaluate the potential oil-in-place within the Gap Mountain prospect, a geo-registered 3-D conceptual geologic model was constructed using 2-D and 3-D geographic information systems (GIS). By using various spatial analysis and modeling tools of 2-D and 3-D GIS and the geologic knowledge of the explorationist, a plausible 3-D geologic model was constructed. For example, formation contacts were extracted from the geologic map as vector data and used to calculate structural dips from the digital elevation model (DEM). Because the conceptual geologic model was geo-registered, other information, such as lease boundaries, hypsography, hydrography, roads, etc., could be integrated directly with the model. These integrated data sets were then used to assess suitability and accessibility of a proposed drill site.

[1]Formerly Dynamic Graphics, Inc., Bakersfield, California, U.S.A.

INTRODUCTION

The Gap Mountain prospect, located in the southern end of the White River Valley, is in a complex geologic region. The Gap Mountain area was mapped in the field, and the geology was presented on a 2-D topographic map. Several structures were identified after mapping was completed, and thus an extensive exploration program was begun in earnest. Gravity acquisition, airborne and ground magnetic surveys, surface geochemical sampling, and telluric measurements were the tools used to grasp a better understanding of the geology and to identify a potential prospect. Much of the data gathered were useful not only in confirming the geologic interpretation but also in identifying a specific drill site. The exercise in conceptualizing the complexity of the Gap Mountain area proved to be extremely challenging and rife with contradictions.

Subsequent field trips to the Gap Mountain area were necessary, with the participants including several structural geologists, who were instrumental in identifying the inconsistencies of the observed surface geology. It became apparent that the 2-D interpretation of this area lacked coherence, and that a 3-D interpretation was necessary in order to (1) project fault planes of known faults in both compressional and extensional configurations, (2) build a stratigraphic section for each separate structural block based on field observations, (3) use the exploration geophysical data in areas where surface outcrop data or wellbore data did not exist, and (4) synthesize the data into a consistent, geometrically valid, and geologically plausible interpretation.

To facilitate the 3-D interpretation, computer-based modeling tools were used to produce a 3-D static geologic model. A static model is a mathematical representation of the current state of the geologic environment. No forward or inverse modeling was undertaken during this project to reconstruct the geologic history.

GEOLOGICAL MODELING AND GIS

Computer-aided data analysis, mapping, and modeling techniques have been used in applied geosciences for many years (Davis, 1973; Loudon, 1979; Jones et al., 1986). A significant number of these methods have been classified within the broader scope of geographical information system (GIS) technology, which extends over a vast range of disciplines (planning, public works, natural resources, etc.). One benefit of formalizing mapping and modeling under GIS is that the data structures, data analysis and modeling tools, and visualization methods have now been clearly defined. Another benefit is that it enhances the communication of how GIS technology can be used to investigate problems of a spatial nature.

It is possible to use the definitions of GIS technology to effectively communicate the approach used to develop the 3-D model of Gap Mountain. As presented by Star and Estes (1990), a GIS must encompass five basic elements:

- Data acquisition (field mapping, geochemical surveys, etc.);
- Pre-processing (data reduction, digitizing, etc.);
- Data management (data storage and retrieval);
- Data manipulation and analysis (exploratory data analysis, mapping, modeling); and,
- Product generation (maps, cross-sections, reports).

The description of what a GIS must contain establishes the systems, procedures, and methodologies used to develop the Gap Mountain model.

GEOLOGICAL SETTING

The Gap Mountain prospect lies within an area referred to as the central Nevada thrust belt. Major compressional events that involved extensive thrusting occurred along a 600 km north-south–trending belt through east central Nevada during the Sevier-Laramide orogeny (Chamberlain and Gillespie, 1993). Along this disturbed belt lies a Mississippian-age source rock fairway that extends from Elko, Nevada, to Las Vegas, Nevada. This source rock is thick (thousands of meters), thermally mature, and oil prone, with total organic carbon values averaging 4–6% (Chamberlain, 1989). The Mesozoic-age thrusting has juxtaposed excellent-quality Devonian-age carbonate reservoirs and Mississippian source rocks, possibly creating large hydrocarbon-filled structures. During Basin and Range extension, which began in the late Tertiary and continues to the present day, relaxation of stress along reactivated thrust faults caused listric normal faults. This includes the often-cited basin bounding normal faults that are largely responsible for the present-day topography in east central Nevada (A. K. Chamberlain, 1996, personal communication).

The Gap Mountain prospect is located in the southern end of the White River Valley, Nevada, and is approximately 60 km to the east-southeast of the Grant Canyon field, located in the Railroad Valley, Nevada (Figure 1). To date, the Grant Canyon field has produced approximately 20 million barrels of oil from 110 ha, accounting for nearly half of the oil production for the state of Nevada. Needless to say, exploration efforts to find more Grant Canyons have been intense at times over the past 15 years, but unfortunately, success has been very limited. Three exploratory wells have been drilled and completed as dry holes just north of the Gap Mountain area by Gulf (now part of Chevron) and Amoco (now BP/Amoco). All three wells, the latest of which was drilled in 1986 by Amoco, encountered good oil and gas shows in the Tertiary volcanics and the Paleozoic sandstones and carbonates. The Gap Mountain prospect was mapped to be updip of these wells, hence the interest in the area.

The Gap Mountain prospect is bounded to the north by the southern Egan Range, to the east by the Schell Creek Range, and to the west by Gap Mountain and Fox Mountain. Remnant thrust sheets have been identified by Tschanz and Pampeyan (1970) around

Fox Mountain and all through the Schell Creek Range. In addition, major east-to-west normal faults/lineaments have been identified by Hurtubise (1994) that bisect the Gap Mountain prospect. According to Hurtubise (1994), the Gap Mountain prospect lies within a region termed the Silver King lineament. This major shear zone trends east to west for approximately 130 km and is about 50 km wide. The Gap Mountain area has experienced at least two episodes of major tectonic events in its geologic history, which has led to complex geology and confusing geologic interpretations for the area.

Based on geologic surface mapping and ground investigations, the Gap Mountain area has been determined to have at least two major thrust faults running through it. The first thrust fault was identified at Gap Mountain and Fox Mountain where Upper Mississippian and Pennsylvanian rocks underlie Devonian-age rocks. The second thrust fault has been inferred from structural deformations that occur in the footwall plate and from mapped thrusts that surface due east of the prospect area (Tschanz and Pampeyan, 1970). Dip-slip and strike-slip faulting occurred after thrusting and have undoubtedly had a major impact on the structural formations that are present in outcrop. The small sub-basin, which comprises the Gap Mountain prospect area, is encircled by a large north-south–trending anticline, which has been breached. Reactivation of thrust faults during Basin and Range extension caused listric normal faults to occur and created this sub-basin. In summary, understanding the complex structural relationship of potential source rock and reservoir rock that resulted from polyphase deformation are critical to the success of the project and was the primary reason for developing the computer-based model.

CONSTRUCTION OF THE CONCEPTUAL GEOLOGICAL MODEL

The data and procedures used in the construction of the conceptual model are presented in terms of the elements required in a GIS, as described in the introduction. At a very high level, the modeling process can be thought of in terms of the following simple workflow procedure:

- Determining what data are needed, their availability, and methods of acquisition
- Transforming the data into usable digital formats and geo-referencing them to a common map projection
- Managing the data and reformatting them so they can be shared among the various software packages being used
- Analyzing, modeling, and fusing the data into a viable 3-D model
- Producing maps and cross-sections, calculating volumetrics, and planning of the exploratory well

DATA ACQUISITION

The data sets from which the 3-D conceptual model was constructed are as follows:

- Four interpretive geologic cross-sections on vellum media
- A geologic bedrock map drafted on a topographic base map and annotated with the location of the geologic cross-sections
- A residual aeromagnetic contour map (vellum)

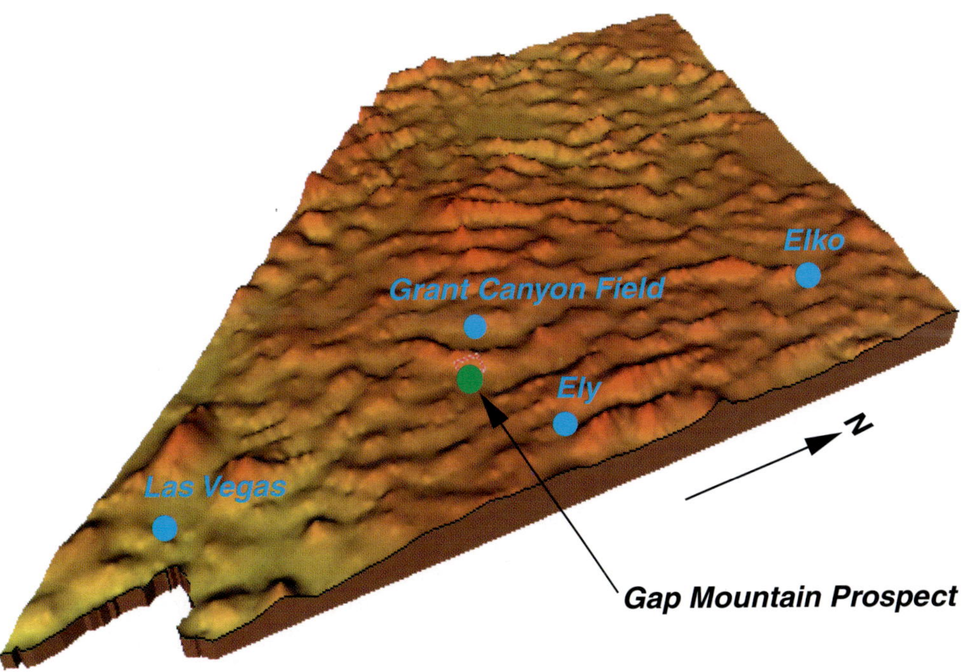

Figure 1—Shaded-relief map of Nevada showing the location of the Gap Mountain prospect.

- Results from geochemical sample analyses maintained as an ASCII data file
- A base map (paper) showing the locations of the geochemical samples
- A paper copy of a map showing lease holdings by Connelly Exploration
- A magnetic reduction anomaly drafted on a topographic base map (paper)
- A table identifying the measured (geologic) sections
- A listing of the depths to key formations observed in several wildcat wells to the north of the prospect area

The cornerstone of the 3-D geologic model was development of a digital elevation model (DEM) over the project area. Not so long ago, getting digital copies of 7.5 min DEMs from the U.S. Geological Survey (USGS) could take several weeks. This process has been reduced significantly since the 7.5 min DEMs can now be ordered online from the USGS Global Land Information System at World Wide Web (WWW) site *<http://edcwww.cr.usgs.gov/webglis>*. The USGS processes the order and then places the files on its *ftp* site for retrieval. For the Gap Mountain project area, four quadrangles (all in Nevada) were ordered: Silver King Well, Silver King Mountain, Timber Mountain Pass (NE), and Gap Mountain.

Several other online data sources were accessed to acquire additional data in digital format. It was desired to have a digital raster graphics (DRG) image that could be draped on the topographic surface of the 3-D model to provide surficial information (culture, hypsography, hydrography, etc.). The four 7.5 minute quadrangle maps previously mentioned were downloaded as DRGs directly from the WWW site *<http://www.nbmg.unr.edu/drg.htm>*. This WWW site is maintained by the Nevada Division of Mines and Geology. All of the 7.5 minute quadrangles for Nevada have been scanned at 1200 dots per inch and are in tagged image file format (TIFF). Each DRG image has an associated "world file," which provides for direct geo-registration within a GIS.

Another data source, available from the Bureau of Land Management (BLM), lists the geo-spatial location, in either Universal Transverse Mercator (UTM) or State Plane coordinates, of the townships in the Public Land Survey System. This data source is called the Geographic Coordinate Database (GCDB) and contains cadastral information for the western region of the United States. The coordinates for all of the sections in the Gap Mountain Prospect area were requested, and the BLM sent a diskette with the coordinate information in digital form. This data set was then processed so that the section boundaries defining the lease holdings by Connelly Exploration could be overlain onto the 3-D model. Details about the GCDB can be found at *<http://wwwsgi.ursus.maine.edu/gisweb/spatdb/urisa/ur94040.html>*. For more information on ordering data from the GCDB, access the WWW site at *<http://www.blm.gov/gcdb/>*.

DATA PREPROCESSING AND REDUCTION

Two software products were used to perform the bulk of the data preprocessing, data analysis, modeling, visualization, and product generation. Arc/Info, a product of Environmental Systems Research Institute Inc. (ESRI), was used to process the DRG images; and EarthVision, a product of Dynamic Graphics Inc. (DGI), was used for digitizing geological maps and cross-sections, processing the USGS DEMs, modeling 2-D surfaces and 3-D volumes, calculating 3-D volumetrics, performing 2-D and 3-D visualization, and generating maps and cross-sections.

Digital Elevation Model and DRG Image

The project area-of-interest (AOI) was defined by the following geographic (longitude-latitude) coordinates:

- Lower-left corner: –115°07′00.15″ 38°09′34.78″
- Upper-left corner: –115°06′45.07″ 38°19′18.50″
- Upper-right corner: –114°56′27.71″ 38°19′08.10″
- Lower-right corner: –114°56′44.15″ 38°09′24.46″

The four USGS 7.5 min quadrangle maps and DRGs covered the following area:

- Lower-left corner (Timber Mountain Pass-NE): –115°07′30″ 38°07′30″
- Upper-left corner (Gap Mountain): –115°07′30″ 38°22′30″
- Upper-right corner (Silver King Well): –114°52′30″ 38°22′30″
- Lower-right corner (Silver King Mountain): –114°52′30″ 38°07′30″

To create the topographic surface from the DEM data, the four DEMs were merged into a single grid, then a smaller grid covering the project area was extracted from the merged grid. The same procedures were followed to create a DRG image for the project—the four DRGs were first mosaicked into a composite image, then a smaller image, based on the AOI for the conceptual model, was extracted.

Bedrock Geology

The bedrock geologic map of the Gap Mountain Prospect was geo-registered and then digitized. For this project, all data were geo-referenced using a UTM map projection. The geologic units were digitized as polygons and the faults as vectors. The modeling methods employed required partitioning the geological map into a fault block map. An important step in this process was identification of the hanging-wall/footwall relationships and how the faults intersected each other. All faults were given names, and dips were estimated and digitized. The digitized geological data were then plotted at the same scale as the original source map, and the two were compared for accuracy. The geographic location of the geochemical samples and cross-section traverses were also digitized off the geological map and verified by plotting and comparing against the source map.

Geologic Cross-Sections

Next, the horizons and faults were digitized from the geological cross-sections by using a seismic line cross-section digitizing program. Although the program uses shotpoints to indicate horizontal offset along the cross-section, for this project, shotpoint values represented distance along the cross-section traverse and depth (instead of time) to the geologic horizon or fault trace. The geological cross-sections were then geo-registered by transforming each shotpoint location to a geographic location by referencing its associated cross-section traverse file that was already geo-registered. In other words, each shotpoint was mapped onto the traverse and an x–y coordinate is returned. The result was a file containing the x–y–z coordinates of all of the horizons and faults drawn on the geologic cross-sections. During digitization, care was taken to distinguish between faults and geologic horizons. Although the horizons and faults were already named during the digitization of the geologic bedrock map, compatible names had to be given to the horizons and faults. This was done so that when the horizons and faults were surface modeled in two dimensions, the data for each horizon and fault from the cross-sections could be combined with their associated surface locations indicated on the bedrock geological map.

Aeromagnetic Map

Because the residual-aeromagnetic map was received on paper, it too had to be digitized. The aeromagnetic map was geo-registered using a geographic-to-UTM transformation. During digitization, the contour lines were automatically projected from geographic spherical coordinates (latitude-longitude) to UTM plane coordinates (x–y). Following digitization, the contour lines were plotted out at the same scale as the source map and verified for completeness and accuracy.

DATA MANAGEMENT

The power of GIS is the linkage between the spatial location of an object and the information (attributes) that describe the properties of the object (Environmental Systems Research Institute Inc., 1991). For example, a strike and dip symbol (a point) on a geologic map has a spatial location (the place where it is measured), and associated with that location are the strike and the dip measurements describing the orientation of a plane within which the point lies. Using a GIS, this point can be queried, and all of the associated descriptive information will be returned. The manner in which the spatial and attribute data are structured and stored depends on the architecture of the GIS being used. For the project described here, the focus is on the GIS toolbox, that is, the tools and methods used to create the model. The data structures employed by the application software are beyond the scope of this discussion.

All of the data acquired, reduced, and geo-registered served as input to the generation of the conceptual geologic model can be classified under the following GIS data types:

- Points—geochemical samples, measured sections, and strike/dip measurements
- Vectors—aeromagnetic contours, fault-traces, and geologic cross-sections
- Polygons—geologic bedrock map, lease boundaries, magnetic reduction anomaly
- Lattices—DEMs, 2-D surface models of faults and horizons, aeromagnetic data
- Rasters (images)—USGS quadrangle maps (DRGs)

The resulting 3-D model consists of a volume that is defined by a polyhedral structure.

DATA ANALYSIS AND MODEL CONSTRUCTION

In general, the construction of a 3-D geologic model is accomplished in three steps: development of the 3-D fault geometry, generation of the geologic structure and stratigraphy, and population of the geologic zones with rock and fluid properties. Following the completion of the 3-D geologic model, a host of outputs (products) can be generated, including volumetric calculations, inputs for reservoir simulation models, maps and cross-sections, and other information to be used in formulating well plans.

Development of the 3-D Fault Geometry

As suggested, three periods of compressional tectonics probably occurred in this region during the Paleozoic and Mesozoic, which were followed by the Tertiary Basin and Range orogeny (extensional tectonics), thus resulting in a complex geologic environment (Connelly et al., 1995). Thus, at least three imbricate low-angle thrust faults were offset by normal faulting. From a 3-D geological modeling perspective, this is an easier geometry to model than if the opposite had occurred (i.e., extension followed by compression). This relative ease in modeling is due to the software package being used to construct the geologic model, and to the number of thrust faults (two) versus the number of normal faults that were modeled (21) in the project area, rather than to the geologic interpretation of deformation history.

Modeling the Normal Faults

The 3-D geologic modeling process requires the development of a 2-D surface model (grid) of every fault in the prospect. This is different from the traditional method of automated 2-D gridding and contouring where vertical faults are represented by line segments and non-vertical faults by polygons. Looking at the fault block map (Figure 2), it is clear that each fault trace was intersected by only one of the geologic cross-sections, and several were not intersected at all. Although a fault surface model can be generated directly from the digitized cross-sections, if there is more than one transect (as in the present case), the limited transect data requires the following approach to be used for each of the normal faults:

- Assigning an average dip to each fault based on the apparent dips seen in cross-section or that had been determined by the explorationist
- Extracting each fault trace as an arc from the 2-D geologic map and passing it through the DEM to estimate the elevation at its vertices
- Calculating an instantaneous strike along each segment of the arc and determining several new points by linearly projecting updip and downdip normal to the strike
- Graphically editing the fault data to remove points that "crossed-over" due to curving along the fault trace
- Generating a 2-D surface model with appropriate review and revision as necessary

After all the 2-D surfaces of the faults were generated, and prior to the construction of the 3-D fault model, the faulting history was defined by a fault hierarchy diagram that described the cross-cutting relationships. An EarthVision program, called the Geologic Structure Builder, used the fault hierarchy to automatically perform all of the 2-D surface intersections and then integrate them into a 3-D fault model. Following the generation of the 3-D fault model, the following steps were performed:

- Cross-sections were generated from the 3-D fault model along the same traverses as the original set of geologic cross-sections, plotted at scale, and visually inspected
- Discrepancies in the fault model were resolved either by reinterpretation of the original cross-sections and/or by modification of the dips

At this stage, the 3-D fault model of the normal faults did not contain knowledge about fault throw or offset. This information was obtained at a later stage with the geological modeling of the horizons. Exceptions occurred when older normal faults were intersected and offset by younger faults, with such offsets being represented by the fault hierarchy. This situation is discussed in more detail in the next section.

Modeling the Thrust Faults

Two imbricate low-angle thrust faults have been mapped in the Gap Mountain area. These are referred to as the "X-Thrust" and "Y-Thrust," with the X-Thrust being older than the Y-Thrust. Although the Y-Thrust crops out within the project area, the X-Thrust does not, which limits the availability of spatial data from

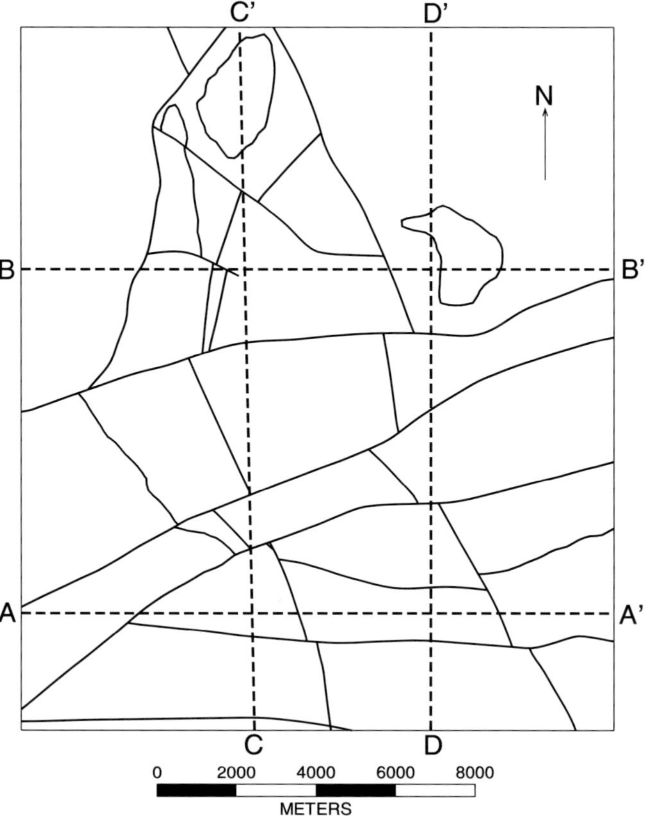

Figure 2—Map showing the traces of the faults at the surface and the locations of the interpretive geologic cross-sections.

the geologic map to help define its geometry. Therefore, it was necessary to rely on the interpretative geologic cross-sections and the imbricate relationship between X-Thrust and Y-Thrust to model it.

The thrust faults were modeled in much the same way as the normal faults, but the process was further complicated because they had been offset by normal faulting. Instead of building a single 2-D surface model for each thrust fault, a 2-D surface model had to be constructed for each thrust fault in each normal fault block. This process produced 22 surfaces for each thrust, for a total of 44 grids (there were 21 normal faults, resulting in 22 fault blocks). The fault hierarchy that defined the normal fault geometry had to be modified prior to building the 3-D fault model of the Y-Thrust. The procedure for generating the model of the Y-Thrust consisted of the following steps:

- Arcs representing the Y-Thrust were extracted from the geologic bedrock maps with elevations being back-interpolated along the vertices (including contacts defining klippes and fensters)
- Fault blocks were separated, and the map data was merged with its associated cross-sections
- 2-D surface models were generated
- A 3-D fault model of the Y-Thrust was constructed along with the normal faults

- Cross-sections were generated, inspected, and adjusted as necessary

The spatial configuration of the Y-Thrust was extremely problematic because it was difficult to honor the placement at outcrop concurrent with its position in the cross-sections; that is, the mapped contact at the surface created a complex geometry that could not be exactly matched by the 2-D surface model when combined with the data from the interpretive cross-sections. In those fault blocks where no fault information was available, a 2-D surface model from an adjoining fault block was rotated, raised, or lowered to honor the perceived offsets between fault blocks.

Likewise, positioning the X-Thrust in three dimensions was accomplished by referencing it to the position of the Y-Thrust in each fault block using 2-D grid mathematical operations (rotations, adding constants, etc.). In the central region of the model, the Y-Thrust was absent due to erosion. In this case, the Y-Thrust was restored from a global 2-D surface model. "Global" means that a 2-D surface was fit to all of the available Y-Thrust structural data as if erosion had never happened. The resulting surface provided information on the regional shape that was used as a guide to place the X-Thrust in somewhat of a parallel configuration to the Y-Thrust. Throughout these operations, it was necessary to retain the perspective that this was only a conceptual model and to acknowledge that reality would ultimately be more complex.

Following the generation of the surfaces describing the X-Thrust, a complete 3-D fault model was generated (Figure 3) and reviewed using interactive 3-D visualization on a graphics workstation. Also, cross-sections were generated at the same scale as the interpretive geologic cross-sections and compared to them. Of the tasks performed, the 3-D fault model proved to be the most difficult to complete to the satisfaction of the team. Since the thrust faults would ultimately determine how the rocks (reservoirs and source rocks) would be juxtaposed, significant time and effort was devoted to this aspect of the project.

As Jones (1992) indicated, the value of working in three dimensions is that it enhances the geologist's interpretative processes. Spatial relationships are simply easier to evaluate in that context than in the context of two-dimensional cross-sections and maps.

Geologic Structure and Stratigraphy

As a critical component of the prospect evaluation, this task generated a table of stratigraphic thicknesses for each of the zones to be included in the 3-D model. This was accomplished by using the measured sections of Chamberlain and Warme (1996), evaluating logged stratigraphy from several wildcats (Gulf "EU" Federal, Amoco #3 White River, and Amoco Sunnyside Well) north of the project area, and measuring the thicknesses of the units as drawn in the four interpretative geological cross-sections.

The following units make up the stratigraphic column used to develop the 3-D conceptual model of the Gap Mountain prospect:

- Alluvium (Quaternary)
- Volcanics (Tertiary)
- Ely (Pennsylvanian)

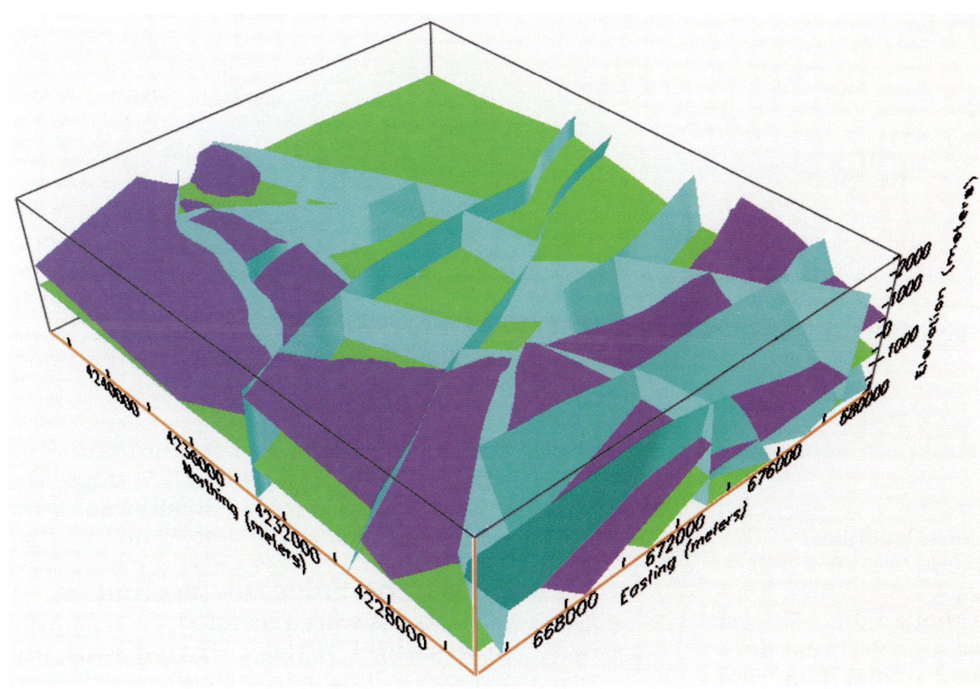

Figure 3—View of the 3-D fault model. Y-Thrust is in magenta, X-Thrust is in green, and normal faults are in cyan. Y-Thrust has been eroded away in the northeast section of the prospect. View is to the northeast.

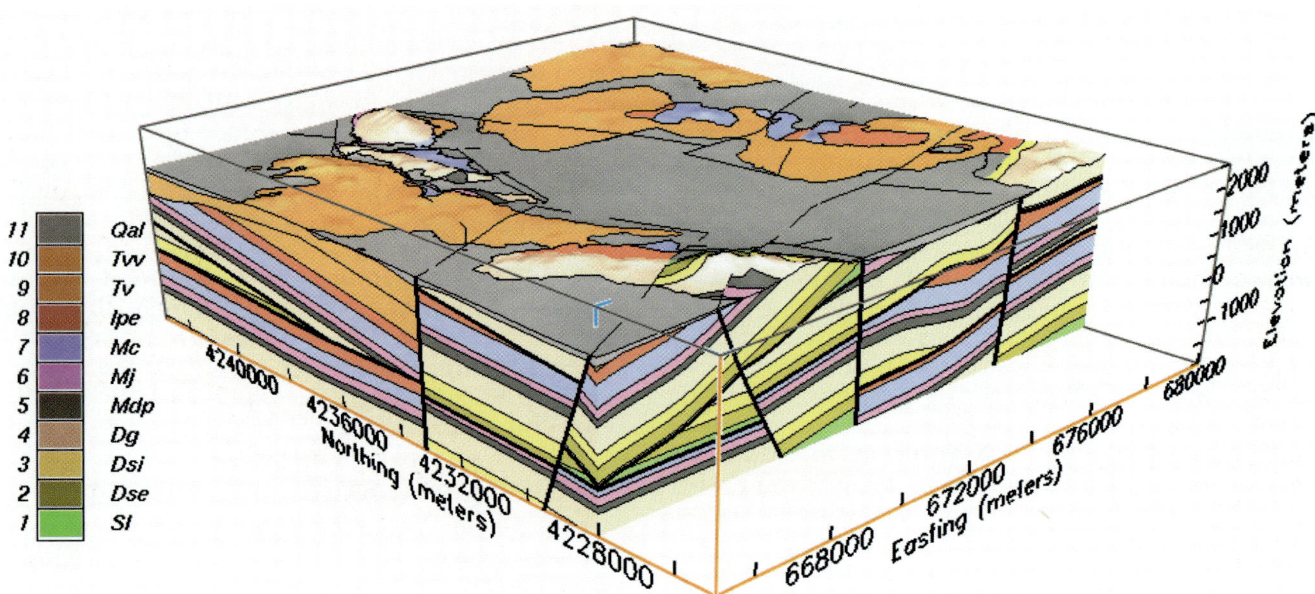

Figure 4—View of the 3-D geologic model along cross section A–A′. View is to the northeast.

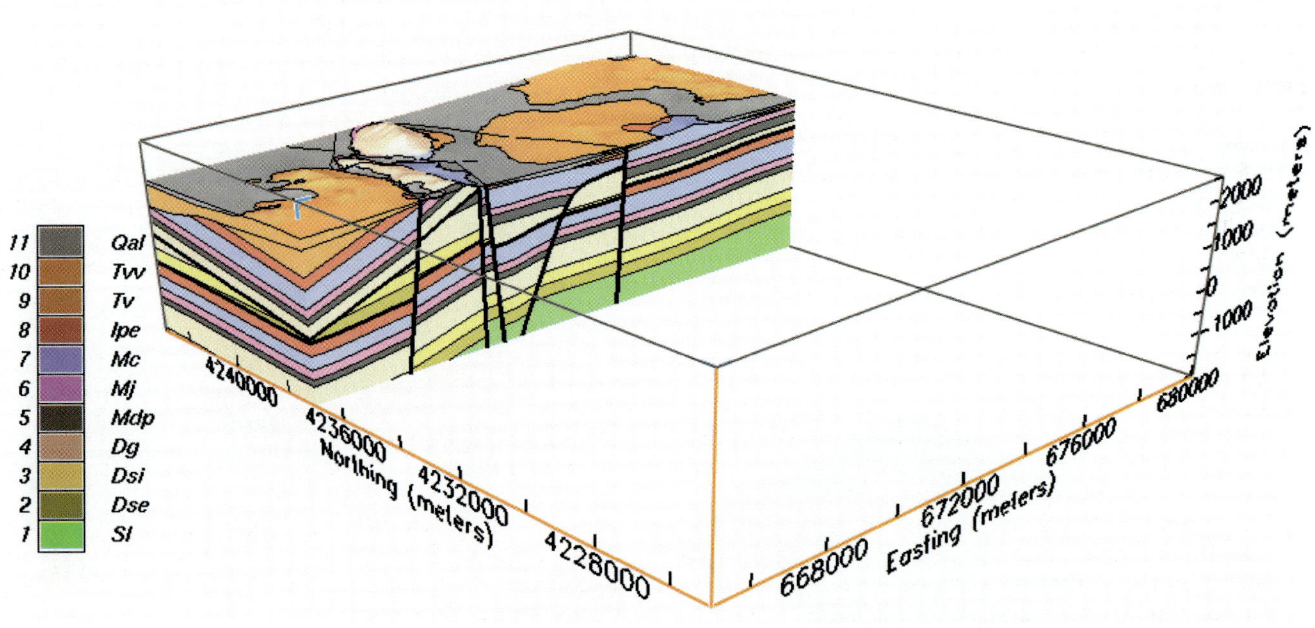

Figure 5—View of the 3-D geologic model along cross section B–B′. View is to the northeast.

- Chainman Shale and Scotty Wash Sandstone (Mississippian)
- Joanna Limestone (Mississippian)
- Pilot Formation (Mississippian–Devonian)
- Guilmette Formation (Devonian)
- Simonson Dolomite (Devonian)
- Sevy Dolomite (Devonian)
- Undifferentiated Silurian rocks

The 3-D geologic structure and stratigraphy were combined with the 3-D fault model in two steps. The Paleozoic section was considered first, followed by the Tertiary volcanics and Quaternary alluvium.

Modeling the Paleozoic Section

Once again, several approaches were formulated and used to generate structural horizons. One approach was

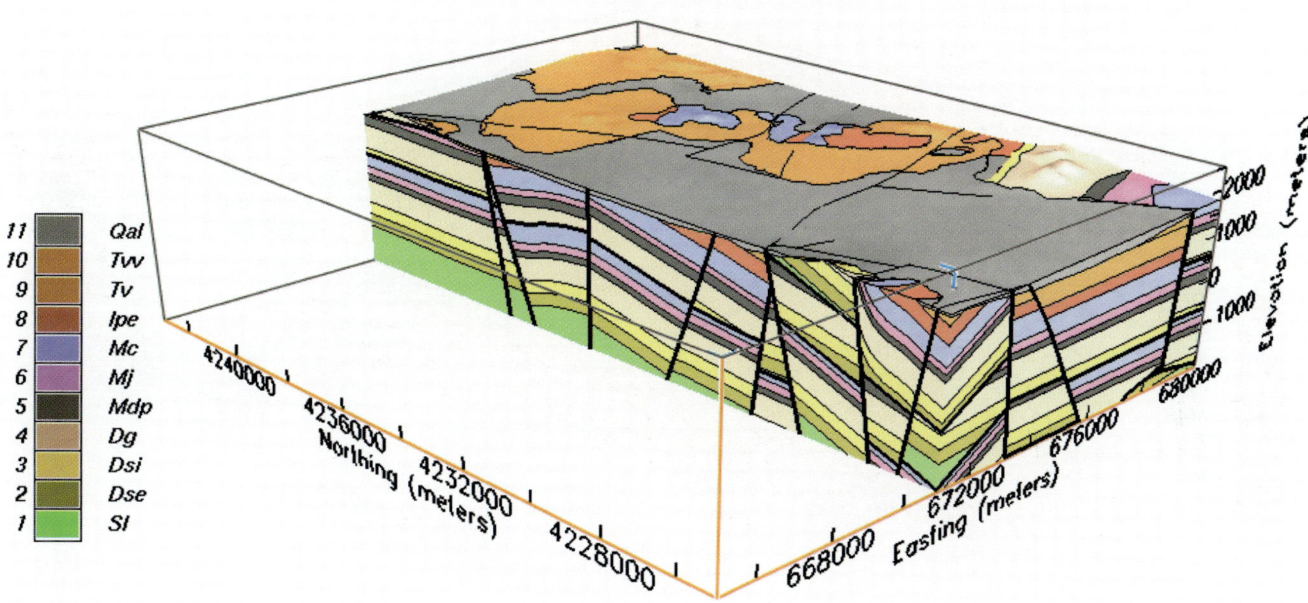

Figure 6—View of the 3-D geologic model along cross section C–C′. View is to the northeast.

used where outcrop data and cross-section data were available, and a different approach was used where neither was available. The former approach encompassed the following steps:

(1) The bedrock-to-bedrock contact data was extracted from the geologic map. It was critical that this contact be bedrock-to-bedrock and not to alluvium or volcanics (otherwise, the wrong geometry would be generated).
(2) The elevation of the contact data was back-interpolated from the DEM and sorted by fault block.
(3) A first-order trend surface was generated through the data. The dip and strike of the trend surface was calculated and compared to the observed dip measurements on the geologic map and to the apparent dips as seen in the geologic cross-sections.
(4) The outcrop data was combined with cross-section data (if available), and a 2-D surface model (grid) was generated through the data. Where cross-section data was limited or not available, only the initial first-order trend surface was used.
(5) The balance of the conformal stack (horizons that did not outcrop) was generated around this reference horizon using the measured section data.
(6) Steps 3–5 were repeated for each of the fault blocks.

As noted, in those cases where outcrop data were not available (especially true below the X-Thrust sheet), another approach was developed. This alternate approach required several assumptions to be made in terms of the relationship of structural tops to the thrust faults. The approach was to take a conformable stack and hang it at a point within the stratigraphic section either above or below the X-Thrust in each fault block. If a reference cross-section wasn't available, a decision was made as to what formation the thrust occurred in or along (i.e., the thrust did not cross formations).

To ensure that compatible offsets across the normal faults were maintained, a series of west-to-east cross-sections were generated every 305 m south-to-north through the 3-D fault model. Using the graphical editor in EarthVision, a reference horizon was developed to provide the control to ensure proper fault offsets. This reference surface was then generated for all of the fault blocks, and the conformable stratigraphic sequence was hung from the reference horizon using grid mathematics. A spatial analysis tool was developed that allowed for the specification of a strike and dip at a point around which a 2-D surface (as a simple plane) could be generated. This application accepted 2-D grid geometry parameters (grid dimensions, spacing, etc.) as input, along with a dip and a dip-azimuth (direction of the dip), and generated the 2-D grid. A 3-D geologic model was then constructed for the Paleozoic rocks alone, and cross-sections were generated and then adjusted to align them with the explorationist's geological concepts.

Tertiary Volcanics and Quaternary Alluvium

The next step addressed erosion of the Paleozoic section to accommodate deposition of the Tertiary volcanic rocks and the Quaternary alluvium in the structural model. The difficult part was determining the 2-D distribution of the volcanic rocks and alluvium and their associated thicknesses. Using the digital geologic map, all of the mapped contacts of the volcanic

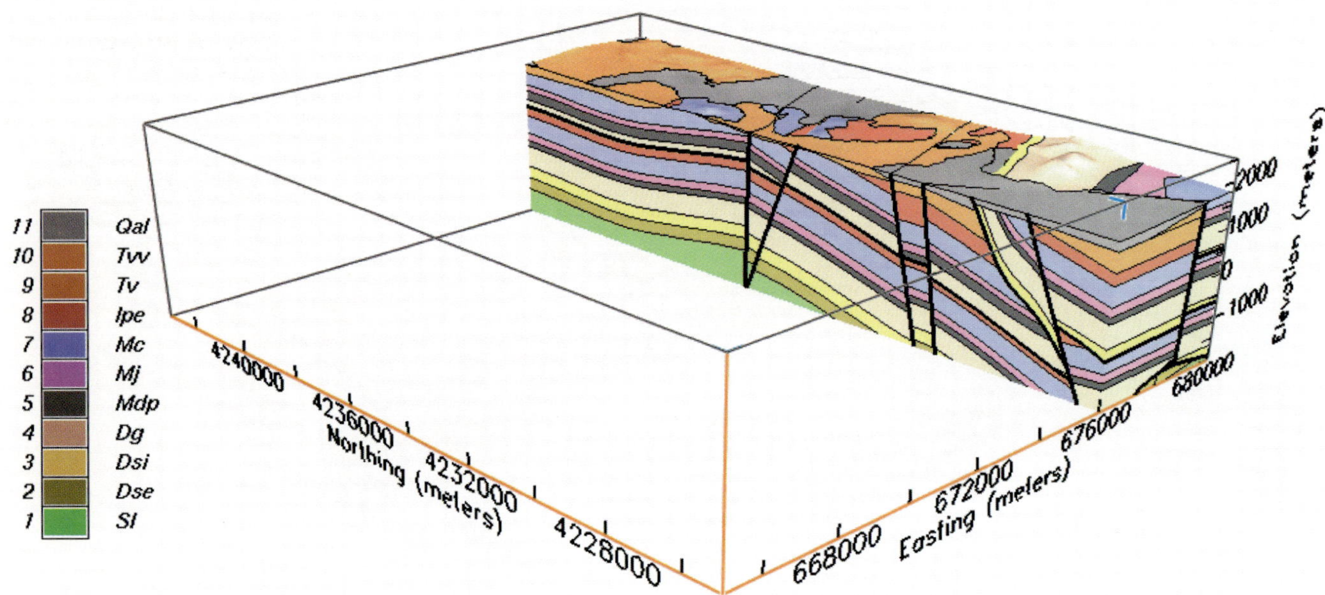

Figure 7—View of the 3-D geologic model along cross section D–D′. View is to the northeast.

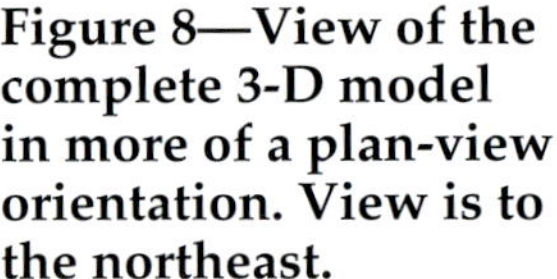
Figure 8—View of the complete 3-D model in more of a plan-view orientation. View is to the northeast.

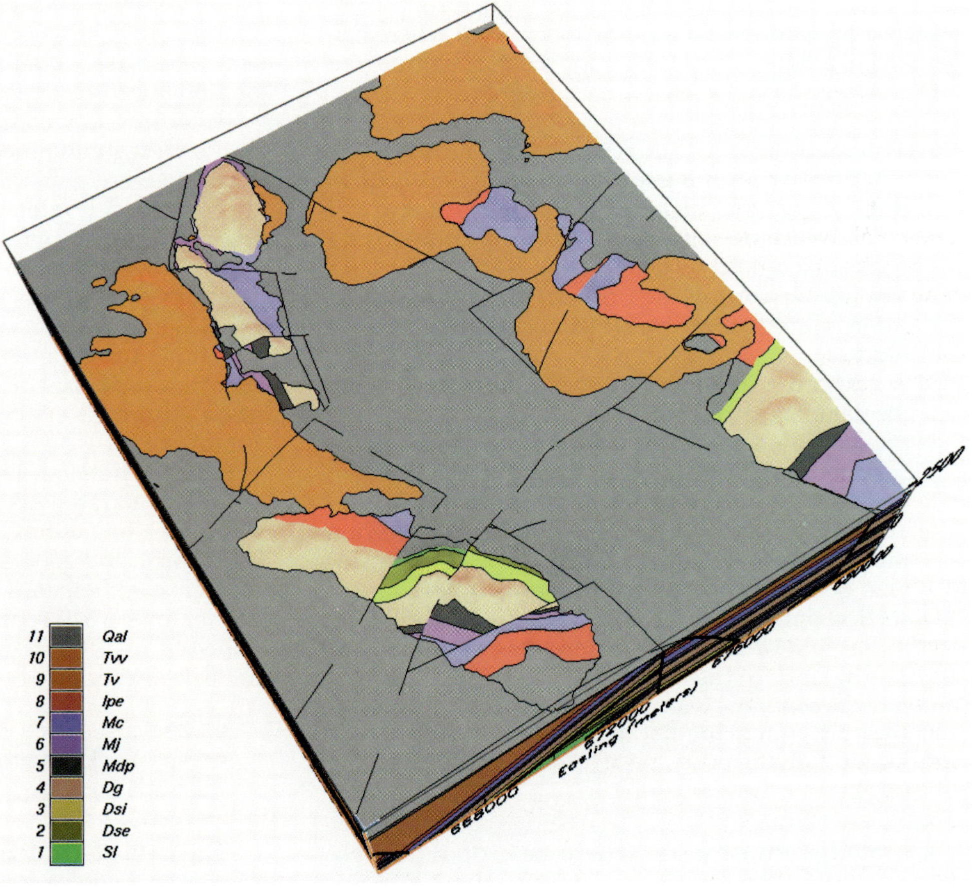

rocks were extracted, and their elevations were interpolated using the DEM. This information provided a limit at the surface. Using the aeromagnetic maps and information from the three wells drilled north of the prospect, the base of the Tertiary volcanics was modeled, allowing it to erode into the underlying Paleozoic

Figure 9—Same view as in Figure 8 with the residual-aeromagnetic map draped over the topographic surface. Positive anomalies (in red) grade toward negative anomalies (in blue). View is to the northeast.

rocks. Following this erosion, a Tertiary volcanics isochore was added. The same process was repeated to generate a 2-D surface model representing the base of the Quaternary alluvium. The resulting model was reviewed for reasonableness by visualizing the model and cutting it at the same location as the interpretative cross-sections (Figures 4–7).

Rock Property Modeling and Volumetric Analysis

The final step in the development of the conceptual model was to populate the geologic horizons with rock properties. Porosity and net-pay thickness were needed to determine the oil-in-place. (Net pay is sometimes calculated as a net-to-gross ratio; that is, the amount of sand divided by the total interval, or zone, thickness.) Having calculated porosity, the amount of fluid occupying the pore-space (oil saturation) could be estimated. Unfortunately, no rock property or oil saturation data were available within the project area with which to estimate these parameters in the prospect on a three-dimensional basis. Therefore, each formation was assigned a constant porosity, net-to-gross ratio, and oil saturation based on data published for analogous settings (Connelly et al., 1995).

The 3-D geological model was now complete (Figure 8) and was used to estimate first-order volumetrics. Volumetric analyses were run using the magnetic reduction anomaly to define the lateral extent of the prospective pool.

SURFACE MODELING OF THE GEOCHEMICAL DATA

Geochemical evaluation was used to further assess the potential of a possible hydrocarbon resource in the Gap Mountain area (Connelly et al., 1995). Because it is possible to overlay various 2-D images directly on the 3-D model using the EarthVision package, 2-D surface models of the following geochemical properties (variables) were made from sample data provided by Connelly Exploration.

- Distribution of propane (hydrocarbon indicator)
- Distribution of Eh (oxidation/reduction potential)
- Distribution of pH (acidity)
- Distribution of conductivity (electrical conductivity)
- Distribution of Q2 Ratio (maturation)

These surface models were developed using a geostatistical approach that involved the estimation of a variogram for each geochemical variable and evaluation of the 2-D distribution using ordinary kriging.

Figure 10—Same view as in Figure 8 with the DRG image draped over the topographic surface. View is to the northeast.

PRODUCT GENERATION

The power of 2-D and 3-D GIS is realized in its visualization and product generation capabilities. Maps, cross-sections, volumetrics, and many other outputs can be generated with relative ease upon completion of the 3-D model. For example, the following products were generated from the Gap Mountain model:

- Structure contour maps on each of the horizons
- Isochore (vertical thickness) maps of each formation (or geological unit)
- Cross-sections at various locations and orientations throughout the modeled region
- Contour maps of each geochemical variable that was modeled
- A contour map and 3-D drape of the residual-aeromagnetic surface (Figure 9)
- A 3-D drape of the DRG on the geologic model (Figure 10)

Following the completion of the 3-D model, Connelly Exploration determined a proposed drill site. For well planning purposes, an initial well path was created. Using this proposed well path and the 3-D model, a well prognosis was calculated. The well prognosis gave an estimate of the expected depth to each of the horizons and fault planes the drill hole was expected to intersect.

CONCLUSIONS

A geo-registered 3-D geological model provides many benefits beyond those of the traditional exercise of constructing 2-D geological maps, structure contour and isochore maps, and cross-sections to describe an exploration prospect. Nonetheless, to develop a viable 3-D geological model, many geometrical problems have to be resolved that would not have to be addressed using the traditional methods. A 3-D model encompasses the complete volume of the prospect, thus further challenging the explorationist. Once completed, however, 3-D visualization can be used to effectively communicate the explorationist's concepts to potential investors, which ideally results in the prospect being drilled.

REFERENCES CITED

Chamberlain, A. K., 1989, Greasy and thirsty Mississippian rocks, Antler Basin, Nevada (abs.). AAPG Bulletin, vol. 73, p. 343.

Chamberlain, A. K., and C. W. Gillespie, 1993, Evidence of Late Mesozoic thrusting, Timpahute Range, south-central Nevada, *in* Structural and Stratigraphic Relationships of Devonian Reservoir Rocks, East-Central Nevada (C. W. Gillespie, ed.). Reno, NV: Nevada Petroleum Society, 1993 Field

Trip Guidebook, p. 139–155.

Chamberlain, A. K., and J. E. Warme, 1996, Devonian sequences and sequence boundaries, Timpahute Range, Nevada, *in* Paleozoic Systems of the Rocky Mountain Region (M. W. Longman, ed.). Tulsa, OK: Society of Sedimentary Geology (SEPM), p. 63–84.

Connelly, M. S., A. H. Pekarek, and R. C. Maly, 1995, New method ideal for frontier plays. The American Oil & Gas Reporter, 38, no. 12, p. 121–126.

Davis, J. C., 1986, Statistics and Data Analysis in Geology, 2nd ed. New York: John Wiley, 646 p.

Environmental Systems Research Institute, 1996, Understanding GIS: The Arc/Info Method (Rev.), Redlands, CA: Environmental Systems Research Institute Inc. (ESRI), 564 p.

Hurtubise, D. O., 1994, Silver King lineament, the missing link of a 500 km east-trending structure in the southern Great Basin, in Structural and Stratigraphic Investigations and Petroleum Potential of Nevada (with Special Emphasis on the South Railroad Valley Producing Trend), Vol. 2 (S. W. Dobbs and W. J. Taylor, eds.). Reno, NV: Nevada Petroleum Society, Annual Conference Proceedings, p. 227–240.

Jones, T. A., 1992, Extension to three dimensions: introduction to the section on 3-D geologic block modeling, *in* Computer Modeling of Geologic Surfaces and Volumes (D. E. Hamilton and T. A. Jones, eds.). Tulsa, OK: AAPG, p. 175–182.

Jones, T. A., D. E. Hamilton, and C. R. Johnson, 1986, Contouring Geologic Surfaces with the Computer. New York: Van Nostrand Reinhold, 314 p.

Loudon, T. V., 1979, Computer Methods in Geology. London: Academic Press, 269 p.

Starr, J., and J. Estes, 1990, Geographic Information Systems: An Introduction. London: Prentice-Hall, 303 p.

Tschanz, C. M., and E. H. Pampeyan, 1970, Geology and Mineral Deposits of Lincoln County, Nevada. Reno, NV: Nevada Bureau of Mines and Geology Bulletin 73, 187 p.

Haertlein, A., and D. Rich, 2000, Evaluating international acquisitions using a spatial database system—A low-cost approach, *in* T. C. Coburn and J. M. Yarus, eds., Geographic information systems in petroleum exploration and development: AAPG Computer Applications in Geology, No 4, p. 151–172.

Chapter 11

Evaluating International Acquisitions Using a Spatial Database System—A Low-Cost Approach

Al Haertlein
Independent Geologist
Spring, Texas, U.S.A.

Dave Rich
President
Geotech Computer Systems, Inc.
Englewood, Colorado, U.S.A.

ABSTRACT

A geographic information system (GIS) can be a powerful tool for analyzing petroleum data. Using GIS successfully on any particular project depends on maintaining a balance between costs and benefits. This paper describes the use of a GIS approach in the evaluation of properties prior to acquisition in South America, the west coast of Africa, and the North Sea. For the areas of interest, the base map data were contained on a CD-ROM. Digital information on the concessions in each area was also obtained, to which proprietary data, such as seismic line locations, were added as appropriate. Maps containing all this information were constructed and used to evaluate concessions on a competitive bid or negotiated sale basis. The GIS system proved to be a cost-effective way to present the data in support of the decision-making process.

INTRODUCTION

The decision to use a geographic information system (GIS) on a particular project is, or at least should be, an economic one. The costs of setting up the system, acquiring or entering data, and creating displays should be less using the GIS approach than for alternative approaches. These other approaches include hand-drafted maps, maps created from scratch using computer-aided drafting (CAD) techniques, and hard-copy base maps purchased from vendors, annotated with proprietary data. If using a GIS approach is the most cost-effective option, then it should be chosen.

A project was recently undertaken to evaluate oil and gas properties in the international areas of South America, the west coast of Africa, and the North Sea. The business plan included conducting property evaluations on a competitive bid or negotiated sale basis and taking a proactive approach to identifying potential acquisition candidates. The basic premise was to identify candidate properties that would have minimum costs, so that time and budget resources could be allocated to evaluating the properties in the due diligence phase of a data room or negotiated sale scenario. Using the capabilities of modern GIS technology, a protocol was developed to screen for acquisition candidates in any country without depleting the resources required for success.

The GIS support required for the evaluation of international oil and gas property acquisitions presents a unique set of problems that vary from country to country. The key data issues are availability, cost, source, accuracy, thoroughness, and format. The solutions to these problems are generally a function of time and expense. It is important to use GIS software that has flexible data management capabilities and is compatible with varying data formats and coordinate projection systems.

Defining the Business Strategy

It is important for international exploration and production (E&P) companies to focus on areas for which the company has opportunities, expertise, and good government relations. Many companies have segregated their properties into "core" properties, which have the closest fit to their business plan, and "non-core" properties, which are all of the rest. The consolidation of properties into core business areas and the divestiture of non-core assets by these companies has resulted in acquisition opportunities for others. These producing properties are generally sold through a solicitation of competitive bids. A proactive approach of submitting an unsolicited offer for a property can be successful if the property is on a company's list of properties to be sold, or if, by virtue of the offer, the company becomes aware of the non-core characteristics of the property. In general, companies are motivated to sell producing properties that meet the non-core criteria. There is a high probability of success for closing purchases of targeted non-core properties on an acceptable economic basis.

The goal is to assess the core or non-core status of properties from the point of view of the current owner. An initial evaluation of an E&P company's intentions for an international concession can be made based on the following:

- Implications of an operating interest versus the interest of a non-operator
- Consideration of the aggregate of the fields in a concession without emphasizing an individual field
- An evaluation of any obvious problems from political, environmental, technical, or product export viewpoints
- The proximity to significant production owned by the same interest owner as the acquisition target
- A review of any recent business activity by the potential seller, such as awards of additional concessions, seismic acquisition, and the results of any drilling activity
- Determination of general production characteristics, such as decline, daily rates, remaining producing reserve potential, and any known undeveloped potential
- A simple analysis of the ratio of production cost to revenue to determine the probable net cash flow
- Review of the production gathering, transportation, and marketing capabilities of the owner
- Information concerning additional in-country downstream business operations that can affect a company's core versus non-core asset determination

Defining the Data Challenge

There are approximately 400 active producing concessions in the areas of South America, Africa, and the North Sea. The average number of separate working interest owners per acreage block is 2 for South America, 1 for Africa, and 5 for the North Sea, resulting in over 900 potential acquisitions that must be screened. Information on just the basic characteristics of a property could result in having to consider over 10,000 data entities. The initial step is to gather as much of this information as possible and organize it into tabular form.

The relative importance of acquisition candidates identified from the tabular data often does not become apparent until the data are examined from the point of view of the spatial relationships. To determine the value of a property and its status as a core or non-core property, it is necessary to analyze the geographic context of the acquisition target and its relationship to other important factors, such as offset operators, pipelines, markets, and additional development potential.

The geographic boundaries of the project described in this article involve three different quadrants of the world, representing a full spectrum of mapping challenges. The land grid and culture, which serve as a backdrop to the field and well data, must be sufficient to allow close inspection and interpretation of the production data. Each geographic area will have varying degrees of development and different levels of data detail for any base map. Concession boundaries, producing field boundaries, wells, seismic surveys, and pipelines will need to be properly located in reference to significant physiological and cultural map features such as coastlines, political boundaries, and major cities.

Defining the GIS/Data Management Solution

The GIS/data management strategy used for this project required the aggregation of various equipment, materials, and information, including appropriate computer hardware, GIS software using data management capabilities, a base map file containing geographic information, a digital database of information about key wells on a worldwide basis, and a digital database of statistics on international concessions. The hardware, software, and data were used together with proprietary information, including seismic line locations, to create the graphical displays used in the evaluations.

The computer hardware for this project consisted of Intel-based IBM–compatible personal computers (PCs) running the Microsoft Windows 95 operating system. This setup was chosen because it provided a reasonable amount of computing power at a low cost, and it supported a broad spectrum of software from which to choose. Ease of use of this hardware-software combination was also a significant factor. Additional hardware included a digitizer used for capturing location information, a plotter for creating large maps, and office printers for making small maps and printing textual data.

The options for the commercial data management software for this project were evaluated based on several initial requirements. These included software quality, availability, cost, PC-based operation, an easy learning curve, and flexibility. In addition to quality and availability, flexibility is a critical feature. Important components of the flexibility aspect included data import and export capabilities, compatibility between different data sources, and the ability to output to large-size plotters, as well as standard office printers.

The GIS software selected for the project was PetroSpase (a product of Geotech Computer Systems, Englewood, CO), which is based on Spase (pronounced "space") (a product of Geotech Computer Systems, Englewood, CO), a spatial database manager. A spatial database manager differs from other types of GIS programs in that it includes a mapping component, a full relational database manager, and a programmable interface integrated into one package. PetroSpase extends the capabilities of Spase by providing a preconfigured format for importing, storing, and displaying base map, well, and seismic information, along with other data. At the same time, it maintains the flexibility and configurability of the native Spase system, which allows individualized data tables to be added to a database, and which uses the industry standard structured query language (SQL) for manipulating the data. This combination of capabilities was important for the project because it was anticipated that a wide variety of different data sources and formats would be needed, thereby requiring the ability to configure the software to work with these formats. It was believed that some formats would be proprietary to various companies and not supported by any commercial software.

The implementation of the business strategy for proactively identifying acquisition candidates used two commercial databases obtained from Petroconsultants (Petroconsultants, S.A., Geneva, Switzerland, now a part of IHS Energy Group), an international energy information company. The first database contains statistics on international concessions, such as ownership interest, operator, and current production, and the other database contains the coordinates of key wells along with any available initial production test information. Both databases are in a spreadsheet format, and they can be queried for acquisition candidates based on various criteria. They can also be accessed for information about properties that are proposed for sale on a competitive bid or negotiated basis, prior to committing the resources needed for the evaluation.

An easy-to-use format of the Digital Chart of the World (DCW), called CVGMap (a product of Cerro Verde Geotech, Golden, CO), was used for this project. DCW was compiled in 1992 from a variety of sources by Environmental Systems Research Institute (ESRI) on contract for the U.S. Defense Mapping Agency, and it is available for use without restriction by U.S. citizens. It contains worldwide coverage of coastlines, rivers, populated places, roads, railroads, hypsography (elevations), and other data digitized at a scale of 1:1,000,000. The level of detail varies due to issues related to U.S. national security. For example, Poland is detailed in the extreme, whereas Gabon is not.

CVGMap provides menus to extract the digital data by selecting a latitude/longitude (lat/long) window. The data is organized into two categories: DCW line data and DCW point and text data. Within each category there are two levels of map attributes having increasing detail. A protocol was created that could be used for any of the countries in this project, allowing a consistent set of basic data to be extracted for each area (Table 1). Using this protocol, the desired data was exported in an ASCII (American Standard Code for Information Interchange) text file using CVGMap's Surfer output format. This simple format was chosen because it was easy to import into the spatial database. CVGMap converted the coordinates from the lat-long coordinates in the DCW data to x–y (Cartesian) coordinates as the file was created, because Surfer, the target of the format, works best in x–y coordinates.

The next step was to create a PetroSpase database for each area evaluated. This option was chosen over the alternative of storing all of the data in one database because it allowed flexibility to be maintained as the project matured from the candidate screening phase to initiating the detailed property evaluations. To import the DCW geographic data into the PetroSpase database, a script was written in the Spase script language (Table 2). The key well information for each area was added through a software feature used to import well data in an ASCII format. The capability of PetroSpase to simultaneously handle data in lat/long and x–y coordinates was needed to overlay the base map data, which was imported in x–y coordinates, with the exploration and other data, which were in lat/long coordinates.

In general, once the computer hardware was set up and the software installed, the following series of steps was performed for each area to be evaluated:

- Extraction of the geographic data from the CVGMap CD
- Creation of the PetroSpase database
- Importing the map data into PetroSpase
- Extraction of the key well data
- Importing the key well data into PetroSpase
- Addition of concession statistics
- Addition of proprietary data
- Creation of the map displays

CASE STUDIES

The GIS system was used to assist in the analysis of concession opportunities in several different geographic regions. The following case studies illustrate the techniques used, and they highlight the positive contribution

Table 1. Protocol for Extracting Geographic Data from the Digital Chart of the World Database to be Imported into the Spatial Database, PetroSpase

Extracting Geographic Data from the Digital Chart of the Word Database

Before you begin fill out the following categories on the area of interest.

Country Area	3-Letter Code	Lat/Long Boundaries South	North	West	East	Map Scale
Example:						
Argentina						
Austral Basin	AUS	-52.5	-48.0	-73.0	-67.5	1:500,000

Note: convert to decimal world coordinate system; **negative** latitudes for **southern** hemisphere and **negative** longitudes for the **western** hemisphere. This means Argentina would be negative latitudes and negative longitudes and the characters such as S and W representing south and west are not included. Degrees are whole numbers, minutes and seconds become tenths.

Example:			
	Northern:	- 48.0 = 48 00` S	
	Southern:	- 52.5 = 52 30` S	
	Western:	- 73.0 =	73 00` W
	Eastern:	- 67.5 =	67 30` W

Insert CVGMAP CD.
Click on the Digital Map icon in the Make Map Group.

Follow the menu

I. The CD-ROM drive is usually "**D**".
II. Hit enter twice
III. The new screen is the **Data Selection** Menu instructions.

A. Arrow down to the "ALL DCW DATA (LINE,.....)" and hit the **F5** key. All of the choices have now been selected.
B. Now using the following **Data Outline** and the **F6** key **unselect** certain choices. The bold highlighted lines in the Data Outline are the items which are to be excluded.
C. Each indention on the outline represents a different data level, as presented in the on screen instructions use the **F3** key and the **F4** key to move to different levels in the outline.

IV. Hit F10 and enter the 3 letter code from above ..."hit enter"
V. Enter Lat/Long boundaries as prompted.
VI. Hit enter twice
VII. Type 0 (zero)
VIII. Type 0
IX. Type N
X. Hit enter 3 times.

made by adding a spatial analysis component to the traditional economic analysis approach.

Gabon, West Africa

Background

Petroleum exploration in Gabon began during a period of colonial rule when the independent nation was part of French Equatorial Africa. Commercial production was established in 1956 with the discovery of Ozouri and Pointe Clairette fields in the coastal area of Port Gentil (Hedberg, 1950, 1951). By the end of 1956, a total of 11 oil discoveries on Aptian salt structures, delineated from gravity and seismic surveys, had been drilled by the Société des Pétroles d'A.E.F. (later renamed ELF/SPAFE). The producing reservoirs were various marine sandstones of Late Cretaceous and Tertiary ages. In 1963, Shell Gabon and Elf discovered the Gamba field in a Lower Cretaceous pre-salt sandstone. In 1997, a daily rate of

Table 1. Continued.

DATA OUTLINE		Level II	Included Data in Bold
10000	DCW line data		
	10100	cultural landmark lines	
	10200	**drainage lines**	
	10300	hypsography lines	
	10400	supplemental hypsography lines	
	10500	**ocean feature lines**	
	10600	physiography lines	
	10700	**political and ocean lines**	
	10800	**populated place lines**	
	10900	road lines	
	11000	railroad lines	
	11100	transportation structure lines	
	11200	utility lines	
20000	DCW point and text data		
	20100	aeronautical points	
	20200	cultural landmark points and text	
	20300	drainage points and text	
	20400	drainage supplemental points	
	20500	hypsography points	
	20600	**ocean feature points and text**	
	20700	**physiography text**	
	20800	**political and ocean points and text**	
	20900	**populated place points**	
	21000	railroad text	
	21100	road text	
	21200	transportation structure points and text	
	21300	utility text	

DATA OUTLINE		Level III	Included Data in Bold
20000	DCW point and text data		
	20900		**populated place points**
	20901	PPP01	populated places (labeled)
	20902	PPP11	populated places (unlabeled)
	20903	**PPP02**	**populated places (for city polygons, labeled)**
	20904	PPP12	populated places (for city polygons, unlabeled)
	20905	PPP03	villages, strategic or landmark places (labeled)
	20906	PPP13	villages, strategic or landmark places (unlabeled)
	20907	PPP04	kampong (labeled)
	20908	PPP14	kampong (unlabeled)
END			

200,000 barrels, or 60% of Gabon's total oil, was produced from pre-salt sandstone completions of the Rabi-Kounga field (U.S. Energy Information Administration, 1997).

Oil export revenue accounts for about 50% of Gabon's gross national product. The Gabonese government is actively seeking new investment from foreign oil companies to maintain its future production. The government's new policy of investment incentives has created the opportunity for exploration on deep-water concessions and for development of fields previously considered to have marginal economics.

Table 2. Script to Import the Extracted Digital Chart of the World Data for a Specified Lat/Long Window into PetroSpase

```
— File Name .........: IM_WORLD
— File Version ......: 1.0
— Spase version .....: 1.34
— Petro Spase version: 2.0
— Author ............: Dave Rich
— Date ..............: 6/25/96
— Modified By .......: Mike Wiley
— Mod Date ..........: 07/12/96
— Mods ..: Added categories: coastline; tracks, trails, footpaths;
—          connectors (drainage); coastal closure line

— Description .:
— Import CVT converted Digital Chart of the World Data
— This is a set of Surfer boundary line and DAT files which are
—  listed in the file NIG-RPT.LOG.
_____________________________________________________________

— Copyright© 1996 by Geotech Computer Systems, Inc.
—                    All Rights Reserved
_____________________________________________________________

—- Dialog boxes and control variables
DECLARE b_ok SMALLINT;
DECLARE b_Cancel SMALLINT;
DECLARE v_msg CHAR 60;

DECLARE fh_0              FILE;              —Control file handle
DECLARE v_fn_0            CHAR 12;           —Control file name
DECLARE v_prefix          CHAR 3;            —File prefix
DECLARE fh_1              FILE;              —BLN file handle
DECLARE v_fn_1            CHAR 12;           —BLN file name
DECLARE v_descr           CHAR 50;           —File contents

DECLARE v_LINE_ID         INT;
DECLARE v_TEXT_ID         INT;
DECLARE v_CATEGORY        MAP_FEATURE_CODES;
DECLARE v_LL_LINE         LINE LAT_LONG;
DECLARE v_XY_LINE         LINE CARTESIAN;
DECLARE v_pt1             POINT LAT_LONG;
DECLARE v_pt2             POINT LAT_LONG;
DECLARE v_count           INT;
DECLARE v_PointCount      INT;
DECLARE v_points          SMALLINT;
DECLARE v_ANNOTATION      CHAR 20;
DECLARE v_COORD           POINT LAT_LONG;
DECLARE v_str             CHAR 99;
DECLARE v_lat             REAL;
DECLARE v_lon             REAL;

DECLARE v_minlat          REAL;
DECLARE v_minlon          REAL;
DECLARE v_maxlat          REAL;
DECLARE v_maxlon          REAL;
DECLARE v_sw              POINT LAT_LONG;
DECLARE v_ne              POINT LAT_LONG;

— Reads a record from log file
DECLARE fmt_0 FORMAT
CHAR  3:14;     —File name
CHAR  18:67;    —Contents
END;

—Point count for BLN files
DECLARE fmt_count FORMAT
INT      1:5;
END;

—BLN points are read free-form - no format needed
```

Table 2. Continued.

```
—Text file
DECLARE fmt_text FORMAT
REAL    1:11;
REAL    12:24;
CHAR    26:45;
END;

DECLARE c_0 CURSOR FOR SELECT MAX(LINE_ID)
FROM FEATURE_LINES;

DECLARE c_1 CURSOR FOR SELECT MAX(TEXT_ID)
FROM FEATURE_TEXT;

— max-min window
DECLARE c_sw CURSOR FOR SELECT COORD_TYPE
FROM CONTROL WHERE NAME='MAIN_SW';

DECLARE c_ne CURSOR FOR SELECT COORD_TYPE
FROM CONTROL WHERE NAME='MAIN_NE';

DECLARE d_0 DIALOG
TITLE 'CVG/DCW Data Import';
TEXT 'Import CVG reformatted DCW Data File' 4 4;
CHAR 3 NOSCROLL 'File prefix' @16 4 18;
BUTTON DEFAULT 'Import' 4 38;
BUTTON 'Exit' 48 38;
END;

DECLARE d_error DIALOG
TITLE 'File Name or Path Error';
TEXT 'Please correct the following problem...' 4 4;
CHAR 60 NOEDIT '' @120 4 18;
BUTTON 'Exit' 48 36;
END;

— main start —

v_prefix:='   ';

FETCH d_0 INTO v_prefix,b_ok, b_cancel;
IF b_cancel=1 THEN QUIT;

— Check the file name
REPEAT
  v_fn_0 := v_prefix+'-RPRT.LOG';
  IF v_fn_0 IS NULL THEN QUIT;

  OPEN fh_0 v_fn_0;
  IF sqlcode <> 0 THEN BEGIN
    v_msg :='Can''t open '+trimstr(v_fn_0);
    FETCH d_error INTO v_msg, b_cancel;
    IF b_cancel=1 THEN BEGIN
      sqlcode:=0;
      QUIT;
    END;
  END;
UNTIL v_fn_0 IS NOT NULL OR b_cancel =1;

IF b_cancel = 1 THEN
   BEGIN
   sqlcode := 0;
   QUIT;
END;
```

Strategy

The evaluation for Gabon followed the general outline for recognizing acquisition opportunities. The focus was on identifying core versus non-core assets and reviewing the potential for developing oil discoveries previously considered marginal or uneconomic.

Table 2. Continued

```
— Init max-min window and initialize CONTROL if necessary
OPEN c_sw;
FETCH c_sw INTO v_sw;
CLOSE c_sw;

OPEN c_ne;
FETCH c_ne INTO v_ne;
CLOSE c_ne;

IF v_sw IS NULL OR v_ne IS NULL THEN BEGIN — not init
  IF v_sw IS NULL THEN BEGIN
    v_sw := POINT(LAT_LONG, 90.0, 180.0);
    DELETE FROM CONTROL WHERE NAME = 'MAIN_SW';
    INSERT INTO CONTROL(NAME,COORD_TYPE) VALUES('MAIN_SW', v_sw);
  END;
  IF v_ne IS NULL THEN BEGIN
    v_ne := POINT(LAT_LONG, -90.0, -180.0);
    DELETE FROM CONTROL WHERE NAME = 'MAIN_NE';
    INSERT INTO CONTROL(NAME,COORD_TYPE) VALUES('MAIN_NE', v_ne);
  END;
END;

v_minlat  := POINT_Y(v_sw);
v_minlon  := POINT_X(v_sw);
v_maxlat  := POINT_Y(v_ne);
v_maxlon  := POINT_X(v_ne);

OPEN c_0;
FETCH c_0 INTO v_LINE_ID;
CLOSE c_0;
IF v_LINE_ID IS NULL THEN v_LINE_ID := 0;

OPEN c_1;
FETCH c_1 INTO v_TEXT_ID;
CLOSE c_1;
IF v_TEXT_ID IS NULL THEN v_TEXT_ID := 0;
—Read control file
READ fh_0 INTO v_fn_1,v_descr USING fmt_0;
REPEAT
  IF FINDSTR(v_fn_1,'BLN') >0 THEN BEGIN        —Line files
    v_descr:=TRIMSTR(v_descr);
    CASE v_descr OF
      'map area boundary (blanking) file': v_CATEGORY:=MF_BOUNDARY;
      'cultural landmark lines': v_CATEGORY:=MF_FEATURE;
      'streams, rivers, channelized rivers': v_CATEGORY:=MF_STREAM;
      'inland shorelines': v_CATEGORY:=MF_STREAM;
      'glacial limits': v_CATEGORY:=MF_STREAM;
      'hypsography lines': v_CATEGORY:=MF_GRAPHIC_BASE;
      'supplemental hypsography lines': v_CATEGORY:=MF_GRAPHIC_BASE;
      'international boundary, dejure': v_CATEGORY:=MF_STATE;
      'populated place lines': v_CATEGORY:=MF_COUNTY;
      'primary or secondary roads or highways': v_CATEGORY:=MF_ROAD;
      'connectors (roads)': v_CATEGORY:=MF_ROAD;
      'connectors (drainage)': v_CATEGORY:=MF_STREAM; — added
      'single track railroads':v_CATEGORY:=MF_RAILROAD;
      'connectors (railroads)':v_CATEGORY:=MF_RAILROAD;
      'transportation structure lines - roads':v_CATEGORY:=MF_ROAD;
      'tracks, trails, or footpaths': v_CATEGORY:=MF_ROAD;        — added
      'transportation structure lines - railroads':v_CATEGORY:=MF_RAILROAD;
      'power transmission lines': v_CATEGORY:=MF_TITLE;
      'above ground pipelines': v_CATEGORY:=MF_PIPELINE;
      'coastline': v_CATEGORY:=MF_HYDROLOGY;        — added
```

A profile of the active companies in Gabon was considered, with the goal of developing a business alliance to execute the strategy. The potential for other favorable opportunities, such as the result of building a relationship with a particular company, was also considered.

Potential acquisition and development targets for the west coast of Africa were categorized into three groups based on different business strategies. These business strategies centered on the current development and operations status of the concession. This three-tier categorization is shown in Table 3.

Table 2. Continued.

```
    'coastal closure line':v_CATEGORY:=MF_HYDROLOGY;          — added
    DEFAULT:v_CATEGORY:=NULL;
  END;

  IF v_CATEGORY IS NOT NULL THEN BEGIN
    OPEN fh_1 v_fn_1;
    IF sqlcode <> 0 THEN BEGIN
      v_msg :='Can''t open '+trimstr(v_fn_0);
      FETCH d_error INTO v_msg,b_cancel;
      IF b_cancel=1 THEN BEGIN
        sqlcode:=0;
        QUIT;
      END;
    END ELSE BEGIN
      READ fh_1 INTO v_PointCount USING fmt_count;
      REPEAT
        v_pt1:=NULL;
        v_pt2:=NULL;
        FOR v_count:=1 TO v_PointCount BEGIN
          READ fh_1 INTO v_lon,v_lat;
          v_pt2:=POINT(LAT_LONG,v_lat,v_lon);
          IF v_pt1 IS NOT NULL THEN BEGIN
            v_LL_LINE := LINE(v_pt1, v_pt2);
            v_XY_LINE := LINE(LL_TO_CART(v_pt1), LL_TO_CART(v_pt2));
            v_LINE_ID := v_LINE_ID + 1;
            INSERT INTO FEATURE_LINES (LINE_ID,CATEGORY,XY_LINE,LL_LINE)
            VALUES (v_LINE_ID, v_CATEGORY, v_XY_LINE, v_LL_LINE);
            END;
            v_pt1 := v_pt2;
            — update max-min window
            IF v_minlat > v_lat THEN v_minlat := v_lat;
            IF v_minlon > v_lon THEN v_minlon := v_lon;
            IF v_maxlat < v_lat THEN v_maxlat := v_lat;
            IF v_maxlon < v_lon THEN v_maxlon := v_lon;
        END;
        READ fh_1 INTO v_PointCount USING fmt_count;
      UNTIL sqlcode=500;
    END;
    CLOSE fh_1;
  END;
END ELSE IF FINDSTR(v_fn_1,'DAT') >0 THEN BEGIN    —Text files
  v_descr:=TRIMSTR(v_descr);
  CASE v_descr OF
    'active civil airports':v_CATEGORY:=MF_FEATURE;
    'active civil and military airports':v_CATEGORY:=MF_FEATURE;
    'active military airports':v_CATEGORY:=MF_FEATURE;
    'other airports':v_CATEGORY:=MF_FEATURE;
    'cultural landmark points':v_CATEGORY:=MF_FEATURE;
    'drainage text':v_CATEGORY:=MF_STREAM;
    'small lakes or glaciers':v_CATEGORY:=MF_STREAM;
    'small inland water islands':v_CATEGORY:=MF_STREAM;
    'spot elevation':v_CATEGORY:=MF_GRAPHIC_BASE;
    'spot elevation, questionable elevation or location':v_CATEGORY:=MF_GRAPHIC_BASE;
    'supplemental hypsography points':v_CATEGORY:=MF_GRAPHIC_BASE;
    'political/ocean text':v_CATEGORY:=MF_TOWNSHIP;
    'populated places (labeled)':v_CATEGORY:=MF_SECTION;
    'populated places (unlabeled)':v_CATEGORY:=MF_SECTION;
    'populated places (for city polygons, labeled)':v_CATEGORY:=MF_SECTION;
    'railroad text':v_CATEGORY:=MF_RAILROAD;
    'transportation structure points - roads':v_CATEGORY:=MF_ROAD;
    'utility text':v_CATEGORY:=MF_PIPELINE;
    DEFAULT:v_CATEGORY:=NULL;
  END;

  IF v_CATEGORY IS NOT NULL THEN BEGIN
    OPEN fh_1 v_fn_1;
    IF sqlcode <> 0 THEN BEGIN
      v_msg :='Can't open '+trimstr(v_fn_0);
```

Table 2. Continued.

```
        FETCH d_error INTO v_msg,b_cancel;
        IF b_cancel=1 THEN BEGIN
          sqlcode:=0;
          QUIT;
        END;
      END ELSE BEGIN
        v_pt1:=NULL;
        v_pt2:=NULL;
        READ fh_1 INTO v_lon,v_lat,v_ANNOTATION USING fmt_text;
        REPEAT
          v_COORD:=POINT(LAT_LONG,v_lat,v_lon);
          v_ANNOTATION:=TRIMSTR(v_ANNOTATION);
          v_ANNOTATION:=SUBSTR(v_ANNOTATION,1,LENSTR(v_ANNOTATION)-1);
          IF v_ANNOTATION IS NOT NULL THEN BEGIN
            v_TEXT_ID := v_TEXT_ID + 1;
            INSERT INTO FEATURE_TEXT (TEXT_ID, CATEGORY, ANNOTATION, COORD)
            VALUES (v_TEXT_ID, v_CATEGORY, v_ANNOTATION, v_COORD);
          END;
          — update max-min window
          IF v_minlat > v_lat THEN v_minlat := v_lat;
          IF v_minlon > v_lon THEN v_minlon := v_lon;
          IF v_maxlat < v_lat THEN v_maxlat := v_lat;
          IF v_maxlon < v_lon THEN v_maxlon := v_lon;
          READ fh_1 INTO v_lon,v_lat,v_ANNOTATION USING fmt_text;
        UNTIL sqlcode=500;
      END;
      CLOSE fh_1;
    END;
  END;
  READ fh_0 INTO v_fn_1,v_descr USING fmt_0;
UNTIL sqlcode = 500;
CLOSE fh_0;
— Update max-min control
UPDATE CONTROL
SET COORD_TYPE = POINT(LAT_LONG, v_minlat, v_minlon)
WHERE NAME = 'MAIN_SW';
UPDATE CONTROL
SET COORD_TYPE = POINT(LAT_LONG, v_maxlat, v_maxlon)
WHERE NAME = 'MAIN_NE';
sqlcode := 0;
— eof —
```

On this basis, several significant issues were highlighted, which were classified as pluses and minuses based on their impact on decisions to pursue opportunities. These issues, which primarily involved operating status, facilities status, and financial conditions, are shown in Table 4. Emphasis was placed on the potential for production enhancement through horizontal drilling.

Information on recent exploration activity in Gabon was then gathered and analyzed. This information is shown in Table 5.

Procedure

The general data management procedure was followed for this evaluation. A new database of the recent activity was created, including new discoveries, older undeveloped discoveries being brought on production, new exploration permits being awarded, new seismic options being granted, and, most important, new divestiture and acquisition transactions. A spatial presentation of the new database was needed to evaluate the impact on any of the acquisition candidates identified from the initial screening process.

The concession statistics database was queried, looking for assets geographically isolated from other producing fields owned by the potential sellers or for properties for which the current operator might consider the cost-to-revenue ratio to be marginal. Candidates were then graded into two groups based on possible production enhancement with horizontal drilling or the potential for developing a valuable business alliance with the other working interest partners in a concession.

A map displaying the key well data (Figure 1) was created to evaluate fields that had not been developed or produced. Reasons for the discoveries not being developed depend on the economic objectives of the operator or the terms of the concession at the time the well was drilled. so a viable development plan had to be determined. Again, the candidates were graded and categorized into two groups based on the possibility for applying production enhancement technologies or the

Table 3. Identification of the Acquisition Business Strategy for the Characteristics of the Potential Target Property

West Coast of Africa

The assessment of potential acquisition and development targets will be categorized into three groups based on the characteristics of the property.

Group 1 Producing Fields

The possibility for taking over as the Operator with the acquisition.
The field is currently producing above the economic limit.
The field is thought to have production enhancement potential.
The government has incentives with favorable terms on any new marginal field development.
The cumulative production is a high percentage of the estimated ultimate recoverable reserves.
The current operator may consider the field to be marginal.

Group 2 Producing Fields with a Non-Operated Interest

Geographically isolated from other producing fields with interest owned by the potential seller.
A joint ownership with the operator could result in other favorable opportunities.

Group 3 Undeveloped Discoveries

The estimated reserves merit the development costs.
Cost saving and production enhancements technologies are applicable.
The government has incentives with favorable terms on any new marginal field development.
A viable development plan can be determined.
The undeveloped opportunity can be incorporated into a strategy for additional acquisitions and or business developments such as electrical power generation.

Table 4. Evaluation of the Opportunity in Gabon

Gabon Opportunity Evaluation

Pluses

Recent acreage relinquishments by Elf and others have resulted in new opportunities.

Elf, Shell, Chevron, and Mobil have reported giant discoveries on prospects drilled in deep water concessions along the West African Coast. Shallow water and onshore opportunities may arise from a change in exploration strategy.

Lower development costs and higher production rates have resulted from the application of horizontal drilling. Beginning in 1994, Elf's production from the Avocette field is expected to increase to 27,000 b/d with the addition of two horizontal wells. Elf has planned 12 additional onshore wells in the Rabi area. Shell Gabon produces over 200,000 bopd from 60 horizontal wells and 50 vertical wells with 440 million barrels of recoverable reserves from Rabi-Kounga field.

Minuses

All of the available released acreage may have been recently awarded to new companies by the government.

The existing production facilities and pipelines for both offshore and onshore are controlled by a combination of Elf and or Shell.

The current operators in Gabon have recently participated in capital expenditures indicating a strategy of expansion.

The success with horizontal drilling and other new technologies has increased the competition for the remaining exploration and exploitation projects.

Table 5. Recent Exploration Activities in Gabon

Recent Activities - Gabon West Africa

New Discovery - Marathon and Sante Fe tested 4,523 bpd, from the Tchatamba -1, located in the Kowe Marine permit. This is reported to be the first economical discovery from the Lower Cretaceous Madiela Limestone.

Horizontal Wells - Beginning in 1994, Elf's production from the Avocette field is expected to increase to 27,000 b/d with the addition of two horizontal wells. Elf has planned 12 additional onshore wells in the Rabi area. Shell Gabon produces over 200,000 bopd from 60 horizontal wells and 50 vertical wells with 440 million barrels of recoverable reserves from Rabi-Kounga field.

New Production - The development of the Hylia, Vanneau, and Pelican East fields is completed with the connection of Hylia to Vanneau.

Chauvco has began production from the onshore Remboue Permit. Production rates are projected to be 7,500 bopd by the end of 1997.

Vaalco and Petrofields are expected to bring the offshore Ekouata field on production in 1997. The estimated recoverable reserves are 7 million barrels.

Agip announced the commencement of the development of Limande field. The daily production rates are estimated to be 10,000 barrels of oil with the ultimate recoverable reserves estimated at 15 million barrels.

New Permits -	Vanco Energy	
	Anton Martin - deep water	
	Astrid Martin - deep water	
	Ocelot	
	Panthere-Nze - Exploitation	
	M'bindji	
	Shell & Amerada	Ineka block
Seismic Opt.	Marathon (100%)	
	Akoumba Marine - deep water	

New Deals - Energy Africa Gabon SA (Engen) is a newly formed company owned 25% by the Republic of Gabon and 75% by a joint venture of which Energy Africa owns 50%. The joint venture received an option on 22 exploration licenses and development rights in three satellite fields to the Rabi-Kounga field and three offshore fields.

opportunity for additional acquisitions.

Several detailed maps showing key well information were created. An example of these maps is shown in Figure 2. Each well is annotated with detailed information, including name, total depth, discovery date, and oil and gas production test rates. The well data were imported into the spatial database from a commercial data file, Petroconsultants Discovery Opportunities, which contains the well information for undeveloped discoveries in the international arena. A portion of this data file is shown in Table 6.

Having the GIS was extremely beneficial to the evaluation of the significance of the recent new business activity by the oil and gas companies. The GIS analysis of the new exploration permits, discoveries, and other industry projects validated the observations in the "minus" column in Table 4. Using the GIS, it was possible to quickly recognize the existence of a competitive situation with companies actively expanding their operations. A decision was made to not invest additional resources in this area, and a minimal amount of time and capital had been expended for this evaluation. The

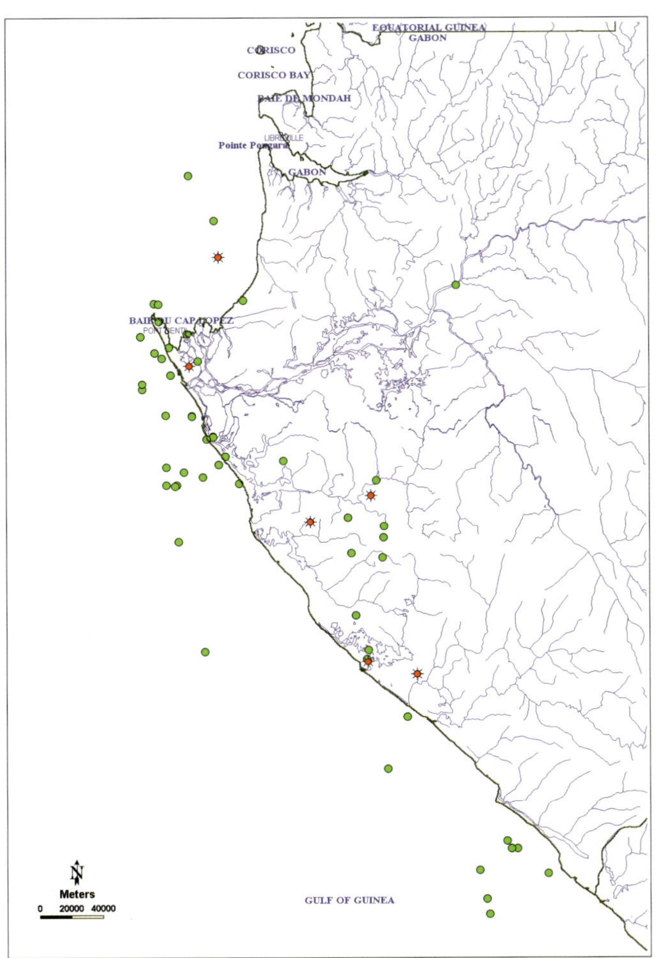

Figure 1—Map showing the undeveloped discoveries for a portion of Gabon.

GIS provided the insight necessary to make a decision regarding allocation of limited resources and,simultaneously it provided valuable data and maps that may be used in the future.

Austral Basin, Argentina

Background

The election of President Dr. Carlos Saul Menem in 1989 was the beginning of a new policy for economic stabilization with a plan of aggressive privatization of government-owned companies. A new incentive for investment in Argentina's oil and gas industry came with the implementation of Plan Argentina in 1991. At the end of 1991, President Menem announced this plan to establish an energy policy based on investment from private Argentine and foreign oil companies. The plan included an exploration licensing round covering 145 tracts in 15 basins, a reduced tax, a maximum royalty at 12%, and the elimination of government participation through Yacimientos Petroliferos Fiscales (YPF) (Keeley et al., 1992). In 1993, privatization of YPF made history as the New York Stock Exchange's largest initial offering, raising over \$3 billion. This success was an important advertisement for the new economic environment being created in Argentina.

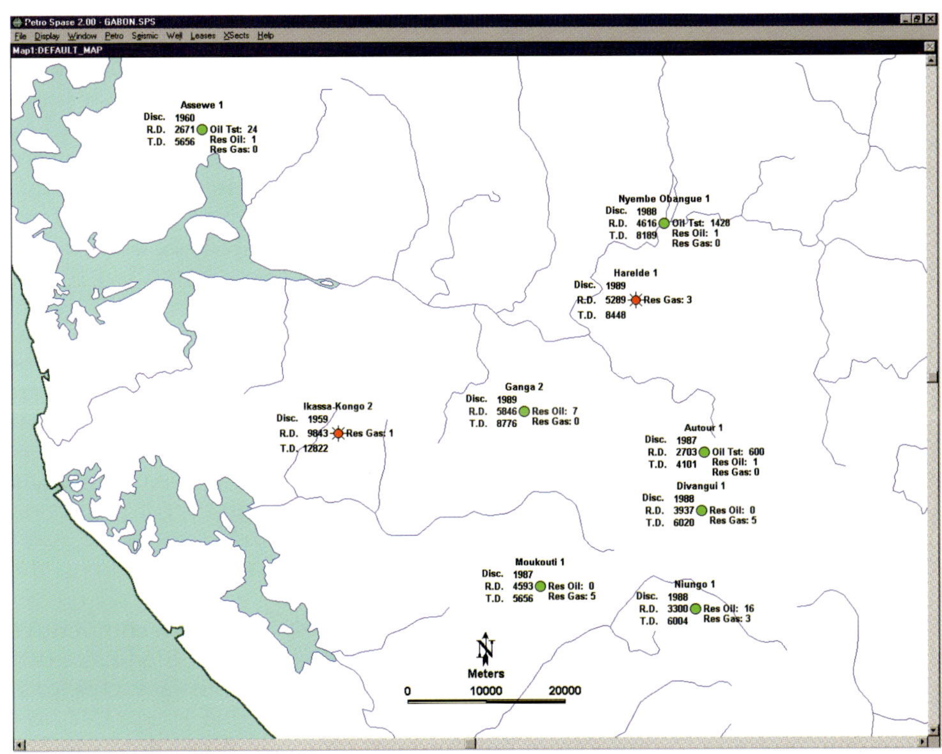

Figure 2—Screen shot of small area of Figure 1 showing the well annotation layer.

Table 6. A Sample of the Petroconsultants Discovery Opportunities Database

Gabon

Discovery Name	Latitude	Longitude	PS	Current Operator	Disc Year	Total Depth	Oil Tst Rte Bopd	Gas Tst Rte Mcfd	Est Rsrv Depth	Reserves Oil Mmbo	Reserves Gas Bcf
Moukouti 1	S 02 00 12 600	E 009 49 26 000	Oil	Elf Gabon SA	1987	5,656			4,593	0.25	
Autour 1	S 01 50 54 254	E 010 00 37 733	Oil	Elf Gabon SA	1987	4,101	600		2,703	1.00	
Niungo 1	S 02 01 48 370	E 010 00 00 140	Oil	Shell Gabon	1988	6,004			3,300	16.00	3.00
Harelde 1	S 01 40 20 000	E 009 56 04 000	Gas	Elf Gabon SA	1989	8,448			5,289		3.00
Ganga 2	S 01 48 00 740	E 009 48 22 600	Oil	Elf Gabon SA	1989	8,776			5,846	7.00	
Ikassa-Kongo 2	S 01 49 29 990	E 009 35 40 140	Gas		1959	12,822			9,843		1.00
Divangui 1	S 01 54 55 170	E 010 00 24 960	Oil		1988	6,020			3,937	0.25	
Nyembe Obangue 1	S 01 34 58 230	E 009 57 57 730	Oil	Elf Gabon SA	1988	8,189	1428		4,616	1.00	

A notable aspect of Argentina's new policy was that owners of an exploration permit would be required to relinquish 50% of the original concession acreage after the first four years to receive a subsequent exploration permit on the remaining acreage (Pucci, 1996). The intent of this policy was to maintain a high level of interest on the part of potential investors; consequently, the retrospective evaluation needed prior to the partial relinquishment of an exploration block resulted in new property acquisition opportunities for both Argentine and foreign oil companies.

Strategy

In this case, the GIS system was used to assist in the evaluation of an acquisition candidate north of the Strait of Magellan (Cuenca De Magallánica) in the Austral Basin (Figure 3). This opportunity came about not as the result of a prior screening process, but of a dedicated evaluation of several different producing properties and associated exploration areas that were offered for sale on a negotiated basis. A successful acquisition depended on accurate and timely assessment of the value of the established reserves, as well as the exploration potential of 14 producing or undeveloped fields and of approximately 4000 km^2 of exploration area.

The hydrocarbon potential is interrelated with the evolution of the basin, which can be divided into three phases: development of a substratum of intense volcanism and block faults by rifting during the Middle to Late Jurassic (Rabinowitz and La Brecque, 1979); subsidence and transgressive sedimentation following the rifting during the Cretaceous; and development of a foredeep basin beginning in the Late Cretaceous into the Tertiary in front of the pending arc collision, which was related to the onset of subduction of advancing Andean thrust sheets (Pindell and Tabbutt, 1994). The major oil accumulations have been discovered within sandstones of the Lower Cretaceous Springhill formation, which overlies the Upper Jurassic Serie Tobifera formation, a volcanic sequence of ignimbrites, breccias, and tuffs (Figure 4). Most of the wells penetrate the top 100 m of the Tobifera formation, because it is often found to be productive, but the reservoir quality has been difficult to predict. The Springhill sand deposition was controlled by the paleotopography of the grabens, tilted fault blocks, and horsts expressed at the surface of the Tobifera formation. The Springhill targets typically include the sandstone reservoir draped over anticlines or, more common to this part of the basin, a stratigraphic pinchout against topographic highs in the Tobifera formation. The Tobifera formation has been found productive on anticlines associated with Springhill production, as well as on anticlines known as "Altos Pelados" (Bald Highs), where the Springhill was not deposited.

In this particular case, the seller presented reserve estimates from two independent reservoir engineering companies. In this process, the production and pressure histories had been digitally compiled, and the estimates of remaining reserves could easily be confirmed

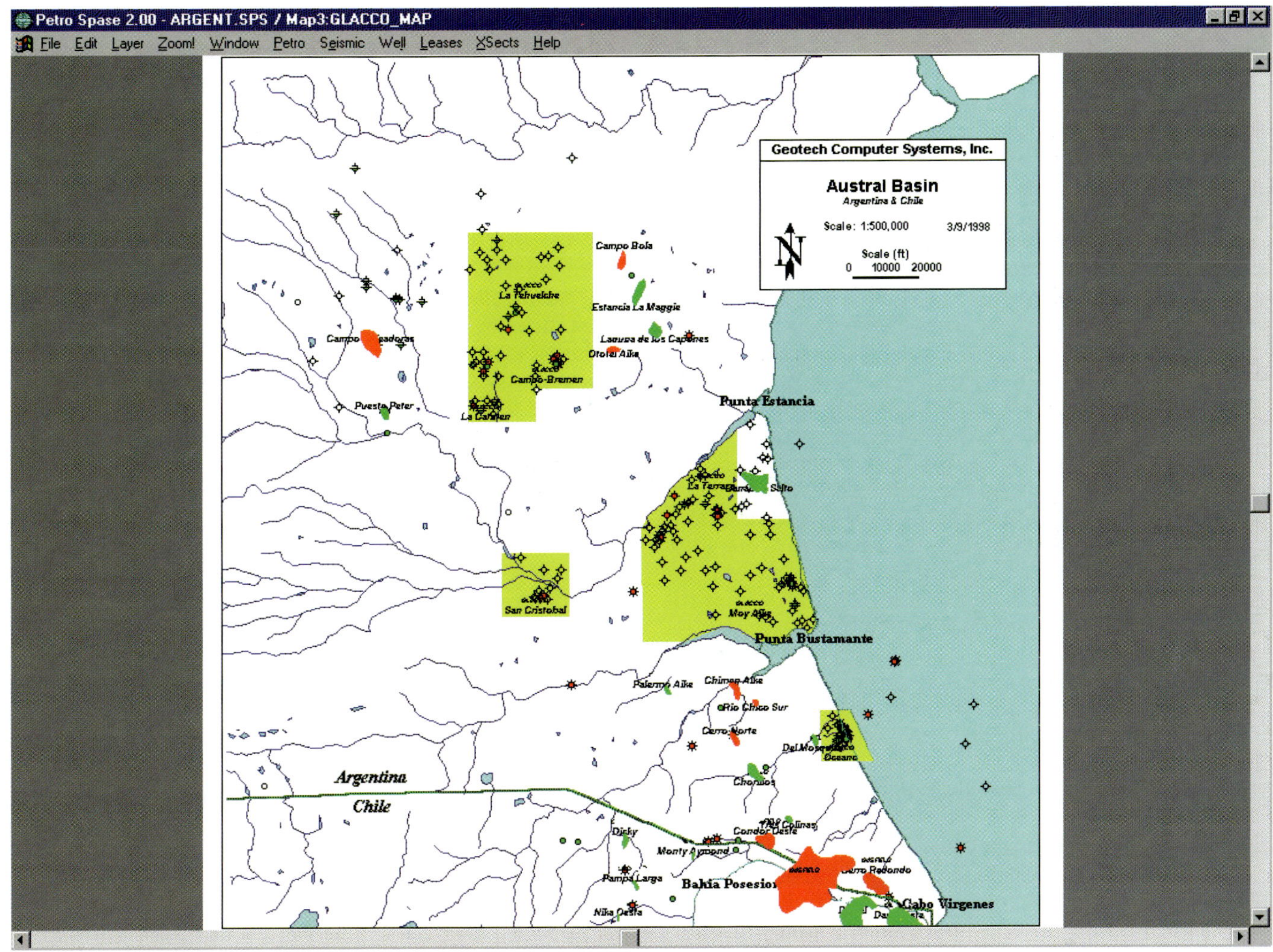

Figure 3—Map showing the producing fields in the northern Austral Basin.

through a decline curve analysis. The seller had also reprocessed approximately 500 km of the 2000 km of available 2-D seismic data, and had provided a digital base map covering approximately 1000 km^2 of the exploration area. The remaining 1500 km of non-reprocessed 2-D data, and approximately 150 well locations with associated production tests and other well history, needed to be quickly incorporated into a usable digital base map.

Procedure

The digital cartographic data covering the active portion of the Austral basin were extracted from the DCW and imported into PetroSpase as previously described in the protocol. At this point, the software features and flexibility became important because it was necessary to integrate data from a variety of sources, including information that encompassed several digital and hardcopy formats.

Before it was possible to begin mapping, it was necessary to add 100 seismic lines (some as long as 15 km) and over 150 additional well spots to the digital base. Using the existing paper base map was not practical because of poor legibility of the shot point numbers. Also, in some areas, there were problems with multiple vintages of 2-D seismic data that had been obtained with different shot point sequences and posted on what appeared to be the same line.

Due to the uniformity of the layout, YPF shot the 2-D data in a consistent pattern of a 2 km by 2 km grid. This process allowed a shortcut to be used during data entry. The seismic lines were added to the spatial database by digitizing the location and shot point for each end of a seismic line. Then the remaining shot points were added through an interpolation feature of the spatial database software.

Roughly 1000 km of 2-D data were added to the database by simply typing in the line identification information and digitizing two points for each of the 100 seismic lines. For seismic data accompanied by digital positional information (the 500 km of reprocessed 2-D seismic data), the standard software feature for importing the SEG-P1 (positional format of the Society of Exploration Geophysicists) file of shot point positions was easily modified through the SQL script editor window to provide for the variations in the Argentine file structure.

The well spots, concession boundaries, pipelines, and offset field outlines were digitized at the same time the seismic was added. To save time, only a brief well name was typed, and the well symbol was selected. After the digitizing was finished, a list of the wells containing a unique identification number, the well name, the well symbol, and the surface location in both *x–y* and lat-long was saved as an ASCII file. The digital data obtained from the seller, consisting of locations and symbols of wells in the concession area, were appended to an existing ASCII file containing well spots and production tests for approximately 125 other key wells scattered across the entire Austral basin. This file was then imported into a spreadsheet, where additional well information and production tests could easily be added. This list was checked against an inventory of electric logs; when a well was missing, the data were added to the spreadsheet from the electric log heading. The additional well data and any available production test information were then imported into the database. The standard software feature for importing well data in ASCII format was sufficiently flexible to allow the well data to be imported without coding.

A regional map was created showing the different producing reservoirs and the trapping style for each. This exhibit was beneficial in presenting the proposed acquisition in the context of the producing fields on trend (Figure 3); it became extremely important in presentations to representatives from the financing partner. Included were the acreage outline, oil and gas pipeline locations, production data for the offset fields, and the location of recent exploration activity.

A plan was designed to allocate resources to manage the remaining data by area based on an assignment of relative importance to the different data sets. For example, several well location discrepancies were known from comparing the two independent engineering reserves studies. A decision was made to research the well file data to verify the surface and bottom hole locations for all of the wells within the boundaries of an existing field. This ensured that the productive area of the original hydrocarbons in place, which was assigned to each field for material balance calculations, would be as accurate as possible.

Upon completion of an audit of the independent reserve reports, an updated reserve report was prepared. The seller, the buyer, and the financing institution agreed to a fair purchase value of the developed reserves. From the negotiations, it became clear that an assessment of the exploration potential and a determination of its value, if any, would be required. Generation of exploration leads now became the new focus of the evaluation. The potential for a new successful exploration program had been demonstrated from the recent drilling activity. The oil reserves for the two exploration areas, Santa Cruz Association Area I and II, had increased from 20.4 to 44.6 million barrels (Pucci, 1996). The increase in new reserves was attributed to the discovery of four new fields.

Because this is a combined structural/stratigraphic petroleum province, both structural and stratigraphic maps were constructed. A map of the structure on top of the Tobifera formation was initially created. This surface represents the paleotopography on which the Springhill sand was deposited. Next, a seismic stratigraphic analysis was performed, which predicted the presence of the Springhill sand based on amplitude phase changes in the 2-D data when integrated with the well control. This analysis was then overlain on the Tobifera structure map, and leads were graded as to their quality and potential. This technique had been used to identify subtle stratigraphic pinchouts from new 2-D seismic data at Estancia La Maggie oil field located in Santa Cruz Association Area I (Gambaretto, 1996). Also, the termination of reflectors against basement highs had been used to predict the Springhill pinchout at La Sara and San Sebastian fields in the southern Austral Basin (Zielinski and Bruchhausen, 1982).

The seismic interpretation technique was applied to the existing discoveries. The support of the reserve estimates with the integrated seismic raised the confidence of the interpretation. In some cases, as with the field shown in Figure 5, risked reserves were added to the value of a field by extending the boundaries of the possible reserves area.

The spatial database made it possible to react quickly to changes in the evaluation needs with changes in the focus of the negotiations.

The North Sea

Background

The 1959 discovery of the giant onshore Gronigen gas field in the Netherlands initiated the North Sea exploration drilling. This drilling then followed through the next decade. The exploration and development history of the North Sea is widely known because petroleum exploration companies drilled over 200 wells before Ekofisk field, the first major oil discovery for the North Sea, was drilled in 1969 (Watson, 1975). In the next decade, the North Sea became a world-class oil and gas producing region with a succession of new giant field discoveries.

Total oil production for the Dutch offshore has declined from a daily average of 65,000 barrels in 1986 to 45,500 barrels in 1996, as reported by the Ministry of Economic Affairs, whereas gas production has continued to increase (U.S. Energy Information Administration, 1997). Operators are faced with lower oil production revenue, higher capital requirements for enhanced recovery programs and, ultimately, expensive field abandonment costs. This situation has resulted in new acquisition opportunities for companies prepared to implement strategies to extend the productive life of existing fields and lower the development costs for shut-in gas discoveries or new fields. The available technologies include horizontal drilling, low-cost subsea completions, and a new generation of floating production platforms.

Strategy

An initial screening for acquisition candidates was based on the ratio of remaining reserves to the estimated ultimate recoverable reserves. A detailed

Age	Series/Age		Formation		Production
Quaternary	Pleistocene to Recent				
Tertiary	Neogene	Pliocene	Santa Cruz		
		Miocene			
	Paleogene	Oligocene	Magallanes		
		Eocene		Campo Boleadoras	Gas & Oil
		Paleocene			
Cretaceous	Upper	Maastrichtian	Upper Inoceramus		
		Campanian			
		Santonian			
		Coniacian		Palermo Aike	
		Turonian			
		Cenomanian			
	Lower	Albian	Margas Verdes		
		Aptian			
		Barremian			
		Hauterivian	Lower Inoceramus		
		Valanginian			
		Berriasian		Springhill	Gas & Oil
Jurassic	Malm	Tithonian		Serie Tobifera	Gas & Oil
		Kimmeridgian			

Figure 4—The post-rift stratigraphic sequence for the Austral Basin.

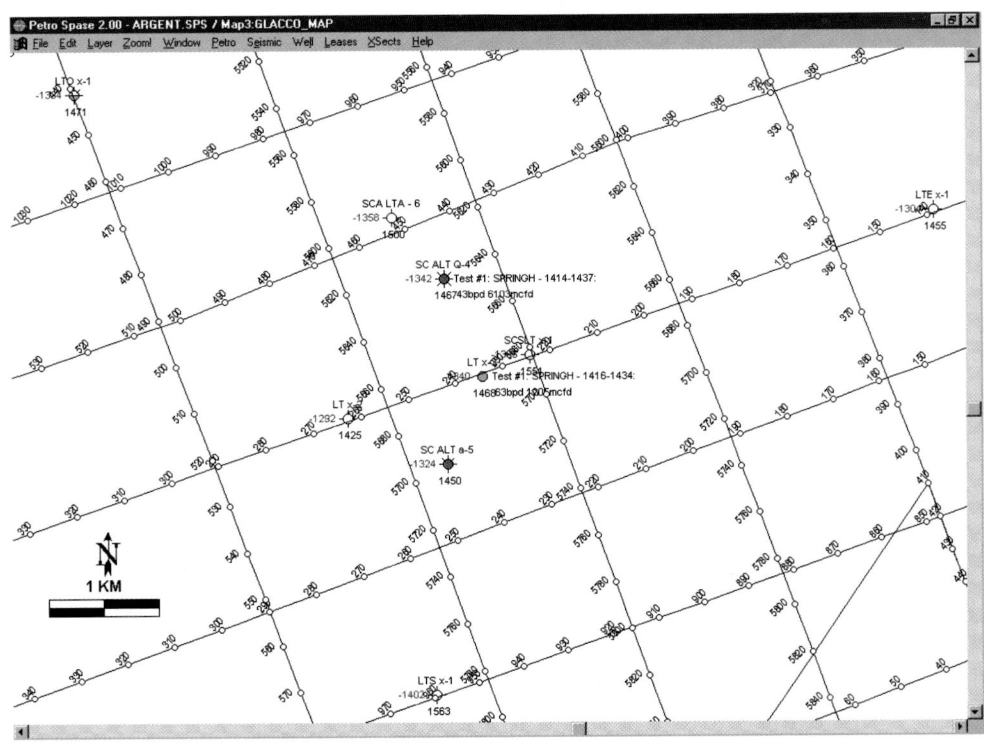

Figure 5—A detailed map showing the seismic grid and well data used in the analysis of a potential acquisition.

evaluation of producing properties based solely on the candidate list from the initial screening would most likely result in dedicating resources to the wrong acquisition lead. The complicating characteristic is the potential alliance between the operator of the acquisition target and the operator of the production platform or pipeline system. A GIS analysis linking producing fields together with the common production facilities shows this otherwise hidden relationship. In some cases, what may appear to be a motivated seller will be a competitor. The operator of an underused production platform will be attempting to acquire additional reserves to fill the vacant capacity left from declining production.

In this case, then, the premise used for the identification of a target is based on the percentage of remain-

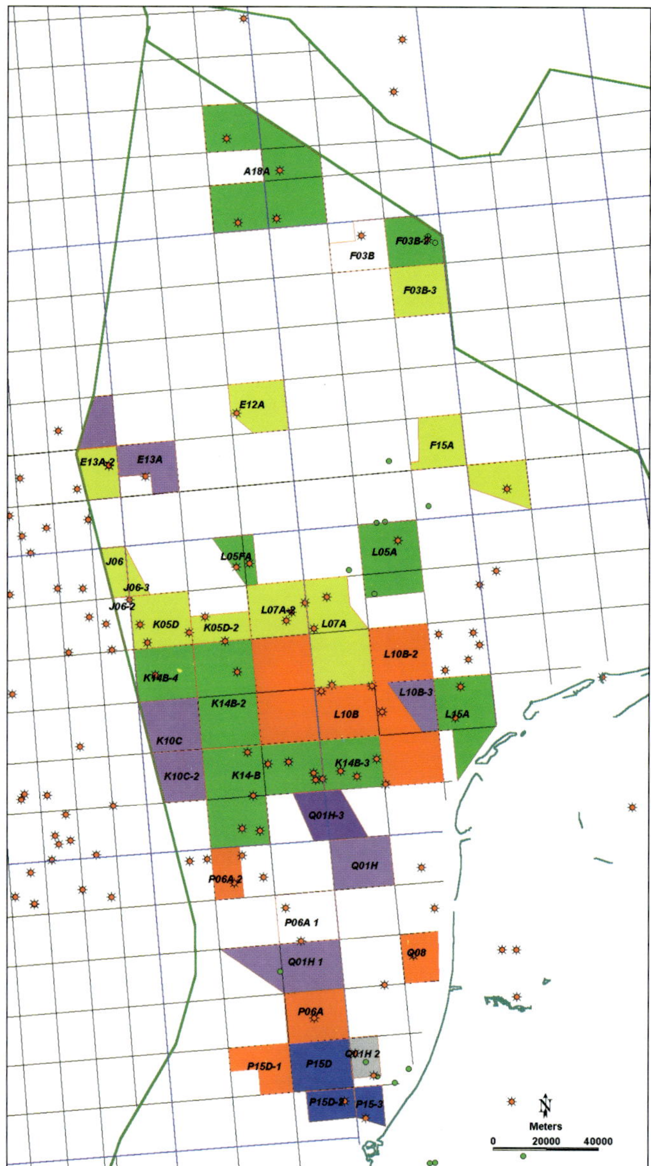

Figure 6—North Sea fields color-coded by operator.

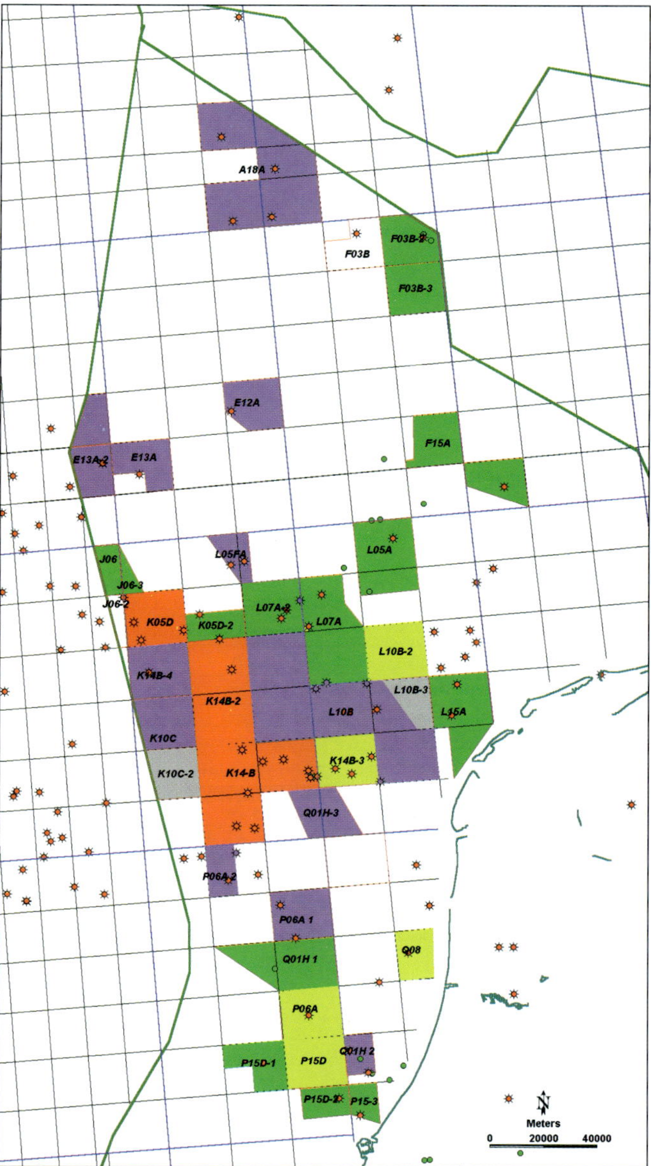

Figure 7—North Sea fields color-coded by production platform tie-in.

ing reserves versus the ultimate reserves, the potential for smaller undeveloped discoveries, and a strategy to capitalize on any additional opportunities from the owners of the production platform or pipeline system (Table 7). Examples of the maps used for GIS analysis are shown in Figures 6–9 (discussion presented in following paragraphs). The value of the data presented in Table 7 is not apparent without the use of the GIS presentation tool.

Procedure

There are approximately 175 producing fields, 30 fields being developed, and over 100 potentially commercial discoveries in the North Sea. For these properties, information on license ownership, production facilities, production history or production test rates, and economic data was gathered to forecast remaining reserves. This approach required the GIS to post various economic attributes that were calculated from the dataset and to link producing properties together by common production facilities without losing the integrity of the distinct ownership for each field.

The availability of accurate information has not been an issue for the North Sea. There are various resources for published data in digital and non-digital formats. This study combined several published non-digital resources for production history, remaining reserve forecasts, and production facility data into the concession statistics database (Table 7).

A DCW base map was generated to outline the North Sea. This map covered the coastline for the Netherlands, Norway, Scotland, and England, as well as the associated offshore international boundaries. An ASCII file containing well spots and production

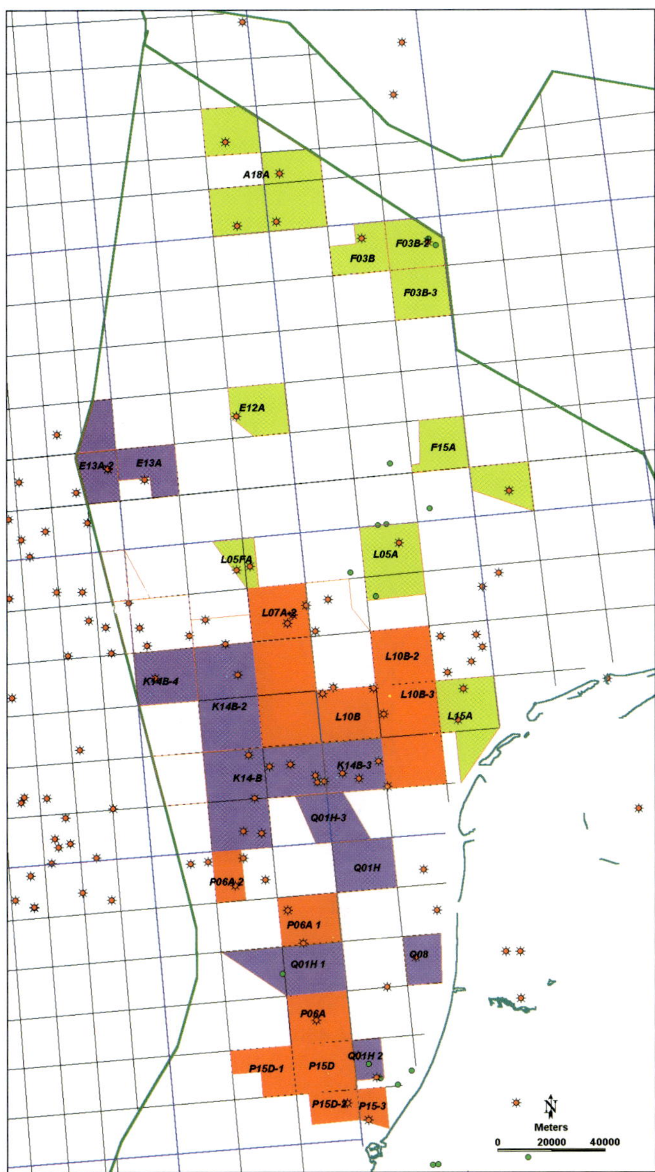

Figure 8—North Sea fields color-coded by pipeline system tie-in.

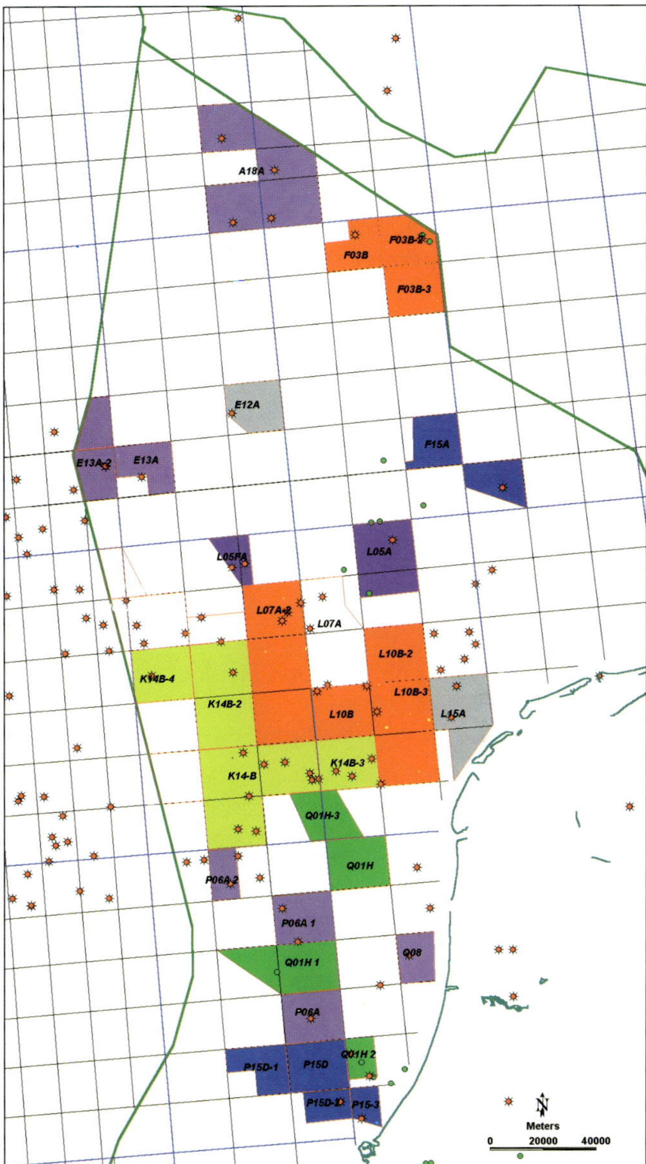

Figure 9—North Sea fields color-coded by estimated remaining reserves as a percentage of the estimated recoverable reserves.

tests for approximately 600 key wells for the North Sea was imported into the database. The well spots established reference points during the digitizing of the grid of offshore blocks.

PetroSpase contained lease management information in two separate tables linked together by a user-supplied scheme, allowing unique lease descriptions (headers) to be tied to single or multiple plats, or multiple headers to be tied to single or multiple plats. The outline of the recognized oil and gas fields, as well as the license boundaries and pipeline systems, were digitized through the lease menu as individual plats. An ASCII file was extracted from the spreadsheet data previously assembled and was imported into the existing header data fields. It was necessary to represent the common ownership in various licenses and common production facilities with the multiple plat entities representing producing fields. A hierarchy was established to represent this many-to-many relationship.

The GIS analysis included the generation of a series of maps highlighting the key characteristics outlined in the initial strategy for identifying a potential acquisition candidate. The oil and gas fields were selected based on the header data fields and presented spatially with different attributes designated by different color schemes (Figures 6–9). These figures show the data color-coded by the following criteria: operator, production platform tie-in, pipeline system tie-in, and estimated remaining reserves as a percentage of ultimate recoverable reserves.

The spatial analysis of the relationships among multiple factors helped identify potential acquisition targets.

Table 7. The Relationships Among Operators, Production Platform Operators, and Pipeline Systems

Netherlands, North Sea

Platform Location	Operator	Terminal Oil/Gas	Pipeline System	Field Name	Location	Operator	Production Status
A18A	NAM	Den Helder	NOGAT	A12 A	A12a	NAM	SI Gas
				B10 A	B10, A12b	NAM	SI Gas
				B13 A	B13	NAM	SI Gas
				B16 A	B16	NAM	SI Gas
				A18 A	A18a	NAM	SI Gas
E13A		Den Helder	WGT	Epidote	E13A	Wintershall	SI Gas
				Andalusite	D12:15	Wintershall	SI Gas
						NAM	
F03B		Tanker	NOGAT	F03 B	F02a	ONGBV	Prod. Oil
		Den Helder		F03 B	F03	NAM	Prod. Oil
				F03 B	F06	Elf	Prod. Oil
				F03 A	F03	NAM	SI Gas
F15A		Den Helder	NOGAT	G16 A	G16a	NAM	SI
				F15 A	F15ad	Elf	Prod. gas
				F14 A	F14a		Undev. SI
				F17 A	F17a	NAM	Unknown
				F18 A	F18a		Unknown
K05D		Den Helder	WGT	Markham	J03b:06	Lasmo	Prod. gas
				Charlie	J03a:06	Elf/Lasmo	
				J03 A	J03a, K01	Elf	
				K04 A	K04b:05a	Elf	
				K05 A	K04b:05a	Elf	
				K04 D	K04a	Elf	
				K04 Z	K04a	Elf	
K10C		Den Helder	WGT	K13 A:D	K13	Wintershall	ABND.
				K10 B	K10a	Wintershall	Prod. Oil
				K10 C	K10a	Wintershall	Prod. gas

CONCLUSION

GIS was used in the screening of producing property acquisition targets, evaluation of the reserves to determine the value of such acquisitions, presentation to financial partners, and for supporting credible negotiations with the sellers. The effort spent carefully specifying the objectives, defining strategies, and identifying the possible data challenges was beneficial in allocating the available time and capital resources.

Protocols that used a spatial relational database manager were developed. The software flexibility and configurability was important in working with various data sources and formats. The DCW was used to obtain the digital geographic data needed for base maps. Digital concession statistics and information on key wells for each evaluated country were used for the initial study, and proprietary data were added based on availability or need. The result was an inventory of acquisition opportunities that could be evaluated from a consistent set of basic maps for any of the countries in this project.

The international arena has proven to be a rewarding area for petroleum exploration. The competition for new prospects has increased as new technologies (deep-water completions, 3-D seismic, etc.) lower the wildcat risk. New technology has transformed the future value of older producing properties, the application of enhanced recovery technologies has extended the life of existing production, and other new technologies have transformed marginal fields or undeveloped discoveries into exploitation targets.

A successful international property evaluation is characterized by timeliness and flexibility. GIS can be a valuable tool in this competitive business environment. The insight gained from presentation of spatial databases can support decisions about which projects to pursue. Given the relatively small commitment of capital and time necessary to initiate a GIS, it is clear that searching the world for oil and gas business opportunities can be accomplished without depleting the resources needed to take full advantage of those opportunities.

REFERENCES CITED

Gambaretto, W., 1996, Advantages of the 3D seismic in the stages of the development at La Maggie, Southern Austral Basin, *in* Proceedings, Vol. 1, XIII Argentinean Geological Congress and III Congress of Exploration of Hydrocarbons, p. 329–342.

Hedberg, H. D., 1950, Developments in Africa in 1949. AAPG Bulletin, vol. 34, p. 1457–1474.

Hedberg, H. D., 1951, Developments in Africa in 1950. AAPG Bulletin, vol. 35, p. 1597–1628.

Keeley, M., S. Hogg, and C. Urien, 1992, Argentina's exploration licensing round covers 145 tracts in 15 basins. Oil & Gas Journal, vol. 90, no. 41 (October 12), p. 85–88.

Pindell, J. L., and K. D. Tabbutt, 1995, Mesozoic–Cenozoic Andean paleogeography and regional controls on hydrocarbon systems, *in* Petroleum Basins of South America (A. J. Tankard, R. Suárez Soruco, and H. J. Welsink, eds.). Tulsa, OK: AAPG, AAPG Memoir 62, p. 101–128.

Pucci, J. C., 1996, Relinquishments in Argentina will boost exploration interest. Oil & Gas Journal, vol. 94, no. 40 (September 30), p. 103–107.

Rabinowitz, P. D., and J. L. La Brecque, 1979, The Mesozoic South Atlantic Ocean and evolution of its continental margins. Journal of Geophysical Research, vol. 84, p. 5973–6002.

U.S. Energy Information Administration, 1997 (August). URL: http://www.eia.doe.gov/

Watson, J. M., 1975, North Sea-major petroleum province. AAPG Bulletin, vol. 59, p. 1098–1112.

Zielinski, G. W., and P. M. Bruchhausen, 1982, Shallow temperatures and thermal regime in the hydrocarbon province of Tierra del Fuego. AAPG Bulletin, vol. 67, p. 166–177.

Hood, K. C., et al., 2000, Use of geographic information systems in hydrocarbon resource assessment and opportunity analysis, *in* T. C. Coburn and J. M. Yarus, eds., Geographic information systems in petroleum exploration and development: AAPG Computer Applications in Geology, No. 4, p. 173–186.

Chapter 12

Use of Geographic Information Systems in Hydrocarbon Resource Assessment and Opportunity Analysis

Kenneth C. Hood
Bernard C. South
F. Dennis Walton
Otha D. Baldwin
William A. Burroughs
Exxon Exploration Company (now part of ExxonMobil)
Houston, Texas, U.S.A.

ABSTRACT

Analysis of undiscovered resource potential is a critical aspect of the petroleum exploration business. The probable number and size of undiscovered oil and gas fields, coupled with economic considerations, forms the basis for determining which areas are most favorable for exploration. Geographic information systems (GIS) provide an excellent tool for analyzing remaining exploration potential and comparing the results among areas. To be effective, the method used must be applicable to settings in which the amount of information available varies widely.

One of the best ways to assess undiscovered potential is in the context of the overall hydrocarbon system. The hydrocarbon systems approach considers the overall history of hydrocarbons, from generation and migration out of primary source intervals to entrapment and retention at reservoir levels. Each hydrocarbon system can be subdivided into one or more exploration plays or families of fields and prospects having similar geologic controls. Geologic maps are constructed for each play element (e.g., source, migration, reservoir, trap/seal) within a hydrocarbon system. Some elements, such as reservoir and seal, must be analyzed separately for each stratigraphic interval. Other elements may be the same for multiple intervals (i.e., several reservoir levels may depend on the distribution and maturity of a single source). In many cases, play element maps may be constructed by combining information from several geologic maps. As an example, a trap adequacy map may be derived from the combination of maps representing structural style and top-seal distribution.

Individual play element maps can be spatially joined to create a play summary map. The play summary map, which encapsulates the best understanding of the geologic controls on hydrocarbon distribution, can be calibrated to historical data (fields and dry structures) to quantify spatial variation in geologic risk. Locations of undrilled prospects can be spatially joined to the play summary map to provide an estimate of the geologic risk for each prospect that is attributable to the regional geology. Counts of known prospects, coupled with an estimate of the number of prospects inferred to exist in unmapped or data-limited areas, can be melded with the geologic risk to estimate the number of future fields. The risked mean volume of undiscovered hydrocarbons for each exploration play can be calculated by multiplying the risked number of future fields times the average size of future fields. Because much of the input into this overall process is based on geologic maps, the assessment results can be kept current as the maps are updated.

Although an understanding of the undiscovered resource potential is a critical piece of an exploration evaluation, it is but one component of opportunity portfolio management. Many areas provide a wide variety of other opportunity types, including discovered undeveloped and enhanced-recovery possibilities. Incorporation of all opportunity types into a GIS provides a highly effective planning tool because the results can be partitioned both spatially and by opportunity type to answer a variety of business questions. For example, undiscovered potential located beneath producing fields (deeper pool objectives) might have a different strategic or business value than undiscovered potential in untested structures. Additionally, assessment results for multiple stratigraphic intervals can be combined to identify areas containing stacked pay potential. A GIS-based analysis tool can be invaluable when attempting to develop a strategic acreage position through acquisition or trade.

INTRODUCTION

Companies engaged in international oil and gas exploration are presented with numerous investment opportunities in diverse geographic and geologic settings. The number of opportunities is so great, in fact, that they cannot all be pursued. As a consequence, opportunities must be evaluated, compared, and placed in a seriatim. The relative success of an exploration program depends in part on the quality of this seriatim because it helps form the basis for resource allocation among the various basins and exploration plays. Generally, an assessment of remaining undiscovered hydrocarbon potential represents a key input into a basin or play seriatim.

Many different schemes have been used to assess remaining undiscovered hydrocarbon potential. Ivanhoe (1986) grouped these methods into three categories: economic estimates, engineering estimates, and geologic estimates. Economic estimates basically assume that the supply of hydrocarbons is infinite and that all basic problems are those of distribution, which can be controlled by money. These methods ignore the geologic controls on hydrocarbon distribution. Engineering estimates rely on the extrapolation of historical data and mathematical constructs such as discovery index (the amount of hydrocarbon discovered per foot of drilling). Such historically based estimates give an increasingly clear picture of what has been done but are of only limited value to prospecting because they do not account for local controls on undiscovered hydrocarbon potential within an area. Geologic estimates attempt to predict the locations and volumes of remaining hydrocarbon potential. These methods are most applicable for areas in which geologic parameters such as structure, stratigraphy, depth, and size of the prospective area are reasonably well known. One method of obtaining geologic estimates of undiscovered potential is through the assessment of exploration plays.

The focus of this paper is the application of geographic information system (GIS) technology to geologic assessment of the remaining undiscovered potential for exploration plays. Much literature has been dedicated to the details of traditional geologic basin and play assessment (e.g., Miller, 1986; White, 1992; Porter, 1995). The approach described here is well suited for exploration plays having moderate to large amounts of historical data to constrain the results. Remaining undiscovered potential is an important factor influencing decisions about the basins in which exploration effort and resources are to be invested. Even for basins with significant undiscovered potential, however, other factors can affect the financial return on an investment. These factors include the size of potential discoveries, the number of wells required to make a discovery, the availability of other opportunity types (such as discovered undeveloped hydrocarbon resources), and the location of resources relative to facilities and to each other. These factors, which are inherently spatial in nature, are well suited for GIS analysis. The application of GIS capabilities allows business analyses to extend beyond traditional exploration play assessment of undiscovered potential into opportunity analysis. Although discussed only briefly here, opportunity analyses represent the ultimate prize available to those who are willing to invest the time and effort required for GIS-based play assessment.

BENEFITS OF GIS IN ASSESSMENT

The phrase "GIS-based assessment" is used here to refer to an assessment in which GIS is employed as the primary data repository and spatial analysis tool. GIS figures into nearly every aspect of the process—defining play boundaries, building field size distributions, developing chance-of-adequacy maps, calibrating play summary maps to historical data, screening and counting closures, displaying and analyzing results, and sundry other details. Nearly all assessment inputs and outputs are directly built from or attached to polygons, lines, or points in GIS datasets. Throughout this paper, references to maps imply the existence of GIS datasets. For example, a polygonal map refers to a GIS dataset that contains polygon topology. Our analysis method requires that key assessment information be organized and built into a suite of GIS datasets and that the GIS application has the capability to accomplish spatial joins on stacks of maps to build new composite products (Figure 1). Each polygon on the play summary map represents a geologic subplay, or an area that is geologically homogeneous at the scale of assessment mapping. Assessment input parameters and output results can be determined for each geologic subplay. For example, the play chance values on the play summary map are calculated by multiplying the play adequacy values attributed to each of the play elements. These stacks of spatially joined maps serve as direct input into established statistical assessment programs. In contrast, many purely statistical assessment techniques (e.g., discovery indexes, extrapolation of historic trends) are one step removed from the exploration map sets. In these techniques, data tables are abstracted from original map sets in one step and then analyzed and revised separately throughout the rest of the assessment process. Because the maps represent the clearest expression of the geologic interpretation of the play, this two-step process occasionally results in unrealistic assessments and obscures any spatial relationships contained within the output results.

One of the more time-intensive aspects of the traditional assessment process has been the creation of maps and the manipulation of map-based information. In the past, information was manually extracted from paper maps and analyzed using conventional statistical methods. The manual nature of these processes resulted in a number of consequences, including the following:

- Different formats necessitated data translation between maps and tabular datasets. Errors, such as data entry, data omission, and drafting, were possible.
- Data simplification, thereby compromising the maximum resolution, frequently was required to facilitate manual analysis.
- The assessment process was designed to answer a relatively narrow set of questions about the undiscovered potential of predefined areas. Adapting assessment results to answer unanticipated questions was potentially cumbersome and often not practical.
- Because the assessment process was time intensive, opportunities to perform the iterations necessary to evaluate sensitivities were limited. Sensitivity cases, when worked, were largely performed by holding the geologic interpretation static and manipulating only the statistical component of the evaluation (e.g., field size distributions).
- Decoupling of assessment calculations from the original maps precluded the extrapolation of assessment results into other areas.

GIS represents a major step forward in providing tools that allow resource assessment to be carried out in a systematic, quantitative, and internally consistent manner. Fundamental work processes can be improved such that analyses can be more thorough. Moreover, the utility of the assessment results is extended because they can be analyzed in combination with any other GIS-based data. Several of the advantages of a GIS assessment process over a manual process are as follows (see discussion in next section):

- *Direct ties between maps and databases:* GIS-based maps are explicitly linked to underlying tabular databases. Pertinent categorical and numerical information is directly associated with map features.
- *Facilitation of the hydrocarbon systems approach:* Because a GIS can determine spatial relationships among many different maps, geologic analyses can be simplified into basic hydrocarbon system components (e.g., reservoir, trap/seal, source/maturation, and migration/timing). These geologic components can be worked independently and the results combined to form the basis of an assessment. Sensitivity cases can be evaluated by using alternative digital geologic maps of any of the hydrocarbon system components.

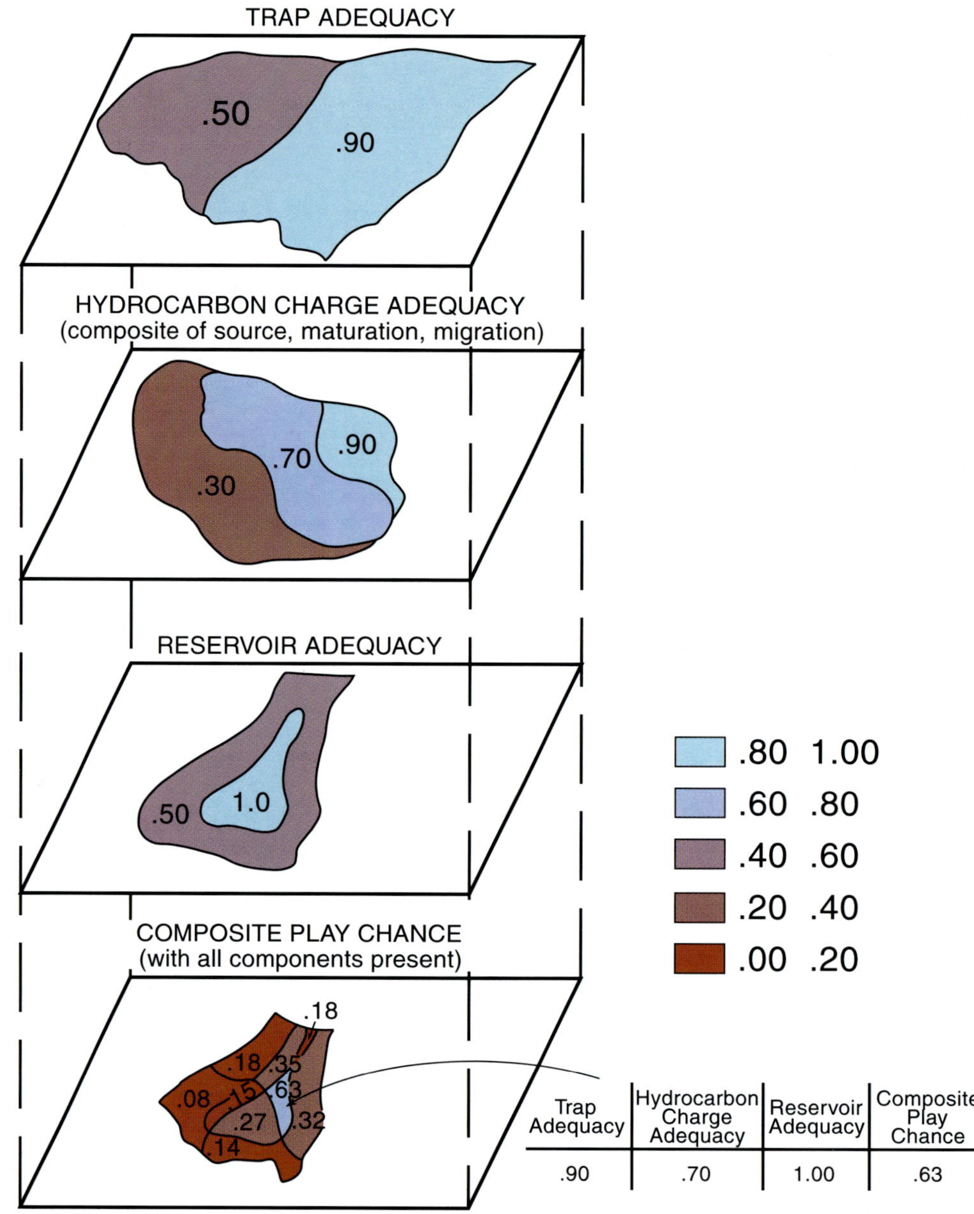

Figure 1—Example showing the spatial joining of individual play element maps to create a play summary map.

- *Better quality control:* By using a GIS, many assessment input parameters can easily be converted into maps or can be automatically abstracted from maps. Visual inspection and analysis of maps by a knowledgeable geoscientist is inherently more efficient and reliable than scrutinizing records in a tabular database.
- *Interactive query and analysis of results:* The link between map display and the underlying databases is the most powerful feature of a GIS. For the explorationist, use of a GIS means that all data associated with a map feature are available for recall, display, and analysis, or can be recombined with other GIS-based datasets to form a basis for further analysis. For instance, a map of hydrocarbon potential could be spatially joined with concession outlines to extract information required to build a simple block or concession seriatim.

ASSESSMENT METHODOLOGY

This section describes a basic methodology used to perform geologic play assessment. Our approach, which is particularly well suited for GIS analysis, is based on prospect counts. Simply put, the method attempts to determine the number of remaining prospects in an exploration play, including prospects that have not yet been identified (postulated prospects), the proportion of those prospects that will ultimately become fields (geologic risk analysis), and the volume of hydrocarbons that will be contained within each new field. The mean resource, risked according to the future number of viable fields (risked mean resource), is calculated by multiplying the number of prospects times the future success ratio for prospects times the average future field size. Other measures of the undiscovered potential, such as mean success case (mean undiscovered resource volume if the play works or contains at least one field) and high-side case (undiscovered resource volume with a 5% exceedance probability from Monte Carlo results), are determined statistically. The current approach uses proprietary software to perform the assessment calculations for each geologic subplay (the basic entity of assessment). This software, which predates the introduction of GIS into the assessment process, explicitly handles dependencies among subplays and uses Monte Carlo simulation to estimate ranges of possible outcomes.

Whereas the assessment method used here is conceptually simple, considerable effort is required to ensure that the values used for the various input parameters are geologically reasonable and are consistent with historical data. Each of the critical parameters can be represented by one or more GIS-based maps, making it possible to explore interrelationships among the various geologic controls. Several relevant aspects of the assessment methodology, as well as key parameters and thresholds, are considered.

Hydrocarbon Systems Approach

Although many business analyses will ultimately depend on knowing the undiscovered potential for the entire explorable geologic section, it is expedient to subdivide the strata into smaller packages that can be analyzed separately. To this end, a hydrocarbon systems approach to assessment has been adopted. The hydrocarbon systems approach attempts to consider the complete history of the hydrocarbons, from the generation and migration out of effective source intervals to entrapment and retention at various reservoir levels. Hydrocarbon systems can be subdivided into one or more exploration plays, defined (after White, 1992) as groups of prospects and related fields having common hydrocarbon source rocks, migration relationships, reservoir formations, seals, and trap types. Key is the fact that the prospects share common elements of geologic risk. In practice, most exploration plays are defined on the basis of a common reservoir interval and trap style (e.g., Eocene structural play). As would be expected, many controls on hydrocarbon distribution are specific to each stratigraphic interval (e.g., reservoir/seal distribution and quality). Other controls may be in common among multiple stratigraphic intervals (e.g., the distribution and maturation history of a single source interval may control the hydrocarbon distribution at multiple reservoir levels). Many controls on hydrocarbon distribution can be represented using topologically defined geologic maps, thus making them amenable for a wide variety of GIS-based analyses.

Geologic Subplays

The basic geologic entity of assessment in the exploration play approach is the geologic subplay. Subplays are represented in a GIS as a series of polygons, each of which has the same family of geologic attributes (e.g., play parameters). Geologic subplays are produced by spatially joining individual play element maps (Figure 1). Each geologic subplay is considered to be geologically homogeneous at the scale of mapping for that exploration play. For each subplay, input parameters, such as number of prospects, prospect chance of adequacy, size distribution of future fields, and commodity ratios, must be determined. Most, or all, of these parameters can be extracted from various geologic input maps. The geologic subplays and the associated assessment parameters are defined for specific stratigraphic intervals. If the stratigraphic interval included in the analysis changes, the subplays and many of the parameters have to be modified accordingly. Data for each subplay are imported into the proprietary assessment software, which calculates the expected number of future fields and risked mean, success case, and high-side resources using conventional statistical procedures.

Geologic Risk

A key assessment input is the geologic likelihood that each prospect will contain a pool of hydrocarbons of at least a specified minimum size (geologic risk). This risk, coupled with the number of prospects (feature counts), directly controls the number of future fields that is expected to be found within each geologic subplay. The geologic risk is represented by both marginal and conditional risk components (e.g., Baker et al., 1986; White, 1992). The marginal probability, or play chance, represents the chance that the exploration play exists at all. For frontier areas, the marginal probability can be a significant component of the overall risk, whereas in established exploration areas the margin probability is commonly close to 1.0. The conditional probability, or future success ratio, represents the proportion of prospects that will ultimately become fields, given that the play exists (accounting for local perturbations in prospect geology). The risked number of future fields for a play is calculated by multiplying the play chance (marginal probability) times the future success ratio (conditional probability) times the number of prospects (feature counts).

In our approach, the play chance and future success ratio are both determined by analyzing the adequacy of each major play element (reservoir, trap/seal, source/maturation, and migration). Because these play elements are assumed to be independent of each other, the composite adequacy can be calculated as the product of the chance of adequacy for each individual play component. Figure 1 illustrates the derivation of play chance by spatially joining the individual play element maps and calculating composite adequacy values as the product of the component parts.

It is important to keep in mind that a calculated risked mean resource for an unestablished exploration play that includes a significant marginal risk does not represent a volume of hydrocarbons that is actually expected to be found. Each play would be expected to contain no resource if the play fails, or a success case volume if the play succeeds. The risked mean resource can be used to aggregate the volume of hydrocarbons that would be expected for a number of potential exploration plays, including both successes and failures. This property of the risked mean resource is important in GIS applications, as will be shown in following paragraphs. For a more detailed discussion of the concepts of marginal and conditional probabilities, and for examples of the application of these two risk components, the reader is referred to the literature in which these concepts have been developed (e.g., Baker et al., 1992; White, 1986; and references therein).

Feature Counts

Feature counts by subplay are required to determine several assessment input values. Features can include not only mapped prospects but also postulated prospects, fields, and structures that have been tested dry (exploration failures). Counts of fields and dry structures are used to calculate historical success ratios, which often form the starting point for estimating future success ratios. To perform feature counts, certain screening thresholds must first be established. One is the closure size that separates features large enough to be systematically identified and counted (large features) from those that are not (small features). This parameter is known as the large trap minimum (LTM), and it represents the smallest size for which an accurate census (count) can be made. Another threshold is the volume of hydrocarbons that separates closures classified as exploration successes (fields) from those classified as exploration failures (dry). This parameter is known as the large field minimum (LFM), which it generally represents the volume of hydrocarbons that would be expected in a successful trap the size of the LTM. Fields with an area smaller than the LTM are not counted. Fields that have hydrocarbon volumes less than the LFM are considered dry for the purposes of assessment feature counts. Both can contain minor volumes of hydrocarbons. These hydrocarbons are represented in a small field fraction, and the associated volumes are determined using one of several statistical processes. Note that the classification of a discovery as an exploration success or failure for the purposes of assessment calculations does not indicate whether it can be economically developed. Economic determinations require that additional local factors, such as water depth, available infrastructure, and distance to facilities, be taken into account. If appropriate, these can be applied as spatial filters after the initial assessment calculations.

Field Size Distributions

To assess the volume of undiscovered hydrocarbons for an exploration play, the sizes of future fields must first be estimated. The future field size distribution (FSD) can be determined in one of several ways, depending on the data available. If all prospects have been assessed, individual prospect sizes can be used to build a distribution. For many plays, however, only a few of the prospects will have been assessed individually. In such cases, existing fields and discoveries that are representative of the remaining prospects can be used. These can be based on previous discoveries within the play for mature plays or on representative fields from analog areas for frontier plays. Only closures that exceed the LTM should be included in the FSD. FSDs are truncated at the lower end by the LFM and, for mature basins, at the upper end by the largest size of field that could potentially remain. Invanhoe (1976a, b, c) addressed the importance of selecting appropriate truncation points for FSDs.

Commonly, fields from several geologic subplays will have similar size distributions and can be combined to form a single FSD. Care must be taken, however, not to combine fields from different structural or stratigraphic regimes that have different size distributions. The interactive query and analysis capabilities of a GIS can be used to select fields graphically based on location within subplays or by using other attributes in related data tables (e.g., structural style or reservoir age) to assemble representative FSDs. Although the details are beyond the scope of this paper, the method used to construct an FSD can have a profound influence on the assessment results. Once the FSDs have been created, log-probability plots can be constructed for the field data, and specific probability cases can be linked to the appropriate geologic subplays and captured as assessment input. For example, Figure 2 illustrates a field size distribution plot for about 1000 reservoirs from multiple stratigraphic intervals. Subsets of these data would be selected to apply to individual exploration plays. Key parameters, such as the minimum and maximum truncation points, the P50 and P16, can be extracted for use as input in the assessment calculations.

Commodity Ratios

In many exploration areas, the economic viability of a hydrocarbon accumulation will depend not only on its size but also on the type of hydrocarbon present (oil versus gas). Where appropriate, the probability of obtaining oil versus gas, or the relative proportion of oil versus gas, can be included as an assessment input. This input can often be provided by polygonal maps that are attributed based on regional trends in the relative abundance of oil and gas (Figure 3). The polygons can be defined by analysis of field data in mature or semi-mature areas, or by source type and source maturity polygons from a source adequacy map. The commodity ratio map can be joined spatially to the play summary map to attach the appropriate information to each geologic subplay.

GIS ASSESSMENT METHODOLOGY—DATA TYPES

Data organization is a critical component of the GIS-based assessment process. The assessment process builds on two basic classes of data: regional play element maps and closure-specific information. The regional play element maps are used to define the geologic interpretation for each assessment interval based on how mappable variations in geology influence the

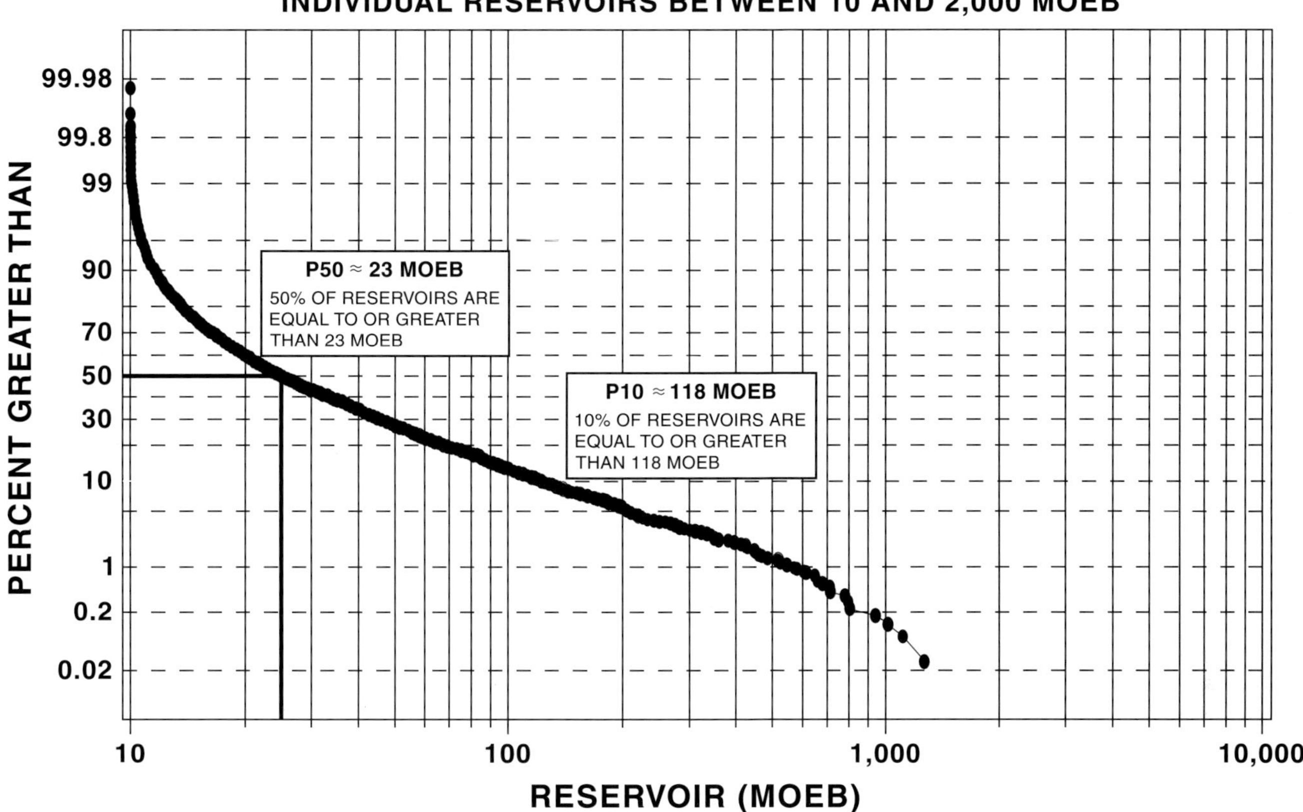

Figure 2—Example field size distribution plot.

distribution of hydrocarbons. The closure-specific information provides the ground truth, namely data about fields, dry structures, and prospects. The calibration of regional play element maps to closure-based historical success information provides a better way to constrain geologic interpretations, as well as an extremely flexible and powerful method to estimate the location and size of future fields.

Regional Play Element Maps

Regional play element maps are the foundation upon which geologic subplays and estimates of geologic assessment parameters are built. Play element maps are typically constructed for each of the major play elements: reservoir, trap/seal, source/maturation, and migration/timing. These elements represent primary controls on the distribution of hydrocarbons within an exploration play. Many different types of data can be incorporated into play element maps, including reservoir facies, source facies, source kitchens, and migration fairways. Play element maps can be simplified or subdivided as appropriate for the local geology and assessment goals. For example, the source distribution, source maturation, and migration play components can be combined into a single hydrocarbon charge map. Conversely, the reservoir play component map could be subdivided into reservoir presence and reservoir quality subcomponent maps. These maps constitute the geologic hypothesis or interpretation of the play or hydrocarbon system. Most of the maps are based on polygons, although grids may be appropriate for some map types.

A play summary map is constructed by spatially joining in a GIS the individual regional play element maps (Figure 1). Areas on the play summary map that have the same set of values for all the spatially joined play element maps are, by definition, geologically homogeneous at the scale of mapping and are termed "geologic subplays." Non-contiguous areas with the same geologic attribute values can also be considered part of a single subplay.

Closure Database

The closure database contains information that is specific to individual closures. Closure information includes such attributes as closure height, closure area, trap type (e.g., four-way closure or fault dependence), exploration status (e.g., undrilled, wet, or flowed oil), and hydrocarbon volume (fields) or undiscovered potential (prospects). Many elements of this dataset are independent of the geologic subplay in which the closures are located. All known closures, whether fields, dry structures, or untested prospects, are included in a single closure database. Grouping closures regardless of their exploration status facilitates comparisons between fields and dry structures and enables knowledge to be extrapolated

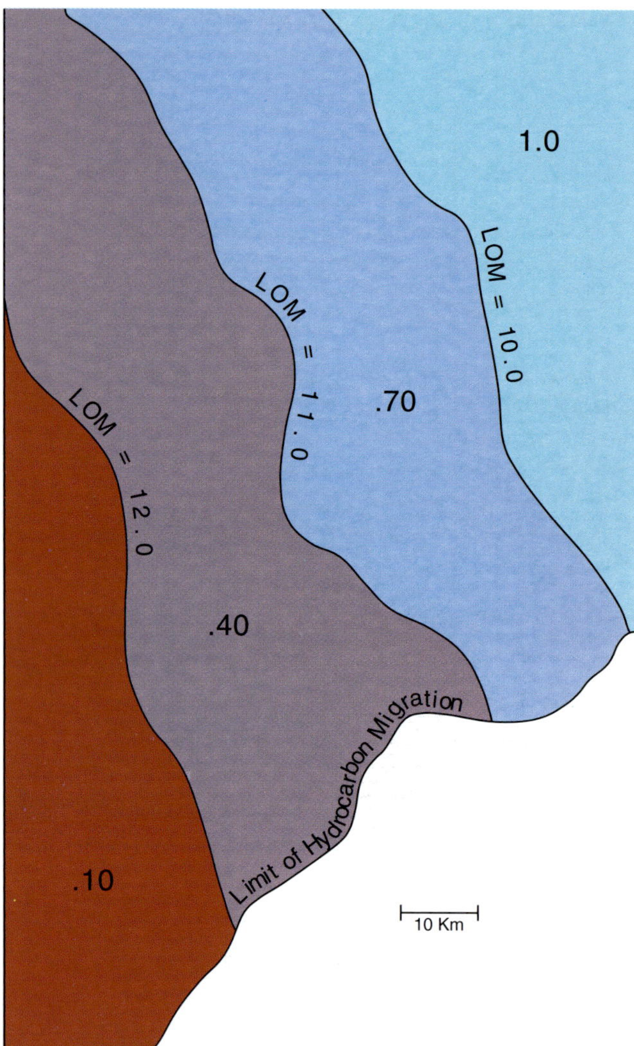

Figure 3—Hypothetical source maturity map that could provide input into the commodity ratio or chance of oil assessment parameters.

to undrilled prospects. Most closure information is tied to specific reservoir intervals. This is important because the size and structural style of a closure can vary with depth, and because reservoir quality can be an important control on the volume and producibility of reservoir hydrocarbons. Because the closure database can be used outside of the assessment process, closure information should be captured in as much detail as is practical (i.e., individual reservoirs/fault block level). Once stratigraphic assessment intervals have been selected, the appropriate assessment input parameters can be obtained by aggregating the more detailed information (e.g., hydrocarbon volume and type, closure area, and exploration status). Detailed information of this sort is typically captured for fields as part of the normal exploration process. To support rigorous play assessments, effort must also be put into capturing comparable information for dry closures. The closure database contains the ground truth of observed exploration results that must be correlated to the regional interpretation (play element maps) for an accurate play analysis and assessment.

The closure database discussed here is built around a general closure model. The basic data entity in the closure model is a "compartment," which is the smallest object to which reserves can be assigned. At a minimum, each compartment has a stratigraphic identifier, an outline, an estimate of the hydrocarbon volume and type (for fields) or undiscovered potential (for prospects), an exploration classification, and an assessment classification. A single closure might have several laterally distinct compartments at the same stratigraphic level, and closures with stacked pays or stacked objectives might have multiple vertically stacked compartments. The complexity and extent of compartments can vary among closures or even at different stratigraphic levels within a single closure, depending on the amount of data available. Compartments within well-established fields might be defined down to the level of individual reservoirs and fault blocks. In contrast, a single compartment within an undrilled prospect or a poorly known field might represent an entire assessment interval. Individual compartments must be limited stratigraphically to a single assessment interval.

The exploration and the assessment classifications assigned to each compartment are both critical to the assessment process. The exploration classification is an observation about the drilling status of the compartment (undrilled, wet, flowed oil, etc.). The assessment status is an interpretation of what the exploration status means for the compartment as it relates to the assessment process. The assessment classification of a compartment could be a success (discovery), a failure (dry), or a prospect (untested). This interpretation is not simply derived from the exploration status. For example, a compartment that contains a show in an updip position might be considered dry, whereas a compartment that contains a show in a downdip position might still be prospective in the updip portion. Proper assignment of the assessment classification draws on the experience of the assessor and a knowledge of the local geology. If a single assessment interval within a closure contains multiple compartments, then the compartment classifications must be aggregated into a single classification for the assessment interval for use as assessment input. We use a hierarchical approach that favors discoveries, then prospects, then dry holes. For example, if an aggregate contains one compartment that is classified as a discovery and two that are classified as dry, the aggregate classification would be a discovery. Each compartment retains the original, as well as the aggregate, classification. One of the powers of this approach is that, after the database is constructed, individual classifications can be compared to aggregate classifications to identify compartments that might represent bypassed opportunities.

Experience has shown that a great deal of care is required in applying the closure database to obtain reliable feature counts. When stratigraphically defined plays stack vertically, a single structure (tested by an individual well) can be classified differently at the various assessment intervals. For example, a single structure tested by one or more wells can be cataloged as dry at a shallow level, a discovery at an intermediate level, and prospective at those levels beneath the limits of drilling. Additionally, a great deal of judgment needs to be exercised to determine if a well represents a true test of a compartment (e.g., penetration of the flank of a structure with a show of hydrocarbons versus penetration of the crest of a structure that lacks significant hydrocarbons).

GIS ASSESSMENT METHODOLOGY—TECHNIQUES

Determining Large Trap Minimum and Large Field Minimum

Establishing the LTM and LFM enables the closure-based assessment approach to be applied to plays with sparse or uneven data coverage. The data distribution may be sufficient to identify all of the large closures for the play or for a representative area within the play (calibration area). If a calibration area is used, then the number of large closures likely to be found in areas of limited mapping can be estimated by applying the closure density from within the calibration area. Prospects that have not yet been identified are termed "postulated prospects." The interactive map query and graphing capabilities of a GIS can be used to help determine the LTM and LFM. By spatially joining the closures to the play summary map, each closure is assigned to a geologic subplay and inherits all of the geologic attributes of that subplay. Frequency distribution plots of closure area for prospects within a geologic subplay, or within a group of geologically similar subplays, provide one method of determining the LTM. The LFM can be estimated by cross-plotting hydrocarbon volume against closure area from historical data, or it can be calculated using representative values for closure height and reservoir properties. The LTM and LFM impact both the calculated historical success ratio and the resulting field size distribution (see following paragraphs).

The use of a GIS allows the construction of a closure database that is independent of discrete thresholds such as the LTM and LFM. LTM and LFM are not considered when assigning the initial assessment classification in the closure database. They are dynamically applied to the database during the feature counting process, so they can be varied from play to play as appropriate for the closure area distribution. This powerful capability allows thresholds to be changed and the features to be dynamically reclassified for sensitivity analysis or with changes in the amount of available data.

Calibrating Chance of Adequacy Maps

Once the regional play element maps and the closure database have been assembled, GIS plays a major role in identifying trends in hydrocarbon distribution. For example, closures that are optimally located within a migration fairway might have a much higher historical success rate than those that are not, thereby validating the interpretation and allowing the extrapolation of favorable geologic trends into less explored areas.

Ideally, the calibration of prospect chance of adequacy to historical success rates will be based on the individual play element maps. Using either rigorous statistical methods or more conceptual approaches, the geology represented on each play element map is classified and calibrated to prospect chance of adequacy. The numeric adequacy value, which is between 0.0 and 1.0, represents the contribution of that play element both to the historical and to the implied future success. The adequacy of each play element often varies geographically across the exploration play. This variation is represented by the polygons on the play element maps (Figure 4). A play summary map is created by spatially joining the four individual play element maps (Figure 5). The likelihood that a closure contains hydrocarbons depends on the adequacy of all four play elements. If a single element fails, then the closure will fail regardless of the quality of the remaining elements. Based on the assumption that the four play elements are independent, the expected success ratio that applies to large closures in each geologic subplay can be calculated as the product of the four play-element adequacy values.

By spatially joining the closures to the play summary map, the regional geology and the associated adequacy values can be treated as attributes of each closure (Figure 6). This geographic linkage enables the chance of adequacy maps to be calibrated to historical data. The appropriate method of calibration depends in part on the amount of data available. For plays with large amounts of historical data, multivariate statistical analysis can be used to extract the relationships among the geology, hydrocarbon distribution, and historical success ratio. For plays with limited amounts of data, adequacy values can be assigned based on exploration concepts about how the geology is expected to affect the hydrocarbon distribution. These concepts can then be compared to historical success rates for consistency and refined in an interactive process. Two measures can be used to evaluate the suitability of the adequacy values: the overall fit to historical data and the discrimination between fields and dry structures. If sufficient data are available, then closure-specific information can also be incorporated into the risking process. For example, four-way closures may have a higher chance of adequacy than fault-dependent traps within the same geologic subplay. The calibration of historical success rates to regional geology requires that the LTM and LFM screening thresholds first be established.

Once the chance of adequacy for geologic subplays on the play summary map has been calibrated to historical data, it can be used to predict the future success ratio of undrilled prospects (Figure 7). The predicted future success ratio for the exploration play can be higher or lower than historical success rates, depending on how the geology of undrilled prospects compares to the geology of closures that have already been drilled. Commonly, the best parts of the exploration

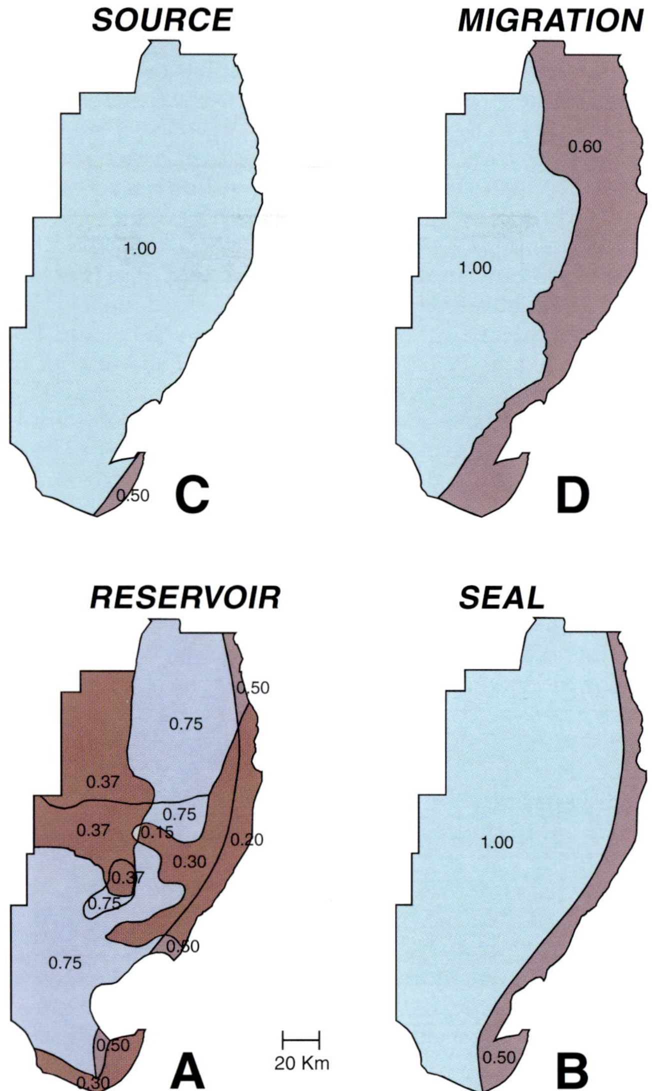

Figure 4—Individual play element maps for a hypothetical play: (A) reservoir adequacy map, (B) seal adequacy map, (C) source adequacy map, and (D) migration adequacy map.

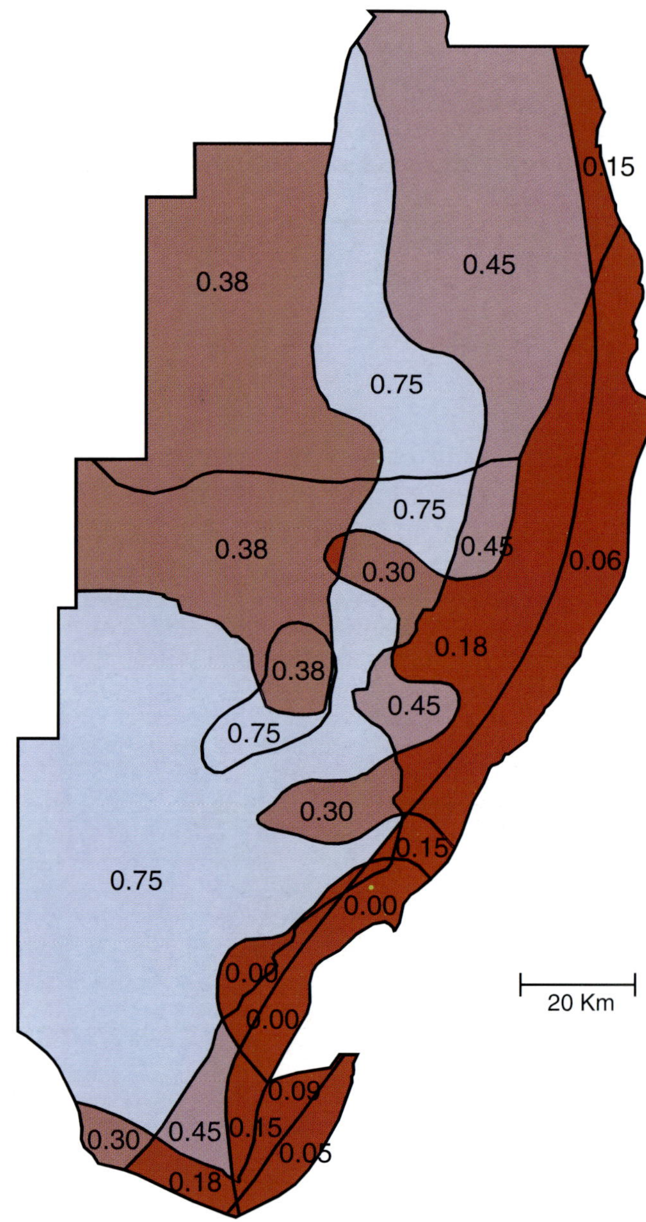

Figure 5—Play summary map calibrated to prospect chance of adequacy, or future success ratio.

play will be drilled first, and the future success ratio will be lower than the historical ratio. Occasionally, the future success ratio could be higher than the historical ratio, such as the case where drilling locations are optimized to test a different play concept at a deeper stratigraphic level.

Automated Feature Counting

Once the play summary map has been constructed and the appropriate threshold parameters (LTM and LFM) established, the closure screening and counting process can be automated using GIS. Drilled structures are screened for assessment classification, LTM, and LFM to establish the historical success rates and to build field size distributions for each geologic subplay. Undrilled structures are screened for assessment classification and LTM to establish known prospect counts for each subplay. Subplay area and ultimate large closure density from a calibration area are used to postulate additional prospects in areas of limited mapping. The future success ratio to be applied to each geologic subplay can be extracted from the calibrated play summary map. Other parameters, such as commodity ratio (oil versus gas), are determined by spatially joining closures or geologic subplays to the appropriate maps. Results from all of these processes are used to build input parameters for each geologic subplay that feed directly into the statistical assessment software.

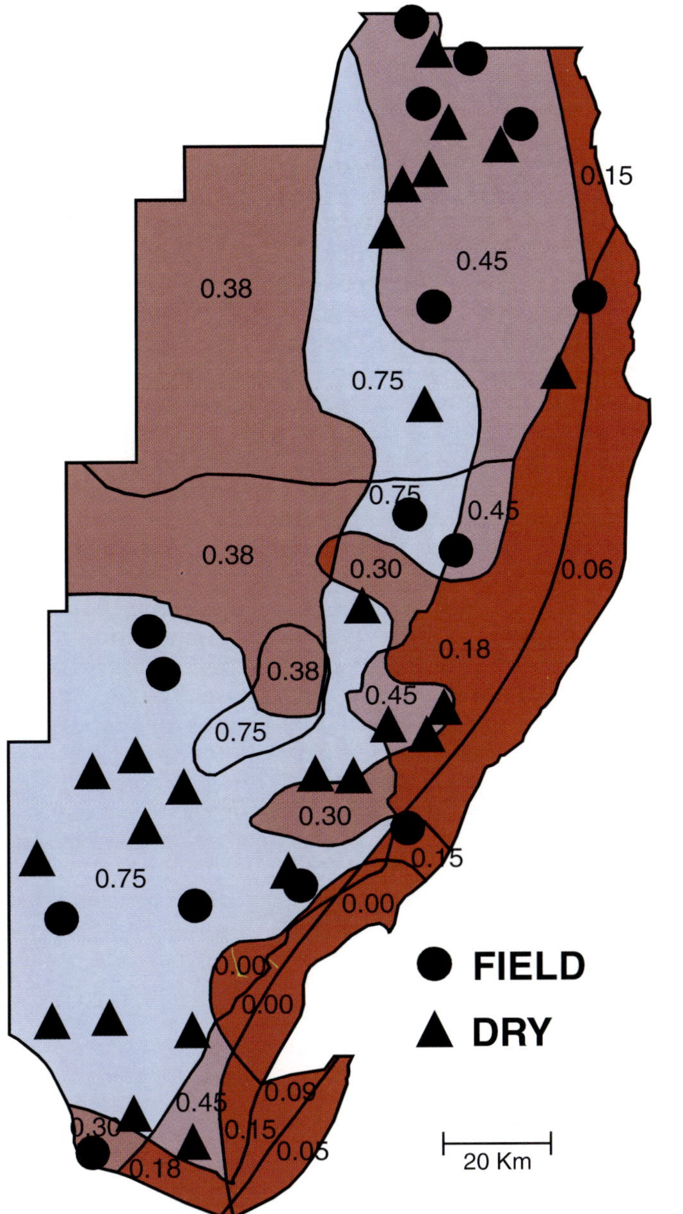

Figure 6—Calibration of prospect chance of adequacy on the play summary map to historical data.

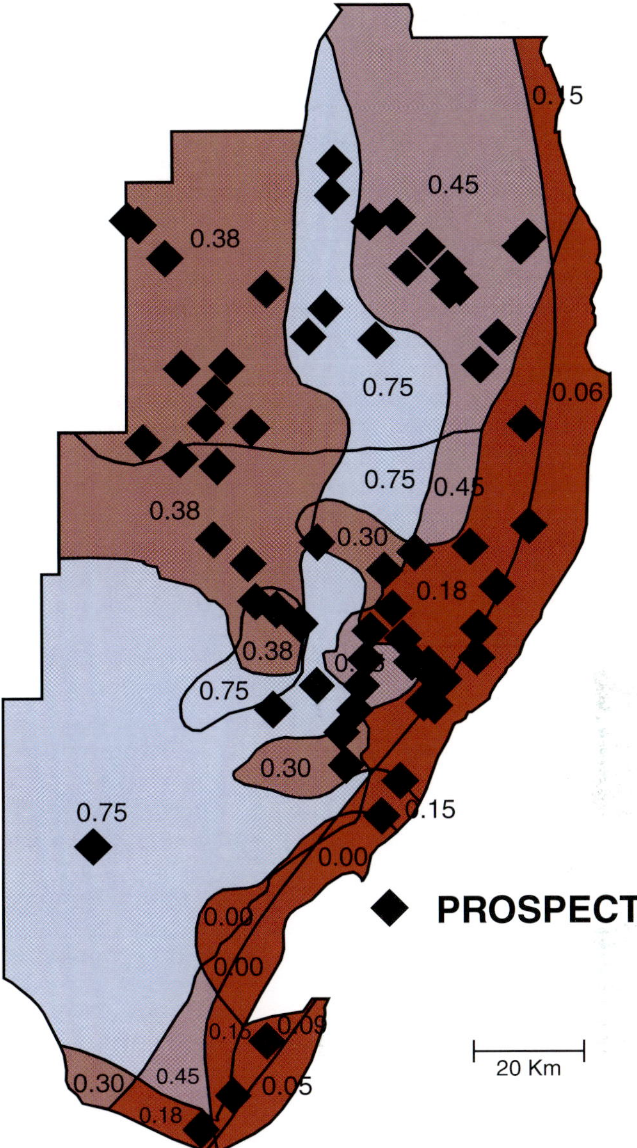

Figure 7—Extrapolation of prospect chance of adequacy to undrilled prospects by spatially joining the prospect locations to the play summary map.

Linking Assessment Results Back to Maps

Once the risked mean, success case, and high-side undiscovered potential for each geologic subplay are calculated, the results are exported from the assessment software to a database and linked back to the play summary map for display, interactive query, and further business analysis. One way to illustrate geographic variations in undiscovered potential is to divide the assessment results for each geologic subplay by the area of the subplay to create an undiscovered potential density map. Such maps allow easy identification of areas that are more prospective due to some combination of closure density, future success ratio, and future field size. If multiple assessment (stratigraphic) intervals have been evaluated, then multiple undiscovered potential density maps can be joined spatially to quantify the remaining potential for the entire stratigraphic column. These "answer" maps provide an objective baseline for the comparison of undiscovered potential between different areas. For example, the undiscovered potential for one or more plays could be apportioned to concession blocks, basins, or countries, and compared one to the other. The interactive querying and analysis capabilities of GIS provide a powerful tool for extracting information from the assessment results for use in making a variety of business decisions.

BEYOND ASSESSMENT TO OPPORTUNITY ANALYSIS

A traditional play assessment provides a quantitative estimate of the remaining undiscovered hydrocarbon potential for one or more exploration plays. Although important, the traditional play assessment is only part of the information that is ultimately required to determine which basins provide the best business opportunities. Not only is the total volume of undiscovered hydrocarbons important, but also the number, size, and availability of potential exploration targets. Because exploration targets can be subdivided stratigraphically for the purposes of assessment, the results must be recombined to obtain an accurate census of the number and size of potential targets. In addition, exploration opportunities represent just one component of an investment portfolio. Other opportunities, such as discovered undeveloped hydrocarbon pools and fields suitable for enhanced recovery techniques, can represent significant business drivers for many basins.

For known closures, the closure database enables undiscovered potential to be combined with information about other opportunity types in a spatially correct, integrated manner. The three-dimensional distribution of hydrocarbons and undiscovered potential in the stratigraphic column is critical to this objective. Because the locations of postulated prospects are unknown, the resource derived from this component of the assessment is generally not suitable to many spatial opportunity analysis techniques.

For plays with large amounts of data, a number of specific opportunity analyses can be performed. Using the closure database, the undiscovered resource calculated using the statistical assessment software for known closures can be allocated back to the proper geographic location and stratigraphic interval. This allows the undiscovered potential for multiple plays to be evaluated in terms of the expected number of fields or the potential that could be tested by a single well. It also allows the undiscovered potential to be separated out into deeper pool objectives in drilled structures versus potential in undrilled closures. As part of the procedure presented here, all discovered hydrocarbon pools are classified as to whether they have been developed. In many areas, undiscovered potential beneath producing fields may have a different business value than undiscovered potential in untested closures. If the producing intervals are operated by others, then the deeper potential may not be available as an exploration target. In contrast, undiscovered potential beneath discovered undeveloped hydrocarbon pools may provide development opportunities with additional exploration potential. Finally, allocating undiscovered potential back to known closures provides block seriatim capabilities that more closely honor the available geologic inputs than undiscovered potential density maps. Because of its ability to handle the spatial and vertical relationships, GIS is critical to extracting these and other business opportunities from a basic resource assessment.

CONCLUSIONS

This paper has shown the adaptation of GIS capabilities to a well-established resource assessment technique. Assessments done in the manner described represent a major step forward in both efficiency of process and in business utility of the final products. As more datasets get built and supported by the organization, a strong element of synergy develops because datasets can be integrated spatially. This flexibility in spatial recombination of data allows ideas and concepts to be explored and tested. For example, consider the use of GIS to study the controls on hydrocarbon distribution. Spatial datasets of depositional environment, seismic attributes, reservoir properties, source rock properties, and source maturation can be joined spatially to mapped field outlines and volumes to determine both qualitative and quantitative relationships. Hydrocarbon occurrence can be analyzed by any mapped attribute (oil volume by depositional environment, source richness, and seismic character; gas volume by source type, source maturation level, and reservoir lithology; etc.). Essentially, this process greatly enhances the conventional search for significant hydrocarbon potential by elucidating geological controls on hydrocarbon distribution.

The use of a GIS system also allows geologic work to be decoupled from other more downstream aspects of exploration and yet, allows datasets to interact efficiently to answer some of those important follow-on questions. This has been one of the themes of this paper. Geologic judgment and spatial analysis are used to deliver an estimate of the hydrocarbon potential, including both the number and volume of potential discoveries, as well as the spatial distribution of the resource. The resultant data can be combined with other spatial datasets to provide business-oriented answers. For instance, spatial interaction of an assessment with a concession offering provides a basis for a seriatim of blocks. As another example, a geologic assessment could be combined with a set of pipeline construction proposals to determine a route that would optimize the pipeline's ability to capture potential hydrocarbon reserves.

Geoscience, at its core, is an inherently spatial discipline in which most data ultimately tie back to maps. GIS will find an increasingly central role in this process. This paper has examined the use of GIS in resource assessment. The spatial component of GIS allows assessment results to be linked with other datasets to support more effective business decisions.

REFERENCES CITED

Baker, R. A., H. M. Gehman, W. R. James, and D. A., White, 1986, Geologic field number and size assessments of oil and gas plays, in Oil and Gas Assessment: Methods and Applications (D. R. Rice, ed.). Tulsa, OK: American Association of Petroleum Geologists, AAPG Studies in Geology 21, p. 25–31.

Invanhoe, L. F., 1976a, Evaluating prospective basins:

Part 1. Oil and Gas Journal, vol. 74, no. 49 (Dec. 6), p. 154–156.

Invanhoe, L. F., 1976b, Evaluating prospective basins: Part 2. Oil and Gas Journal, vol. 74, no. 50 (Dec. 13), p. 108–110.

Invanhoe, L. F., 1976c, Evaluating prospective basins: Part 3. Oil and Gas Journal, vol. 74, no. 51 (Dec. 20), p. 82–84.

Invanhoe, L. F., 1986, Oil discovery index rates and projected discoveries of the free world, *in* Oil and Gas Assessment: Methods and Applications (D. R. Rice, ed.). Tulsa, OK: American Association of Petroleum Geologists, AAPG Studies in Geology 21, p. 159–178.

Miller, B.M. 1986, Resource appraisal methods: choice and outcome, *in* Oil and Gas Assessment: Methods and Applications (D. R. Rice, ed.). Tulsa, OK: American Association of Petroleum Geologists, AAPG Studies in Geology 21, p. 1–23.

Porter, E., 1995, Are we running out of oil? Chicago, IL: American Petroleum Institute, Policy Analysis and Strategic Planning Department, API Discussion Paper #81, 56 p.

White, D. A., 1992, Selecting and assessing plays, *in* The Business of Petroleum Exploration: Treatise of Petroleum Geology (N. H. Foster and E. A. Beaumont, eds.). Tulsa, OK: American Association of Petroleum Geology, Handbook of Petroleum Geology, p. 87–94.

Barrell, K. A., 2000, Conducting a field study with GIS: Port Hudson field, Tuscaloosa trend, east Baton Rouge Parish, Louisiana, *in* T. C. Coburn and J. M. Yarus, eds., Geographic information systems in petroleum exploration and development: AAPG Computer Applications in Geology, No. 4, p. 187–194.

Chapter 13

Conducting a Field Study with GIS: Port Hudson Field, Tuscaloosa Trend, East Baton Rouge Parish, Louisiana

Kirk A. Barrell
Geodynamic Solutions, Inc.
Houston, Texas, U. S. A.

ABSTRACT

A multi-discipline team was assembled to reevaluate Amoco (now BP/Amoco) Production Company's Tuscaloosa producing assets. Port Hudson field, the largest field in the trend, was targeted for a reservoir evaluation. A geographic information system (GIS) was designed and implemented to manage, analyze, and present spatial and tabular datasets for the project. Geographic datasets covering the entire Tuscaloosa trend were digitized and input into a centralized database. The GIS was used to integrate multiple disciplines, databases, and software applications. The software technology that was employed by the team included geophysical interpretation, log analysis, and spreadsheet programs. The GIS was designed to be the tool for integrating, accessing, and analyzing data spatially. The GIS approach presented an entirely new, unique, and efficient business process. This paper represents a case study of using GIS technology to perform an exploitation field study.

INTRODUCTION

Port Hudson field is located in east Baton Rouge Parish, Louisiana. Amoco Production Company (now BP/Amoco) discovered Port Hudson field in 1977. The field produces from the Upper Cretaceous Tuscaloosa Formation and, to date, has produced over 500 billion ft^3 of gas. The field is a complex faulted anticline occurring over a piercement salt dome.

A multi-discipline team was assembled to reevaluate Amoco Production Company's Tuscaloosa producing assets. Port Hudson field, the largest field in the trend, was targeted for a reservoir evaluation. The objective of the project was two-fold:

- Define the original and remaining hydrocarbons in the field; and,
- Create and implement a plan to recover the remaining reserves.

It was determined that the latest state-of-the-art technology would be employed to achieve the project objectives. A 3D seismic survey was acquired over the field to image this complex structure.

GIS TECHNOLOGY

Geographic information systems (GIS) is a rapidly emerging technology that enables the spatial analysis of datasets. Once the geographic relationships of data are determined, many types of analysis can be performed on individual or multiple map (data) layers. The maps from the GIS are a representation of one or more map layers displayed on top of one another. Layer-to-layer queries and iterative and dynamic evaluations are possible. Symbology is the ability to use color, size, thickness, or style

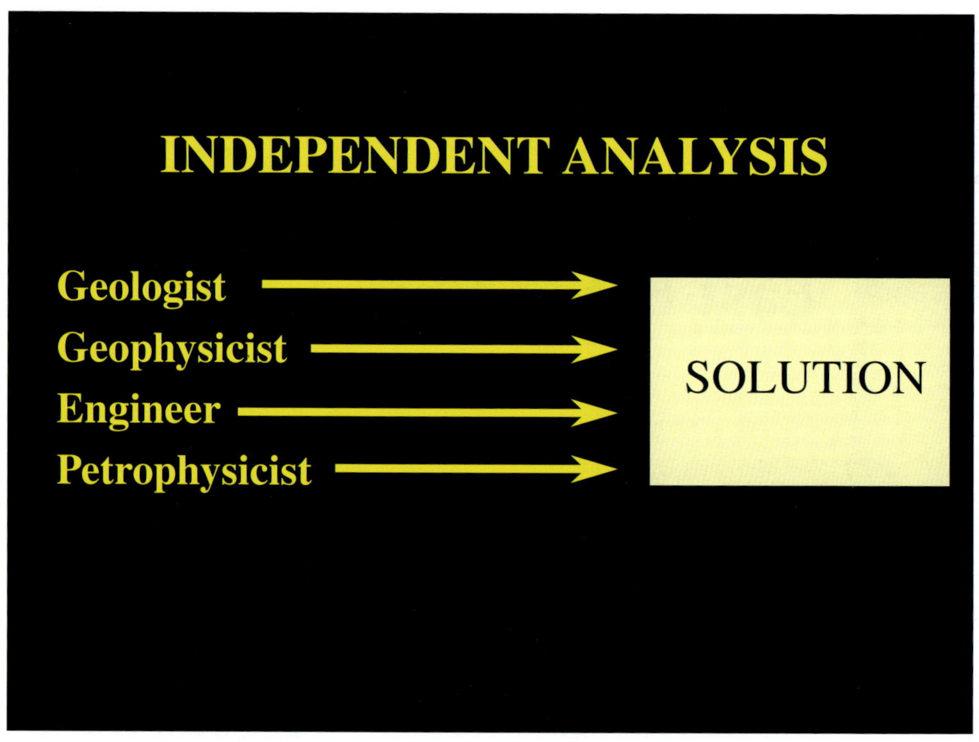

Figure 1—A traditional field study evaluation process.

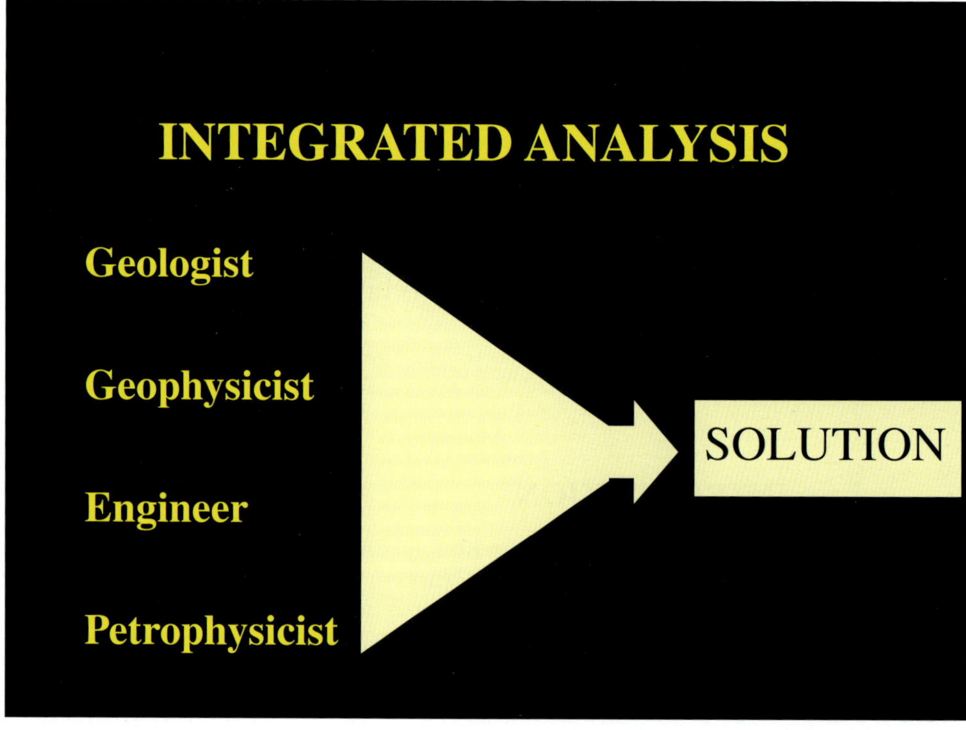

Figure 2—A field study evaluation process using a geographic information system (GIS).

to represent an attribute of the features from the map layer. Using digital data greatly improves the accuracy of the map. GIS technology can greatly improve the integration of people, data, and software. Recent decreases in cost and complexity have made GIS more accessible to a much larger user base.

EVALUATION PROCESS

Multi-Discipline Team Approach

In 1991, at the onset of the Port Hudson field evaluation, Amoco Production Company was evaluating

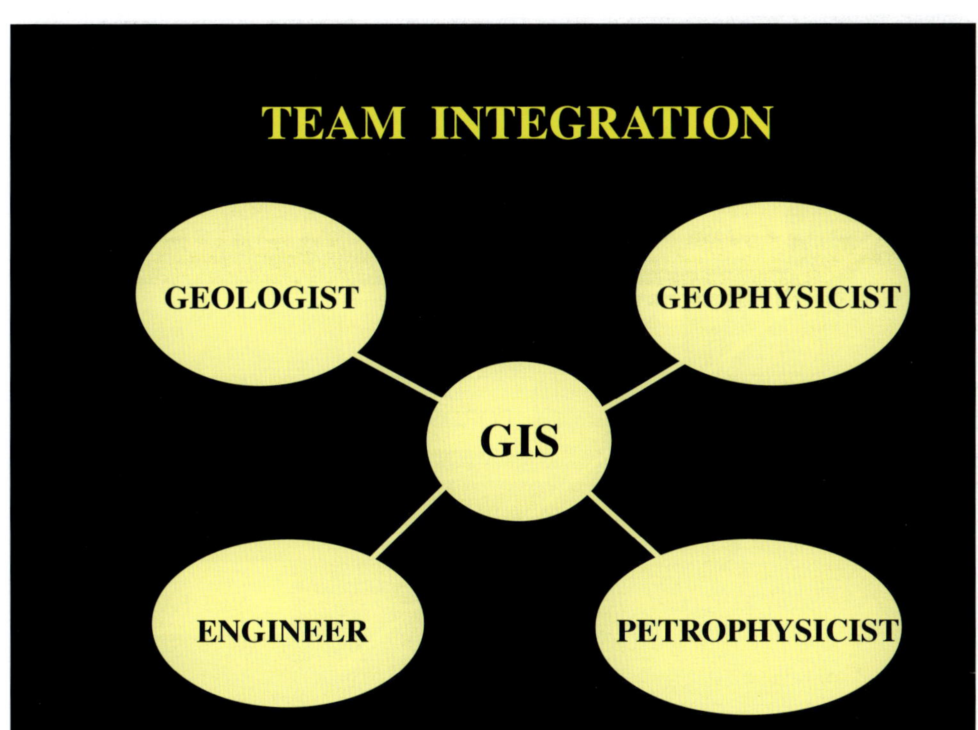

Figure 3—Multi-discipline team integration through the use of geographic informations systems (GIS).

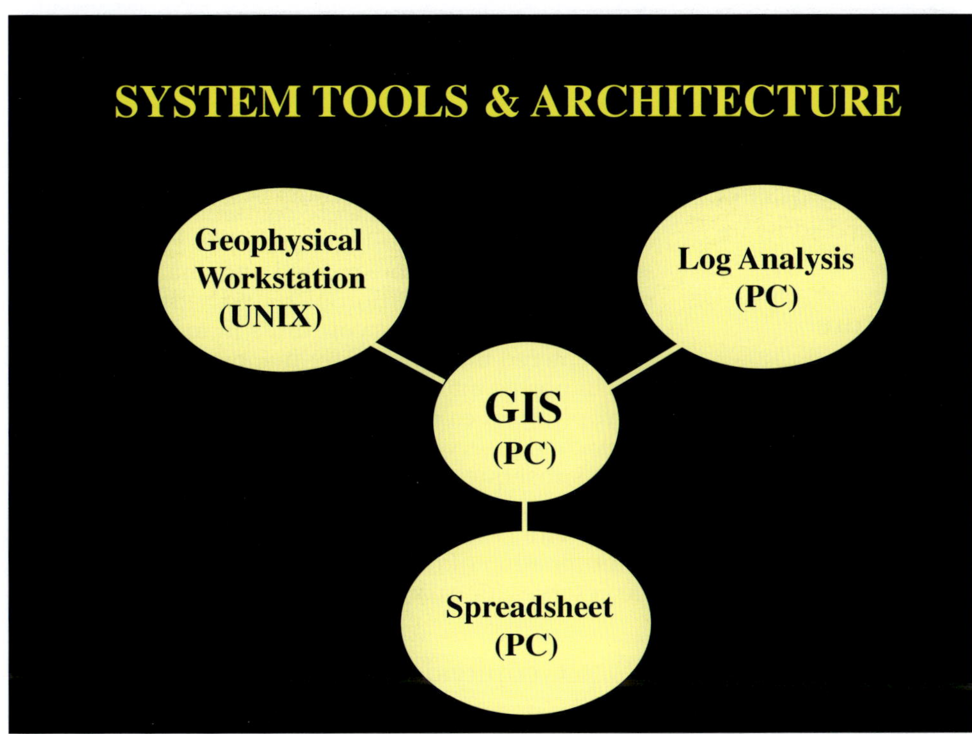

Figure 4—System architecture and tools for a geographic information system (GIS).

business processes across the corporation. During this time, the multi-discipline team approach for reservoir evaluation was considered.

Geologists, geophysicists, engineers, and petrophysicists usually perform exploitation evaluation. Historically, petroleum exploitation has been conducted through independent analysis of multiple individuals from different disciplines (Figure 1). Following the reservoir analysis process, the individuals come together to integrate the independent evaluations into a final solution. As illustrated in Figure 2, the recent trend in the industry is to form multi-discipline teams and approach

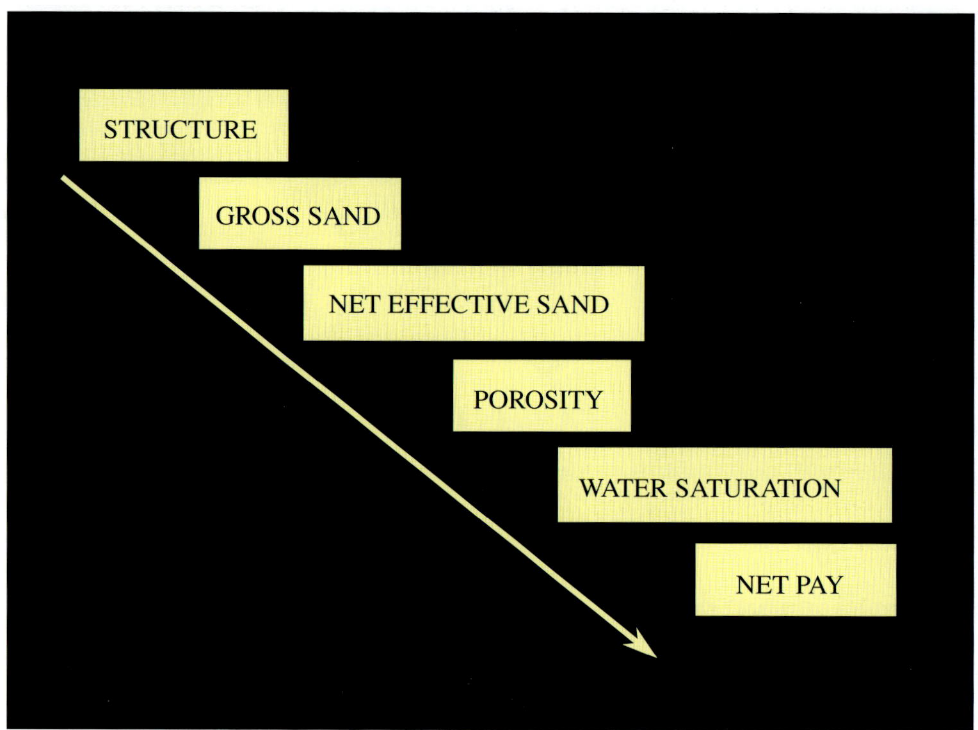

Figure 5—Field study mapping process.

the analysis using an integrated approach (Sneider, 1986).

A multi-discipline team was assembled to perform the Port Hudson field reservoir evaluation. The team consisted of a geologist, a geophysicist, a reservoir engineer, and a petrophysicist. The GIS was used to integrate multiple disciplines, databases, and software applications (Figure 3).

System Tools and Architecture

One of the goals of implementing the GIS was to integrate existing software applications. The software technology that was employed by the team included geophysical interpretation, log analysis, and spreadsheet programs. The GIS was designed to be the tool for integrating, accessing, and analyzing data spatially (Figure 4). All geophysical software resided on a UNIX workstation, whereas the log analysis and spreadsheet software resided on a personal computer operating under Microsoft Windows. The 3-D seismic survey was interpreted on the geophysical workstation, and map layers were exported into the GIS. Geophysical map layers included top-of-sand structure contours, top-of-sand faults, and sand isochron contours. Log analysis was performed by the petrophysicist, and calculated results were imported into the GIS as database tables. These tables were relationally joined to the well map layers, which enabled the log calculations to be presented and analyzed spatially on the map. Reservoir net pay contour areas were directly calculated by the GIS. These values were exported into a spreadsheet for reservoir volumetric calculations.

Geological Evaluation Process

During the reservoir evaluation, a significant amount of geological research and analysis was performed prior to using the GIS. Conventional core measurements were evaluated to determine reservoir rock types. These rock types were combined with well log data to create a rock-to-log correlation. Well log cross-sections were then constructed and analyzed, which resulted in the definition of 16 individual reservoir units. Tops of the reservoir units in measured depth were entered into the GIS for mapping. The tops were also imported into the log analysis application to allow petrophysical calculations to be performed on each individual reservoir unit. Reservoir unit calculations were subsequently imported back into the GIS for mapping of multiple attributes on all 16 reservoir units. The result of the mapping exercise was the determination of original gas in place and remaining gas reserves for each individual reservoir unit.

Geophysical Evaluation Process

Using the geophysical workstation, synthetic seismograms were incorporated into the evaluation. Tops were imported from the GIS and merged into the synthetic seismograms. Fieldwide correlatable, continuous time reflectors were defined and mapped. Faults and stratigraphic features were also identified and mapped. Time structure and isochronal maps were generated on the definable time horizons. The time structure and isochronal maps were converted into depth and thickness. The structure and isopach maps were then exported to the GIS.

Petrophysical Evaluation Process

Rock types identified by the geologist were input into the petrophysical evaluation software. As noted, these data were combined with the well logs to determine a rock-to-log correlation index. The correlation index consisted of a probability tool that enabled the petrophysicist to predict rock types over intervals where conventional core was not available. Knowing rock types for each 1 ft interval in the well enabled the petrophysicist to apply different log parameters to each rock type for calculation. The result was a more accurate determination of the presence of hydrocarbons. Petrophysical values, such as average water saturation, net effective sand, and net pay were exported into the GIS for each of the reservoir units. For purposes of sensitivity analysis, a different set of values was calculated for each of the three different porosity values.

Reservoir Engineering Evaluation Process

Each well was evaluated to determine original and current water levels in the reservoirs. These values were input into the GIS for spatial analysis. Production analysis was performed to predict future hydrocarbon recoveries. Cumulative production was first assessed and assigned to each perforated interval and was then allocated to each reservoir unit. Following the mapping process in the GIS, net pay contour areas were exported into a spreadsheet. Reservoir volumes were then calculated from the net pay contour areas. These calculated volumes were compared with cumulative production volumes and originally calculated ultimate reserve recoveries. Numerous iterations were then performed on the map layers in the GIS to obtain reasonable net pay contour areas.

Mapping Process

The various data sets were imported and integrated into the GIS. The goal of the mapping process was to determine the original and remaining hydrocarbon volumes in the field. This process consisted of generating six different maps on each of the 16 reservoir units (Figure 5). Numerous iterations were performed on the maps as new data were generated and imported from the 3-D seismic and petrophysical data sets.

In the GIS, subsea top values were calculated from the tops of the reservoir unit given in measured depth. From these values, reservoir structure maps were created on each unit (Figure 6). Because the 3-D seismic could not resolve down to the reservoir unit level, the top-of-structure maps from the 3-D seismic survey were integrated to serve as a guideline and to ensure accuracy. From the petrophysical data, gross reservoir unit isopach maps were generated. These maps illustrated the orientation of, and variance in, thickness of the sand body.

Net effective sand maps, representing the feet of effective reservoir rock, were then generated to display the spatial variance in reservoir thickness (Figure 7). These maps present the amount of potential reservoir rock across the field. Using the filtering capability of the GIS, net effective sand maps were generated for the three different porosity values. This exercise enabled the team to assess the change in reservoir thickness as affected by different porosity. Average porosity of the reservoir unit was then generated to determine the spatial variance in porosity across the structure (Figure 8). Average water saturation maps were also generated to determine the original and current water levels in the water-drive reservoirs (Figure 9).

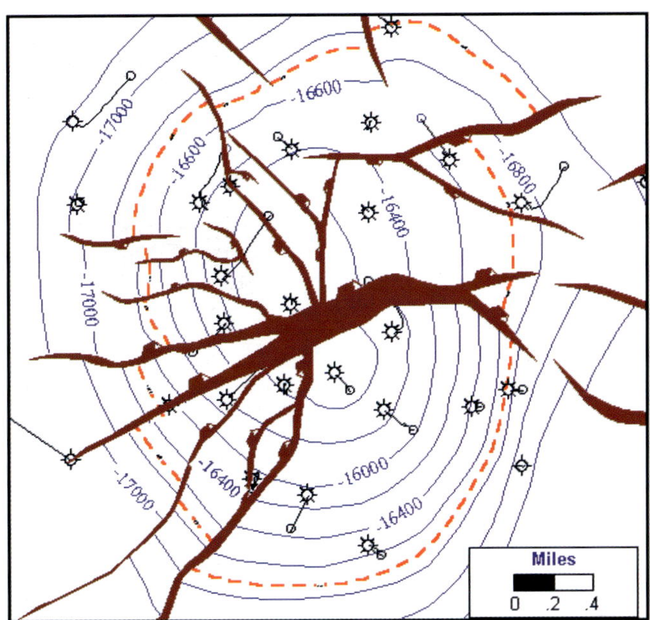

Figure 6—Structure map on the B2 reservoir unit.

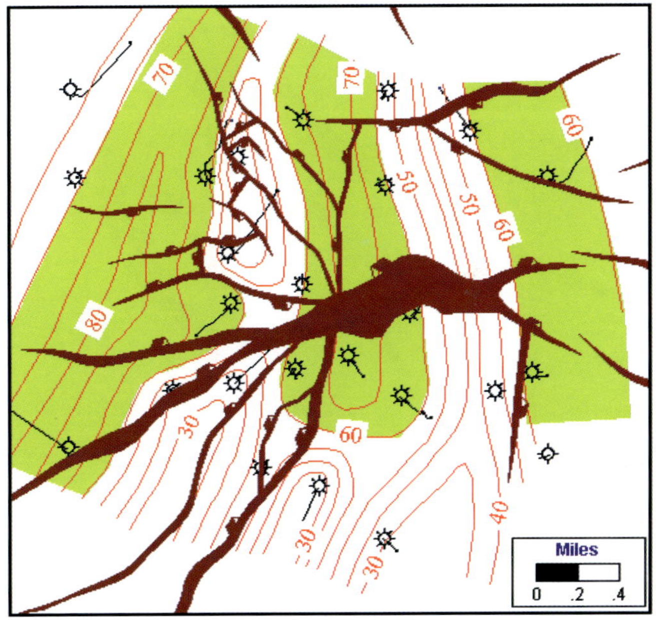

Figure 7—Net effective sand map on the B2 reservoir unit.

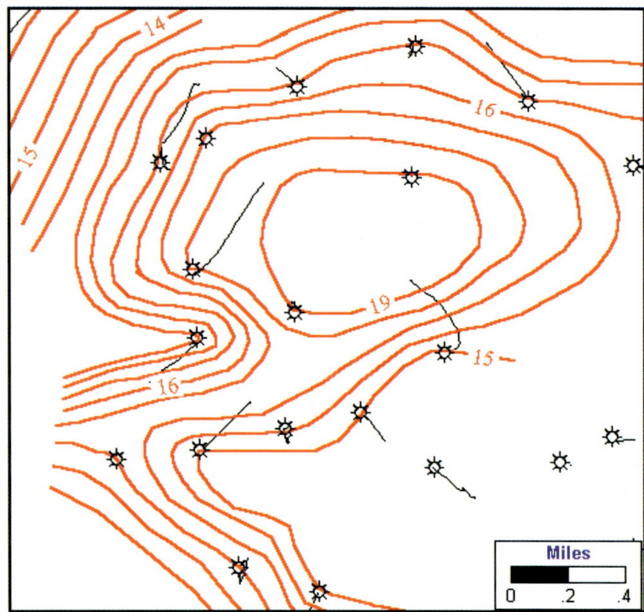

Figure 8—Average porosity map on the B reservoir units.

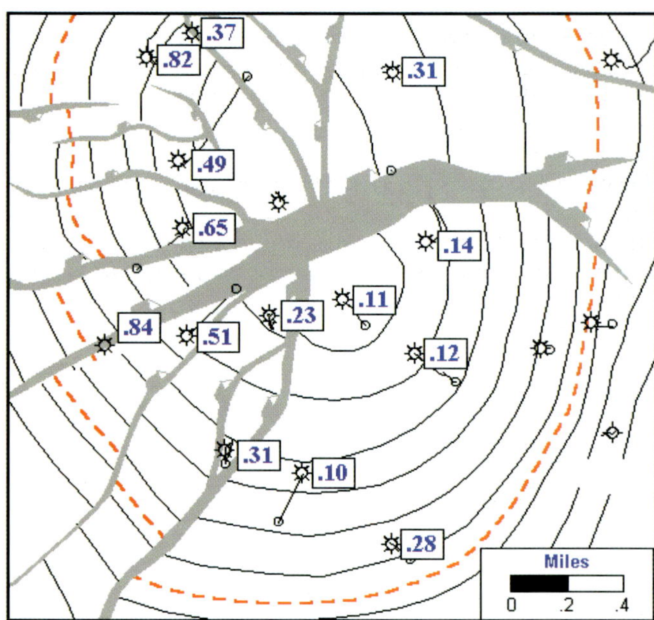

Figure 9—Average water saturation map on the B2 reservoir unit.

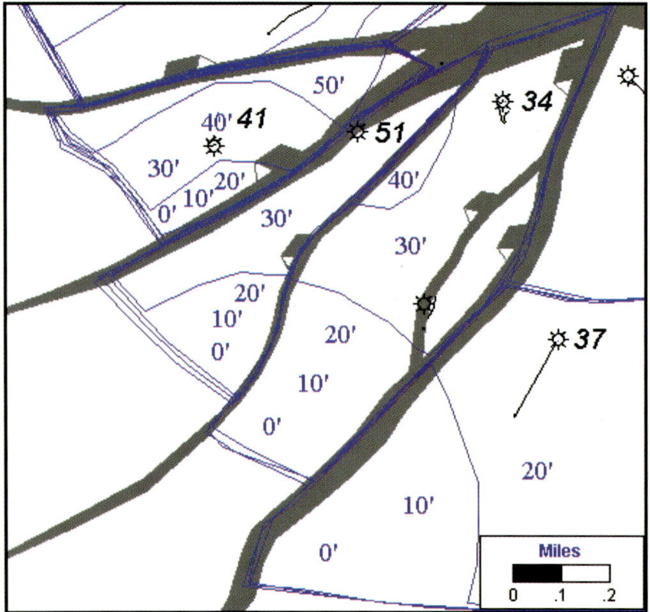

Figure 10—Net pay map on the A8 reservoir unit.

The final and most important map generated was the net pay map (Figure 10). The numbers on the map represent the net pay value for each contour. The net pay map was created dynamically in the GIS by overlaying the top-of-reservoir structure, base-of-reservoir structure, and the net effective sand map. These contour maps were also integrated with the fault locations from the 3-D seismic to ensure accurate delineation of the reservoirs. Net pay contours were defined and constructed as polygonal map features that enabled the GIS to automatically calculate the area under each individual contour. These area values were imported into the reservoir engineer's spreadsheet, where the reservoir volume was calculated. Calculated reservoir volumes were then compared to the production analysis and the original calculated ultimate reservoir hydrocarbon recovery. Individual maps were reexamined in cases where the calculated volume did not align with cumulative production or current reservoir water levels.

DATABASE

The GIS database consisted of two types of data: map layers and tabular databases. In the GIS, map layers are dynamically linked to tabular data tables. This link allowed for the posting and spatial analysis of any attribute in the tabular databases.

Three types of spatial data can be stored in a GIS: points, lines, and polygons. Points are represented by a single pair of latitude and longitude values, or x–y coordinates. Examples of points are well locations. Two or more pairs of x–y coordinates are used to represent lines. An example of a line layer is a gas pipeline. Three or more pairs of x–y coordinates are used to represent polygons, the most complex data type, where the first and last pairs represent the same coordinates.

In this project, the map layers were grouped into four categories: well, reservoir, 3-D seismic, and culture. The well data were stored as three map layers: surface location, bottom hole location, and the directional wellbore path. The directional wellbore path represents the direction of the wellbore from the surface to the bottom hole location as projected onto the earth's surface. The reservoir map layers consisted of the following contour types: structure, gross sand, net effective sand, and net pay. The reservoir layers also included the original and current reservoir water levels. Polygonal layers included the sand depositional fairway and areas where erosion had occurred. The 3-D seismic layers included structure contours, isopach contours, and fault locations. The culture layers included section boundaries, township-range boundaries, unit boundaries, and the locations of surface gas plants and processing facilities.

The tabular databases included three types of data: well header, petrophysical, and reservoir unit tops. The well header table contained general information about the well, such as operator, lease, well number, total depth, spud date, and completion date. There were many petrophysical tables and attributes as a result of the calculated values for three different reservoir porosity values. The table containing tops of the reservoir units included both tops and bases for all 16 reservoirs, as well as associated gross isopach values that were calculated in the GIS.

CONCLUSION

The application of GIS technology to this reservoir evaluation project resulted in many substantial benefits. The use of GIS technology greatly enhanced the analysis process and significantly decreased the project cycle time. This reservoir evaluation, as all others, involved many iterations. Digital datasets integrated into a GIS enabled the team to perform numerous "what-if" analyses. Dynamic editing and updating of map layers resulted in rapid retrieval of new information. The ability to overlay several map layers automated the net pay map-generation process. Using digital data sets provided a more accurate method for generating reservoir maps. Automatic calculation of the net pay contour areas significantly reduced the time required to assess reservoir volumes. Finally, GIS enabled the team to share data and analyses, which resulted in an integrated interpretation. The ability of the GIS to integrate data from the geophysical and petrophysical applications provided a seamless mechanism for analyzing the entire data set and resulted in a more efficient business and project analysis process.

ACKNOWLEDGMENTS

I wish to express my appreciation to Amoco Production Company for providing me the opportunity to publish the results of this study. This project involved the creative contributions of many people, and without the use of the multi-discipline approach, these results would never have been achieved. I especially would like to acknowledge the many people involved in this evaluation, specifically Scott Young, Chip Story, Terry Manning, Eric Djuvik, Kelly Tumino, Odelia Gurrola, Amy Horan, Dave Wright, Frank Sheppard, Pat McGrievy, Harold Cansler, Virginia Callaway, Chris Roth, Gary Caldwell, Carl Butler, John Jeffries, and Julie Spoth.

REFERENCES CITED

Barrell, K. A., 1997, Sequence Stratigraphy and Structural Trap Styles of The Tuscaloosa Trend. Transactions, Gulf Coast Association of Geological Societies, vol. 47, p. 27–34.

Sneider, R. M., 1986, The synergistic team approach—its value to petroleum exploration and development in the next decade. Proceedings, Eighth Australian Geological Convention, Adelaide, Australia, p. 16–19.

Safley, L. E., and M. L. Fowler, 2000, Waterflood optimization using GIS techniques in Citonelle field, Mobile County, Alabama, U.S.A., *in* T. C. Coburn and J. M. Yarus, eds., Geographic information systems in petroleum exploration and development: AAPG Computer Applications in Geology, No. 4., p. 195–204.

Chapter 14

Waterflood Optimization Using GIS Techniques in Citronelle Field, Mobile County, Alabama, U.S.A.

L. Eugene Safley
J. K. Research Associates
Las Vegas, Nevada, U.S.A.

Michael L. Fowler[1]
Independent Geologist
Tulsa, Oklahoma, U.S.A.

ABSTRACT

Citronelle field, an aging domestic giant oil field in south Alabama now in its mature waterflood stage, was selected for participation in the U.S. Department of Energy's Reservoir Management Demonstration Program. A reservoir management strategy was developed to extend the life of the field by improving profitability through improvements in waterflood sweep efficiency and increased production.

Early in the project, waterflood optimization was identified as the primary focus for strategy development. Because the field contains over 450 wells operated by more than 100 individual well operators, the amount of data for analysis was large and of varying quality. The approach chosen was to maximize the use of existing data and select a focus area for improved reservoir characterization.

A geographic information system (GIS) was used to organize, present, and interpret existing geologic and engineering data. In addition, techniques were developed to optimize the waterflood on a local scale using a flow unit approach based on previous geological studies. Various production and injection well geometries were developed to identify potential wells for recompletion and conversion to injection. Additional areas in the field with recovery potential where the flow unit approach might be applied were identified.

This project demonstrates an approach to reservoir management strategy development using a GIS that efficiently organizes and presents existing well and reservoir data. Efforts undertaken in this project have the potential to ascertain significant recoverable reserves that previous mapping efforts have found difficult, if not impossible, to identify. The approach to strategy development described may be representative for large, mature to supermature domestic reservoirs approaching their economic limits under supervision of independent oil operators.

[1]Formerly BDM Petroleum Technologies, Bartlesville, Oklahoma

INTRODUCTION

Many operators have an incomplete understanding of reservoir management, thinking of it only as a high-tech, high-dollar venture, and giving little consideration to simpler, lower cost tools and techniques. Low profit margins are often viewed as adequate justification for not taking action to improve recovery or profitability. Independent operators may not be technically experienced and could be reluctant to apply technologies that they do not fully understand. The net result is that reservoir management is viewed as a risk to be avoided.

Recognizing the potential resulting from suboptimal reservoir management practices, the U.S. Department of Energy (DOE) developed a reservoir management demonstration program to increase reservoir management understanding. Three cost-shared reservoir management demonstration projects were initiated in Ohio, Alabama, and Montana (Figure 1). The goal of the cooperative government-industry project in Mobile County, Alabama, was to develop a comprehensive reservoir management strategy to improve the operational economics and to optimize the oil recovery from Citronelle field. A reservoir management team was formed with professionals from the Main Citronelle Unit operator and BDM-Oklahoma Inc., and charged with developing the reservoir management strategy.

In addition to having limited resources available for plan design, the team was faced with a vast amount of variable quality and insufficiently documented data that had been accumulated over the nearly 50-yr life of the field. Also unique to this project was the geological complexity of this fluvial reservoir with its 240-m-thick (800-ft-thick) gross pay interval in the Cretaceous Rodessa Formation. A focused approach to waterflood optimization was chosen, which included the use of a geographic information system (GIS) and improved reservoir characterization techniques that could be adapted to the field as a whole. Figure 2 illustrates the steps taken to arrive at the appropriate reservoir management decisions.

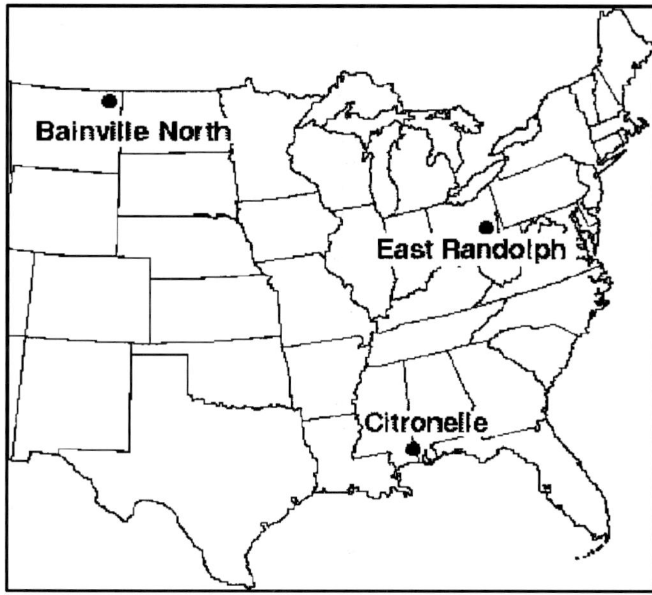

Figure 1—Locations of the three cost-shared reservoir management demonstration projects initiated by the U.S. Department of Energy and BDM-Oklahoma, Inc.

GIS CHARACTERISTICS

Burrough (1991) defined a GIS as a powerful set of tools for collecting, storing, retrieving, transforming, and displaying spatial data from the real world for a particular set of purposes. GIS software generates computer maps and links spatial, geographically referenced data stored in electronic databases to visual representations of features on these maps. The particular GIS hardware and software combination selected depends on the number of users, the size and extent of the system, the capabilities of the applications, the characteristics of the data, and the role of GIS within the organization (Somers, 1994). A large number of specific GIS software packages are available that support structured query languages. Drawing tools should be capable of output in a compatible exchange format such as drawing exchange format (DXF) or initial graphics exchange specification (IGES).

For the petroleum industry, the types of data that can be linked to the maps include geological, geophysical, geographical, and engineering data. The items included in the data sets must fall in mappable "themes" (e.g., well and seismic data) to be usable by current GIS software (ESRI, 1997). The data sets are based on "entities" (e.g., wells, pipelines, and rock types) of geographic features, which are collections of map "feature classes" (e.g., points, lines, areas, and collections of similar features). The labels used to define what a feature represents are called "attributes," and geographic features can have an unlimited number and variety of them. Each feature included in a GIS database must be specified by location on the earth's surface, by what the feature represents, and by its spatial relationship to the other surrounding features. The treatment of the spatial relationships of features (topology) constitutes a fundamental difference between modern GISs and most other types of computer graphics systems.

The types of data required for the GIS depend on the desired application and can be added to as the system grows. Control files interpret each commercial vendor's format to convert ASCII data into a usable format. If not available in GIS formats, data must be digitized or converted from existing datasets into the appropriate GIS formats. Conversion of data from paper maps, handwritten notes, and other digital forms into GIS formats can be time consuming and requires careful planning. Map scales, accuracies, and resolutions depend on how the GIS will be used and on how that accuracy affects analysis. The larger databases, higher accuracies, and finer resolutions require high-performance computing and carry an obvious

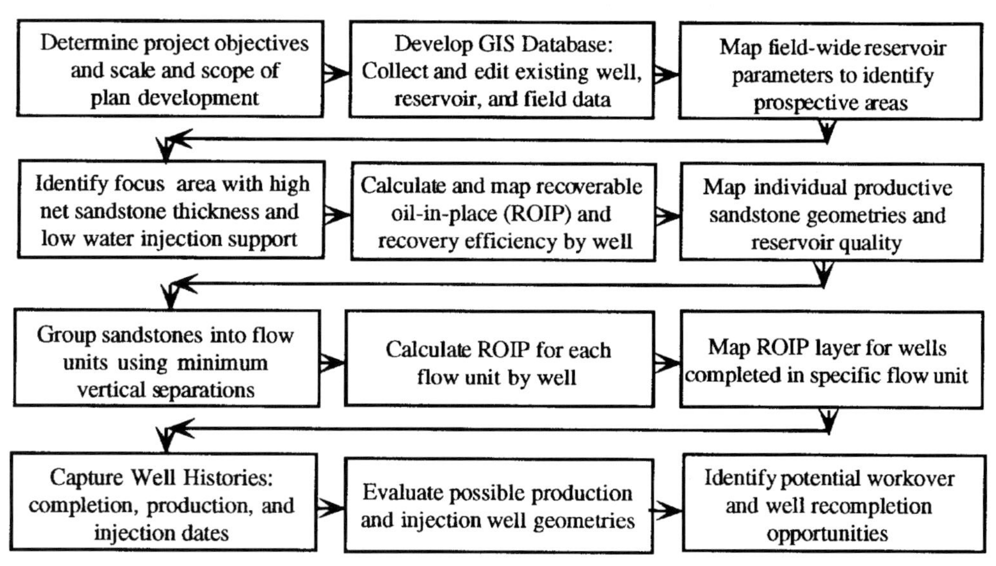

Figure 2—Flow chart illustrating the steps involved in waterflood optimization for Citronelle field.

disadvantage of increased computing resources and higher costs (Armstrong, 1994).

The real power of a GIS is its ability to integrate various data layers and to perform calculations among them (Burrough, 1991). Each grid cell on a map is referenced by a row and column number and represents an area on the surface of the earth and the average value of the attribute that is mapped. The resolution of the data cell is the relationship between the size of the grid cell and the area represented on the earth's surface. Each data cell may be displayed with the surrounding cells to produce a data layer. The data represented by the grid cell can be mathematically combined with material from other data layers to generate new information from a dataset.

The computer-generated map is simply a visual representation of information. GIS links drawn entities with descriptions and pointers to other data sets. With a mouse click on a graphical entity (e.g., a particular point on a computer display), GIS links the drawn entity to the GIS theme and database (e.g., a particular well or rock type). This technology is used to integrate and overlay vector and raster data, photographs, scanned documents, satellite images, drawings produced with computer-aided drafting (CAD), and multimedia data types with diverse textual and visual descriptions. Effective use of GIS requires careful planning on how the databases are organized, what information is contained in them, and how the links are constructed.

HISTORY OF CITRONELLE FIELD

Since its discovery in 1955, Citronelle field has been producing from Donovan sandstones of the Cretaceous Rodessa Formation at a depth greater than 3000 m (10,000 ft). This domestic "giant" oil field, located over a deep-seated salt intrusion, was developed, and essentially remains, on 40-ac (16-ha) spacing, covering 16,400 ac (6640 ha) with 468 wells (Figure 3). The field is divided into four operating units with more than 100 individual well operators.

Initial reservoir pressure (in excess of 5000 psi) declined rapidly after discovery, a result of the fluid expansion drive mechanism. Water injection to stabilize the rapid pressure decline began in 1961 and has had the additional benefit of sweeping oil to producing wells, but it has performed inefficiently due to the high reservoir heterogeneity. Daily oil production peaked at 17,600 bbl in 1965. By early 1995, 15,000 bbl of water were being injected daily into 50 injection wells to produce 3600 bbl of oil from 175 active producers. Cumulative recovery through 1997 was over 160 million bbl of oil, 1.2 billion ft^3 of gas, and 120 million bbl of water.

RESERVOIR CHARACTERISTICS

The domed, circular structure of the Rodessa Formation at Citronelle field is typical of Gulf Coast reservoirs located over deep salt piercement features. About 75–90 m (250–300 ft) of closure are observed on the productive Donovan sandstones in the field. A 240-m-thick (800-ft-thick) gross pay interval contains at least 42 productive sandstone zones that form more than 300 separate reservoirs, each with the highly variable distribution of porosity and permeability characteristic of fluvial deposition. The Donovan sandstones are interpreted as a stacked series of meander belt deposits trending in a roughly northeast–southwest direction. Meandering stream deposits are typically limited in overall size and effectively isolated from each other and from other permeable deposits by overbank clays. Based on core analysis, the sandstones have a gradational grain size from coarser at the base to finer at the top and a distinctive vertical sequence of bedding structures (massive beds at the base, medium-scale cross beds, horizontal laminations, and ripple beds at the top).

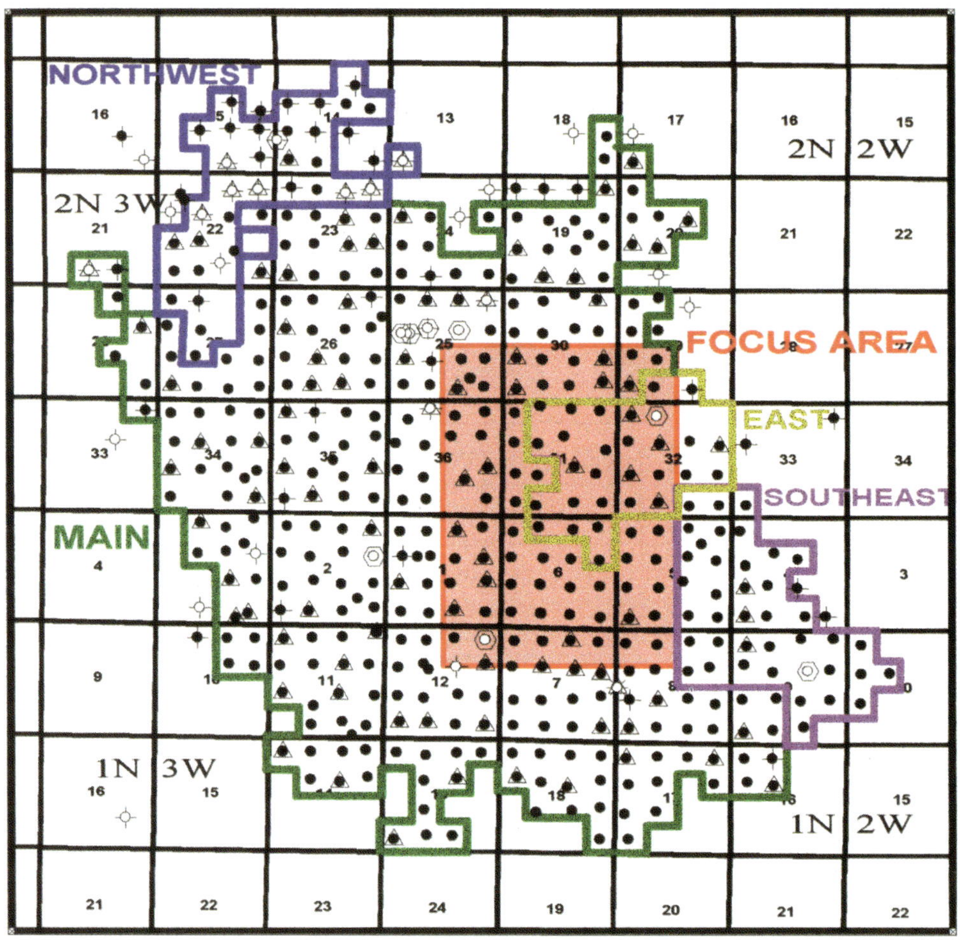

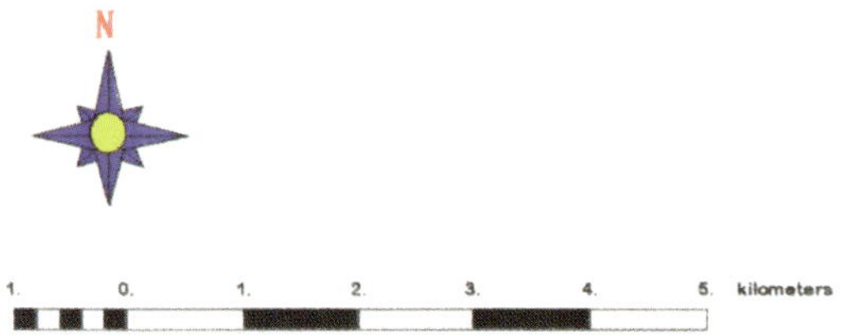

Figure 3—Base map of Citronelle field, Mobile County, Alabama, showing producing and injection well locations, unit boundaries, and outline of project focus area.

DATABASE DEVELOPMENT

Analysis of the historical performance of Citronelle field required the detailed review of production and injection data from individual wells and operating units. Because completion, production, injection, and reservoir data were unavailable from data vendors, the necessary information was obtained directly from the unit operators and the State of Alabama Oil and Gas Board. Monthly production and injection data were available from the early 1970s on, and the cumulative injection and production values reflect the life of each well. Almost all the reservoir descriptive information was manually entered over a period of two man-months from hardcopy reports, memos, and computer printouts accumulated over the 40+ yr history of the field. Data were considered more accurate and consistent when they could be associated with field studies that extensively relied on cores and wireline logs.

A customizable, relational database was used for efficient storage and retrieval of various types of data, particularly because a large number of data records were present. The items in the database also became available for sharing with other software applications across multiple platforms and computer networks. In this project, a database using GeoGraphix software, consisting of more than 100 discrete fields, organized critical data into logical tables. Each record listed a particular description, type (i.e., alphanumeric, integer, or floating decimal), number of characters, and position within the database. Land coordinate data were stored as latitude and longitude values. To display this information, the data had to be converted into Cartesian x–y coordinates by using a map projection. The well data were referenced by the well's API number and included well history and reservoir data. Well histories provided each well's location, production and injection status, drilling history, and completion data.

Reservoir data included the formation tops and pay thickness values for each of the mapped sandstone zones. Various other reservoir parameters obtained from core and log analyses, including porosity, permeability, water saturation, and productivity indices for each sandstone encountered in each well, were entered into the database. All of the data in the database could be posted on the maps or used to generate bubble plots and display layers.

When a display layer was created, a binary vector file was built in the project directory. When a file was selected for display, the entire file was loaded into memory and represented a snapshot of the data in time. Each map could contain any number of elements. An element represents a single graphic unit and may contain up to 25 layers, a metafile, or a bitmap. Communication with elements stored in the database was extended to other Windows applications through the use of dynamic data exchange (DDE). By clicking on a well symbol on any of the layers, the well history and reservoir parameters could be accessed in the GIS or other spreadsheets and databases.

Data could be output as hardcopy reports or graphics, or exported electronically within the database or to other Windows applications. Reports could be generated using an output definition system to specify textual elements for output, or as ASCII files for import into other applications. Graphics files contained all the parameters necessary to reconstruct and display the graphical layout by referencing the particular layers without duplicating the actual files.

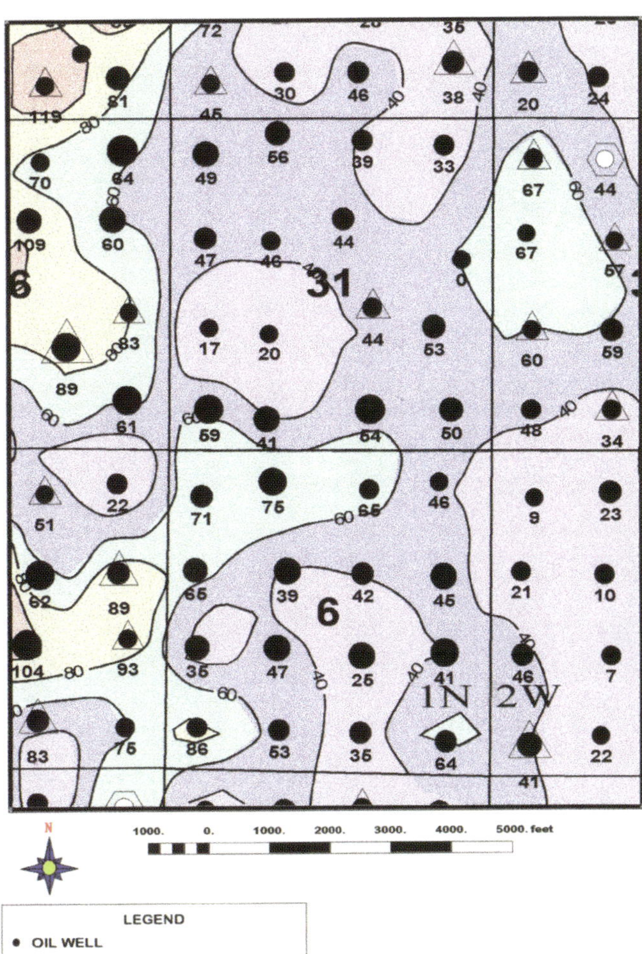

Figure 4—Contour map of net sandstone thickness of the Donovan sandstones using a 10% porosity cutoff. Well symbols are bubbled on cumulative oil produced or the cumulative water injected. Contour interval is 20 ft.

FIELD-WIDE ANALYSIS

Areas within Citronelle field where efforts to optimize the waterflood could be performed needed to be identified. A series of graphical layers were generated to store and display logically related reservoir data. This effort established the capability to map any data element for all of the wells in the field, whether the map was created for the entire field or for a group of wells in a focus area. Various well symbols were used to depict the current well status, with the symbol type or size reflecting the relative amount of cumulative oil produced or water injected.

Contour layers were generated and interpreted using various gridding and contouring algorithms to reflect the current understanding of reservoir conditions. Contour layers of reservoir parameters included the reservoir structure, net sandstone thickness, amount of oil originally present, cumulative amount produced, amount of reserves potentially remaining to be produced, and the amount of water injected. By depicting the interaction of various factors through manipulation of mappable layers using various filters and statistical tables, team members were able to identify key areas on which to focus attention.

FOCUS AREA ANALYSIS

Rather than analyze the entire field, the most economical approach was to concentrate on specific areas of the field where significant problems or opportunities could be identified. It was assumed that solutions developed in these areas would have the greatest and fastest impact on profitability per dollar expended. A focus area spanning the boundary between the Main and East units of the field was selected for improving waterflood recovery (Figure 3). Historically, there had been little mutual or coordinated effort on the part of the unit operators to optimize the waterflood in this area. Active wells were present, with several shut-in wells available for recompletion or conversion to injectors.

In this boundary area, reservoir conditions are favorable for increased recovery based on the parameters used in fieldwide mapping. Figure 4 shows that the net sandstone thickness (contour interval 20 ft) of the Donovan sandstones ranges from 12–24 m (40–80 ft) using a 10% porosity cutoff. Note the areas of high net sandstone thickness in Sec. 31 T2N R2W and Sec. 6

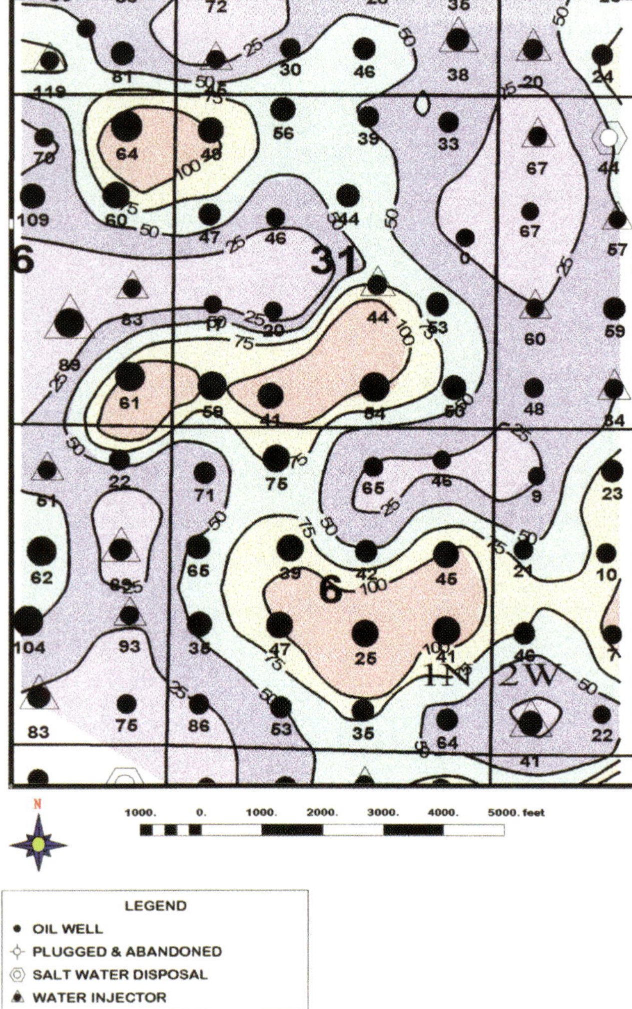

Figure 5—Contour map of recovery efficiency (cumulative oil production divided by recoverable oil-in-place by well) for the focus area. Contour interval is 25%.

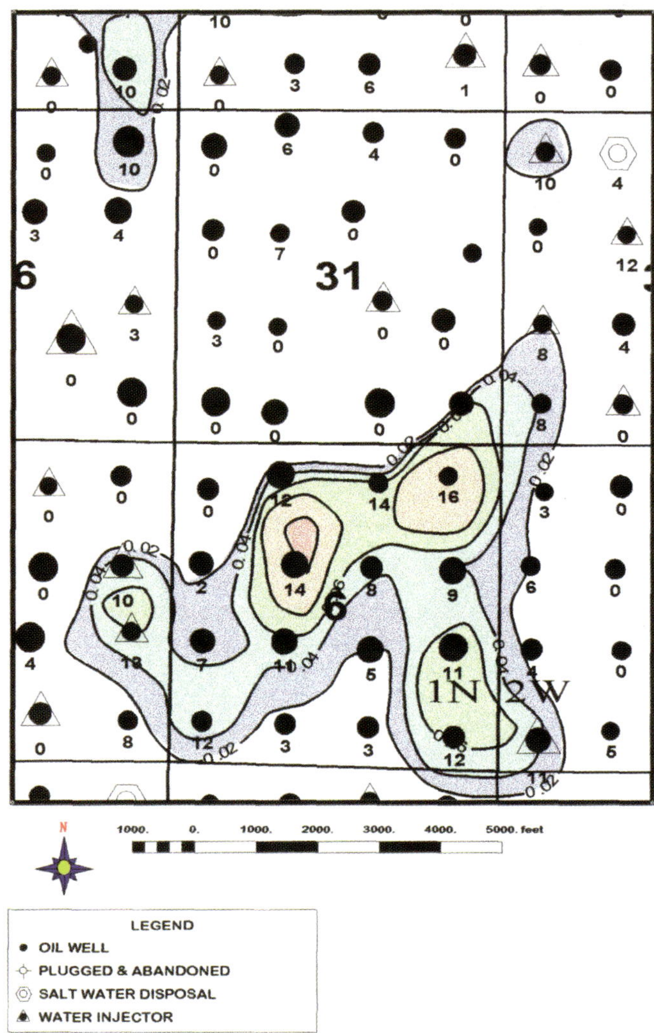

Figure 6—Contour map of productivity index values for lower Donovan sandstone layer 38A illustrating the isolated, sinuous, meandering geometry of the fluvial sandstones. Contour interval is 0.02.

T1N R2W and the absence of injection wells. High cumulative oil production can be identified by the larger sized well symbols. Areas of high net sandstone thickness and smaller producing and injection well symbols suggest areas with potential unrecovered reserves.

The recoverable oil-in-place (ROIP) layer was generated by surface-to-surface volumetric calculations for each 40-ac (16-ha) drainage area (polygon), using the thickness layers of the known productive sandstones, and the average porosity and water saturation of the productive sandstones in the well. The reservoir properties were averaged over the well's drainage area using sandstone distribution maps and calculations from previous reservoir studies.

The recovery efficiency layer was generated by taking the cumulative oil production for each well divided by the calculated ROIP for each well's drainage area. Over the entire field, the effectiveness of injection support is revealed by the presence of a donut-shaped ring of wells with less than 50% recovery efficiency circling the center of the field. Inside and outside this ring, recovery efficiency goes above 50%, and above 100% in some areas. The high recoveries for some areas suggests contribution from sandstones not previously considered as pay or a larger pay sandstone distribution in the 40-ac (16-ha) drainage area. Note that the focus area has benefited from the movement of oil toward the field margins, where recovery efficiencies for wells with high net sandstone thickness are above 50% (Figure 5); however, because water injection has been minimal, potential reserves remain.

INDIVIDUAL SANDSTONE GEOMETRIES

Individual sandstone layers were generated for the productive intervals in the focus area based on detailed log correlation. Maps of net sandstone thickness

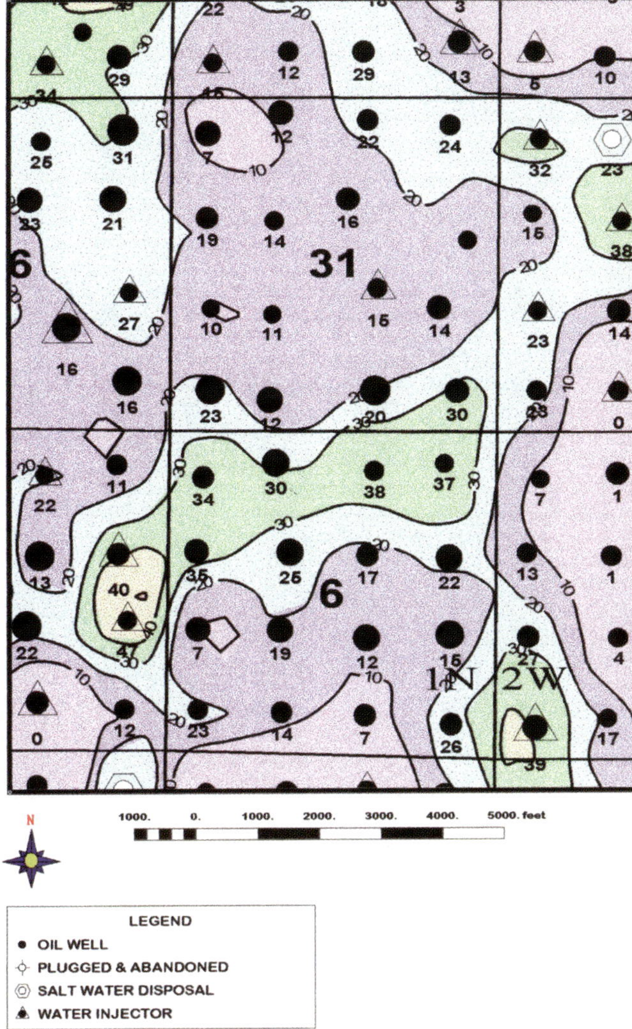

Figure 7—Contour map of net sandstone thickness of flow unit 15 containing the eight productive sandstone layers of the lower Donovan. Contour interval is 10 ft.

indicated that many of the fluvial sandstone deposits were limited in distribution and were effectively isolated from other permeable sandstone layers, both vertically and horizontally. Numerical filters were used to identify which wells had been completed in the mapped sandstone layers. This analysis helped to identify potential opportunities where wells had not been completed in mapped, productive sandstone intervals.

Reservoir quality was depicted using permeability or productivity indices obtained from previous core-log modeling studies. For example, the map of productivity index values for sandstone layer 38A illustrates the isolated, sinuous, meandering geometry of the fluvial sandstones (Figure 6). Note the presence of the thick, reservoir-quality deposit that has a high calculated ROIP in Sec. 6. Reservoir-quality maps for each of the productive sandstones aided the identification of specific wells that could provide improved sweep efficiency, as well as sandstones that could act as waterflood thief zones because of high water injection support.

FLOW UNITS

Identifying groups of sandstones that perform as a single fluid-flow unit was important for local waterflood optimization. Because wide vertical intervals have traditionally been fracture stimulated, there is reason to believe that many sandstones with close vertical association may be in communication behind pipe. To define packages of sandstones that might perform as single flow units, a vertical separation of at least 10 m (30 ft) of shale between sandstones was used as a criterion for effective isolation. Wells within the focus area were examined to identify flow units containing the greatest net sandstone thickness. Approximately 20 flow units were identified using geological cross-sections oriented east–west, roughly perpendicular to the trend of fluvial deposition in Citronelle field.

To illustrate the complexity of the analysis process, Figure 7 shows the net sandstone thickness of flow unit 15 containing the eight sandstone layers of the lower Donovan. Net sandstone thickness ranges from 0–12 m (0–40 ft) and appears widely distributed with a meandering trend across the focus area. Individual sandstones, however, have much more varied distributions and lack continuity across the area, reflecting shifting depositional patterns and orientations. Evaluation of production and injection responses must consider which productive sandstones are involved in the corresponding flow unit, what is their distribution, and whether corresponding zones are open in producers and injectors to enable the operator to identify optimum injection strategies.

Areas where significant quantities of recoverable oil were associated with these fluid-flow units were identified to focus water injection support. Calculation of volumetrics for key flow units involved layer-to-layer manipulation of the summed pay thickness for sandstone zones in the flow unit, the reservoir drainage area associated with each well, and averages of productivity index, porosity, and water saturation. A series of maps were generated with the wells bubbled on cumulative production and injection, with overlays of ROIP by flow unit to assess the waterflood performance.

Figure 8 shows the distribution of ROIP in barrels for the 40-ac (16-ha) tracts associated with flow unit 15. An area of high ROIP trends northeast–southwest across the northern half of Sec. 6. Although the area has recovery efficiency for the Donovan sandstones of over 50% (shown on Figure 5), water injection support has been minimal.

WELL HISTORIES

Well completion histories were collected for each pay zone from selected wells within the focus area. Using filters to identify wells matching specific completion date

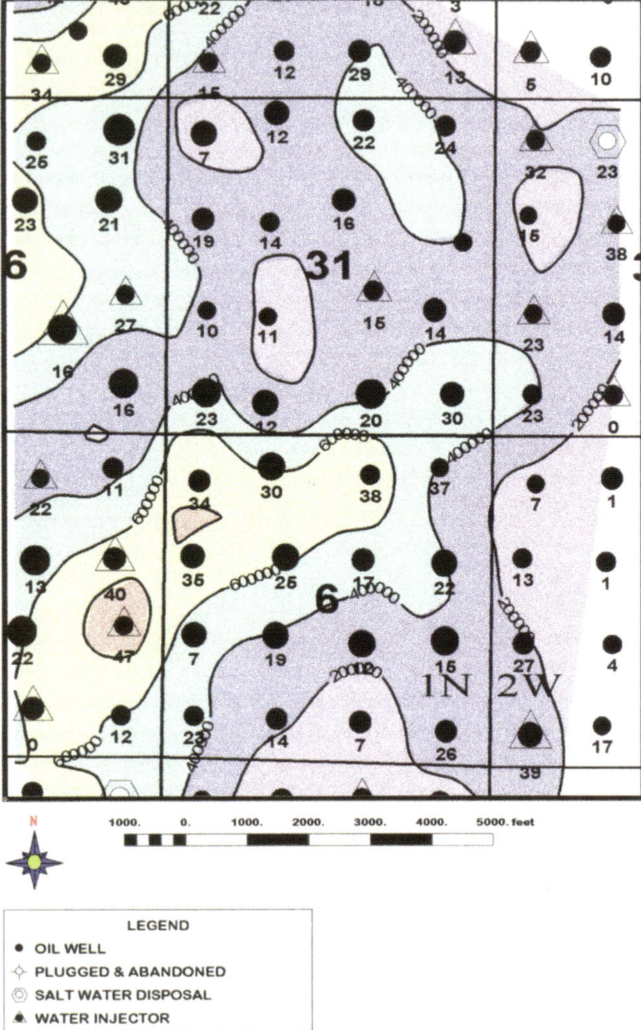

Figure 8—Contour map of recoverable oil-in-place (in barrels) for the 40-ac (16 ha) tracts associated with flow unit 15 in the focus area. Contour interval is 200,000 barrels.

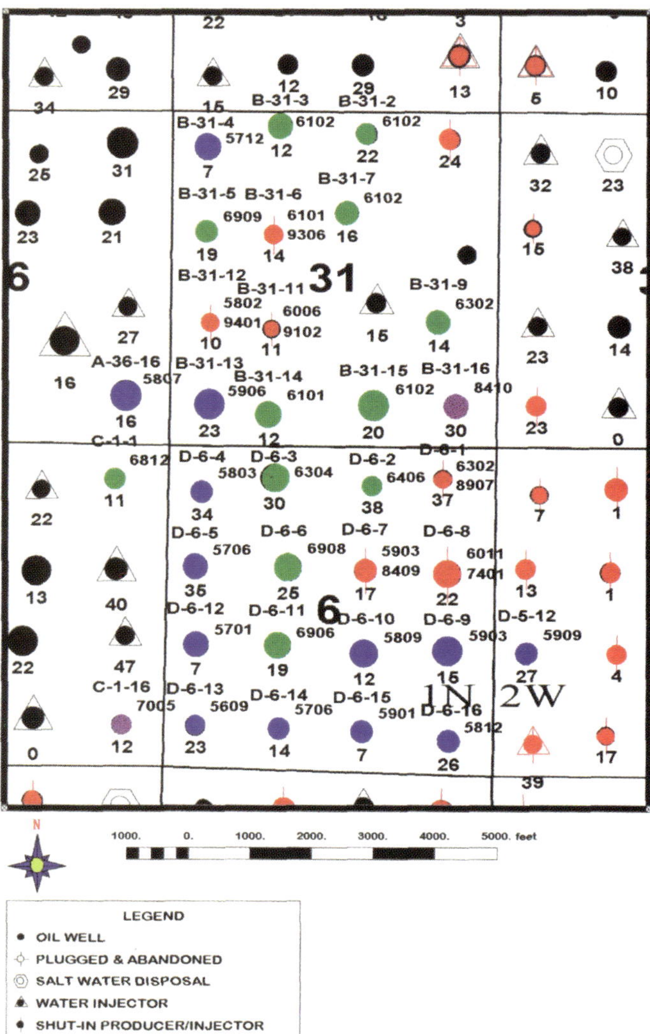

Figure 9—Color-coded well histories for flow unit 15. Wells completed prior to 1960 are shown in blue, wells completed between 1960 and 1970 are shown in green, and wells completed after 1970 are shown in magenta. Currently shut-in wells are shown in red. The completion, shut-in, water injection, and water injection shut-in dates (YYMM) are posted clockwise, consecutively around the well symbol.

criteria, the well histories were incorporated in the geologic mapping. Most of the wells in the field were drilled in the late 1950s and early 1960s. On Figure 9, wells completed in flow unit 15 prior to 1960 are shown as blue well symbols, wells completed between 1960 and 1970 are shown as green, and wells completed after 1970 are shown as magenta. Currently shut-in wells are shown in red. The completion, shut-in, water injection, and water injection shut-in dates (YYMM) for the well are posted clockwise, consecutively around the well symbol. This graphical display helped team members to effectively link the well completion histories and cumulative production and injection data with the defined flow units.

The wells encountered numerous continuous and discontinuous sandstone intervals. The thickest Donovan sandstones were generally the first to be completed. These sandstones contained the largest recoverable reserves but not necessarily the highest porosity and permeability. Initially, the upper and lower Donovan intervals were not commingled in the producing wellbores, so early production records present a history of the performance of these two intervals independently. The individual sandstone zones within these two main intervals were all completed and fracture-treated, so the allocation of production to the individual sandstones or flow units is difficult.

The waterflood was initiated in the early 1960s using a random type of flood pattern with the production from the upper and lower Donovan commingled. Injection was initiated by recompleting previously producing zones in existing wellbores. When the thickest sandstones watered out, they were isolated by cement squeezing, and additional thinner, sandstone zones were opened and produced. Injection surveys were run, and injection into the various intervals was allocated over time using these surveys and by using the calculated productivity index value for each sandstone zone. Throughout the life of the field, freshwater injection wells and producers were shut in and reactivated based on the progression of the flood fronts in those particular sandstones penetrated by the wellbores.

POTENTIAL WORKOVERS

Because each well penetrated multiple flow units, several possible production and injection well geometries were developed with the aid of the individual sandstone and flow unit mapping. This was done to help assess the maximum recoverable oil-in-place (ROIP). Based on these geometries, the following wells (shown on Figure 9) in this unit-boundary area were identified as the best potential candidates for conversion to injection: B-31-5 and B-31-13 in Sec. 31 T2N R2W, and D-6-3, D-6-7, and D-6-8 in Sec. 6 T1N R2W. Two of the wells, D-6-7 and D-6-8, were shut in at the time of this analysis. Top ranked among these from a recoverable oil-in-place perspective (Figure 8) were the B-31-13, D-6-3, and D-6-7 wells. Additional analysis of the condition of wellbore facilities would have to be performed before a final selection could be made.

In terms of distribution of net sandstone thickness, cumulative production, ROIP, and recovery efficiency, additional areas of the field may also merit closer examination. The greatest numbers of these areas, however, are located near the perimeter of the field where targets are smaller (i.e., net pay thickness, cumulative production, and ROIP are lower, and pay quality is generally lower). A finer contour interval may be required for adequate definition of recovery anomalies. Details of production and injection history in these areas should be reviewed before proceeding to a flow unit assessment. This task will be of critical importance to the functionality of the GIS. Another factor to consider is the general paucity of active wells in many of these areas at the present time.

CONCLUSIONS

The experience gained in characterizing the reservoir and improving waterflood sweep efficiency at Citronelle field illustrates the complexity of assimilating an immense amount of data of various qualities and vintages. Considering that the maps initially available were 30-yr-old, hand-drafted sand maps, it was a major project accomplishment to have constructed a system capable of mapping and updating 482 wells and hundreds of reservoirs on a well-by-well and sand-by-sand basis. This was made possible by the availability of relatively inexpensive, desktop GIS software and the availability of easy user-interface builders capable of performing the large database, analytical, and graphical display tasks. The strategic analyses involved three aspects: (1) the development of a relational database to facilitate fieldwide and local analyses, (2) development of flow-unit analysis techniques that could be applied on a local scale, and (3) generation of graphical display layers of reservoir parameters to identify and rank promising areas and individual wells for recompletion and water injection support.

Estimates of incremental recovery from implementing the procedures range from 5 to 10 million bbl (1.3 to 2.5% of original oil-in-place assuming an OOIP of 400 million bbl). Targeted well recompletions and conversions are required to evaluate the merits of this effort and to identify significant recoverable reserves that previous mapping efforts have found difficult to ascertain. For this and any other similar project, using a GIS to depict the interaction of various geologic and engineering factors through map layers and statistical tables substantially enhances operators' understanding of the potential impacts of specific well conversions or increased water injection support.

REFERENCES CITED

Armstrong, M. P., 1994, GIS and high performance computing. Proceedings, GIS/LIS Annual Conference. Bethesda, MD: American Congress on Surveying and Mapping (ACSM), p. 4–13.

Burrough, P. A., 1991, The development of intelligent geographical information systems, Proceedings, 2nd European Conference and Exhibition on Geographic Information Systems (EGIS '91). Brussels: EGIS Foundation, vol. 1, p. 165–174

ESRI (Environmental Systems Research Institute), 1997, How GIS Works, *in* About GIS. Redlands, CA: Environmental Systems Research Institute (ESRI), Inc., <http:www.esri.com/library/gis/index.html>.

Somers, R., 1994, GIS organization and staffing, Proceedings, URISA Annual Conference. Park Ridge, IL: Urban and Regional Information Systems Association, p. 41–52.

Day, R. A., et al., 2000, Reservoir characterization of a stratigraphically and structurally complex Gulf of Suez oil field using three-dimensional modeling and GIS techniques, *in* T. C. Coburn and J. M. Yarus, eds., Geographic information systems in petroleum exploration and development: AAPG Computer Applications in Geology, No. 4, p. 205–212.

Chapter 15

Reservoir Characterization of a Stratigraphically and Structurally Complex Gulf of Suez Oil Field Using Three-Dimensional Modeling and GIS Techniques

Ron A. Day[1]
BP/Amoco
Aberdeen, Scotland

Khaled Talaat
Reservoir and Field Development Dept., Qatar General Petroleum Corporation
Doha, Qatar

Karen S. Hoffman
Dynamic Graphics, Inc.
Houston, Texas, U.S.A.

ABSTRACT

Many of the oil fields in the Gulf of Suez operated by GUPCO (a joint partnership between the Egyptian General Petroleum Company and Amoco Corporation's Egypt Business Unit) are mature, stratigraphically and structurally complex fields. Three-dimensional reservoir characterization models that include GIS query and extraction capabilities provide an advantage over traditional methods in ongoing efforts to exploit reserves in an efficient and cost-effective manner. The use of such technology by GUPCO's multidisciplinary asset management teams has resulted in successful exploitation of fields that were previously regarded as too complex to yield economic results. Application of the technology to one such field located in the southern Gulf of Suez is described here.

INTRODUCTION

GUPCO, a joint partnership between the Egyptian General Petroleum Company and Amoco Corporation's Egypt Business Unit, operates a significant proportion of the oil fields in the Gulf of Suez. To date, these fields have produced in excess of 3 billion barrels of oil. Many of these oil fields are mature, and they are stratigraphically and structurally complex. Extending the lives of the fields requires continued assessment of the potential of previously undeveloped and under-exploited reservoirs. GUPCO has created multidisciplinary, multinational asset management teams comprised of geophysicists, geologists, petrophysicists, and reservoir engineers. It is the task of these teams to conduct field studies and propose new developments in order to fully exploit the reservoirs and any existing production infrastructure.

[1]Formerly Amoco Trinidad Oil Company, Houston, Texas

One such field designated for study lies in the southern Gulf of Suez (Figure 1). This field had already produced over 100 million barrels of oil from pre-Miocene (Mesozoic) and Oligocene–Miocene reservoirs. As these reservoirs had become depleted, exploitation efforts were focused on determining the potential of the shallower, Miocene sandstones. This reservoir is not as structurally complex as the older reservoirs, but it is stratigraphically much more complex, and the presence of good quality reservoir rock was thought to be much more difficult to predict.

The Miocene sandstones had been penetrated about a dozen times, and wireline electric logs had initially suggested the presence of oil in porous reservoir rock, primarily in the northern part of the field. Completions had been attempted in a few of these wells, but after initially observing good production rates, production rapidly declined. The cause of the decline was not fully understood, but the quality and technique of the well completions, along with rock quality, were considered possible factors. Intragranular and intercrystalline microporosity with poor permeability was suspected to be locally present and difficult to differentiate from better-quality reservoir based solely on well log data. The poor performance gave little indication that this could be a significant reservoir, but a field study, hinged on improved characterization of the lateral and vertical variation of reservoir properties, nonetheless was undertaken to determine its potential.

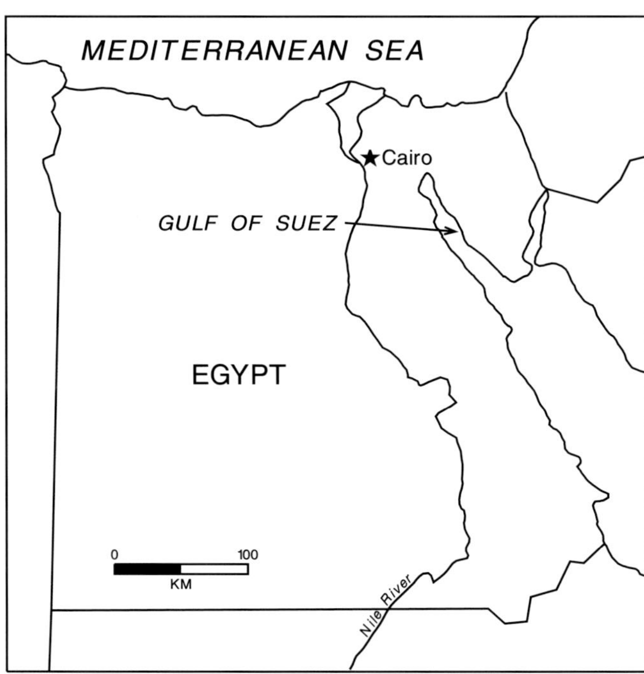

Figure 1—Location map of the general area of the field.

FIELD STUDY

The first part of the study was devoted to a detailed petrophysical analysis. This analysis concluded that the reservoir was typically immature and that, in fact, microporosity, mostly in the form of occlusion of primary porosity by intercrystalline dolomite, was locally significant and likely the primary culprit of the reservoir's poor performance to date. It was also observed that salt had precipitated from the overlying halites, most likely after perforation of the wells, which could contribute to poor production performance.

Further investigation of the available well log porosity data indicated a very large range of values both within wells and areally across the field (Figure 2). Some wells exhibited a distinctly bimodal porosity distribution, whereas others had a broader, more even distribution. This variability had to be taken into account in any reservoir characterization scenario.

Difficulties inherent to geophysical data in this part of the Gulf of Suez (steeply inclined layered halites, anhydrites, and shales that produce reverberation between layers, resulting in the generation of strong multiples) meant that, although they might help with the interpretation of structure at this level, seismic attributes could not be used to guide the trend analysis and interpolation of petrophysical properties throughout the field. Therefore, it was determined that a three-dimensional reservoir characterization model would have to be developed based on well log data alone.

The first step in the creation of this model was to build the structural framework. Using SeisWorks (SeisWorks is a product of Landmark Graphics Corporation), various geophysical interpretations were developed in the depth domain, including several fault traces and one horizon surface. The data supporting these interpretations were subsequently transferred to EarthVision (EarthVision is a product of Dynamic Graphics Incorporated), where they were adjusted to tie together the existing wells. In this same phase, a fault hierarchy was determined and an initial fault model was constructed (Figure 3).

An adequate petrophysical portrayal of the reservoir required the structural horizon that had been geophysically interpreted to be subdivided into four different zones (layers). These zones were based on the presence of several characteristic anhydrite markers. Because so few well penetrations of the Miocene interval existed to supply the necessary data (some fault blocks had no penetrations at all), the method of conformal mapping was used. At each well location, a true stratigraphic thickness (TST) was calculated, and a grid of TST was generated. This grid was then superimposed on the fault blocks (even those containing no well penetrations) to create a new surface. Such a process is quite different from simply adding two grids together because the three-dimensional methodology accounts for the lateral movement of faults with depth. The resulting fault/horizon intersections are thus not simply vertical shifts, allowing the three-dimensional integrity of the model to be maintained (Figure 4).

Statistical multiple regression techniques were employed to relate the available core measurements to well log responses and to derive prediction equations

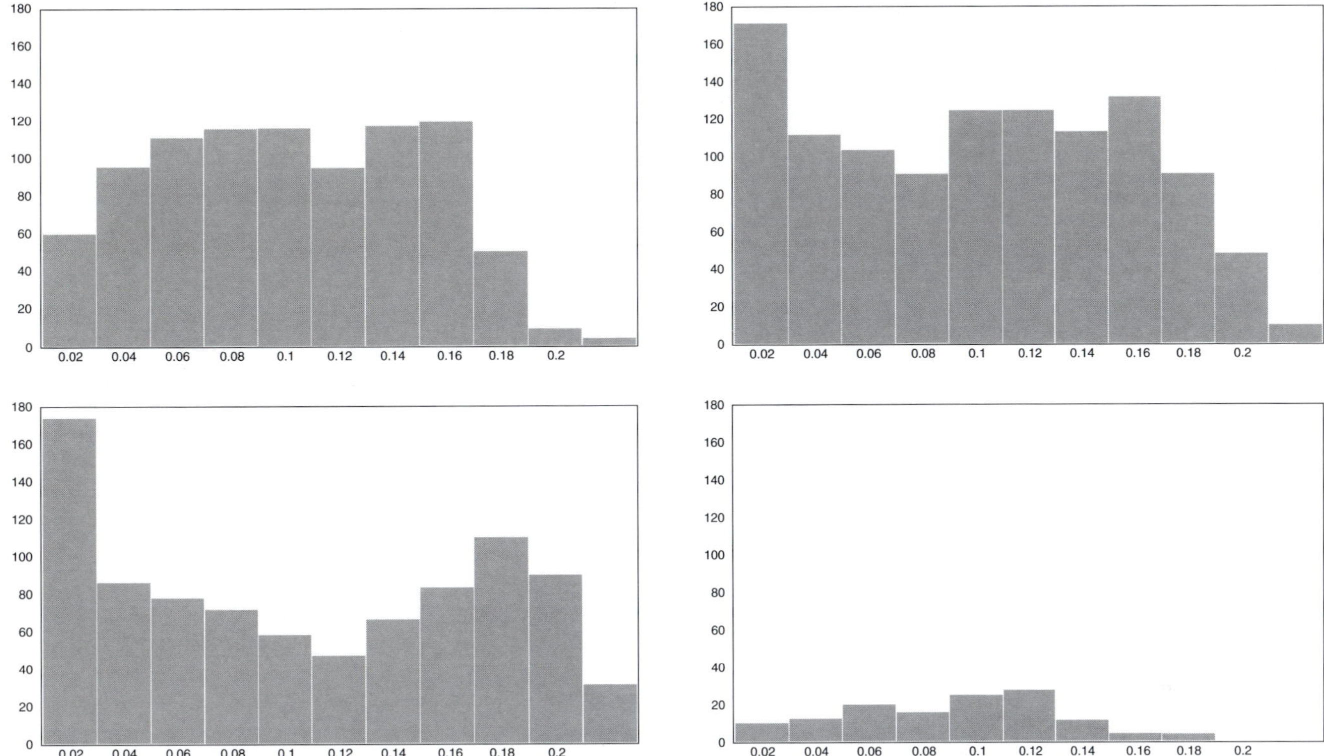

Figure 2—Four representative histograms indicating the difference in the distributions of porosity at well locations within the reservoir.

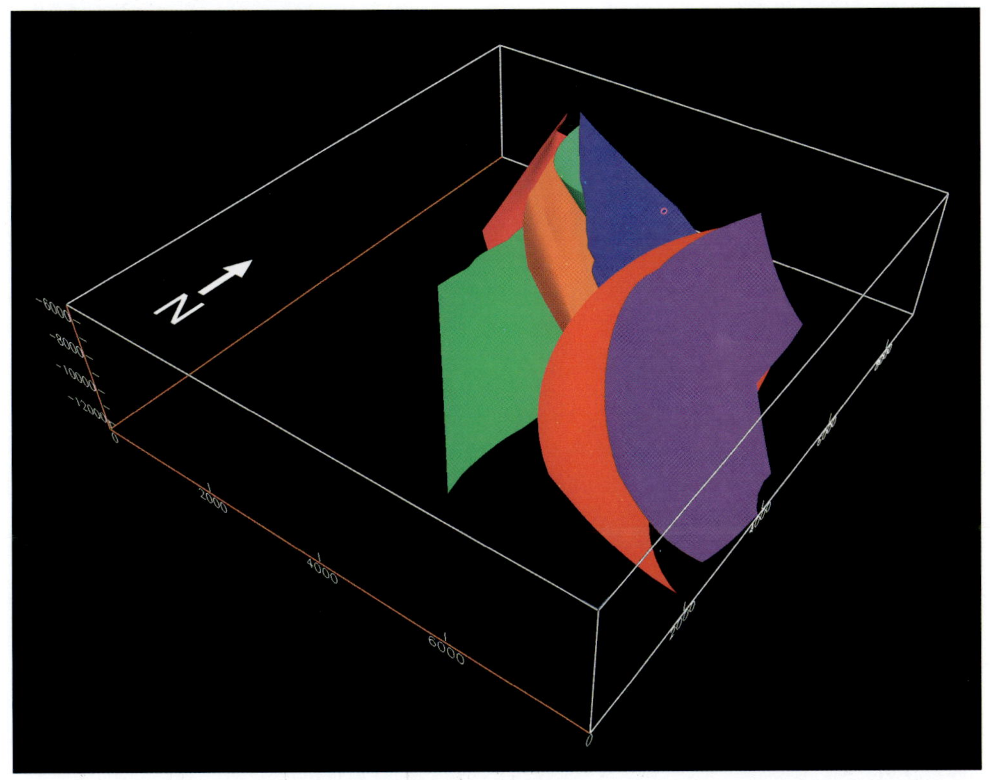

Figure 3—The fault surfaces model. Vertical exaggeration 1×.

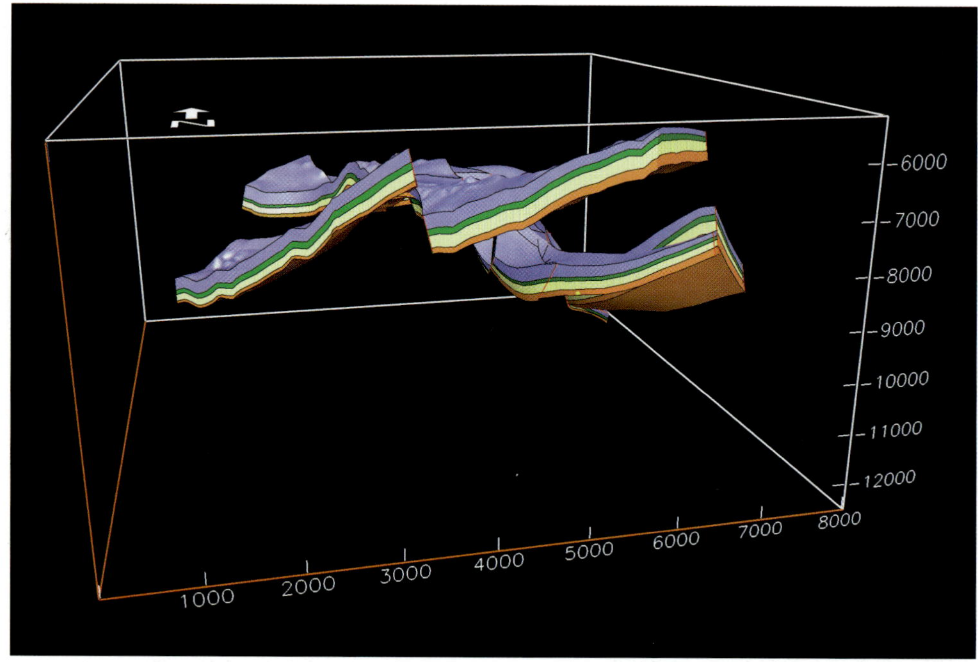

Figure 4—A model showing the four reservoir layers. Vertical exaggeration 2×.

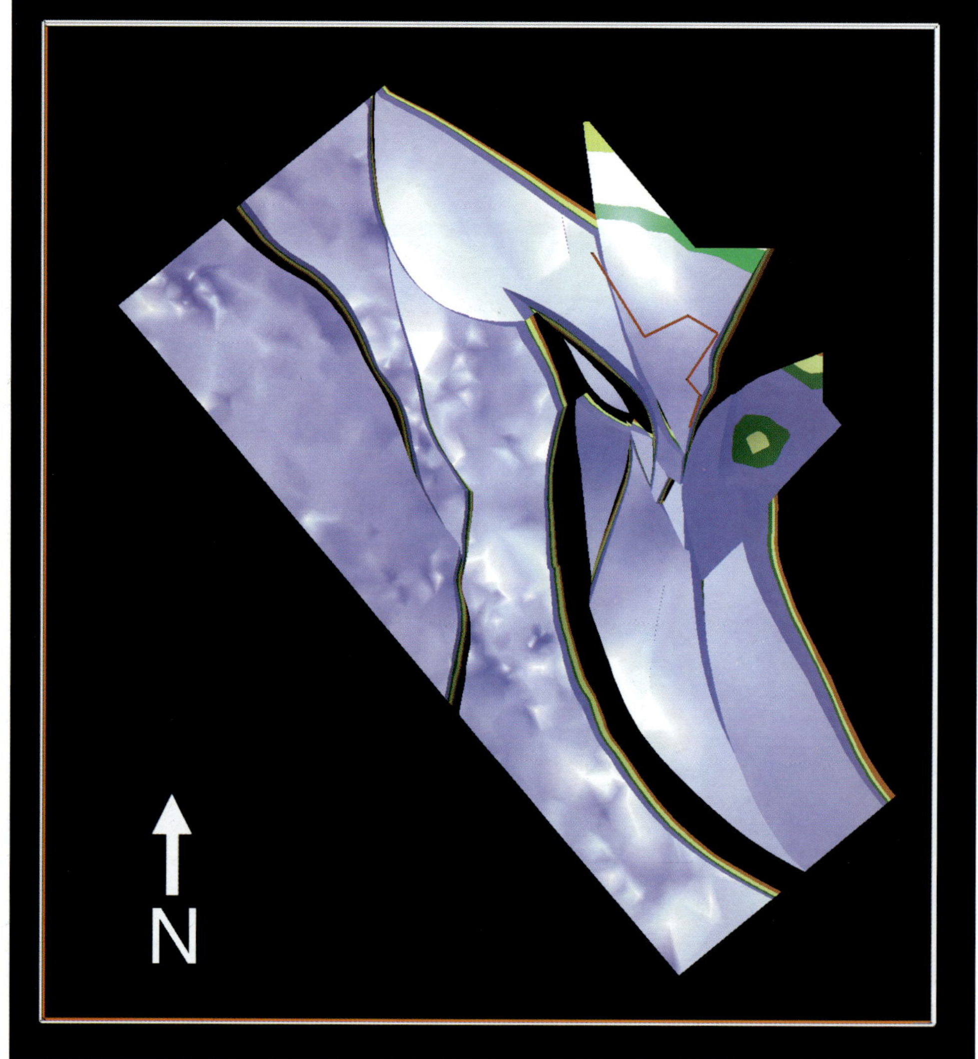

Figure 5—A map view of the three-dimensional model showing the location of the fence diagram.

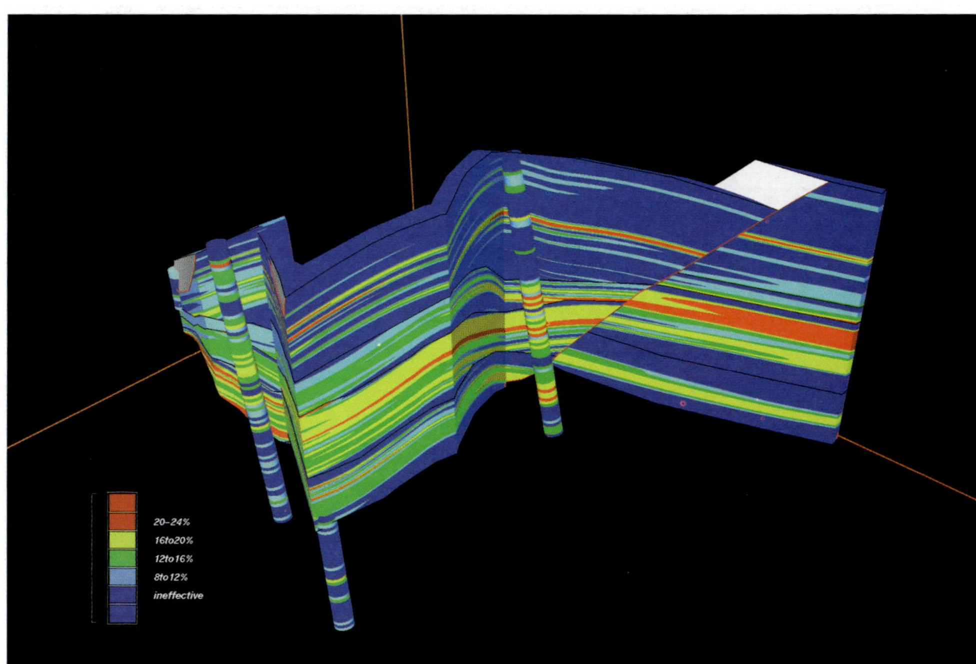

Figure 6—A three-dimensional fence diagram from the porosity model, viewed from the northeast. Porosity data from wells are indicated as tubular columns. Red lines indicate faults.

for various reservoir properties, in particular, porosity and permeability. This step resulted in the calculation of porosity and permeability values on half-foot spacing within all the wells (even the non-cored wells) that had penetrated the Miocene reservoir interval.

The next step was to construct a spatial representation of porosity and permeability in the reservoir for purposes of imputing values of porosity and permeability to locations away from the well control. This imputation process involved interpolation between wells and extrapolation beyond them using the three-dimensional algorithms available in EarthVision. Due to the highly faulted nature of this field, a gridding technique was required that treats faults as "soft" boundaries. The top of the reservoir was chosen as a datum, and then all grid nodes were shifted to this datum, effectively eliminating the vertical displacement of the faults; consequently, interpolation/extrapolation of porosity and permeability at each grid node was accomplished in this pseudo-stratigraphic space. Having constructed grids of porosity and permeability for each fault block, they were back-shifted to reflect the present-day setting based on the respective structural reference surfaces. The results of this process are shown in Figures 5 and 6. High-porosity portions in the reservoir can be seen extending from one fault block to another.

At this point, porosity and permeability cutoffs (0.8% for porosity; 1 md for permeability) were applied to yield effective porosity and permeability, the values of which were averaged within each zone. Maps of average effective permeability and true vertical thickness (TVT) maps of average effective porosity (phi**h* maps) were then prepared for each of the zones within the reservoir (Figure 7). As a result of applying the porosity and permeability cutoffs, the maps produced with this technique excluded areas where the statistical regression equations predicted values associated with microporosity. The maps indicated attractive reservoir quality adjacent to existing wells, and it was hoped they would suggest other areas having good reservoir quality that could be exploited.

Initially, the results were discouraging because several "bulls-eye" features were noted on both the porosity and permeability maps and thought to be spurious results of the three-dimensional model. Hence, to test the validity of the modeling and mapping technique, a separate three-dimensional representation of fractional quartz was also developed for each of the zones within the reservoir. Converting these to two-dimensional views of quartz thickness for each zone showed the existence of lobes of fractional quartz that were consistent with the deltaic sedimentation patterns of the reservoir. Hence, the "bulls-eye" features on the porosity and permeability maps were judged to be reasonable. With the method thus validated, the team recommended that a well be drilled through one of the areas where good-quality reservoir was indicated by the model. A well was selected that was already planned for a deeper and lower risk target, and which the model identified as having a prospective high-quality pay zone.

Reservoir volumes were calculated from the porosity model within the 150-ac (60 ha) drainage areas suggested by the team's reservoir engineers. Because reservoir porosity was shown to vary greatly in both lateral and vertical directions, volumes based on average porosity (the two-dimensional maps) were not felt to accurately represent the reservoir volume. In this case, volumes were instead calculated directly from the three-dimensional model, using 2% porosity bins. Three-dimensional GIS query techniques were used to accomplish this task. These results were used by the reservoir engineers to predict the exploitable volume

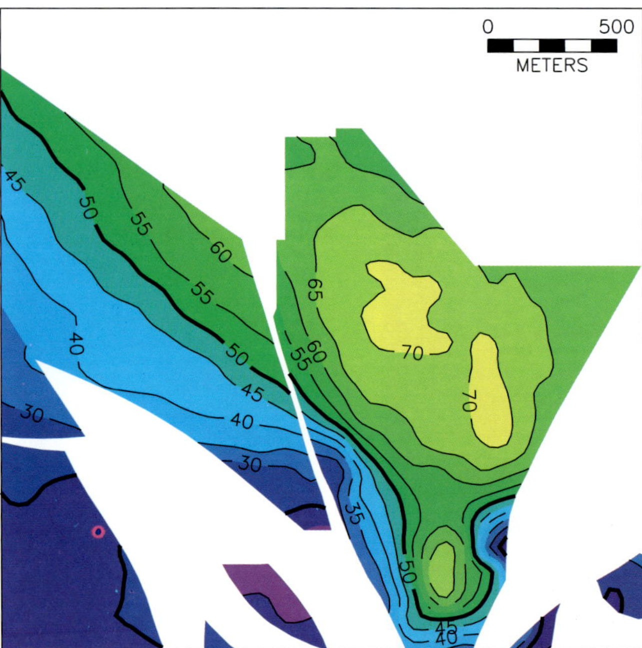

Figure 7—True vertical thickness (TVT) porosity map of the northern part of the field. Thickness was calculated based on petrophysical cutoffs of 0.8% porosity and 1 md permeability.

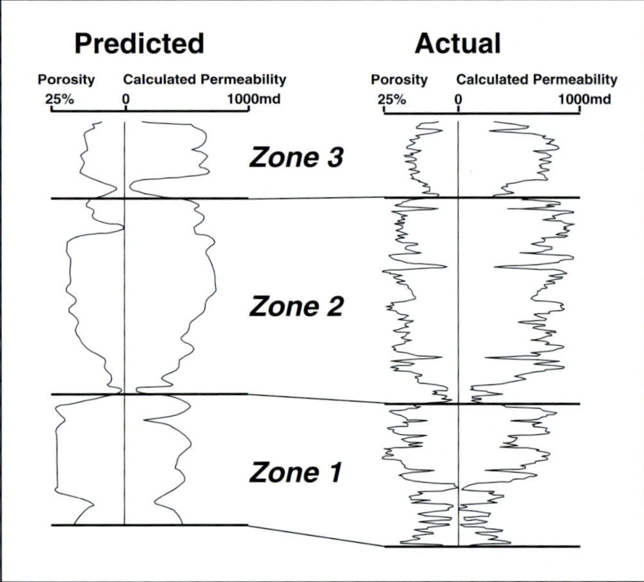

Figure 8—Comparison of the predicted and actual logs for the well drilled on the basis of reservoir characterization model. Permeability was determined using a statistical multiple regression equation.

of hydrocarbons around the proposed well, along with a production profile. The total volume was significantly higher than that predicted by the traditional, average porosity technique, which increased the upside potential for the well.

Also using three-dimensional GIS query techniques, the proposed well path was intersected with the three-dimensional porosity and permeability models to derive "predicted" log curves. This process involved the selection and manipulation of points and lines in three-dimensional space and matching their coordinates with the grid points, nodes, or cells that comprised the depictions of the reservoir properties. How closely the results matched the actual values and the performance of the well would be the ultimate test of the quality of the petrophysical and reservoir characterization models. Consequently, further work and development was put on hold until the well was completed.

DRILLING RESULTS

As noted, the well drilled to test the model had dual objectives to split the risk. This well was initially given a low chance of success in the Miocene reservoir, but as a planned dual-completion, it provided enough upside potential to be attractive when combined with a lower risk, deeper target. As predicted by the model, the well encountered reservoir quality better than adjacent wells that had never been tested for this reservoir. The initial production rates were among the best that have been seen in this area, and the decline rate was significantly better than for previous completions. After sufficient time, the production performance was judged to be acceptable and indicated additional potential to be exploited in as-yet-untested parts of the field.

In addition, actual well log responses matched the predicted values very closely (Figure 8). Not only are the overall shapes of the curves very similar, but the magnitudes of the values are also not that different. The predicted logs are smoother than the actual logs primarily because the resolution of the model was not as fine as that of the recorded logs. This close match, along with the very positive drilling results, validated the model. In addition, the well did not produce water during its lifetime, which also helped to substantiate the initial petrophysical interpretations.

The overall porosity model was updated with the new data, although the added well control did not significantly change the maps (Figure 9). Additional successful wells have since been drilled (based on these techniques), and the reservoir is under continuing development.

CONCLUSIONS

The detailed petrophysical analysis and reservoir characterization based on the procedures described here provided the basis for a new approach to exploiting a reservoir that had been thought to be uneconomical. The asset management team, working with a three-dimensional model common to all the disciplines and having access to three-dimensional GIS query capabilities, was able to plan a well that targeted what was initially thought to be a high-risk objective. The success of this and subsequent wells has shown that the team's techniques and approaches are valid,

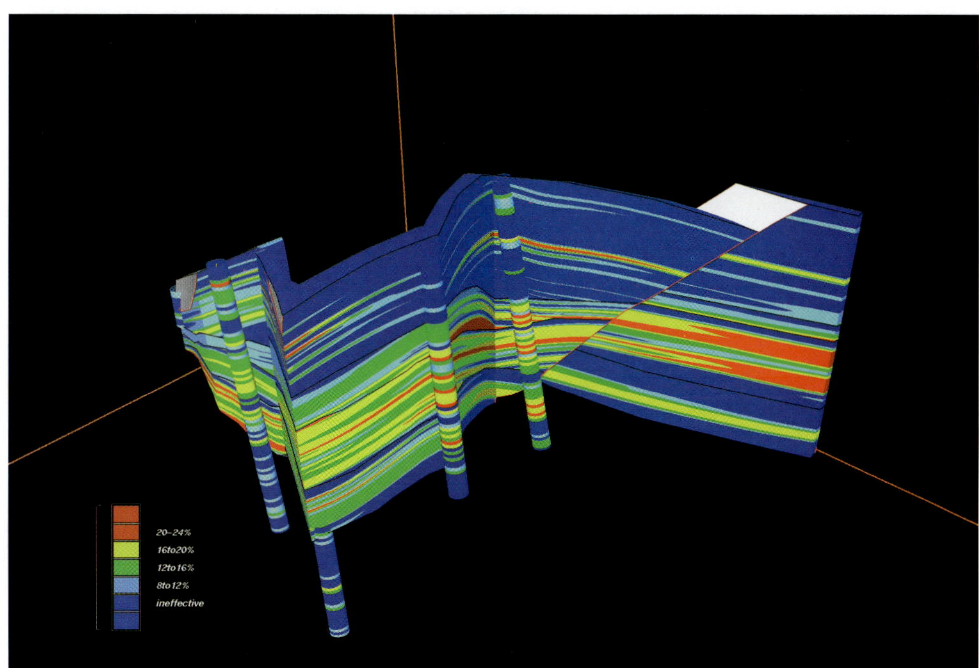

Figure 9—A three-dimensional fence diagram of the porosity model updated with data from the new well. Note the similarity to Figure 6.

leading to significant corporate revenue enhancement.

The successful exploitation of this reservoir could not have been accomplished as efficiently, nor with as much confidence, without the use of the three-dimensional modeling and GIS tools. These tools allowed the asset management team to work together using an efficient task flow to identify targets and predict their performance in a rather systematic fashion, despite the complexity of the reservoir. All intermediate tasks and products were completed or generated in substantially less time than more conventional approaches (two-dimensional maps of reservoir properties, calculation of net-to-gross interval ratios, computation of average interval thickness, etc.) would have taken, and predictions and projections based on the established methodology have proven to be accurate.

ACKNOWLEDGEMENT

The authors gratefully acknowledge the assistance of Tim Coburn in compiling the final version of this paper.

Supporting Roles and Applications

Not all of the benefits of using geographic information systems (GIS) in petroleum exploration and development come from their direct application to geological modeling, resource assessment, seismic planning, or similar high-profile endeavors. The robust nature of their data- and information-related tools make GIS an excellent choice of technology to support the extensive range of activities, disciplines, and personnel that must be brought together in a timely and effective manner to achieve exploration or exploitation success. Consequently, this fourth section showcases the versatility of GIS as an informational, communications, and organizational backbone for the overall operational structure that is unique to petroleum exploration and development.

The section contains four chapters, the first of which comes from Khalid Amin Khan of K-tron Research. In **Integrated Geo-Systems—A Computational Environment for Integrated Management, Analysis, and Presentation of Petroleum Industry Data**, Khan describes the central role GIS can play in integrating and managing various kinds and sources of petroleum-related data. Example applications include base mapping, visualization of three-dimensional subsurface models, and interactive seismic survey planning and design. The second chapter is entitled **Using the World Wide Web to Manage Exploration and Production Team Data**, by Charles Fried from Amoco Exploration and Production Technology. In this chapter, Fried characterizes the use of GIS technology in an Internet environment as an effective means of capturing and maintaining new data, as well as legacy information, exchanging data among various corporate offices, and communicating among team members stationed in diverse locations. An example of a project from Venezuela is provided. In the third chapter, Kirk Barrell of Geodynamic Solutions describes how GIS can be effectively employed as the focal point around which multidisciplinary exploration or exploitation teams can be organized. The synergy of the team is developed on the basis of a common, integrated database that is managed, maintained, processed, and visualized using GIS tools. Barrell's chapter, which is entitled **GIS—The Exploration and Exploitation Tool**, contains several examples from onshore and offshore areas of the Gulf of Mexico region. Finally, Charles Barnwell of Environmental Systems Research Institute (ESRI) describes the use of GIS technology in developing the GeoExplorer product for the U.S. Geological Survey. GeoExplorer is a desktop computer application designed to identify and distribute federal geological and earth sciences data. Barnwell's report on this project, particularly as it applies to the Livengood Quadrangle of Central Alaska, is contained in the chapter entitled **The USGS GeoExplorer Project: Using New GIS Technology for Scientific Data Publishing**.

Khan, K. A., 2000, Integrated geo-systems—A computational environment for integrated management, analysis, and presentation of petroleum industry data, *in* T. C. Coburn and J. M. Yarus, eds., Geographic information systems in petroleum exploration and development: AAPG Computer Applications in Geology, No. 4, p. 215–226.

Chapter 16

Integrated Geo-Systems—A Computational Environment for Integrated Management, Analysis, and Presentation of Petroleum Industry Data

Khalid Amin Khan
K-tron Research, Incorporated
Islamabad, Pakistan

ABSTRACT

Petroleum exploration and production (E&P) involve complex techniques for processing, analyzing, and interpreting large volumes of diverse data. To get the most out of this data resource, it is necessary to use effective data management tools for integrated analysis. This paper presents an integrated approach in which geographic information systems (GIS) technology plays a central role, along with other technologies, in integrated management, analysis, and presentation of various types of E&P data sets. A GIS provides a visual front-end as an intuitive user interface that interacts with the array of databases for data extraction. The GIS is supported by powerful graphical tools for integrated display of results from various methods that complement each other, thereby presenting a variety of complex information using simple visualization models. GIS architecture is linked with an increasing amount of geoscience application software through its common data formats for displaying their computed results.

INTRODUCTION

Today's exploration and production (E&P) environment, being information intensive, is highly dependent on computational support. In the last two decades, there has been a tremendous increase in the complexity, diversity, and volume of E&P data (Khan, 1995). With the advent of new methods, the procedures involved in processing these data sets to get the required results have become so complex that a single application software package cannot perform all the tasks that are necessary (Ward et al., 1996); consequently, specialized software applications have emerged for specific tasks. All commercial geological and geophysical application software have their own unique databases and/or data file format (Bellotti and Dershowitz, 1991). In contrast, the E&P functions require all application software involved in data acquisition, analysis, storage, visualization, and report generation to effectively communicate with other software through the sharing of data. This situation is not unlike those that have occurred in various other industries or research settings in which large volumes of geoscientific information have needed to be managed and integrated with

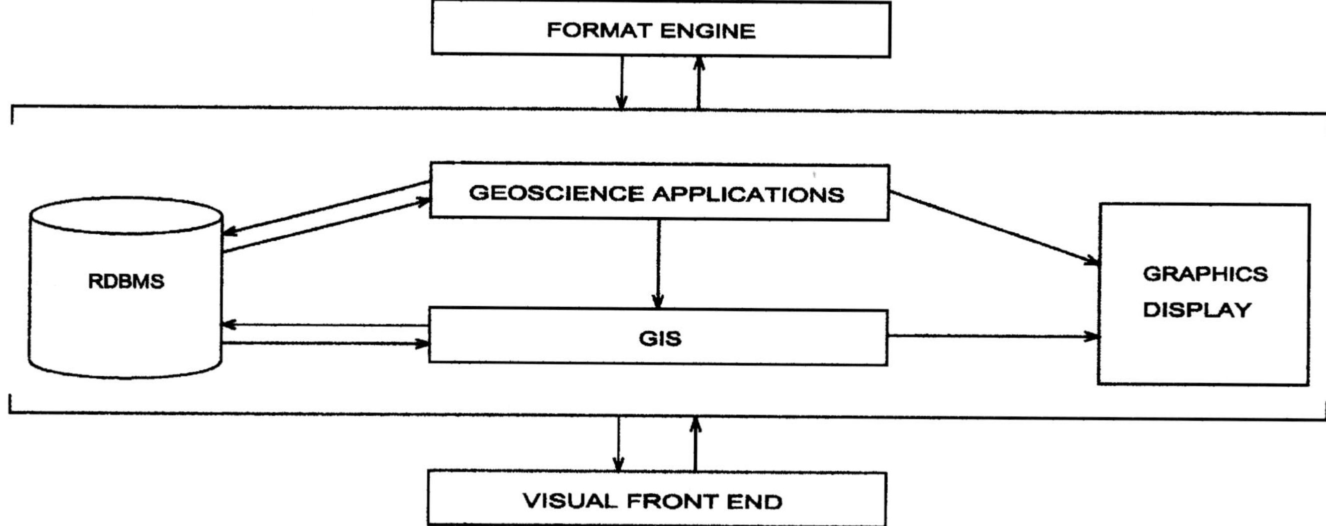

Figure 1—The conceptual architecture of the system showing the geographic information system (GIS) coupled with the other main components of the system.

various application software packages using file conversion formats. Consider, for example, projects from the mining industry (Baumewerd-Ahlmann et al., 1991), climate research (Stonebraker, 1994), global change (Hachem et al., 1993), environmental studies (Khan, 1994), hydrology and resistivity (Barker et al., 1996), and ground water contamination (Khan and Khan, 1995).

Under these circumstances, a major concern within the petroleum industry is the development of standard protocols and formats for database sharing and data transfer among various applications. The main aim behind this data sharing is to integrate data, both at the numerical and graphical levels, leading to more thorough and effective interpretation. Most geoscientific data are space-variant with respect to the earth's geographic coordinates, and they are usually plotted in some graphical form. Making use of this characteristic, GIS technology has come to the forefront and is playing a vital role in integrated management and presentation of E&P data.

Keeping these trends in mind, there is a need to develop a system that can support the following objectives:

- Integrated data management
- A tight connection among geoscience application software, through standard internal formats, for integrated processing, analyzing, and interpreting
- Flexible importing and exporting of data to and from various application software packages
- Coupling of the GIS with a graphical user interface (GUI) and other graphical tools for innovative display and map generation

To achieve these objectives, one such E&P computational environment having a GIS as its central system for managing and integrating various geographically dispersed data has been developed for the petroleum industry. A prototype subset of the system, called integrated geo systems (IGS), was developed earlier as a test project for Oil and Gas Development Company Limited of Pakistan (Banaras et al., 1996).

SYSTEM ARCHITECTURE

The conceptual architecture of the system, depicting its main components, is presented in Figure 1. All important features of this architecture are discussed in the following paragraphs.

Relational Database Management System

One of the main objectives of this environment is integrated data management, for which an extensive E&P relational database management system (RDBMS) has been developed. It consists of an array of databases for storage of diverse information comprising seismic lines, navigation data, exploration and mining concession boundaries, well locations, core data, formation tops, petrophysical logs, lease information, production information, observer reports along with geometry data, stacking velocities, statics, and seismic data cartridge library information.

In implementing an effective data storage and retrieval strategy, this information has been classified into three data types based on the raw and processed data volumes and processing times. A separate mode of data storage, retrieval, and input to the GIS has been adopted for each data type as shown in Figure 2. Data type 1 requires no processing, and therefore raw data are stored in the database, which, after retrieval, feeds directly to the GIS. Data type 2 involves long processing times or reduced processed data volumes. The raw data are first processed using one or more application software packages, and the final processed output,

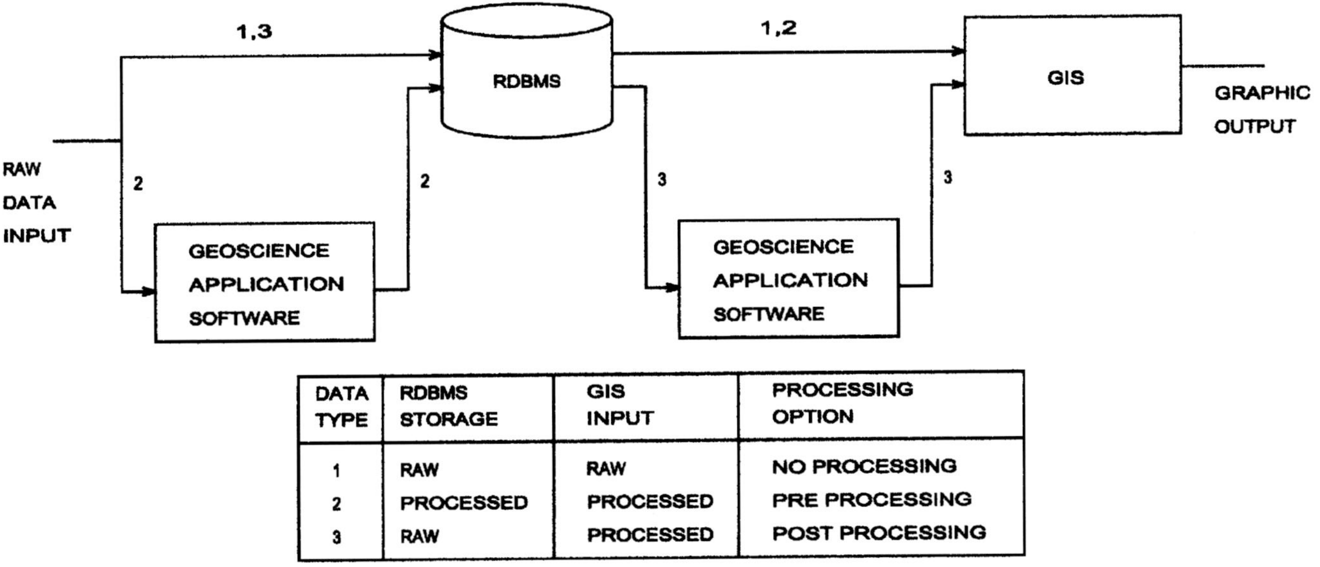

DATA TYPE	RDBMS STORAGE	GIS INPUT	PROCESSING OPTION
1	RAW	RAW	NO PROCESSING
2	PROCESSED	PROCESSED	PRE PROCESSING
3	RAW	PROCESSED	POST PROCESSING

Figure 2—Modes of data storage, retrieval, and input to geographic information system (GIS) for the three types of data. The data type number is given on its data flow path. RDBMS = relational database management system.

which may be smaller in volume compared to the raw input, is stored in the database. The processed data are retrieved and input directly into the GIS. Such processing and storage is part of a preprocessing strategy. The main ideas behind this strategy are a reduction of database storage space, quick data retrieval for GIS display, or both. In contrast, data type 3 involves shorter processing times with negligible changes in the sizes of the raw and processed data volumes. In this case, the raw data are initially stored in the database and, only after retrieval, are processed using one or more application software packages with the final results being sent to the GIS for display. Such storage and retrieval is part of a postprocessing strategy.

Format Engine and Internal Formats

E&P data exist in a number of formats and reside on a heterogeneous array of computers attached to extensive information networks or remotely located field equipment. To facilitate communication across all these platforms, a programmable format engine (Figure 3) has been developed. This engine acts as a file format conversion bridge for data input, database storage, and importing/exporting of data to other application software packages. Format definition files encompassing all industry standard formats, as well as the specialized formats for some major petroleum software, are provided. Using these files, IGS-based applications can be loosely connected with other application software. In addition, the system has a device interface through which it can directly import data from a variety of hardware devices, ranging from global positioning system (GPS) receivers to digitizers.

The vast data resource input into the system must be processed in a number of ways, so several processing tools have been provided. These tools allow the sequential application of a number of data manipulation operators, which may be part of separate application software packages. To facilitate processing in such a way, the formatting engine converts data in external formats into the environment's internal formats, and vice versa. The IGS environment comprises a number of standardized internal formats for various data types through which IGS-based applications are tightly connected for integrated analysis.

Geoscience Application Software

An ever-expanding part of IGS is its internal geoscience application software packages, which consist of a wide range of applications for report generation, quality control, gravity and magnetic modeling, petrophysical log analysis, synthetic seismograms, seismic refraction and statics computation, velocity analysis, and 3-D seismic planning. All these applications communicate with each other through the internal formats, and results generated from them are sent to the GIS for display in a variety of forms. The results can also be exported to other free-standing application software packages through the formatting engine.

GIS and Graphics System

At the heart of IGS is its GIS and the associated graphical tools, which are, perhaps, the most visible aspects of the system. The GIS provides a GUI through which the user can interactively extract a variety of E&P information from the RDBMS by using its spatio-temporal database engine. A detailed block diagram of this part of the system, along with some application software, is shown in Figure 4. The environment is separated into two parts: the essential part comprising the GIS and

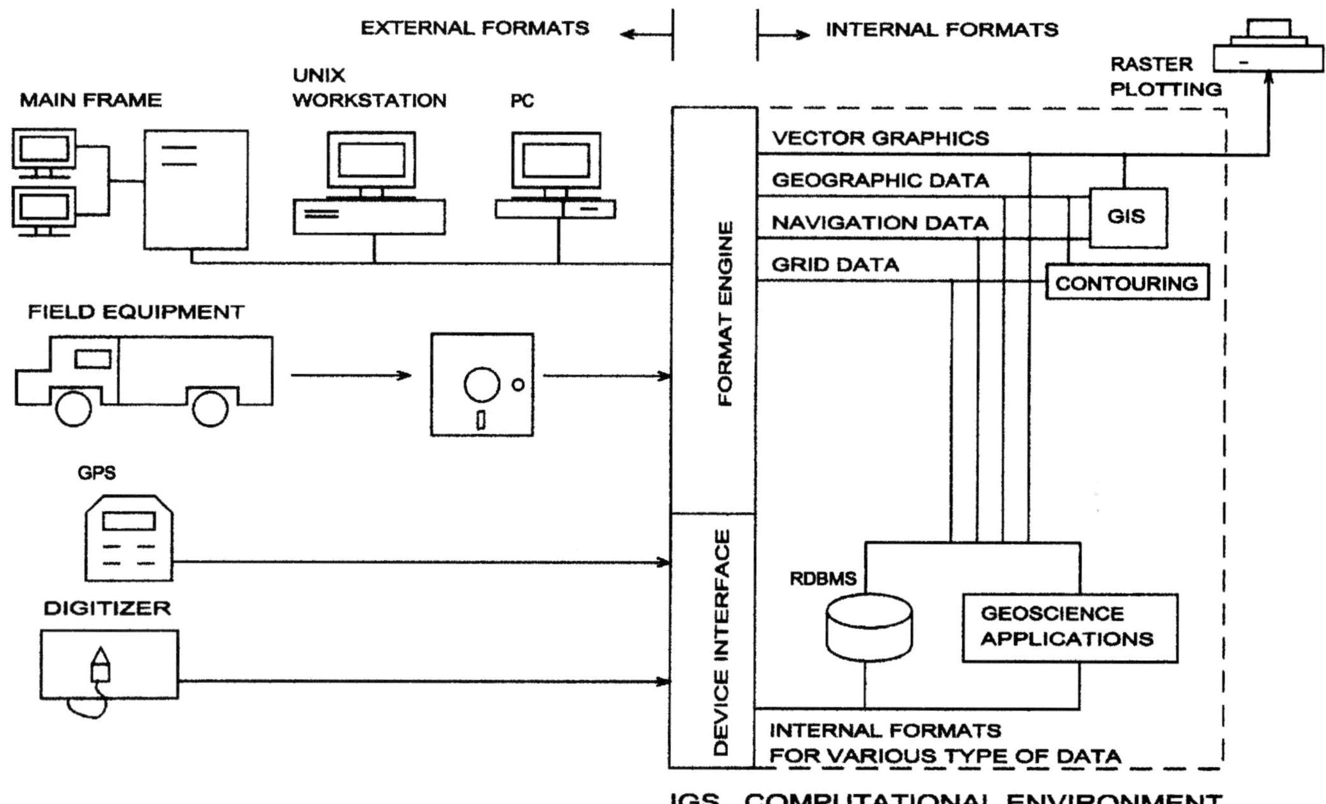

Figure 3—The format engine as a data communication bridge between applications based in the integrated geo-system (IGS) and free-standing independent software applications. The bridge encompasses a number of formats used on a variety of computing platforms. RDBMS = relational database management system; GPS = global positioning system; GIS = geographic information system.

related components, and the optional part comprising a bundle of geoscience application software.

The GIS consists of a GIS engine for extracting information from the database and processing it into geographic objects for display and visualization by the graphics system. The GIS also includes several tools for interactively calculating the area of mapped polygons (concessions), tying points between seismic lines, indexing numbers of the topographic sheets for the mapped area, and several useful report generation utilities that present the data in tabular form. In addition to the E&P data, the GIS is backed by an extensive library of cultural information comprising provincial and district administrative boundaries, cities, towns, roads, and waterways. This information helps in planning seismic surveys and selecting camp sites.

An important component of the GIS is its geodetic system, which transforms geographic data into various map projections. Geodetic transformation parameter files for all Lambert Conic Conformal and Universal Transverse Mercator projections are provided. The system contains a number of geometric operators that can be directly applied to large streams of navigation data. It acts as an essential input interface to the GIS. All applications requiring computation and presentation of data on any map projection send their data to the GIS through the geodetic system.

It has been observed that most geoscientific data end in some graphical form with or without a geographic reference. Consequently, among the essential components of the graphical part of IGS is the gridding, contouring, and 3-D surface-generation software, which is directly interfaced with the GIS as one of its major tools. Data related to any geographically dispersed raw or computed parameter obtained from any geoscience application software can be processed into contour maps, which, in turn can be directly displayed by the GIS. The gridding program contains several filters, such as polynomial surface fitting, edge enhancement, and sun shading, as well as low- and high-cut filters for extracting various features from the 2-D gridded data (Cooper, 1997). Geological structures and rock attribute distributions display anisotropic behavior with significant directional trends due to tectonic

Figure 4—(opposite page) Detailed block diagram of an integrated geo-system (IGS) showing the geographic information system (GIS), graphics capabilities, and other tools as the essential components, and the geoscience software applications as the optional components. RDBMS = relational database management system.

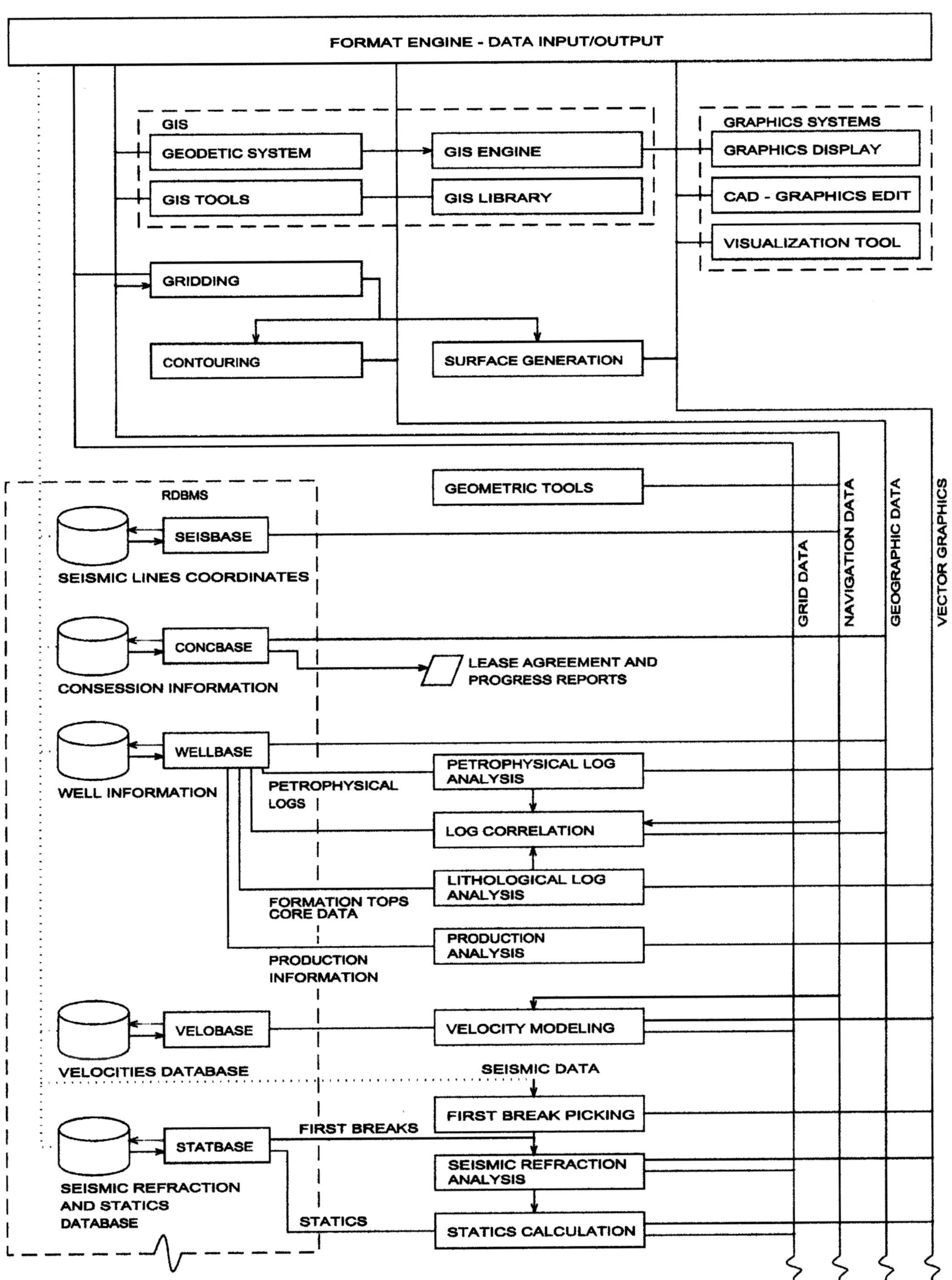
FORMAT ENGINE - DATA INPUT/OUTPUT
GIS
GEODETIC SYSTEM
GIS ENGINE
GIS TOOLS
GIS LIBRARY
GRAPHICS SYSTEMS
GRAPHICS DISPLAY
CAD - GRAPHICS EDIT
VISUALIZATION TOOL
GRIDDING
CONTOURING
SURFACE GENERATION
GEOMETRIC TOOLS
RDBMS
SEISBASE
SEISMIC LINES COORDINATES
CONCBASE
CONSESSION INFORMATION
LEASE AGREEMENT AND PROGRESS REPORTS
WELLBASE
WELL INFORMATION
PETROPHYSICAL LOGS
PETROPHYSICAL LOG ANALYSIS
LOG CORRELATION
LITHOLOGICAL LOG ANALYSIS
FORMATION TOPS CORE DATA
PRODUCTION ANALYSIS
PRODUCTION INFORMATION
VELOBASE
VELOCITY MODELING
VELOCITIES DATABASE
SEISMIC DATA
FIRST BREAK PICKING
FIRST BREAKS
STATBASE
SEISMIC REFRACTION ANALYSIS
SEISMIC REFRACTION AND STATICS DATABASE
STATICS
STATICS CALCULATION
GRID DATA
NAVIGATION DATA
GEOGRAPHIC DATA
VECTOR GRAPHICS

events (Kushnir and Yarus, 1992). To impose such geologic interpretation, distance transforms are applied to the gridded data (Zoraster, 1996). Most of these operators applied to gridded data are routinely applied by GIS to image processing of satellite data. The gridded data can be directly sent to the GIS as a color attribute map or contoured and, subsequently, sent to the GIS. The GIS has interactive tools for extracting data along a profile between two points.

The GIS is coupled with a powerful graphics system consisting of a graphics display, a computer-aided drafting (CAD) system, and a 3-D visualization tool. The detailed structure of the system presented in Figure 4 indicates that four types of data—grid data, navigation data, geographic data, and vector graphics data—are commonly used throughout the system. All such data can be visually displayed using the GIS and its associated graphics tools. Applications that require scaled presentations in specific map projections send their data directly to the GIS. Geographically scattered data are displayed as an attribute map by the GIS, or they are gridded and contoured for display by the GIS. Similarly, the navigation and geometric data are also sent to the GIS through the geodetic system. Applications that generate graphs, cross-sections, fence diagrams, and other displays of subsurface models communicate directly with the graphics display features or the CAD system for presentation and graphic editing, whereas interpreted 3-D models are loaded into the visualization system for simulation and analysis.

The GIS, in conjunction with the E&P RDBMS, the graphics system, and the contouring suite, forms an integrated application for managing and displaying petroleum industry data. It is a powerful tool for spatial data representation where data stored from all sectors of the petroleum industry, along with cultural information from the GIS library, can be integrated for innovative display and map generation.

APPLICATION EXAMPLES

IGS encompasses a broad spectrum of E&P data handling, processing, visualization, and modeling applications, each designed to support a specialized task. Detailed discussion of all of these applications is beyond the scope of this chapter; however, the following section briefly presents some selected examples that highlight the GIS-related features of IGS. Each of these examples focuses on a specialized aspect of the GIS, highlighting its capabilities in innovative data search, integrated graphics presentation, visualization, application software connectivity, and real-time mapping.

Integrated Graphics Presentation

One of the basic applications of GIS to E&P data has to do with generating base maps that indicate the locations of seismic lines, wells, and concession boundaries. IGS, with its abundance of data and applications, can graphically integrate a variety of information and present it on a base map. Figure 5 presents a map produced by IGS having basic information such as seismic lines, wells, and concession boundaries, along with statics contours, seismic refraction pickets, and tie points between seismic lines. Several data layers are superimposed to generate a single, graphically integrated map. Figure 5 also contains the output from some of the GIS tools contained in IGS, such as interpolated statics along a seismic line and tabular results from tie point and other report-generating utilities.

3-D Visualization

E&P interpretation involves comparative study of various interrelated parameters. IGS provides 3-D visualization tools for generating simple visualization models that help in understanding parameter variation and anomalies. Figure 6 presents a visualization model showing the locations of seismic lines, contours of statics, topography, interpreted thickness of weathered layers, and interpreted velocity. It clearly indicates that the clusters in the contours of statics can be attributed to the variation in the thickness of weathered layers and velocity, which, in turn, follow the topography.

Interactive Seismic Survey Planning

GIS can be coupled with other geoscience applications for interactive seismic survey planning and design. Figure 7 illustrates the interaction of GIS and another software application. The GIS extracts the required information from the database and displays it with the graphics system. It also provides an interactive user interface for graphic input. Based on the analysis of existing information displayed by the GIS, the user can input new parameters and design criteria that are transferred to the coupled application. The secondary software application processes the user input and sends its computed output to the GIS, which displays it along with the original information.

Figure 8 shows a map that contains 2-D seismic lines and time contours as initial information. After analyzing the displayed information, the user can interactively mark the boundary of the geological target area over which a 3-D seismic survey is to be conducted. This information is subsequently transferred to the 3-D seismic planning software, which computes the survey area boundary on the basis of migration aperture. Similarly, swath parameters and the geometry of shot receiver spacing can also be interactively input to obtain various fold zones. The advantage of such an interactive system is that the user can continuously change the parameters until the desired results

Figure 5—(opposite page) Base map indicating location of seismic lines, wells, concession boundary, tie points, refraction pickets, and statics contours along with interpolated statics for line KTN-9 and tabular results from some computational utilities. GIS = geographic information system.

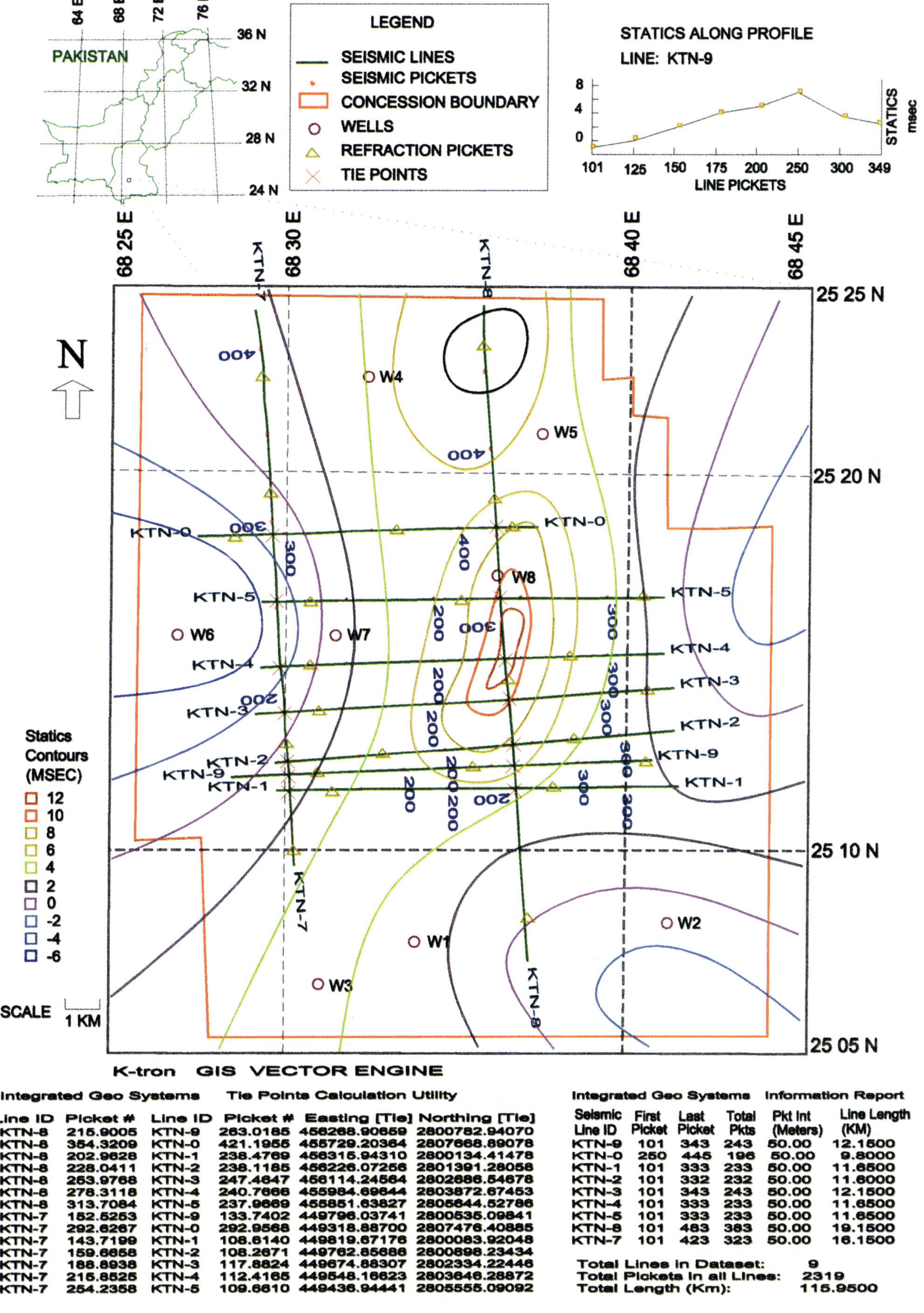

Integrated Geo Systems   Tie Points Calculation Utility

Line ID	Picket #	Line ID	Picket #	Easting [Tie]	Northing [Tie]
KTN-8	215.9005	KTN-9	263.0185	456268.90659	2800782.94070
KTN-8	354.3209	KTN-0	421.1955	455729.20364	2807668.89078
KTN-8	202.9628	KTN-1	238.4769	456315.94310	2800134.41478
KTN-8	228.0411	KTN-2	238.1185	456226.07256	2801391.28058
KTN-8	253.9768	KTN-3	247.4647	456114.24564	2802686.54678
KTN-8	278.3118	KTN-4	240.7666	455984.69644	2803872.67453
KTN-8	313.7084	KTN-5	237.9669	455851.63827	2805644.52786
KTN-7	152.5253	KTN-9	133.7402	449796.03741	2800535.09841
KTN-7	292.6267	KTN-0	292.9568	449318.88700	2807476.40885
KTN-7	143.7199	KTN-1	108.6140	449819.67176	2800083.92048
KTN-7	159.6658	KTN-2	108.2671	449762.85686	2800898.23434
KTN-7	188.8938	KTN-3	117.8824	449674.88307	2802334.22446
KTN-7	215.8525	KTN-4	112.4165	449548.16623	2803646.28872
KTN-7	254.2358	KTN-5	109.6610	449436.94441	2805555.09092

Integrated Geo Systems   Information Report

Seismic Line ID	First Picket	Last Picket	Total Pkts	Pkt Int (Meters)	Line Length (KM)
KTN-9	101	343	243	50.00	12.1500
KTN-0	250	445	196	50.00	9.8000
KTN-1	101	333	233	50.00	11.6500
KTN-2	101	332	232	50.00	11.6000
KTN-3	101	343	243	50.00	12.1500
KTN-4	101	333	233	50.00	11.6500
KTN-5	101	333	233	50.00	11.6500
KTN-8	101	483	383	50.00	19.1500
KTN-7	101	423	323	50.00	16.1500

Total Lines in Dataset: 9
Total Pickets in all Lines: 2319
Total Length (Km): 115.9500

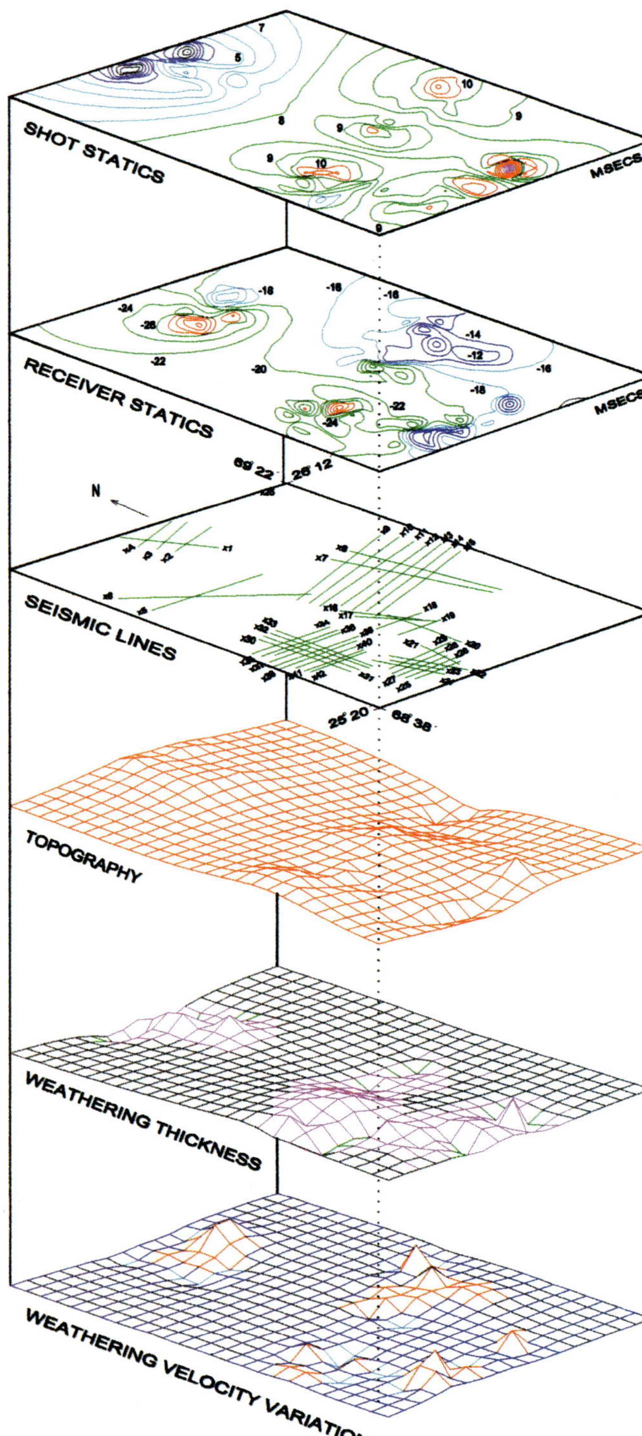

Figure 6—3-D statics visualization model indicating a tilted seismic line base map, shot and receiver statics contours, and surfaces showing the variation in topography and weathered layer parameters.

are obtained. Once the survey has been planned, the final information can be referred to the database for storage.

Innovative Database Search

Most GIS applications deal with the management and display of 2-D dispersed data; however, there is a distinct need for 3-D selection involving the incorporation of a vertical component in the geographic attributes (Rogers, 1996). IGS has a spatio-temporal database engine that, in addition to conventional lateral geographic searching, facilitates searching for attributes at depth, as well as other selection criteria. Figure 9 illustrates a simple example in which a geographic area has been searched for all wells that produce from a Middle Jurassic formation. Such selection can be useful in mapping the formation tops and variation of subsurface structures.

Real-Time Geographic Mapping

The GIS can directly input data from a GPS receiver connected to a computer's serial port for real-time data acquisition and monitoring. Figure 10 shows an onscreen display of a map containing previously digitized information and an indication of current location. This is useful in marking well points on existing seismic lines in the database.

NEW DIRECTIONS

To keep pace with the new trends in information technology and geoscientific methods, IGS is being continuously updated. New tools are being added to the GIS, existing geoscience applications are being revised, and several new applications are being added. Work is in progress to include a toolbox-based decision support system (DSS) in the next release of IGS (Walls, 1996). DSS would more fully exploit the large E&P information resource and help in decision making in a competitive and risky business environment.

CONCLUSIONS

An E&P computational environment comprises efficient and effective tools for automated storage, retrieval, integration, analysis, and presentation of such diverse and plentiful E&P data. Under the umbrella of this environment, various geoscience application software can share the E&P data resource for thorough and robust results, which can be graphically integrated for presentation in a GIS.

REFERENCES CITED

Banaras, C. M., M. A. Khan, and K. A. Khan, 1996, Use of database management systems, GIS and CAD in geophysical investigations, *in* Proceedings, 2nd Meeting of the Environmental and Engineering Geophysical Society (European Section), Nantes, France, p. 234–237.

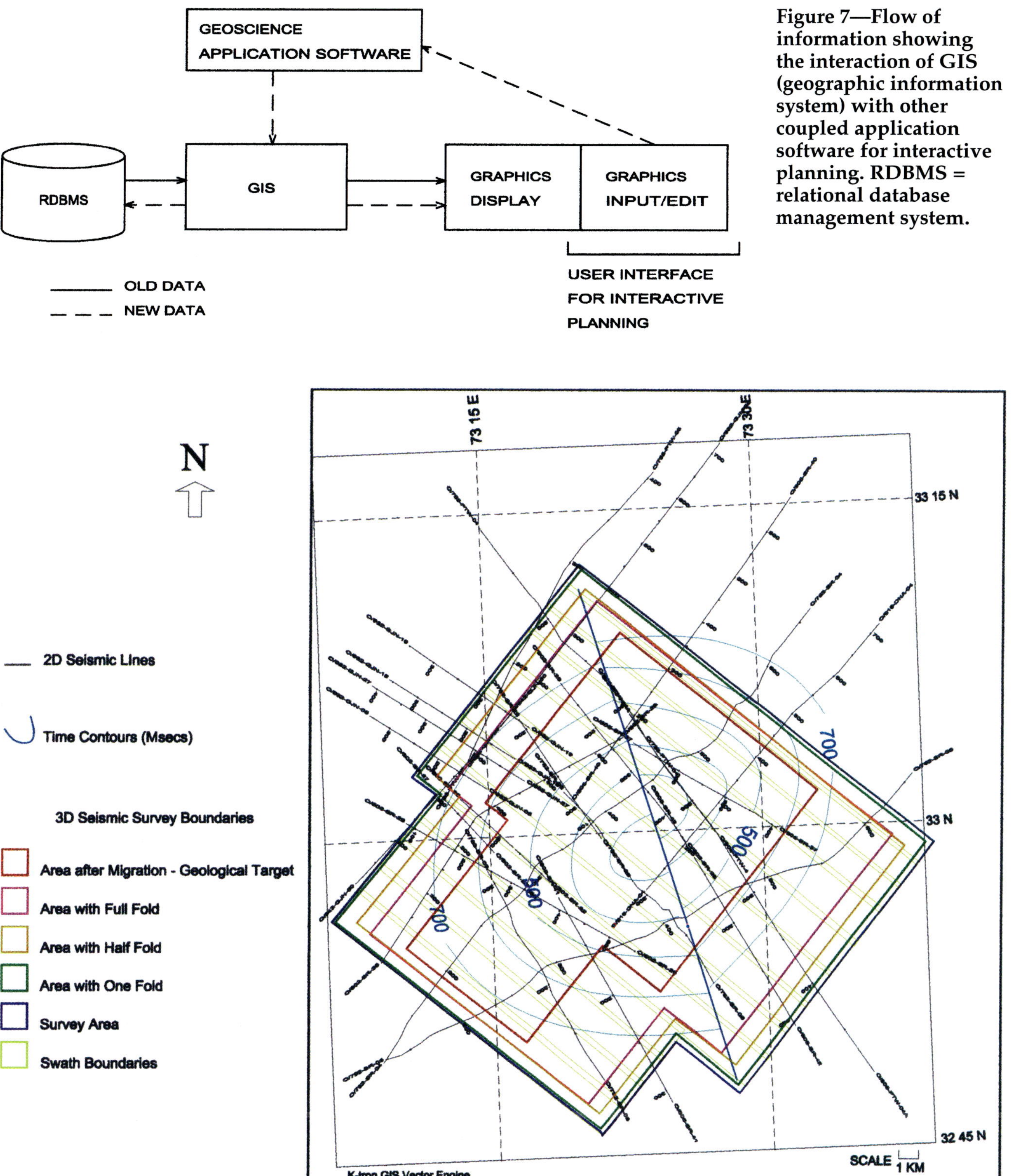

Figure 7—Flow of information showing the interaction of GIS (geographic information system) with other coupled application software for interactive planning. RDBMS = relational database management system.

Figure 8—Base map showing the initial information as 2-D seismic lines and time contours, along with the planned final information as various 3-D seismic survey boundaries.

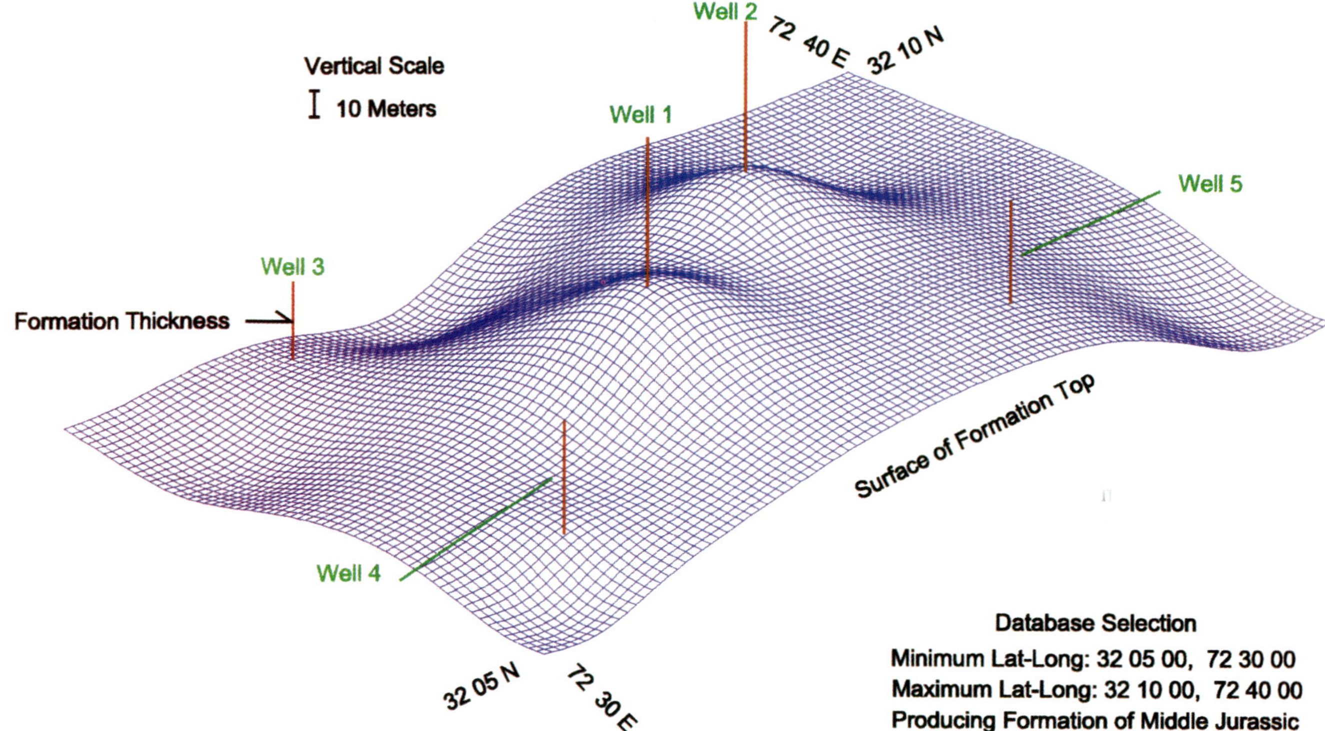

Figure 9—Surface of a producing formation top. The data are extracted from the wells' database by specifying the geographic area along with the formation attribute.

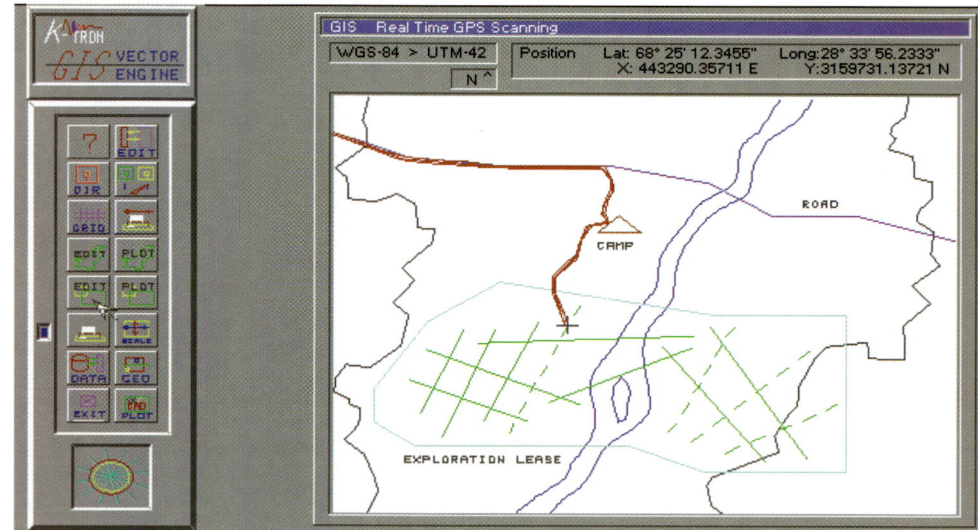

Figure 10—Computer monitor screen capture showing a digitized map along with the user's current location. The user's movement path is indicated by the thick red line, and the current position is indicated by the cursor. Geographic coordinates, as well as grid coordinates, in the selected map projection are also simultaneously displayed.

Barker, R. D., I. Blunk, and I. Smith, 1996, Geophysical considerations in the design of UK National Resistivity Sounding Database. First Break, vol. 14, p. 45–53.

Baumewerd-Ahlmann, A., A. B. Cremers, G. Kruger, J. Leonhardt, L. Plumer, and R. Waschkowski, 1991, An information system for the mining industry, *in* Proceedings, Database and Expert Systems Applications: DEXA '91 (D. Karagiannis, ed.). Berlin: Springer-Verlag, p. 86–91.

Bellotti, M. J., and W. Dershowitz, 1991, Hydrogeological investigations: data and information management. Computers & Geosciences, vol. 17, p. 1119–1136.

Cooper, G. R. J., 1997, GravMap and Pfproc: software for filtering geophysical map data. Computers & Geosciences, vol. 23, p. 91–101.

Hachem, H., M. Gennert, and M. Ward, 1993, The Gaea system: a spatio-temporal database system for global change studies, *in* Proceedings, Annual Meeting of the American Association for Advancement of Science. Boston, MA, p. 84–89.

Khan, K. A., 1994, An expert system based on geophysical techniques for environmental control, *in* Proceedings, Symposium on Environmental Geology, Peshawar University, Peshawar, Pakistan, p. 97–100.

Khan, K. A., 1995, Seismic and its supercomputing challenges. Pakistan Journal of Hydrocarbon Research, vol. 8, p. 59–61.

Khan, K. A., and M. A. Khan, 1995, Integrated system based on automatic VES curve matching and contouring algorithm for investigating groundwater and its contamination, *in* Proceedings, 1st Meeting of the Environmental and Engineering Geophysical Society (European Section), Torino, Italy, p. 480–483.

Kushnir, G., and J. M. Yarus, 1992, Modeling anisotropy in computer mapping of geologic data, *in* Computer Modeling of Geologic Surfaces and Volumes (D. E. Hamilton and T. Jones, eds.). Tulsa, OK: American Association of Petroleum Geologists, p. 75–92.

Rogers, B. R., 1996, Three dimensional selection and its potential application in graphics and image processing of geoscience data. Computers & Geosciences, vol. 22, p. 1187–1189.

Stonebraker, M., 1994, Sequoia 2000—a reflection on the first three years, *in* Proceedings, 7th International Working Conference on Scientific and Statistical Database Management, Charlottesville, VA, p. 108–116.

Walls, M. R., 1996, Developing an exploration decision support system (EDSS): a strategy for combining information and analytics. Nonrenewable Resources, vol. 5, p. 181–192.

Ward, M. O., W. L. Power, and P. Ketelaar, 1996, A computational environment for the management, processing and analysis of geological data. Computers & Geosciences, vol. 22, p. 1123–1131.

Zoraster, S., 1996, Imposing geological interpretations on computer-generated contours using distance transformations. Mathematical Geology, vol. 28, p. 969–985.

Fried, C., 2000, Using the World Wide Web to manage exploration and production team data, in T. C. Coburn and J. M. Yarus, eds., Geographic information systems in petroleum exploration and development: AAPG Computer Applications in Geology, No. 4, p. 227–236.

Chapter 17

Using the World Wide Web to Manage Exploration and Production Team Data

Charles Fried
Exploration and Production Technology Group, BP/Amoco
Houston, Texas, U.S.A.

ABSTRACT

Efficient management of all the data necessary to conduct large-scale exploration projects can be a massive and challenging undertaking. This chapter details the inefficiencies of traditional data-handling procedures, discusses some new and beneficial technological advances in information management, and outlines the application of a much-improved strategy to administer corporate data resources. The new process integrates geographic information systems (GIS) with the capabilities of the World Wide Web to facilitate a truer, more interconnected team approach. Its successful application is succinctly illustrated by an account of BP/Amoco's Venezuela Project.

INTRODUCTION: GIS AND DATA INTEGRATION

Two recent technological enhancements of geographic information systems (GIS) promise to integrate GIS with a broad variety of data used in exploration and production, and at the same time make GIS applications easier to use and support. Although these are not new algorithms or new GIS functions, the improvements that are promised by these developments are important.

First, the maturation of GIS spatial data extensions (SDE) for relational database management systems will finally enable the full integration of GIS data with corporate data. This will eliminate a significant barrier to the adoption of GIS applications in exploration and production, and reduce the burden of support for GIS application developers and users. Second, the advent of GIS Internet map server (IMS) technology promises to allow GIS to integrate with the diversity of data available on the World Wide Web. Internet distribution of GIS applications can materially reduce the support burdens for GIS application developers, as well. An example of how these technologies have and will be applied is described in the following paragraphs.

HISTORICAL PROBLEMS ASSOCIATED WITH GIS

A brief summary of how GIS has been used in typical exploration and production situations serves to illustrate some of the failings of GIS in the past and establishes the context for significance of SDE and IMS.

Stand-Alone Applications/Stand-Alone Data

GIS has generally been used as a stand-alone technology. Data for GIS applications must be imported into the GIS software where they will be used. This is fine for ad hoc, one-time, casual data sets. They can be easily imported into a GIS environment and do not require periodic updates from multiple sources or synchronization with other data stores. However, much more data are contained in the corporate data stores.

These databases are most commonly relational in nature and often exist as variants of one of several standard data models, the tables of which contain geographic coordinates. These coordinates are extracted and reformatted for use by the GIS. To improve performance, a subset of commonly used attributes is also extracted and placed in the GIS environment. Once replicated, the coordinates and attributes must be synchronized with the source tables, which are updated frequently. This can represent a formidable challenge with the large and dynamic data volumes common in exploration and production settings.

Such corporate databases are supported by corporate support organizations, which employ database administrators who are charged with protecting corporate data resources. These administrators are required to ensure that the data are backed up, that they be compiled according to a standard data model, that they be kept up-to-date, and that they be housed in corporately standardized database management systems. Such stringent requirements are at odds with the introduction of a nonstandard database manager (the proprietary one used by the GIS platform), which requires data replication. On top of that, the database administrators lack the training to support GIS data and data storage.

Historically, the result has been that the corporate support organization resists and avoids supporting GIS databases. The burden of extraction, maintenance, and synchronization then falls on the developers or users of the GIS applications. The result is a major data integration problem.

Specialized Spatial Analysis Functions

Over the years, GIS has been used primarily to accomplish two functions: spatial analysis and automated mapping. In spatial analysis, GIS experts, both those trained and those clever enough to construct sophisticated GIS modeling programs, have built unique and complex specialized systems of spatial technology. Although the intent has been to develop new analytical techniques, what has often resulted has been a series of complicated functions and usable by small groups of individuals having niche applications. The more typical users (such as geologists and engineers) are either not technologically savvy or they already have enough complicated applications to master and are not enamored of yet another difficult one to learn.

Automated Mapping Functions

A GIS has also been employed to create hard-copy maps of information in the database. It is certainly capable of performing this function, which can be reduced to fairly simple and accessible applications. However, this does not really exploit the most interesting spatial analytical capabilities of the technology.

Support Burdens

Support for these types of GIS applications, already constrained by data integration problems, is further complicated by the requirement to support them on multiple hardware platforms. A large number of users may have PCs, whereas another large group has UNIX workstations. Differences in path names, length of file names, graphical user interfaces, operating system calls, etc. make system and application support more difficult than they should be.

RECENT SIGNIFICANT ADVANCEMENTS

As noted, the introduction of GIS spatial data extensions to relational database management systems (DBMSs) and the appearance of IMS extensions for a GIS are both meaningful because they directly address the data integration issues previously mentioned.

Spatial Data Extensions to Relational DBMS

Spatial data extensions consist of methods for encoding geographic coordinates in relational tables. Application program interfaces to these new entities provide for GIS access to the coordinates. Essentially, the relational data model can be spatially enabled in such a way that the GIS environment can access these data directly.

The advantage of these methods is that geographic coordinates can now be stored and integrated in the same relational database as the attributes of the features mapped by the GIS, and it can be done with acceptable performance. Periodic update and maintenance of various data sets can likewise be accomplished all at one time within the confines of a single DBMS. Because all the data are stored in the same corporate standard DBMS, the influential guardians of corporate data do not have to address deviations from corporate standards or data replication and synchronization.

Because special data used by GIS applications can now be integrated into a relational database and can be supported by the same organization that maintains all other corporate data, the burden of self-support previously placed on GIS users or developers is eliminated. GIS software environments still contain ad hoc data updating facilities that provide for the flexibility commonly needed for quick response to one-time tasks, but large, long-term, world wide GIS applications can now be accessed across all corporate boundaries without requiring their users or developers to provide individualized or specialized support.

Internet Map Servers

IMSs are extensions of GIS platforms that permit views of geographic data to be delivered to the Web. Any user that can retrieve the Web page containing a map can view it. The map is dynamic, however. Layers can be added or subtracted from the display, the area can be zoomed to display areas of interest, and, most important, entities on the map can be used as hyperlinks to other data that are accessible on the Web. The accessible information includes relational attribute data, as well as Web pages and even Web searches by content.

IMPACT OF GIS INTEGRATION WITH WEB DATA

The integration of a GIS and the Web is significant because the Web itself is so significant. It also has some interesting technological implications that should help reduce GIS support requirements.

How the Web Complements GIS

The Web as a "database" complements traditional databases in many ways, as shown in Table 1. Neither the Web nor a GIS is a replacement for the other. The combination of the two is similar to the synergistic relationship between GIS capabilities and relational database capabilities. Thus, GIS queries can now be combined with Web queries and with database queries. Each type of query can be used to feed one of the other types.

The uniquely flexible characteristics of the Web as a database, when used as an intranet, make it useful for sharing information within a team. Memos, presentations, reports, spreadsheets, e-mail traffic, and other diverse elements can be filed. All team members can do this themselves, without the aid of support staff. The index, updated automatically, makes it easy to search the mass of documents with little additional effort. Moreover, these searches are available to GIS Internet applications. As an example, a user could click on a field on a map to initiate a Web search for all documents that mention that field name. This might be a feature of an activity monitor system. A different activity monitor function might be to display on a map all wells that have been drilled below 10,000 ft within the last 12 months. This can be done using a traditional GIS function. Four common types of searches that can be accomplished via the Internet are shown in Table 2.

Advances in Ease of Use

Typical GIS developers have considerable experience with aptitude for complex applications. They favor products with considerable functionality and flexibility rather than simplicity. The resulting applications are downright scary to casual users. The arrays of buttons and options do not encourage use by the average person. Thus, many GIS applications lack user acceptance despite functionality and performance.

For example, the typical user of an activity monitor system may be a negotiator, manager, or other casual user who has little knowledge of the underlying GIS concepts and who uses the system only infrequently or on a limited basis. Having to become refamiliarized with how the system works every time it is initiated is a disadvantage for these individuals and may lead to nonuse. It is critical to note that the purpose of the system is defeated if the users are discouraged from employing it, even if more capabilities are available. Even geoscientists with considerable computer experience are often unaccustomed to the concepts used in a GIS or the vocabulary of a GIS.

The point-and-click interface of the Web, however, is very familiar. Web-based GIS applications tend to be simpler and can be implemented as a series of relatively simple pages, each with a focused set of functions. Although care must be taken to maintain simplicity in the GIS Internet application interface, if it is done well, it is accessible and nonthreatening to any user. This situation greatly improves the ease of use of these applications.

Reduction of System Resource Requirements

Given the all-too-rapid evolution of hardware and software, most corporate environments encompass a variety of desktop platforms. UNIX workstations are used for seismic interpretation, whereas PCs are most

Table 1. Comparison of Relational Databases and Web Databases

GIS or Relational Database	Web Database
Highly structured	Unstructured
Difficult to use	Easy to use
Requires specialist support	Easy to distribute and support
Relational/geographic	Graphic/textual
Client hardware intensive	Minimal client hardware and software requirements

Table 2. Complementary Search Methods

Search Method	Example
By content	Find all documents that contain the words "API gravity" authored by Jones
By database	Find all wells drilled after 1990 that are deeper than 10,000 feet
By map	Display all wells, seismic lines, and fields in the area of Concession 3-A
By menu	Navigate with "point and click" to Venezuela page, then select link to Project 4

commonly used for all other work. Most individuals have only one or the other of these machines. Some computers have lots of memory and fast processors but others have relatively small resources. GIS applications that must run on all of these platforms are difficult to build, and they represent only a lowest-common-denominator solution. The difficulty of making the same data simultaneously available to GIS applications running under multiple operating systems (e.g., UNIX, Windows 95, and Windows 3.1) can also be a challenge.

To use a GIS application via an IMS, all the client needs is a Web browser, which today is generally already installed on the computer. Consequently, the system requirements of the computer are reduced and platform independence is increased, which, again, means that the burden of supporting GIS applications is reduced.

Inclusion of Metadata

Because it is so easy to publish a wide variety of diverse documents and information using the Web, information about the data (or metadata) can be more easily included. Such information, typically consisting of word processing files, charts, graphics, and other varied material, cannot be included in relational or GIS databases with the coordinate and attribute data. However, this information can be added to the Web site, indexed, and thus more easily made available to users because the Web allows word processing, spreadsheet, and other documents to be published without conversion.

THE BP/AMOCO VENEZUELA PROJECT

The Web was used as a medium for information sharing and data management in BP/Amoco's Venezuela project. The history and results of this exercise illustrate the advantages and disadvantages of the GIS approach outlined.

Experiences with BP/Amoco Venezuela

No two projects are the same, yet many share common characteristics. Typical of such ventures, BP/Amoco's Venezuela project encompassed numerous technical and nontechnical data sets that existed in a variety of formats and that had to be shared among a large group of workers in more than one location over a span of years. In this respect, the solutions that were found to be effective may be relevant to any large group having similar requirements, regardless of the relationship to petroleum exploration and production.

As an initial step, an intranet Web was constructed for BP/Amoco's Houston-based Venezuela team. The first priority was to distribute data related to a bid round. The establishment of a means for information exchange was a strategic objective.

A GIS interface was part of this Web. It was considered an important component of the data distribution mechanism. Venezuela lacks the type of land survey system common in the United States, and often the only way to communicate the location of a well or other feature is a map. Thus, the quantity of maps involved, and the ease with which they can be produced, represented an example of how well the overall GIS approach could address these issues.

Historical Perspective

A brief history of BP/Amoco's Venezuela operations may help to explain the data management challenges. BP/Amoco was active for several decades in Venezuela prior to nationalization, which occurred in January 1976. At that time, with no hope of returning to Venezuela, BP/Amoco staff packed all materials in the office and shipped them back to the United States. This material included a diverse collection of unique original records, many of which were scientific or technical in nature, including original well records, maps, seismic sections, and well logs. Some 200 boxes of information were shipped to BP/Amoco offices in Chicago, then later transferred to Houston.

In the early 1990s, apparently unsatisfied with the level of investment and returns that a monopolized national oil company could generate alone, Venezuela began to re-open some opportunities to foreign exploration and production concerns. BP/Amoco's interest was renewed, and geologists, geophysicists, engineers, and others eager to resume the chase exhumed the boxes from their dusty resting places in offsite storage.

The Data Quagmire

Original, yellowed documents were retrieved, restudied, and filed in desks and drawers in offices. Seismic maps were digitized and the data mapped by computer. Well logs were digitized and assembled into cross-sections. Some of these data were entered into the BP/Amoco computer master files, along with other information that had been routinely collected by scouts, vendors, and other commercial entities. Despite the best intentions of everyone involved, the data management issues became nightmarish. For example, due to the casual nature in which well locations had been identified, data relating to the same entity was often filed under different keys, and therefore a single comprehensive collection of such information could not be produced. Also, because local land surveys and coordinate systems commonly used in Venezuela were not completely understood, coordinates for these various entities often included variances of many kilometers, even for the same well! To further complicate matters, additional data officially released by Venezuela in data packages for bid rounds over several years was added to the pile with often unreliable quality control. Finally, as individuals transferred out of the group, they cleaned out their offices, leaving many documents in a data room. Although the room was periodically inventoried, subsequent removals and additions of material were not recorded. Then BP/Amoco won several blocks in bid rounds. Partnerships were formed. Data were traded. Plans to shoot seismic were made, and the data were acquired, processed, and interpreted. Environmental studies were carried out. An office was opened in Caracas. Wells were committed. BP/Amoco was back in Venezuela. Suddenly, or at least, eventually, the situation had changed from casual to critical, and the real state of the data was an issue.

Over time, the data room was inventoried several times. As a result, some of the critical study areas were organized and cleaned up. Yet, in the heat of battle of bid rounds, all efforts were focused on short-term priorities, and the data room cycle repeated itself. Comprehensive database cleanup demanded major efforts, and maintaining the data room required new procedures. Boxes remained in offsite storage containing unknown quantities and qualities of still-yellowing original and irreplaceable documents.

Originally, data were organized roughly into "East Venezuela" and "West Venezuela," with the data from each region occupying respective sides of the room. However, the distinction was not exact, because with so many individuals working on the project (some of whom lacked total familiarity with Venezuela), and with Venezuela itself having so many confusing place names, materials often got misplaced. For example "Quiamare," "Quiamare-La Cieba," and "Quirequire" are three distinctly different areas, but information about them was frequently interchanged. Due to lack of attention (e.g., individuals transferred out of the group simply dumped materials from their offices into the data room), the structure of the files deteriorated. Thus, it was difficult to locate the proper information, even with the requisite intimate knowledge of Venezuela place names and history.

Several data managers cycled through the group. Yet as more new workers joined the group, both in Houston and Caracas, confusion with the databases and in the data room intensified. As new teams were formed, including regional studies groups dedicated to looking at data all over the country, the data quality issues mushroomed and became seemingly insurmountable.

Yet another bid round was scheduled for 1997. This was to be a marginal production bid round with 20 fields offered, including the first discovery in Venezuela—the Mene Grande field, discovered in 1914 and still producing. The quantity of data involved was staggering: thousands of wells, hundreds of square miles of 3-D seismic, thousands of miles of 2-D seismic, hundreds of pages of scanned documents, and thousands of ASCII files, MS Office documents, CAD drawings, and digitized logs. To facilitate evaluation of the offering, all data was to be digital and encoded on 8 mm Exabyte tapes.

Although the data provided in the offering were well organized, they still had to be printed, labeled, filed, or distributed, if they were to be handled as they had been in the past. Although several vendors stood ready to provide the service of printing documents, due to the sheer volume of material the effort to receive, inventory, check, punch, bind, and label even a subset of the files, once printed, would require many man-weeks.

As staggering as this part of the project seemed, it paled against the need to integrate the data from the offering with information already available internally to BP/Amoco, especially given the state of that information. The variety of data and their formats (both old and new) that had to be integrated included the following:

- Documentation files: Excel, Word, PowerPoint, paper
- Logs: LAS (Log ASCII Standard, a commonly used format for digitized logs), ASCII (American Standard Code for Information Interchange, used for plain text files), TIFF (Tagged Image File Format, a commonly used scanned-image format), Oracle, paper
- Maps: CAD (Computer-aided design/drawing, for digital drafting files), ASCII, Arc/Info, TIFF, Oracle, paper
- Seismic: UKOOA (United Kingdom Offshore Oil Scouts Association, a commonly used format for seismic navigation data), ASCII, SEG-Y (Society of Exploration Geophysicists-Y, a commonly used format for seismic trace information), Excel, TIFF, Oracle, paper
- Well reports: TIFF, ASCII, Oracle, paper
- Production data: Excel, ASCII, TIFF, Oracle, paper

Resources available to address data management issues were constrained by the lack of time. Each day of delay in delivery of the data reduced the time available for analysis. Only 90 total days were available, and the last 30 were expected to be needed to reconcile and review results, develop strategies, and obtain approval for bidding. This left only about two months for technical work. Even one day of delay would be a significant portion of the allotted interval. Thus, highly efficient methods of data delivery were sought. The Web was suggested as a potential method on the observation that adolescents use it to publish and access information without adult assistance.

Based on the model of the Web, a new "self-serve" data management concept was conceived (see Figure 1) having the following objectives:

- Users can file documents and update data directly without conversion or intermediate steps
- Indexing is done automatically
- Users can print and directly download documents and data without conversion or intermediate steps
- Instructions are available via the same system
- Metadata are also available

This idea held great promise, but there was still a lot of work to do, both from the standpoints of the data themselves, and the computer technology necessary to enable the data. Two efforts were mounted to address these challenges.

(1) The paper data were inventoried again. Original documents were archived in BP/Amoco's records. The catalog of this data, in database form, was

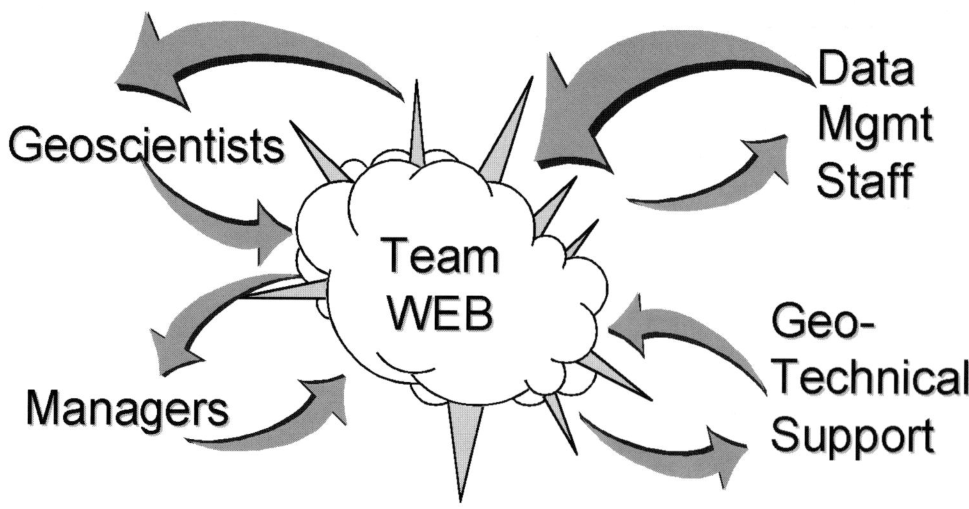

Figure 1—A new data management concept.

updated. Although this was not a trivial effort, it was nondiscretionary. Without an accurate inventory of the paper records, accurate analysis was impossible.

(2) A Web-based, self-serve system was established to provide direct user access to the data, including the inventory of paper records. This system was also designed to permit users to add without onerous conversions or training. Information from the bid round data packages was also to be added with minimal conversion and effort.

Two Web servers (computers) were configured for the Venezuela data. One was a Windows NT server running Microsoft Internet Information Server (IIS) software. The other was a UNIX server running Apache's Web server. All office documents were placed on the NT server, where the IIS could index them automatically. All binary files, such as TIFF images, digital logs in LAS format, CAD drawings, etc., were placed on the UNIX server. Pointers from Web pages on the NT server provided the index to these data. A series of commonly requested database queries returned lists of wells or seismic line links to the data contained on the two servers.

An advantage of the NT server over the UNIX server was that PC-based word processing files, spreadsheets, presentations, etc. could be filed there without effort. Once filed on the Web server, which appeared to the user as a file server, these files could be automatically indexed by content within minutes. An advantage of the UNIX server was that the data from the bid round, consisting of tapes containing UNIX archive files, could be easily dumped without conversion. Consequently, both types of servers were essential, and they acted in conjunction to serve up files via Web links to all users in the group. In addition, the NT server hosted the ArcView IMS software (developed by Environmental Systems Research Institute, or ESRI) providing geographic links to both text files on the NT server and binary files on the UNIX server.

The strength of this combined system was not in the GIS, database, or content queries alone, but in the fact that the queries turned links into a large mass of technical data. This made an otherwise difficult body of information retrievable from any machine, by any user, via a familiar interface. Users could zoom in on an area of interest, and then click on wells, seismic lines, blocks, or fields to search the Web for inventories of paper documents or references to those entities.

As an example of the kinds of searches that can be conducted, consider the information shown in the graphic provided in Figure 2. In this example, a keyword Web search for "API Gravity" returned a PowerPoint presentation on that topic, authored by one of the professionals on the Venezuela team. This document was directly filed on the Web server by its author and was available immediately to all interested individuals. The exact location and name of the file is not critical because the index server knows its name and where it is located. Potential users see the title supplied by its author and a summary of the document. The other document returned is an addendum to the contract supplied as part of the bid round data package, which was placed on the server by the team data manager. The Web server has integrated both sources of documents, although they were collected independently. Other kinds of diverse documents and data can be returned by similar GIS applications and database queries.

Special Features of the Web-Based Data Management System

To the user, perhaps the most important aspect of this data management system is its user interface; that is, what the user sees and manipulates on a computer monitor. Because most prospective users would be familiar with maps, an IMS was used to

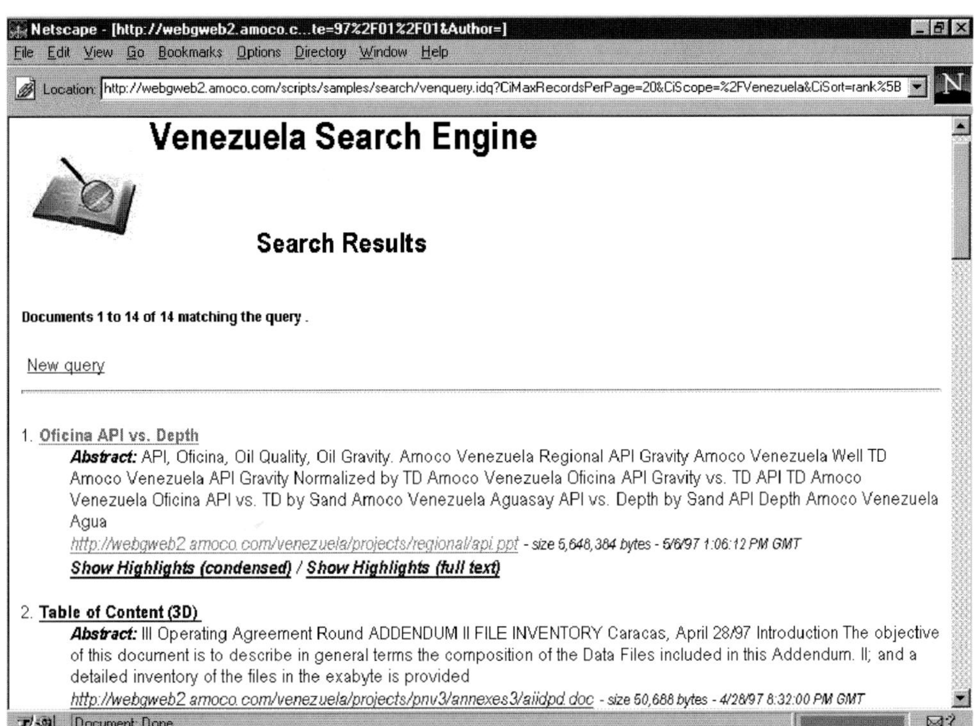

Figure 2—Results of a query searching for the words "API Gravity," which has returned two documents.

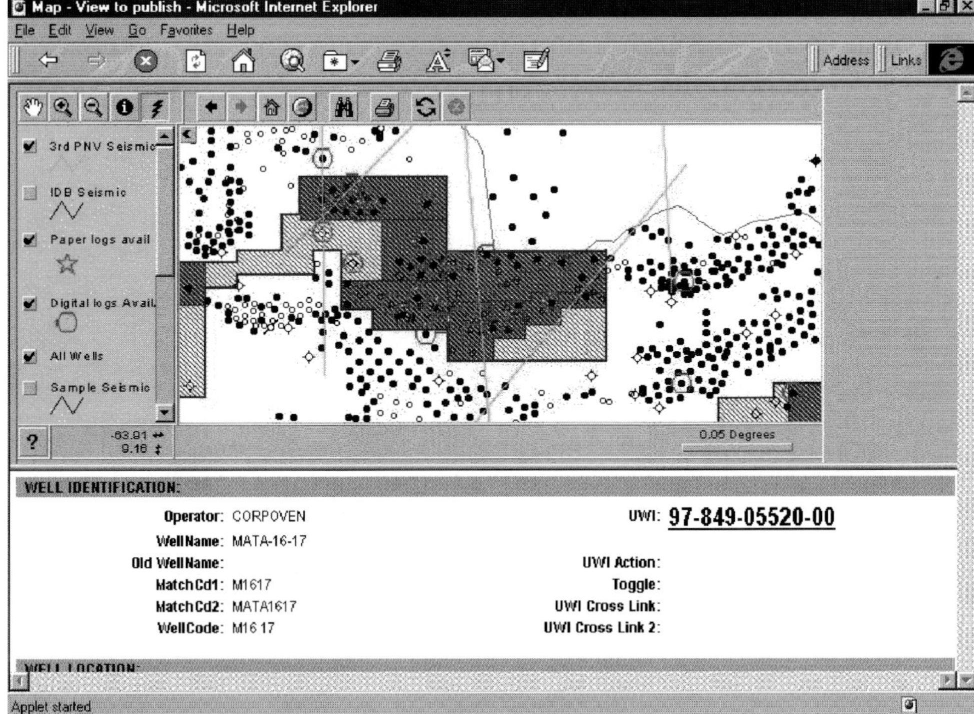

Figure 3—The results of zooming into a small area. Each well, concession, and seismic line on the map is a hyperlink to additional information.

provide links to the diversity of data made available on the Web. Figure 3 contains a screen snapshot of the GIS interface, where the view has been zoomed into a small area of interest. As the area becomes smaller, additional detail that does not appear at the regional scale begins to appear. Suppressing display of all wells, for example, at regional scale improves map legibility and drawing speed. Each concession, seismic line, and well on the map is a link to more Web data about that feature. These links can take various forms: database access (a portion of one such retrieval as a scout ticket is shown in the lower frame of this screen snapshot), Web pages, or content searches. These same searches can also be used independently or in various combinations with each other to provide a highly flexible and integrated data store.

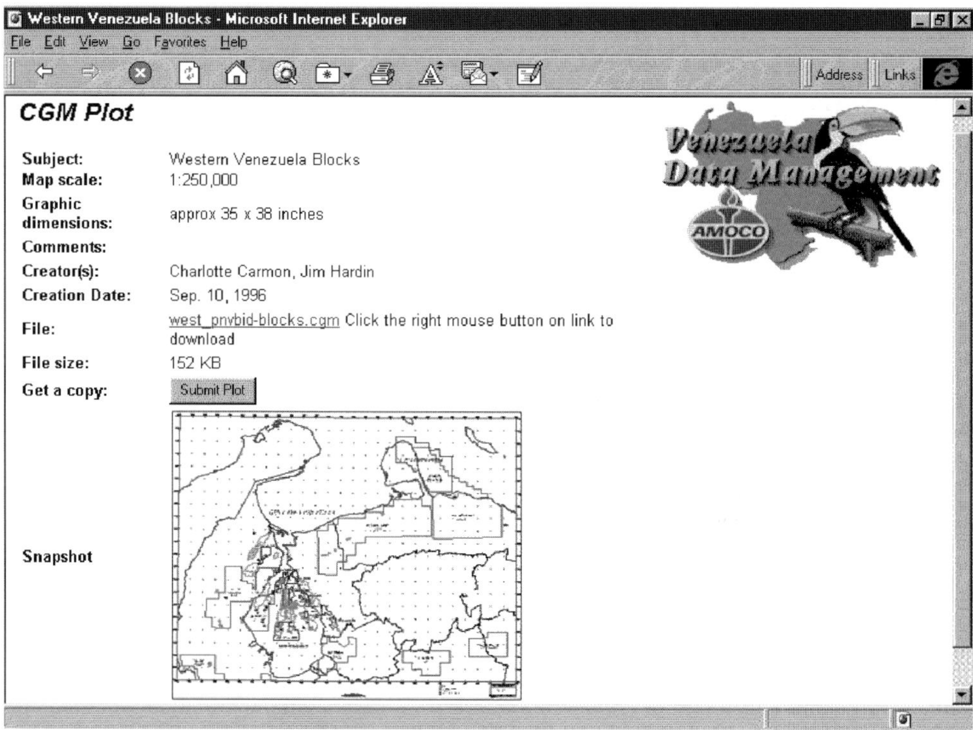

Figure 4—A page from the catalog of maps and plots stored in Computer Graphics Metafile (CGM) format.

Table 3. Closing the Productivity Gap

	Old way	New Way
Seismic Technical Interpretation	Seismic lines are processed, plotted, and interpreted with colored pencils. Interpretations are manually transferred to maps.	Seismic is processed and then interpreted on-screen with the data flowing into a database where multiple workers in multiple locations can map it instantly.
	Changes require the full cycle, limiting the number of iterations.	Changes can propagate quickly through the system, allowing multiple iterations or testing of multiple hypotheses.
Business Technical and Nontechnical Analysis	Most data exist in hard-copy form, and they must be reproduced, distributed, inventoried, and archived by data management technologists. Analysis is hampered by the logistics of paper handling.	Data are digital, distributed instantly to multiple locations, indexed automatically, and searchable. All appropriate workers can easily and completely get to the data in a variety of ways. The analytical results can flow back into the system without delay.

Notice that the number of GIS buttons (second row, below those for the Web browser) are few in number and function. This is neither an automated map production system, nor a spatial analysis system, but a data integration and query tool. Such a focused and limited design for this particular page makes its capability accessible to a full range of users.

Another important aspect of the Web-based data management system is its ability to store the results of specialized GIS analyses and maps. The digital files (in standard Computer Graphics Metafile, or CGM, format) were placed in a catalog that was indexed by a creator-supplied description. A Web page form to facilitate submission of graphics to the catalog was developed. This application builds a Web page with the information supplied, copies the CGM to the Web server, and generates a "postcard" image of the file for reference. The resulting Web page contains a button to allow the user to send a copy of the graphic to a local plotter. An example of one of these pages, in screen snapshot, is shown in Figure 4.

Finally, using the intranet as a database holds some

promise as a new way to increase the productivity of the team. This is particularly relevant, for example, in the case of seismic interpretion. Table 3 compares the interrelated processes of data management and seismic interpretation under both the conventional approach and the new approach suggested by the concepts outlined in this chapter.

SUMMARY

Recent GIS technological advances promise dramatic new capabilities for data integration. These improvements result in a boost in efficiency, allowing GIS to work more smoothly in the orderly and regimented corporate data world, as well as in the free and easy, diverse world of the Internet. The effectiveness of this integration has been proven in a real project, the system for which has been in use for several years. The next step is to apply spatial data extensions to relational databases to reduce the burden of support for the GIS components.

The advantages (and disadvantages) of the corporate data world are well known. The pros and cons of the Internet as a data integration tool are still being explored; however, it already seems clear that the Internet provides some powerful, easy-to-use, and easy-to-support new techniques for distributing and sharing data. These techniques can be used in the context of an exploration and production team, just as they can be used in the World Wide Web. The promise is that because of greater breadth and flexibility, this development can revolutionize the productivity of the entire team rather than of a few individuals, such as seismic interpreters. The unique and valuable capabilities of GIS are leveraged by integration with the Web.

Barrell, K. A., 2000, GIS: The exploration and exploitation tool, *in* T. C. Coburn and J. M. Yarus, eds., Geographic information systems in petroleum exploration and development: AAPG Computer Applications in Geology, No. 4, p. 237–248.

Chapter 18

GIS: The Exploration and Exploitation Tool

Kirk A. Barrell
Geodynamic Solutions, Inc.
Houston, Texas, U.S.A.

ABSTRACT

In recent years, the significant increase in performance and decrease of prices of the personal computer (PC) platform have accelerated the growth of geographic information system (GIS) usage. GIS is a very dynamic technology enabling the user to display a map of any location in the world and rescale that map instantaneously. The key to the success of any exploration or exploitation team is the integration of the members, the database, and the multiple software functions. GIS is a particularly effective means of providing functionality for all of the disciplines represented on the team. GIS cannot replace all existing software applications, but it can be used to integrate and link other programs. Although GIS can be effectively applied in various petroleum industry settings, its use in exploration and exploitation is of particular interest. One specific exploration application involves the creation of reconnaissance maps. Uses of GIS in exploitation projects are perhaps more varied because exploitation evaluation typically deals with more extensive data sets than those typically used in exploration settings. Exploitation approaches are generally applied to mature producing areas where well control is dense, whereas exploration projects may not involve any wells at all. GIS is a particularly effective technology that enables exploration and exploitation teams to share information, analyze data in new ways, and integrate the evaluation process.

INTRODUCTION: THE BENEFITS OF GIS

A geographic information system (GIS) is a powerful technological tool that can be used in the problem-solving process facing any exploration and exploitation team. A GIS can provide the team with a whole new way of analyzing, visualizing, and integrating data. GIS technology is now available across all computer platforms, including the Internet. In recent years, its cost has decreased significantly, whereas functionality has been dramatically enhanced. GIS can specifically benefit exploration and exploitation teams in the following ways:

- Integration of the contributions of various team members through cross-discipline interpretation and software functionality
- Provision of new methods for visualizing data sets through the use of symbology
- Dynamic mapping of digital databases
- Presentation of data in various forms, such as maps, charts, data tables, and query results
- Sharing of, integration of, and access to centralized databases via computer networks or the Internet
- Linkage of multiple software applications
- Technology use across multiple computer platforms

- Enhanced portability of technology and data via laptop and handheld computers
- Proliferation of new tools, which are affordable and easy to use

In recent years, the significant increase in performance and decrease in prices of the personal computer (PC) platform has accelerated the growth of GIS usage. GIS technology has become very affordable and easy to use. The true power of GIS is that it presents a new way to analyze databases. GIS enables the analyst to visualize the data as they are represented spatially. Analysis of the spatial relationships of digital databases can present new pictures of the data that never could be assembled by analyzing tabular representation of data sets.

GIS is a very dynamic technology, and it enables the user to display a map of any location in the world and to rescale that map instantaneously. Interpreters can zoom into a specific area of interest by simply defining the four geographic coordinate boundaries of the map. GIS and digital map data sets create a dynamic combination whose functionality is unlimited in their combined capabilities.

Symbology is a classification method used in GIS to represent data or individual map features as varying sizes, thicknesses, colors, or styles. In any GIS, there are three types of data: points, lines, and polygons. Each type of data can be represented by different types of symbology. Color, size, and style can represent point data, such as well locations. Color, thickness, and style can represent line data, such as pipeline locations. Polygon-fill color or style can represent polygonal data, such as offshore block boundaries.

Symbology on a map presents three types of results: trends or patterns, anomalies, or a random representation. One of the most vivid results of a GIS map is when the symbolized data present a recognizable trend or pattern. The explorationist then must search for an explanation for the trend or pattern. In many cases, the trend or pattern is a result of the underlying geology. Anomalies become very apparent on symbolized maps.

Either due to color, size, thickness, or style symbology, a feature or features on the map clearly represent themselves as anomalies. Anomalies must then be further investigated to establish explanations for their occurrences. If a trend, pattern, or anomaly is not represented, then the data will exhibit a random pattern with no new pictures or ideas being presented; however, even random data patterns can be informative.

A common GIS technique is to filter data to restrict an analysis or presentation to only the information that satisfies some specific criterion. Filtering data is very effective in areas with dense data control, enabling some of the data to be removed from consideration, if only temporarily. Filtering is an excellent tool to be used, for example, in field exploitation settings, especially when many wells have been drilled. As an example, a filter could be applied to "total depth drilled" to remove all of the shallow wells from a map.

The key to the success in any multidisciplinary effort is the integration of the members, the databases, and the multiple software functions. GIS is a particularly effective means of providing functionality for all of the disciplines on the team. Most GIS systems can present data via maps, as well as in charts and tabular database tables. Multiple databases can be integrated via the GIS; and once they are integrated and centralized, every member of the team can share the same database and data sets across networks or the Internet.

GIS cannot replace the capabilities of all existing software, but it can be used to integrate and link various programs. A GIS can also access databases that are being used in other applications, and individual features in the GIS database can be linked to files from other software packages.

Today's multidisciplinary teams use various computer platforms such as UNIX workstations and PCs operating under Windows. Cross-platform GIS functionality enables the team members to continue to use their individual machines. Also, with the significant increase in speed and power of laptop PCs, GIS functionality is even more portable. The multidisciplinary team can now take a GIS on the road to the well site, lease sale, or out-of-town meeting without fear of losing access to data or established geocomputing routines. In addition, Internet-based GIS now enables spatial databases to be accessible from anywhere in the world.

USER GROUP

The implementation of multidisciplinary teams to solve exploration and production problems has become a recent trend in the petroleum industry. A multidisciplinary or synergistic team consists of two or more individuals of diverse technical expertise working toward solving common problems. The team shares a common leader, goals, workspace, technology, and databases. The multidisciplinary or synergistic team concept is based on the premise that individuals of varying technical backgrounds working closely together will produce a significantly better final product over a shorter cycle time (Sneider, 1986).

A typical petroleum exploration or exploitation team might consist of one or more geologists, geophysicists, engineers, petrophysicists, landmen, and geotechnicians. Each discipline approaches the problem-solving process in different ways. In addition to the diversity of disciplines and problem-solving approaches, petroleum industry teams, particularly those involved in exploration and development, have historically used and relied on a wide variety of computer software, hardware, and databases, which has added to the need to seek common ground and synergy. Because of the very nature of the problems involved, multidisciplinary teams have been especially effective in the exploitation arena where the skills of individuals from many disciplines are used to piece together complex geological and reservoir solutions.

APPLICATIONS OF GIS

As previously noted, GIS can be used in both the exploration and exploitation environments. The specific

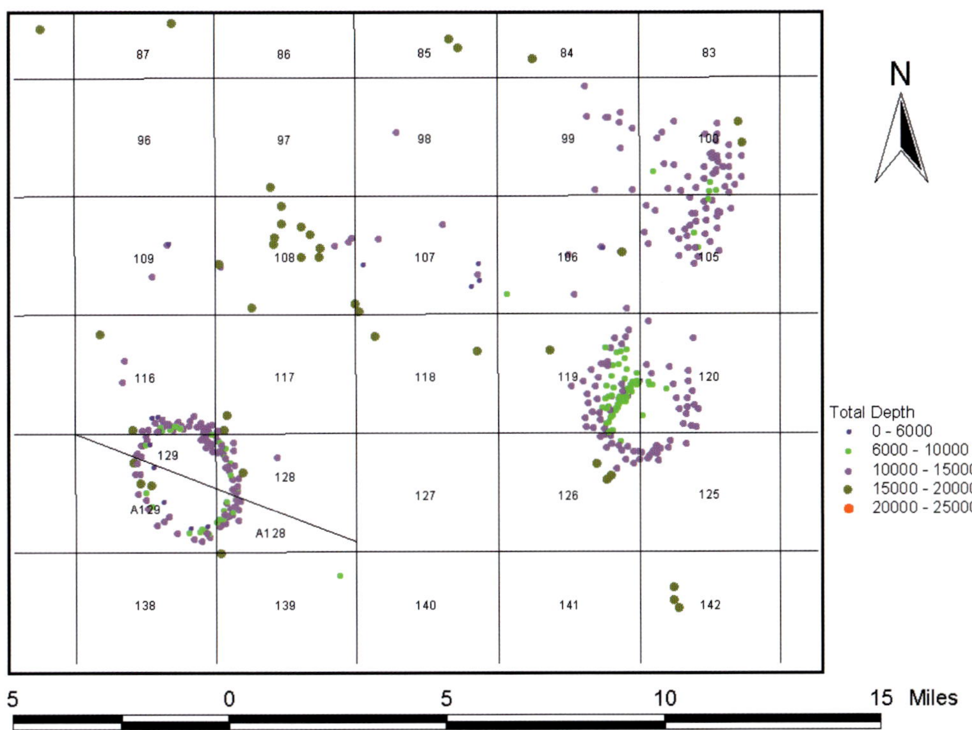

Figure 1—A Gulf of Mexico base map of wells classified by total depth drilled (includes data supplied by Petroleum Information/Dwights and Whitestar Corporation).

processes and applications of the technology will vary depending on the task or project and on the type and size of the databases involved. Although GIS can be effectively applied throughout petroleum industry settings, its implementation in exploration and exploitation is of particular interest.

Exploration Applications

GIS is a powerful tool for petroleum exploration, particularly with regard to exploration mapping. Such mapping is usually performed across large geographic areas, where many data sets or map layers are used in the analysis of hydrocarbon potential. Raster data, such as aerial photos or satellite imagery, can be incorporated with vector data, and surface culture, such as hydrography, elevation contours, and topographic landmarks or points of interest, can be presented. Where appropriate, coordinates from land surveys, such as section, township, and range, can be integrated with well data.

In exploration applications, one of the first steps taken is the creation of reconnaissance maps. Such maps are typically produced relative to the available well or well layers, with the sizes of the well data sets being dependent on the maturity of the area or basin of interest. Reconnaissance maps of mature exploration areas can present distinct trends, patterns, and anomalies.

Maps of the following features and characteristics should be included in the first phase of any exploration reconnaissance mapping program:

- Total depth
- Spud or completion date
- Well operator identification
- Producing formation
- Cumulative production
- Initial potential
- Well classification

Such maps are discussed in more detail in following paragraphs. A map of total depth can quickly identify where shallow and deep drilling has occurred over a basin, trend, or field, and apparent trends, patterns, and anomalies can be identified. Figure 1 is a total depth map for an area in the Gulf of Mexico for which well layers have been classified according to driller's total depth of the well. In this example, the symbology creates circular patterns in which the deeper wells are indicated on the flank of the salt dome structures. Deeper drilling potential can be identified in between salt dome structures.

Spud date maps represent a chronological history. Figure 2 is a spud date map for an area in northern Louisiana. Using this map, it can be determined when and where drilling has taken place. Such a map is useful when an interpreter has just started evaluating a new trend or basin because it represents a chronological history of the area. Figure 3 shows a similar map filtered in such a way that only wells drilled in the 1990s are represented. This type of map is very effective for evaluating recent activity in an area of interest because remote drilling locations and areas of dense drilling are apparent.

Operator maps contain symbology that identifies and denotes various well operators. Figure 4 is an operator map for an area in the Gulf of Mexico showing the location of wells operated by Shell Oil, Mobil

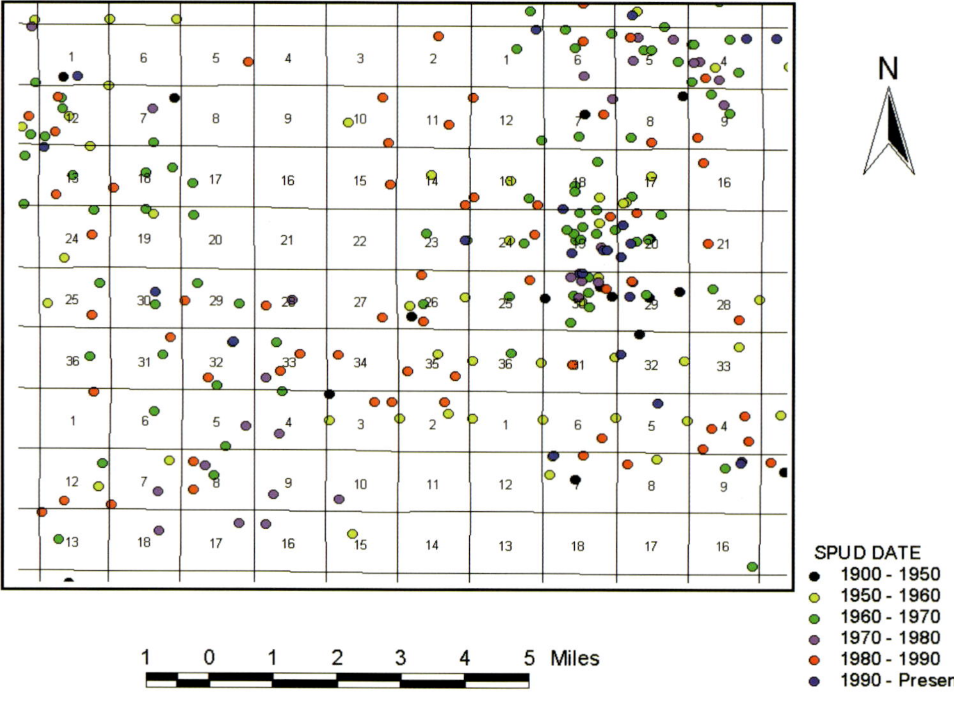

Figure 2—A northern Louisiana base map of wells classified by spud date drilled (includes data supplied by Petroleum Information/Dwights and Whitestar Corporation).

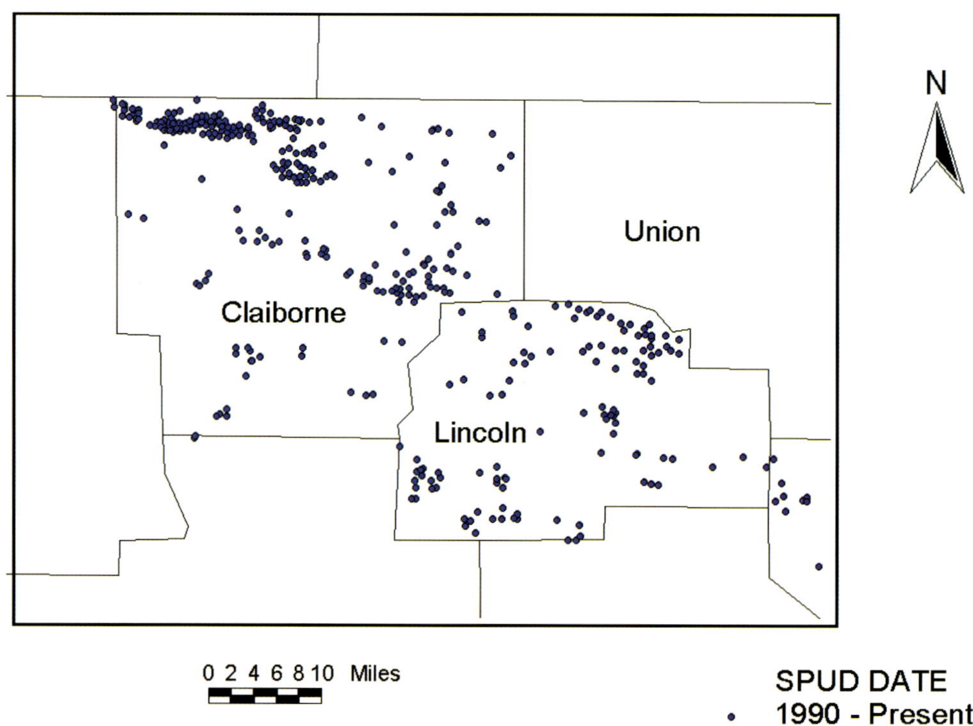

Figure 3—A northern Louisiana base map of wells filtered by spud date to display only wells drilled after 1990 (includes data supplied by Petroleum Information/ Dwights).

Oil (now ExxonMobil), and Phillips Petroleum. These maps can be used to locate where competing companies are drilling wells. They can also be used to support the process of evaluating producing properties for acquisition purposes. The map in Figure 4 has been filtered to show only the wells drilled by the three specific operators.

When evaluating a new trend or basin, one of the first steps necessary is to determine which geologic formations produce hydrocarbons and where the production is located. If the production has been assigned to a formation name or code in the database, the production can then be mapped or symbolized according to its producing formation. Figure 5 is a production map of the Cotton Valley, Hosston, and Smackover horizons in northern Louisiana. A strike-trending pattern following deep, strike-trending fault blocks in the

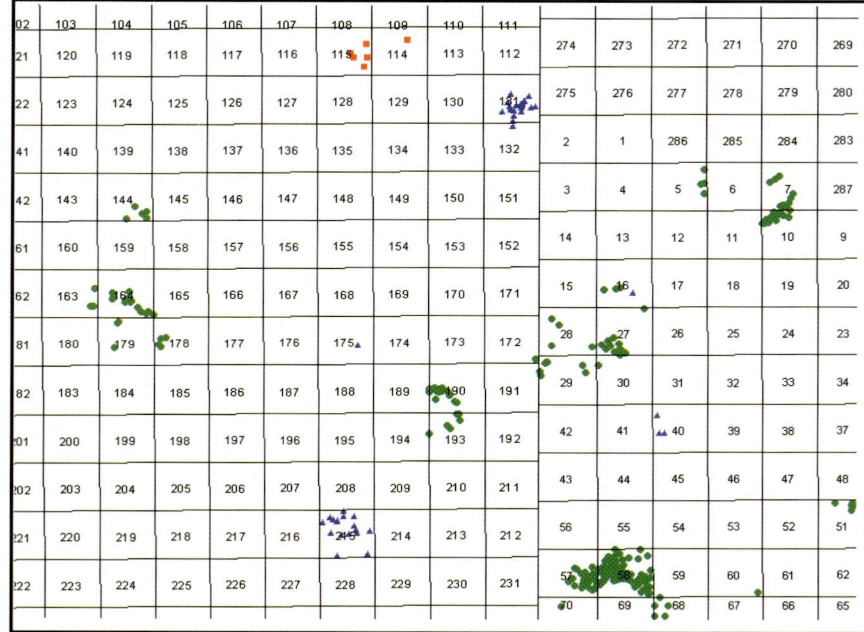

Operator
MOBIL OIL
PHILLIPS OIL
SHELL OIL

Figure 4—A Gulf of Mexico base map of wells restricted to three well operators (includes data supplied by Petroleum Information/Dwights and Whitestar Corporation).

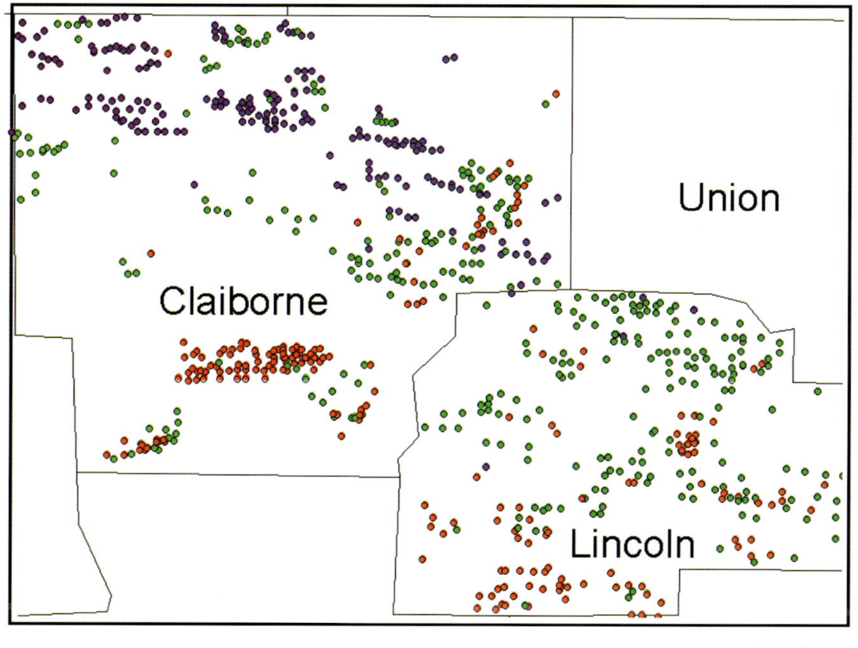

0 2 4 6 8 10 Miles

PRODUCING FORMATION
COTTON VALLEY
HOSSTON
SMACKOVER

Figure 5—A northern Louisiana base map of wells restricted to three producing formations (includes data supplied by Petroleum Information/ Dwights).

Smackover Formation is represented by the symbology. As the map shows, the production from the Hosston Formation occurs in dense well patterns or clusters, and production from the Cotton Valley Formation occurs randomly across the map without any clear trend or pattern.

Once production has been identified, it can be integrated with a geological structure map (Figure 6). When the geological structure and production are integrated, the structural positions of individual wells can be evaluated. Figure 6 shows that the Hosston production typically occurs within a structural closure,

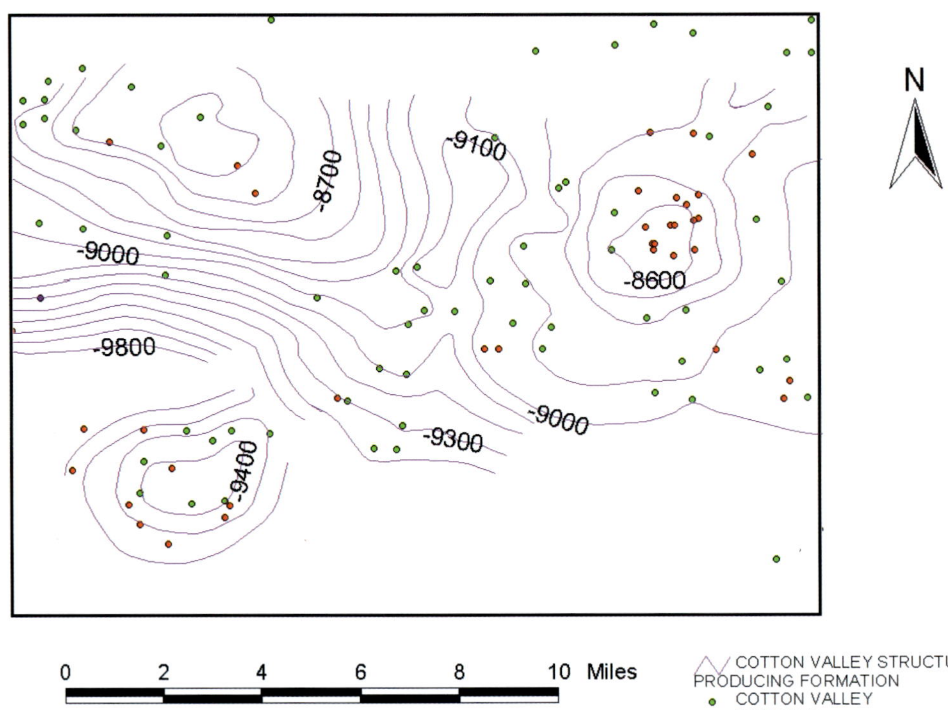

Figure 6—A northern Louisiana Cotton Valley structure map integrated with wells restricted to three producing formations (includes data supplied by Petroleum Information/ Dwights).

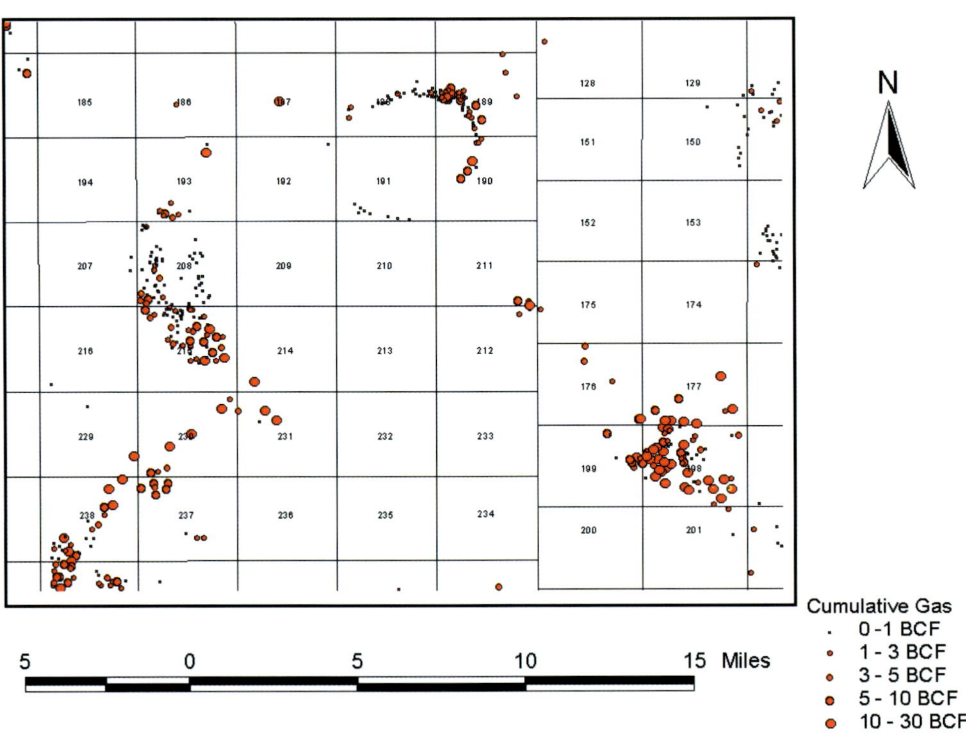

Figure 7—A Gulf of Mexico well base map of wells classified by cumulative gas production (includes data supplied by Petroleum Information/Dwights and by Whitestar Corporation).

whereas the Cotton Valley production also occurs within closures but somewhat randomly across the structure. Such a map can be used to quickly identify undrilled or untested areas.

A second step in evaluating production is to assess the cumulative total production in various wells. Spatial representation of cumulative well production can be very revealing, and trends, patterns, and anomalies can become apparent. Figure 7, a map of cumulative gas production in Gulf of Mexico wells, exhibits both unique patterns and anomalies. The most apparent pattern is a circular one representing

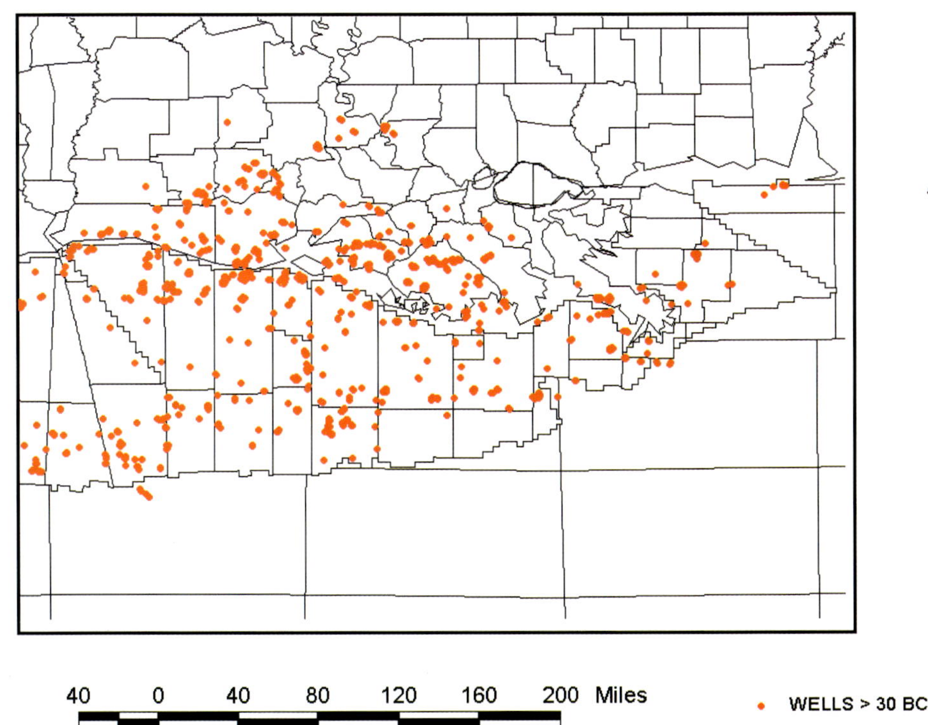

Figure 8—A Gulf of Mexico well base map restricted to wells whose cumulative gas production exceeds 30 billion ft^3 (includes data supplied by Petroleum Information/Dwights and by Whitestar Corporation).

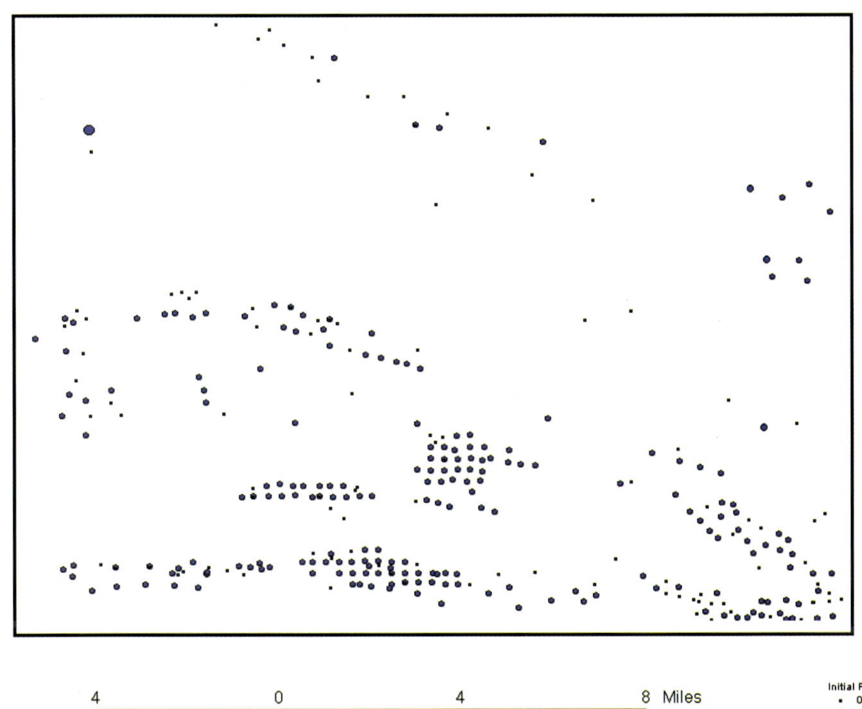

Figure 9—A southern Louisiana base map showing wells classified according to initial oil potential (includes data supplied by Petroleum Information/Dwights).

production on the flank of a salt dome in blocks 188, 189, 190, and 191. Most interesting is the fact that the amount of production around the salt dome varies significantly. For example, the southwestern flank of the dome is much less productive than the northeastern flank.

Because of the variation in production, the map in Figure 7 warrants further investigation. Blocks 187 and 193 each contain two wells that are apparently anomalous. These wells are prolific producers but are not offset by many wells. In contrast, the southeastern corner of block 211 has significant production over a

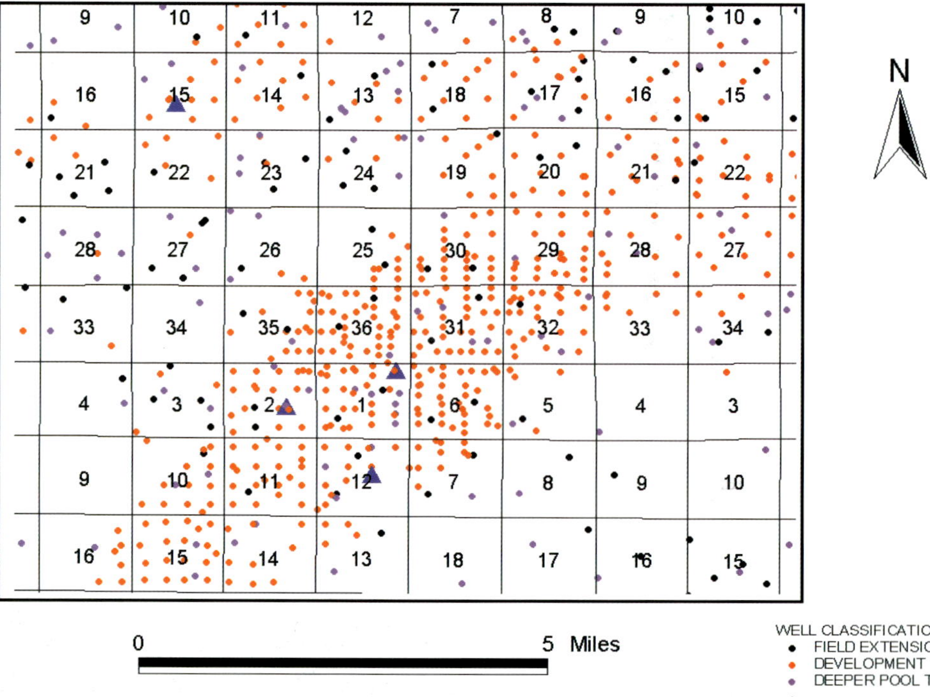

Figure 10—A northern Louisiana base map of wells by classification (includes data supplied by Petroleum Information/Dwights).

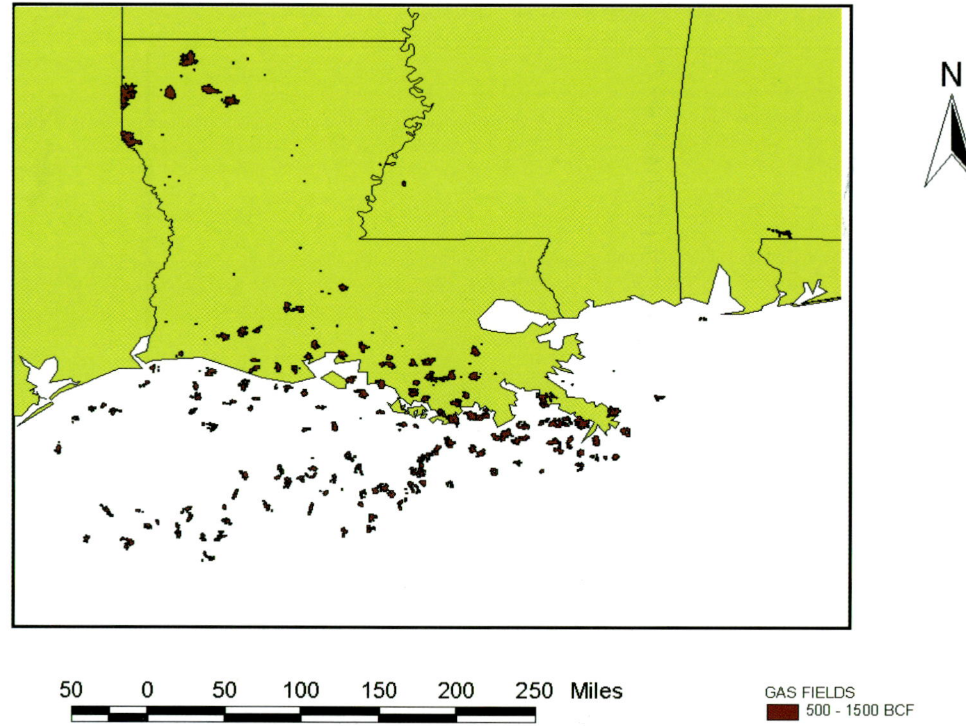

Figure 11—A Louisiana and Gulf of Mexico field map showing fields that have produced over 500 billion ft^3 of gas (includes data supplied by Petroleum Information/Dwights).

small geographic area, possibly suggesting that there may be more potential reservoirs such as these.

Cumulative production can also be evaluated over an entire basin. The map in Figure 8 shows all wells that have produced over 30 billion ft^3 of gas in southern Louisiana and the offshore Louisiana shelf. The symbology reveals clusters of prolific producing wells, along with wells exhibiting anomalous production levels in remote areas. Large geographic areas, which are void of any significant well production, are also evident. Consequently, a map produced at the basin or trend scale can be extremely informative and can reveal new areas for exploration or data acquisition.

Once cumulative production has been evaluated,

well rates or initial potentials should be analyzed. Figure 9, an oil initial potential map for an area in southern Louisiana, shows both consistent patterns and a significant anomaly. The pattern is a west-to-southeast strike trend likely resulting from down-to-the-basin expansion faulting. The wells in the southern one-half of the map are more closely spaced than those in the northern one-half. However, a significant anomaly is indicated in the northwestern corner of the map, where there is a well with high initial potential offset by only one other well. This area warrants further investigation. Initial potentials in the southern portion are fairly consistent.

When evaluating a new trend or basin, it is informative to know how wells were classified when they were drilled. Typical classifications might be wildcat, development, deeper pool test, or field extension (Figure 10). A well classification map can illustrate where the wildcats have been drilled and how fields have been extended and developed through time. Figure 10 indicates that four wildcats were drilled in this area and significant development of the field followed. Many deeper pool tests were attempted across the entire area.

GIS can be an effective exploration tool at either the basin, trend, or field level because it is not limited to a specific geographic area or map scale. Figure 11 is an example of a map created to evaluate the gas field sizes in the southeastern Gulf Coast Basin. This map is filtered on cumulative gas production, and only fields that have produced over 500 billion ft^3 are presented. Such a map would illustrate a basin-scale evaluation. The location of producing fields can additionally be integrated with regional geological features to present a basin-level picture (Figure 12). The impact of geological features can then be evaluated relative to the size and location of existing fields.

Raster data, such as aerial photographs and satellite imagery, can provide powerful backdrops to cartographic map layers consisting of field outlines and well locations. The integration of a raster backdrop enables the interpreter to compare surface expression with the subsurface. For example, digital aerial photographs can be very useful for the design of 3-D seismic surveys when superimposed on a conventional map. Figure 13 illustrates the integration of LandSat imagery and well data for the design of a 3-D seismic survey. In this instance, the hydrography, or water bodies, can be evaluated for their impact on seismic acquisition, and future well locations can be sited based on the proximity to such entities.

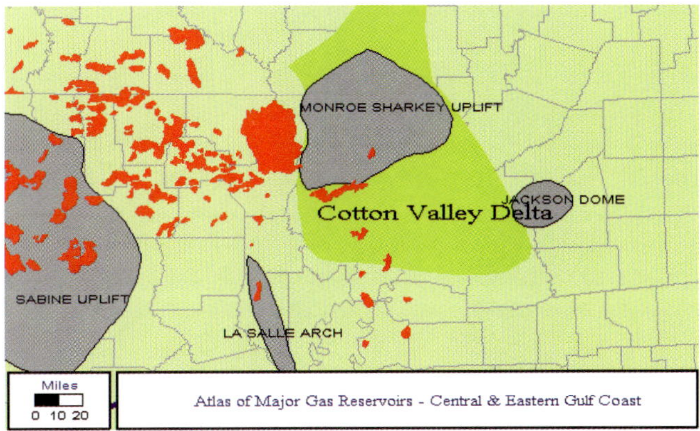

Figure 12—A northern Louisiana basin map presenting major gas fields and regional geology (after Gas Research Institute, 1998).

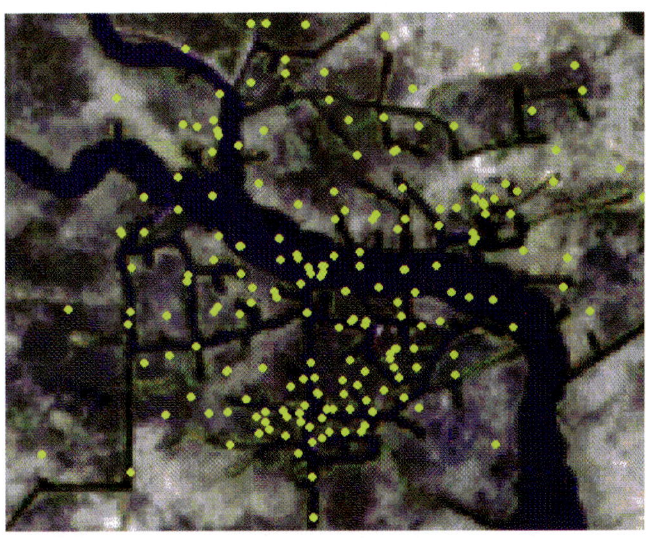

Figure 13—LandSat image superimposed on a southern Louisiana base map of wells (includes data supplied by Petroleum Information/Dwights).

Exploitation Applications

Exploitation evaluation typically deals with much larger datasets than those used in exploration settings. Exploitation approaches are generally applied to mature producing areas where well control is dense. As previously noted, the filtering capability of GIS can be useful in this environment because multiple types of maps on multiple horizons are generally required, and information from many diverse databases must be incorporated. In this situation, GIS technology can be more than just a mapping tool—it can be a powerful database management tool.

In field exploitation settings, integrating and overlaying multiple types of maps is essential. Figure 14 is a map from an area in southern Louisiana that incorporates three different map (data) layers: well locations, structure on top of the Oligocene Camerina sand, and fault locations. Note how database attributes, such as fault names or subsea formation tops, can be easily displayed as labels on the map. GIS functionality is a powerful technological tool that greatly extends the geoscientist's analysis capabilities beyond a conventional paper-based geologic map.

Another type of map frequently used in exploitation is the sand isopach map. Figure 15 is a typical map that has been constructed using GIS that integrates three

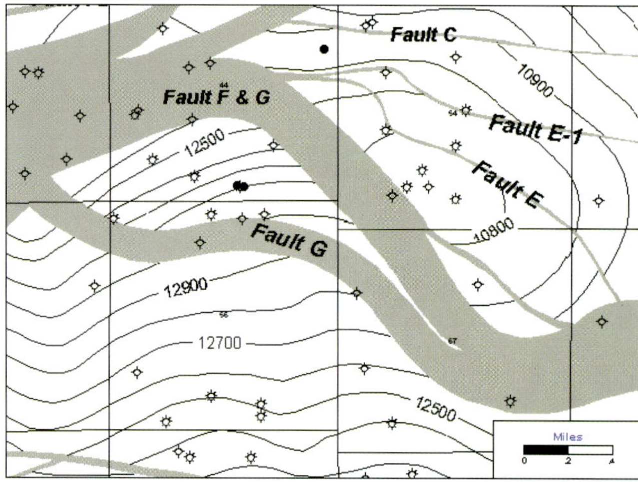

Figure 14—A southern Louisiana structure map on the Oligocene Camerina sand (includes data supplied by Petroleum Information/Dwights).

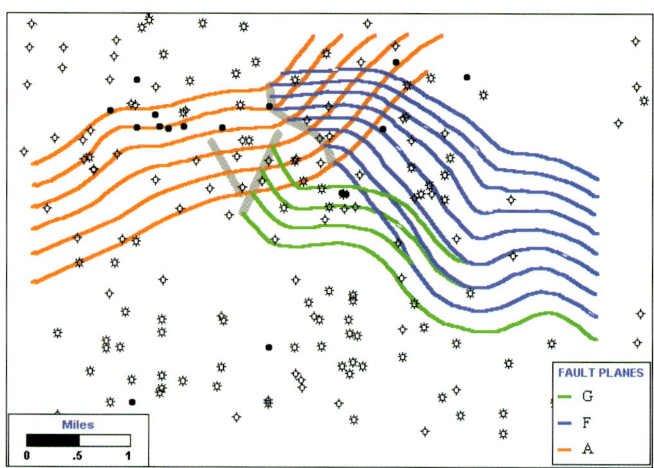

Figure 16—A southern Louisiana map of fault planes classified by fault name (includes data supplied by Petroleum Information/Dwights).

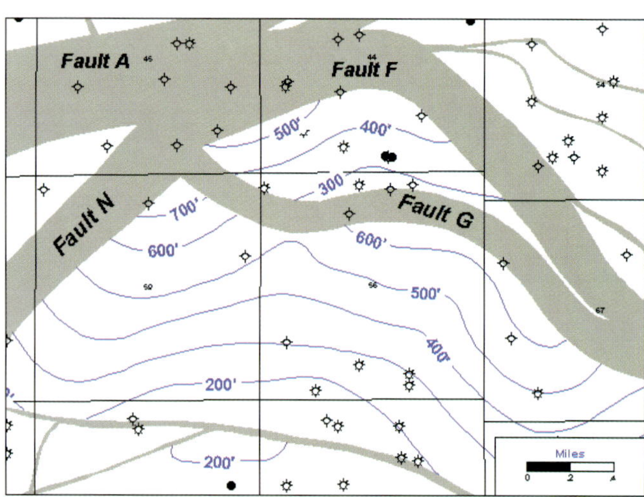

Figure 15—A southern Louisiana Oligocene Camerina sand, isopach map (includes data supplied by Petroleum Information/Dwights).

different map layers: well locations, sand isopach contours, and fault locations. Using such a map, spatial relationships between wells and sand thickness can be evaluated. The direct integration of the three data layers is nearly instantaneous using GIS; and more important, it is repeatable, leading to more productive use of the analyst's time than does conventional mapping.

Fault plane maps are also frequently generated in the exploitation process. These maps are typically very complex and again require the integration of several data layers. Figure 16 illustrates the results of having created a fault plane map using GIS where the symbology distinguishes individual fault plane contours. Note how the color symbology creates a much clearer picture of the intertwined fault contours. In the past, fault plane maps were created by overlaying multiple hardcopy maps on a light table; but with GIS, this exercise is performed dynamically on the computer. Using additional computational techniques, the resulting fault plane map is subsequently intersected with the structure contours to create the faults.

Assessment of reservoir volume and drainage area is an important part of the exploitation process. Figure 17 is a map of a gas field developed on 40-ac spacing in which the tools of GIS have been used to create buffers, or circular rings, around individual wells. Buffers are circles calculated around a point that create a visual tool for clearly identifying possibly undrained or overdrilled reservoir acreage. Having created such a map, the well spacing can be evaluated to determine whether the reservoir has been efficiently produced.

Calculating reservoir volumetric size can be achieved using GIS functionality by dynamically integrating and overlaying structure and net sand maps to produce a net pay contour map (Figure 18). The net pay contours are created as polygonal features so that the area of the polygon can be automatically calculated by the GIS. Such maps can be edited on-screen, which leads to dynamic calculation and what-if analysis of new net pay contour areas.

CONCLUSION

GIS is a particularly effective technology that enables petroleum exploration and exploitation teams to share information, analyze data in new ways, and integrate the evaluation process. Multidisciplinary teams are now present in most exploration and exploitation departments. Technological tools, such as GIS, play a vital role in the team-based approach to upstream problem solving. In particular, the application of map symbology enables geoscientists from various disciplines to more productively visualize data, examine spatial relationships, and compare and integrate interpretations. Such an approach enhances the overall problem-solving process.

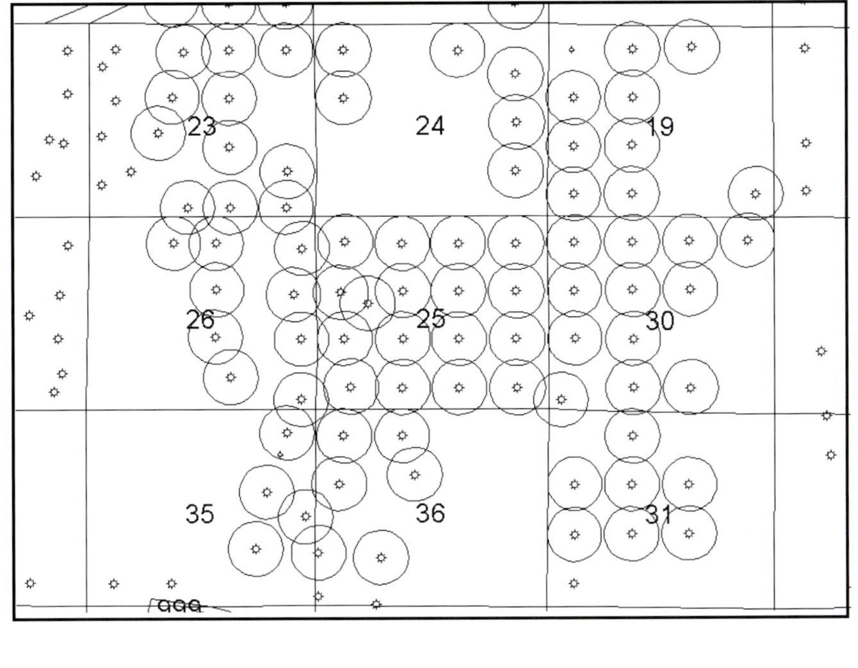

Figure 17—A northern Louisiana base map of wells showing 40 ac drainage buffers (includes data supplied by Petroleum Information/Dwights).

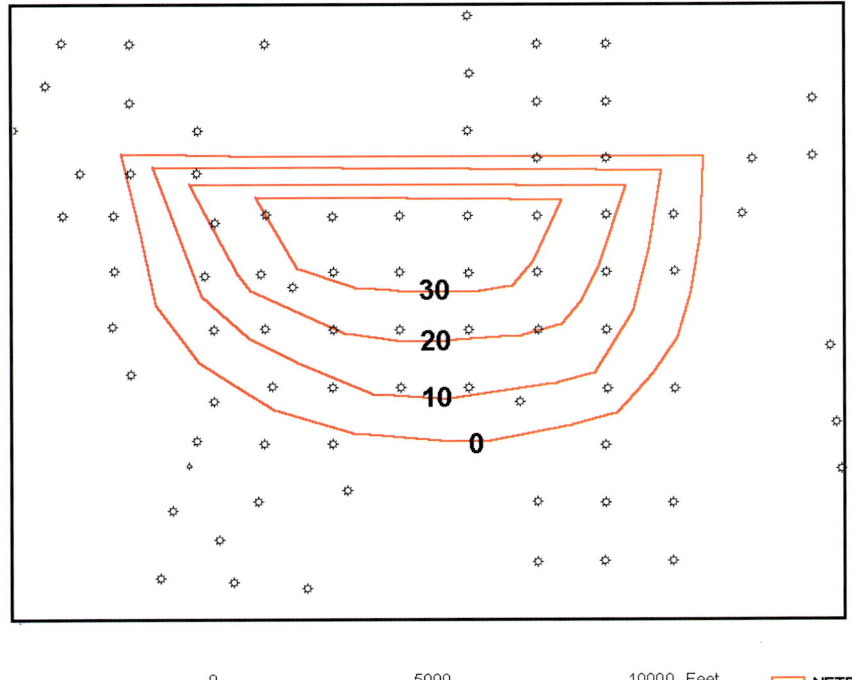

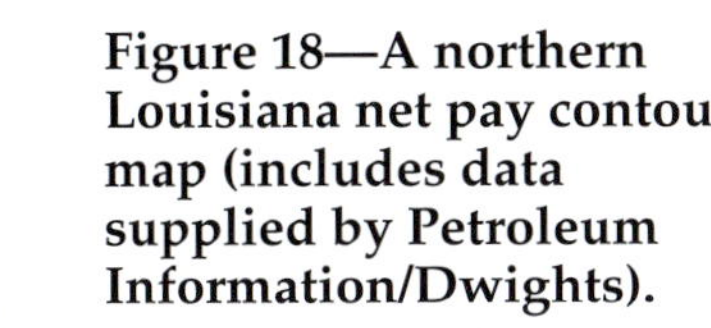

Figure 18—A northern Louisiana net pay contour map (includes data supplied by Petroleum Information/Dwights).

REFERENCE CITED

Sneider, R.M., 1986, The synergistic team approach: Its value to petroleum exploration and development in the next decade. Proceedings, Eighth Australian Geological Convention, Adelaide, Australia, p. 16–19.

Barnwell, C. E., 2000, The USGS GeoExplorer project: using new GIS technology for scientific data publishing, *in* T. C. Coburn and J. M. Yarus, eds., Geographic information systems in petroleum exploration and development: AAPG Computer Applications in Geology, No. 4, p. 249–260.

Chapter 19

The USGS GeoExplorer Project: Using New GIS Technology for Scientific Data Publishing

Charles E. Barnwell[1]
Municipality of Anchorage
Anchorage, Alaska, U.S.A.

ABSTRACT

This chapter relates the cooperative efforts of the Environmental Systems Research Institute, Inc. (ESRI) and the U.S. Geological Survey (USGS) to use modern geographic information systems (GIS) technology to distribute federal geological and earth resources data, along with the tools necessary to display and analyze such data. Largely constructed using the functionality of ESRI's MapObjects software, Livengood GeoExplorer is the name given to the desktop GIS product developed through this joint effort. Options for developing a desktop GIS product are discussed, as are the applications adapted for use in this specific project.

The Livengood GeoExplorer product is one example of modern GIS technology that takes advantage of component-based technology and object-oriented programming for data distribution. The various decisions and activities associated with the development of GeoExplorer are described, and examples of the kinds of map-based data displays that can be produced are presented. The process undertaken in this project is directly applicable to petroleum assessment in various fields, regions, basins, or regimes for which a desktop product that integrates spatial data management and contemporary map display capabilities is desired.

INTRODUCTION

One of the major challenges facing everyone who deals with data—particularly spatial data—involves distribution and visualization (Bernhardsen, 1999). It is relatively easy to develop a CD-ROM containing huge volumes of information, yet it is often difficult for users to display and assimilate such data in a simple manner that facilitates interpretation and understanding.

Spatial data are somewhat different from other kinds of information, and they are often relegated to the corners of organizations where professionals generate maps. Such data are typically hard to access and are equally difficult to use in formal analyses. Fortunately,

[1]Formerly with ESRI, Inc.

the advent of geographic information systems (GIS) technology has brought about an improved ability to access and analyze spatial data in digital form, which, in turn, has been a major step toward better understanding of geoscience problems. Spatial information systems based on technology such as GIS provide an excellent means of overcoming the shortcomings of simple graphic representations because they facilitate the storage and integration of diverse knowledge within a single operational entity or system.

The U.S. Geological Survey (USGS) is an organization that relies heavily on spatial information. In particular, it is in the business of delivering this information to the public. Institutions such as the USGS that work with resource data are moving away from the use and production of static maps and toward the use of interactive multimedia systems to create dynamic spatial displays.

A recent focus of the USGS is to provide scientific and spatial data to the public via the World Wide Web (Internet) and CD-ROM technology. The reasons for this focus are speed and efficiency. CD-ROM technology and the Internet are two profoundly effective and powerful means for information sharing, but how best to use them remains a challenge.

The Environmental Systems Research Institute (ESRI) and the USGS recently cooperated to develop a desktop GIS product called the Livengood GeoExplorer, which focuses on a single quadrangle in the interior part of the state of Alaska. The technology chosen to implement this product allows for the development of applications that are interactive and can incorporate data contained in a wide range of media (e.g., video, sound, and images).

The Livengood Quadrangle is a highly prospective area for minerals and has been the subject of numerous government-sponsored geologic studies since the early 1900s. A wealth of geoscience information about this region is available from the USGS. Additionally, since 1992, a significant amount of the data have been digitized in GIS format (McCammon, 1997). Thus, the Livengood Quadrangle provides an excellent area in which to prototype a GIS project (Figure 1).

One of the goals of this project was to ensure that the ultimate product capitalized on the most up-to-date GIS functionality. A second goal was to develop a true desktop application that could be seamlessly integrated with other desktop applications. Given the disparate nature of the geoscience data in the study area, it was concluded that the most appropriate approach was a general object-oriented one (Chung, 1995; Kang, 1995; Orfali and Edwards, 1996). Consistent with the object-oriented approach, a component-based programming strategy was determined to be best suited for the project. Hence, to accomplish the project goals, GeoExplorer was constructed using a standardized mainstream object architecture, in particular, Microsoft Corporation's Component Object Model

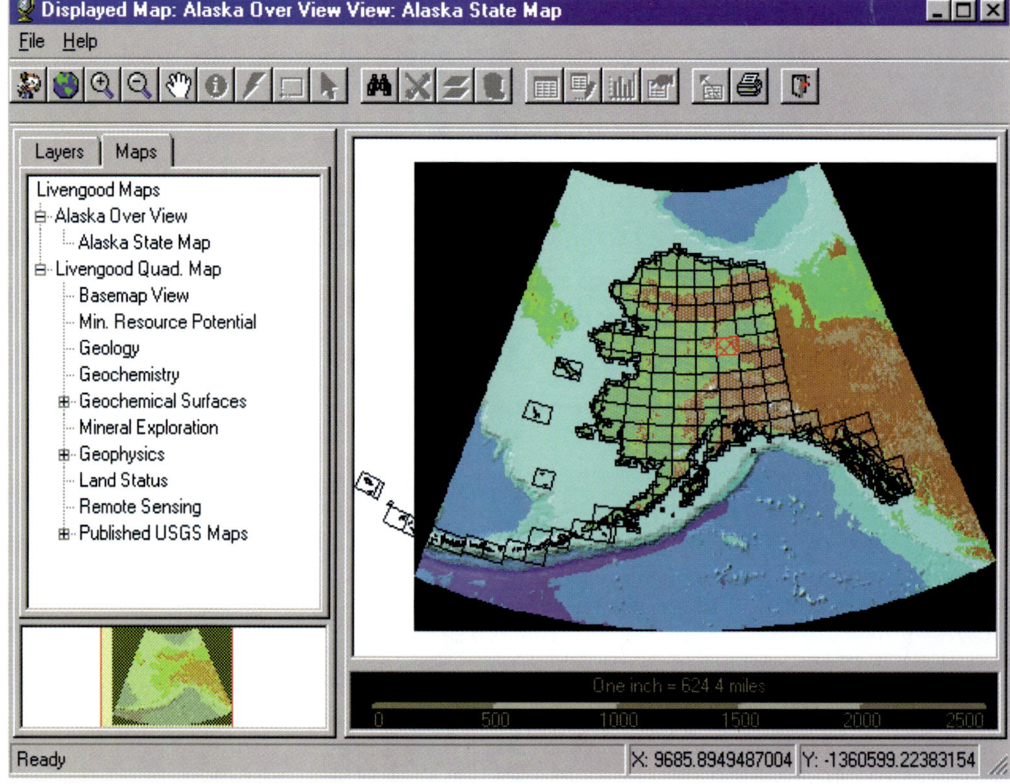

Figure 1—Opening view in the Livengood GeoExplorer, showing the state of Alaska with the Livengood area highlighted in red in the center of the map.

(COM) standard. The object-oriented approach was used in programming, in graphic user interface (GUI) design, and in organization of the data.

Given the USGS focus, a major consideration at the outset of the project was whether to publish data on a CD without software to display them, or to include a stand-alone run-time product to allow the user to "tour" the data and to map them using GIS functionality (Asche and Herrmann, 1994). One of the advantages of GIS technology is that it provides a view of the relationships of many types of data. By layering data in a common geo-referenced grid, one can view a particular set of data and determine their accuracy and validity. This alone is valuable in the sense that once a large set of disparate data is commonly geo-referenced, the user can perform analyses from various points of view. Consequently, an alternative consideration is whether to provide the users with prepackaged views of the data to facilitate rapid interpretation.

Another issue involves the use of metadata. Metadata, or a description of the data, have become an important topic, particularly in regard to the distribution of government data. The Federal Government Data Committee (FGDC) has established detailed standards for metadata (Somers, 1997), but a simple way was needed to provide metadata and to build an inventory of available data layers. To address this concern, a database component was programmed into the application that both controls and stores the information related to view and layer properties. Microsoft Access was the database software selected because it runs under Windows and because Microsoft provides components that work with Visual Basic. An Access database is an easy way to manage the data being displayed. Having such an internal database allows metadata to be easily stored and the properties of data layers and views to be easily managed.

Developing a stand-alone application that can be distributed on CD-ROM requires the following: creation of a system that is fairly easy to learn, making the data accessible to the application itself, and to users who may choose other programs; and providing clear documentation. These requirements are discussed in more detail in the following paragraphs, as are many of the technological decisions that contributed to the development of GeoExplorer.

DESKTOP GIS AND DATA PUBLISHING

Desktop GIS applications are rapidly evolving in the marketplace. The major GIS technology vendors are developing and selling desktop GIS software programs that work on a variety of machines, including PCs, Macintoshes, and UNIX-based workstations. Most recently, the trend has been to expand the robustness of these systems, so that many of them are beginning to have capabilities that approximate those of professional GIS packages, such as Arc/Info.

For this project, a number of GIS software products were evaluated, resulting in the selection of ArcView and MapObjects. Both are desktop GIS programs, and both offer advantages and disadvantages for a CD-ROM run-time environment. ArcView, ESRI's first entry into desktop GIS, includes many GIS capabilities. Its key advantages, the ones that have the greatest appeal in the mainstream market, are its easy-to-use, standard GUI, and its wide range of analytical functions.

Similarly, the appeal of MapObjects lies in its powerful set of mapping and GIS components that are designed for application development professionals. Unlike ArcView, however, MapObjects is not an end-user product. It is built on Microsoft Corporation's ActiveX programming language and provides a collection of programmable ActiveX automation objects that facilitate the embedding of mapping and GIS capabilities into various applications.

The use of ActiveX is advantageous because it is fully integrated into many desktop office software programs available today. MapObjects can be used with any developmental programming environment that supports custom controls, such as Visual Basic, Delphi, Visual C++, PowerBuilder, Access, Visual FoxPro, and others. All of these are considered to be visual development languages because they allow an application programmer to use a variety of different controls to construct integrated applications in the Windows environment. For example, using Visual Basic, a programmer can create a single application consisting of components of Excel, Access, and MapObjects functions.

Once it was determined that an object-oriented approach would work best for this project, selection of MapObjects was a logical choice. It reduces the amount of programming needed to deliver a customized solution that includes mapping and GIS functionality, and once the application is built using COM standards, the various application components are reusable in Windows-based environments. MapObjects applications also provide the desired level of performance, an important consideration for run-time GIS operations delivered on CD-ROM media. Because MapObjects works in a variety of standard development environments, application developers can choose from a wide variety of interchangeable third-party components. Consequently, MapObjects provides the technology to create a variety of applications that cannot be produced with other products.

MULTIMEDIA ASPECTS

Today's GIS and Internet browsers support a variety of data and formats (CAD, coverages, shapefiles, postscript files, tabular data of many types, image files, text, images, video, and sound). Each GIS software program handles the display of different data in its own unique way. As has been suggested, MapObjects efficiently enables the integration and seamless sharing of data from multiple media under GIS. For example, hotlinking allows the user to specify (by pointing and clicking) a location on the map to retrieve a photo or graph for that site (Andrews and Tilton, 1993). Such hotlinking can be used to display photos,

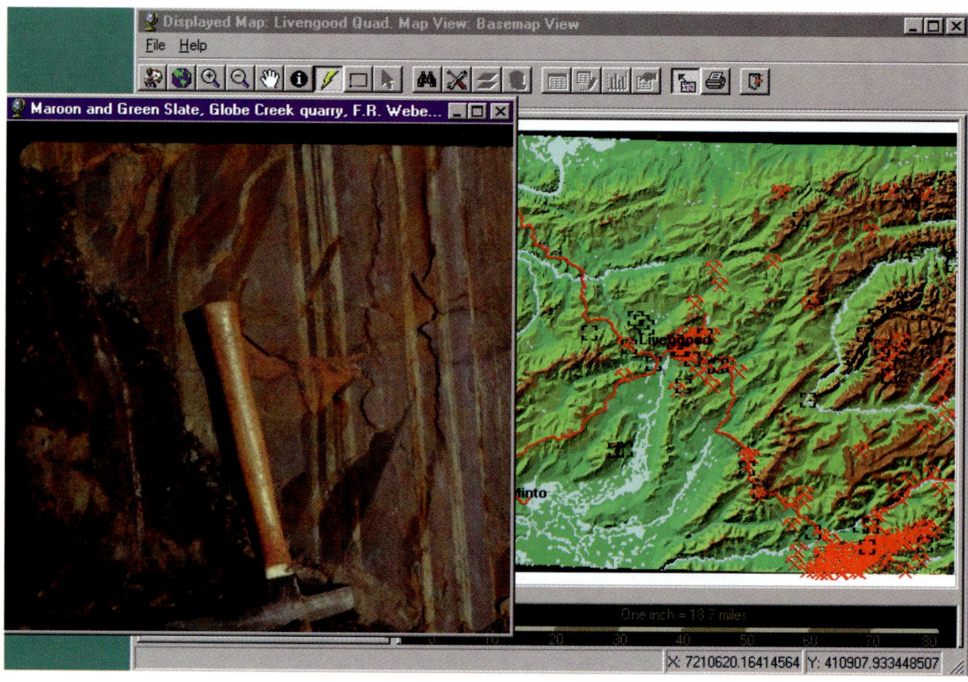

Figure 2—Hotlinked photo to location on a map. This could also be another Windows-compatible application (e.g., a video) that displays when the hotlink is activated.

videos, sound, or information from other media, as illustrated in Figure 2.

DATABASE CONSIDERATIONS

In its truest sense, a database can be used to implement a variety of technical applications. GIS is becoming increasingly more integrated with the database concept, particularly as it derives from relational database management systems. The desire to formulate and execute ad-hoc database queries is the motivation for almost all interactive query languages. However, the usefulness of these languages in developing *spatial* information systems is fairly limited. Spatial information systems can and must be responsive to user queries that involve notions such as neighborhood membership, and they must be able to select, restrict, or extract data based on criteria such as distance and direction.

In GIS applications, the same data are often presented in various graphical fashions depending on the perspectives of different users. For example, geochemists, petroleum geologists, and mining companies all have differing perspectives of, and uses for, satellite images. One person regards such images as a means of identifying structural features in the subsurface, whereas another person uses them to search for relationships between chemical concentrations and mineral deposits. As part of an information system, the images must be presented in combination with other data in such a way as to enhance interpretation from all perspectives. Establishing and maintaining a logical order in which data are combined and portrayed are also important to effective interpretation.

Databases that support mapping products, such as those incorporated into, or accessed by, GIS applications, must be designed in such a way as to allow map content to be quickly modified. A traditional paper map is a static product—once it is printed it cannot be updated. Maps produced with GIS applications, however, are highly interactive, their essential communication medium being the computer screen. Consequently, objects contained in the supporting databases must be represented in ways that allow them to be added, removed, or modified essentially "on the fly" without having to start over. The ability to combine various database queries is also important, as is the ability to integrate spatial data from various sources in a way that appears to be most suitable to each user's needs. Users must be able to easily add data, query the data, review information about data, and refresh an on-screen map using standard Windows menu items. GeoExplorer provides all these capabilities through the functionality of MapObjects in conjunction with the built-in Access database.

MORE ON THE HANDLING OF METADATA

As described, metadata, or information describing the data, are always a major issue in GIS development. In the case of this project, a component (available free from Microsoft) is embedded in the software that facilitates viewing and navigation of metadata via documents formatted with HyperText Markup Language (HTML). Such documents are linked to both views and themes (data layers), with the linkage being accomplished by storing the path to the document in an Access database. The view in Figure 3 was created by highlighting "mineral resources" and depressing the description button, instructing the browser to open an HTML document containing information about the view. This is particularly useful because HTML documents can

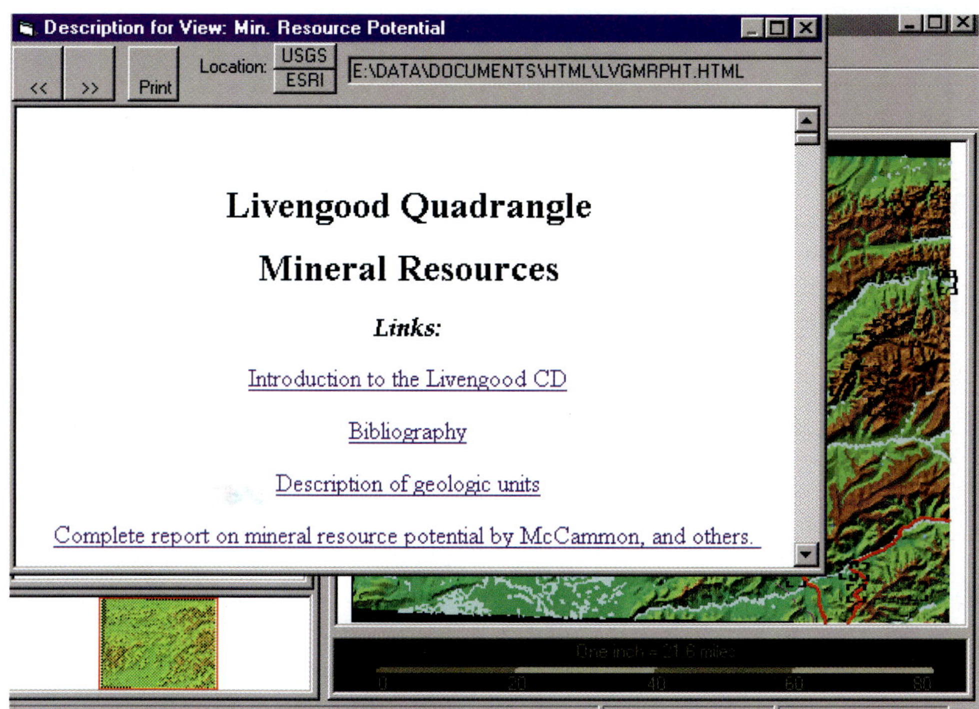

Figure 3—The embedded Web browser allows the user to view related HTML documents, as well as access Web sites containing relevant information about the GIS data or area of interest. HTML = Hypertext Markup Language.

contain links to additional documents; moreover, the use of an embedded browser allows direct access to selected Web sites.

HARDWARE AND OPERATING SYSTEM REQUIREMENTS

An important consideration for the USGS and its constituency was to have a CD-ROM product that was usable on a variety of hardware. PCs, UNIX-based workstations, and Macintoshes alike were dispersed throughout the agency and were employed by various users (both internal and external) of their data. However, the decision to build the data delivery software around MapObjects forced an automatic restriction on the hardware platform. MapObjects (and, therefore, applications constructed around MapObjects) requires a 32-bit Windows environment (i.e., Windows 95 or Windows NT 3.5.1 or later) and hardware that is compatible with Windows 95. Therefore, it was decided to make the data readable by all platforms, restricting the run-time GIS application to only PCs or other Windows-based computing environments. The recommended configuration is a Pentium processor with a minimum of 16 MB of random access memory (RAM).

PROGRAMMING SOFTWARE REQUIREMENTS

As previously stated, MapObjects requires the use of a programming language that supports the ActiveX custom control (OCX). Some of these languages are Visual Basic, Delphi, Visual C++, PowerBuilder, Access, Visual FoxPro, and others. Version 5 of Visual Basic was selected for this particular project.

DATA FORMAT CONSIDERATIONS

MapObjects applications can accommodate shapefiles, a variety of geo-referenced images, spatial database engine (SDE) databases, and Arc/Info coverages. A wealth of preexisting data from the Livengood Quadrangle is available in these formats, including a wide variety of geologic, geophysical, and cultural information.

AN OBJECT-ORIENTED APPROACH TO PROGRAMMING

Much has been made of the object-oriented programming approach as it relates to the development of GIS applications. Recently, many software vendors have invested heavily in the development of GIS software that uses the object-oriented approach. The decision to employ object-oriented programming for this project was based on the nature of the data themselves, a desire to incorporate the most up-to-date technology, and a need to conform to standards. As has been suggested, the development of truly object-oriented GIS applications enables the integration of data having unique characteristics, independently of how they are formatted.

Typically, information is loaded onto a CD with little concern for a truly logical directory structure or the user's familiarity with the data. This is particularly true with geoscience information, which is highly disparate in nature (e.g., the CD produced for this project alone

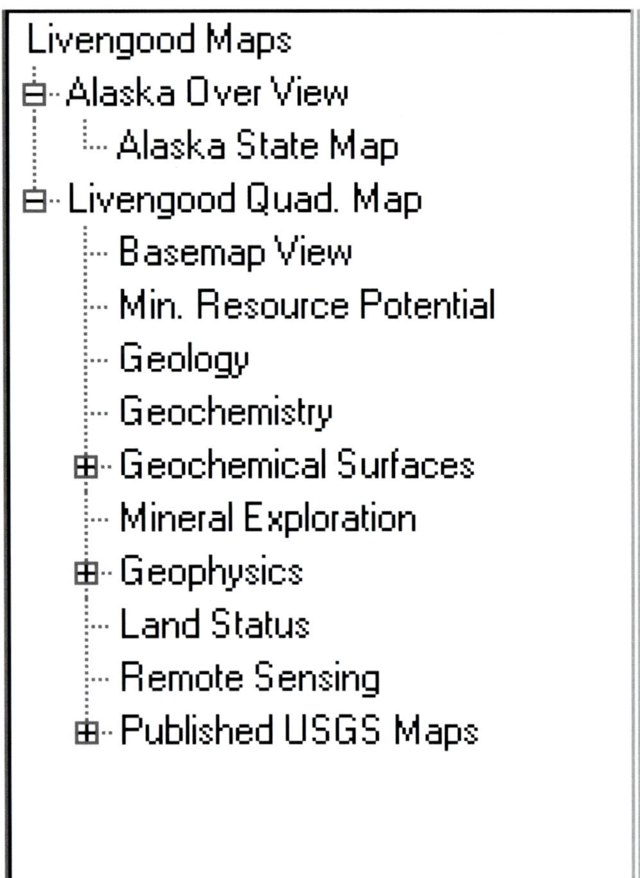

Figure 4—Depiction of the various thematic data available on the CD-ROM. The data are arranged hierarchically for logical viewing and organization.

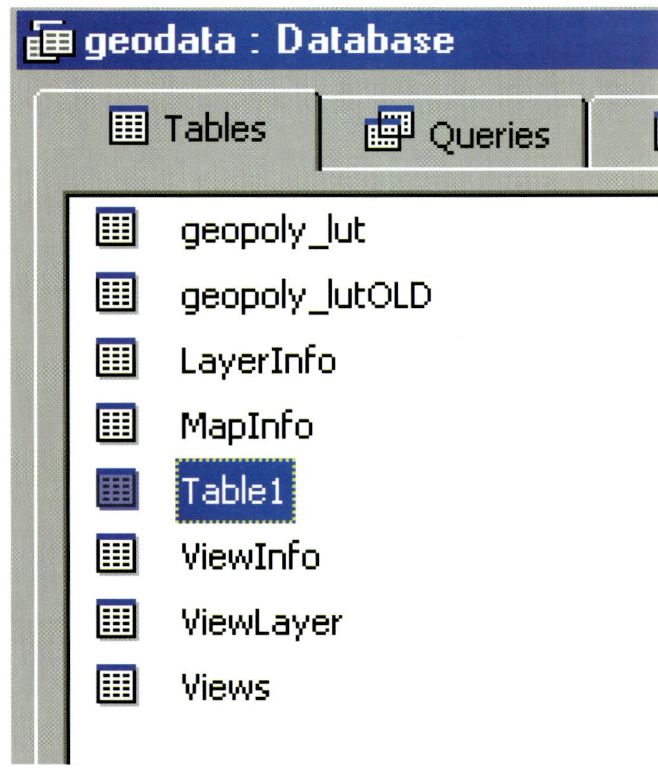

Figure 5—Various tables in the Microsoft Access database of the Livengood application. This database controls descriptions and properties of the views and layers present in the application.

contains various geochemical, geophysical, paleontological, and lithologic data). The strategy taken in this project, however, was to organize and make the data available thematically, building on the object-oriented approach.

Upon opening the CD from the computer desktop, the data delivery software presents the user with a "table of contents" frame that lists all available views or maps. These views can be considered to be "objects" representing specific themes, such as geophysics or geochemistry, with each one containing a unique set of GIS data layers. The user can then select the particular view with which to start (Figure 4).

In GeoExplorer, the hierarchical data structure is managed by the Access database, which controls the views selected at any given time and the layers contained in a particular view. This arrangement is governed by a series of linked database tables (see Figure 5).

Using this hierarchical view structure, it is possible to more easily and logically select a view of interest and to extend the hierarchy by adding subviews to the list. Figure 6, for example, depicts the geophysics view, with subviews corresponding to various categories of geophysical data. The Access table named "layerinfo" (see Figure 7) contains all of the information describing the properties of the various layers, as well as the actual controls on scale, color, etc.

VIEWS

A view is essentially a multidimensional picture (e.g., a map) of data portrayed from a specific standpoint. Each view may contain several data components that are interrelated. To draw an analogy, a view is similar to a painting or a scene from a play that gives proportion to all aspects of the situation being represented. In a geoscience setting, views might be represented by the hydrology or the seismicity of a particular geographic region or regime, with hydrology and seismicity each reflecting various aspects that contribute to those concepts. Some of the important views of the Livengood Quadrangle provided in GeoExplorer are discussed in more detail in the following paragraphs.

Geology

The geology view contains a predefined set of GIS layers representing various geological units. The colors used to display each unit are derived from the layerinfo table, which, in turn, derives the proper display and drawing characteristics from a lookup table of colors in the Access database (shown in Figure 9). Figure

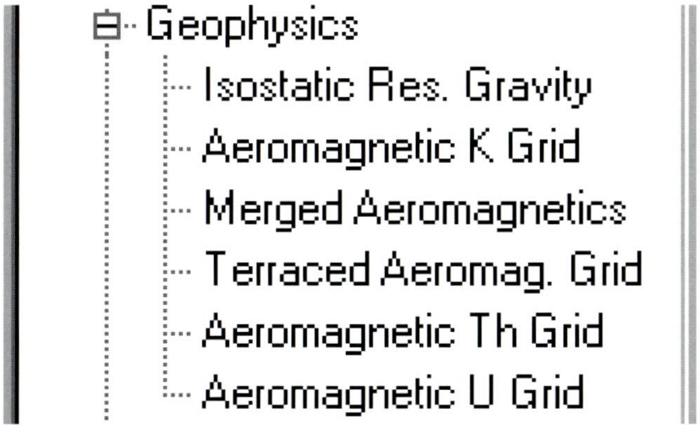

Geophysics
- Isostatic Res. Gravity
- Aeromagnetic K Grid
- Merged Aeromagnetics
- Terraced Aeromag. Grid
- Aeromagnetic Th Grid
- Aeromagnetic U Grid

Figure 6—Subviews in the hierarchical view structure as developed in the application.

8 portrays the lookup table of colors and Figure 9 presents a resulting geologic map.

Geochemistry View

GeoExplorer accesses and manages geochemical data as GIS shapefiles that can be viewed singly or together in the geochemistry view, or that can be superimposed on any other GIS layer. Figure 10 is an example of geochemical data sampling locations that have been superimposed on the landscape of the western Livengood area that can be used in an investigation of mineralization. To construct this picture, a query was first executed to restrict the data to samples having concentrations greater than 100 parts per billion (ppb). The resulting table contains only those sample values (and their associated locations) satisfying this restriction. The theme-on-theme function (an operational component developed with MapObjects) then allows additional features to be identified that are in close proximity to these sample locations and to be displayed as another layer. For example, occurrences of granite within 1000 m of the mineral of interest are shown in yellow. Theme-on-theme is an analytical function unique to GeoExplorer developed with MapObjects. It allows two thematic layers to be superimposed and intersected for purposes of producing a subset of data displayed on the map or outputting a separate subset of data.

Land Status

Land status (i.e., ownership, governance, leaseholding, etc.) is always a critical factor in exploration for petroleum or any mineral resource. In the Livengood area, land status is a particularly important consideration because federal, state, and native governments own most of the region. Using GeoExplorer, simply clicking on any part of the map will retrieve a window containing information on land status. "Map tips" (a MapObjects function) displays a label for each sector of interest that is derived from field attributes. For example, in Figure 11, "BLM" (signifying Bureau of Land Management) is shown in the upper left of the view. The synthetic aperture radar (SAR) image used as a base layer in GeoExplorer is a May, 1997 scene captured by Canada RadarSat, which was processed and geo-referenced using the Imagine software developed by ERDAS, Inc.

LayerInfo : Table

LayerID	MapName	Display	SymbolSize	MaxScaleFactı	MinScaleFact	TextSizeFactı	LabelRendered	LayerName	Shapefi
13	Quadmap	400	4	2	0	8	☑	Towns	towns.shp
16	Quadmap	395	14	2	0		☐	Airports	airports.shp
57	Quadmap	390	12	2	0		☐	Photos	photoloc.shp
37	Quadmap	380	16	0.75	0		☐	Fossils	paleoutm.sh
17	Quadmap	370	40	0.75	0		☐	Mines By Type	mils049.shp
34	Quadmap	360	14	2	0		☐	Mines & Prospects	mils049.shp
20	Quadmap	350	12	2	0	0	☐	Gold Rich Samples	goldmf.shp
21	Quadmap	340	9	0.3	0	0	☐	Sediment Samples	livseds.shp
22	Quadmap	340	9	0.3	0	0	☐	Rock Samples	livrxs.shp
23	Quadmap	338	9	0.3	0	0	☐	Concentrate Samples	livcons.shp
11	Quadmap	336	2	0	0		☐	Roads	roads.shp
8	Quadmap	335	0	0.35	0		☐	Detailed Streams	hydrolin.shp
9	Quadmap	330	0	0	0.351		☐	Big Streams	bigstrm.shp
14	Quadmap	329	0	0.15	0		☐	Township	secpol.shp
12	Quadmap	329	0	0.2	0		☐	Contours	contour.shp
10	Quadmap	329	0	2	0		☐	Pipeline	pipes.shp
31	Quadmap	329	2	2	0	0	☐	Faults	faults.shp
6	Quadmap	328	0	2	0		☐	Water Bodies	hydropol.shp
15	Quadmap	325	2	0	0		☐	USGS Quads	in2miutm.shp
24	Quadmap	320	0	2	0	0	☐	Placers	placers.shp
25	Quadmap	310	0	2	0	0	☐	Granites	granites.shp
58	Quadmap	305	0	2	0	0	☐	Fault Buffers	hab1c.shp

Record: 1 of 75

Figure 7—Illustration of the "layerinfo" table depicting the information about the GIS data layers (themes) stored in the database.

geopoly_lut : Table

ID	Value	Style	Color
3	water	0	16777197
4	Qa	0	11927551
5	Qg	0	5896191
6	Qlc	0	14090239
7	Qsu	0	14942179
8	QTg	0	16578559
9	Tvs	0	65535
10	Tb	0	13064703
11	Tgp	0	2362096
12	Tqm	0	10050047
13	Tm	0	16554495
14	TKg	0	1717247
15	TKm	0	16734719
16	Kmo	0	3365375
17	Km	0	3407830
18	Ka	0	16528383
19	Ks	0	16718591

Figure 8—Color lookup table. This table manages colors assigned to geologic units in the display.

Figure 12 shows the same information with the locations of mines and faults superimposed.

USER INTERFACE

GIS development has often concentrated on database design and on the system's internal aspects, such as methods for data storage, access to multidimensional data, and structures for modeling spatial information. Increasingly, however, more attention is being paid to the interface through which users can actually access information. One of the first objectives of the GeoExplorer project was to create an interface that employed common desktop standards and was extremely user friendly. The target audience spanned a wide range of users with differing degrees of computer skill. This order of development—designing and implementing the interface first—is the reverse of the approach that is typically taken.

Similar to static maps, interactive multimedia maps are graphic instruments for analyzing and communicating geographically related data. Designing interactive maps therefore involves the tackling of issues typical of any cartographic communication process, such as information content, visualization, and data extraction. At the same time, the relevant principles of multimedia (interactivity, navigation, and transparency) have to be honored. The challenge of interactive map design is to satisfy the constraints of both sets of objectives in such a way that information can be intuitively and effectively visualized.

Establishing and maintaining consistency is a major question that must be addressed. It requires establishment of a set of conventions for each part of the application, specifically as it entails visual presentation, screen design, information access and exploration, and database navigation. The use of MapObjects allows this task to be accomplished in a relatively rapid and consistent fashion, resulting in the design of interactive maps that convey the options of data access and navigation to the user.

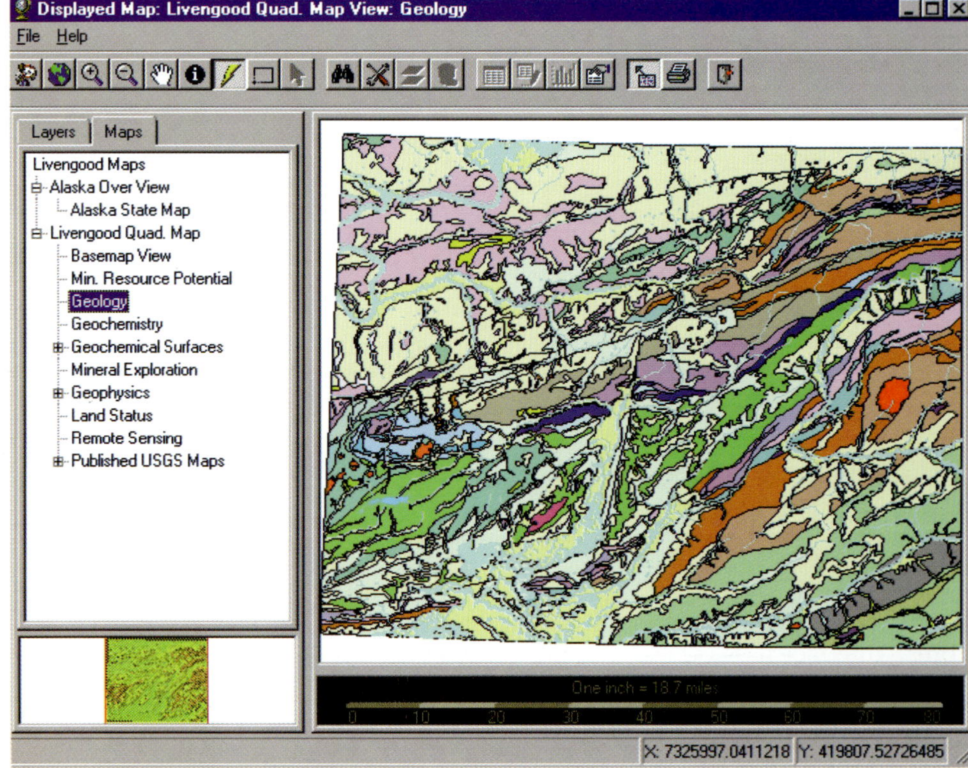

Figure 9—Geologic map for the Livengood area as shown in the geology view.

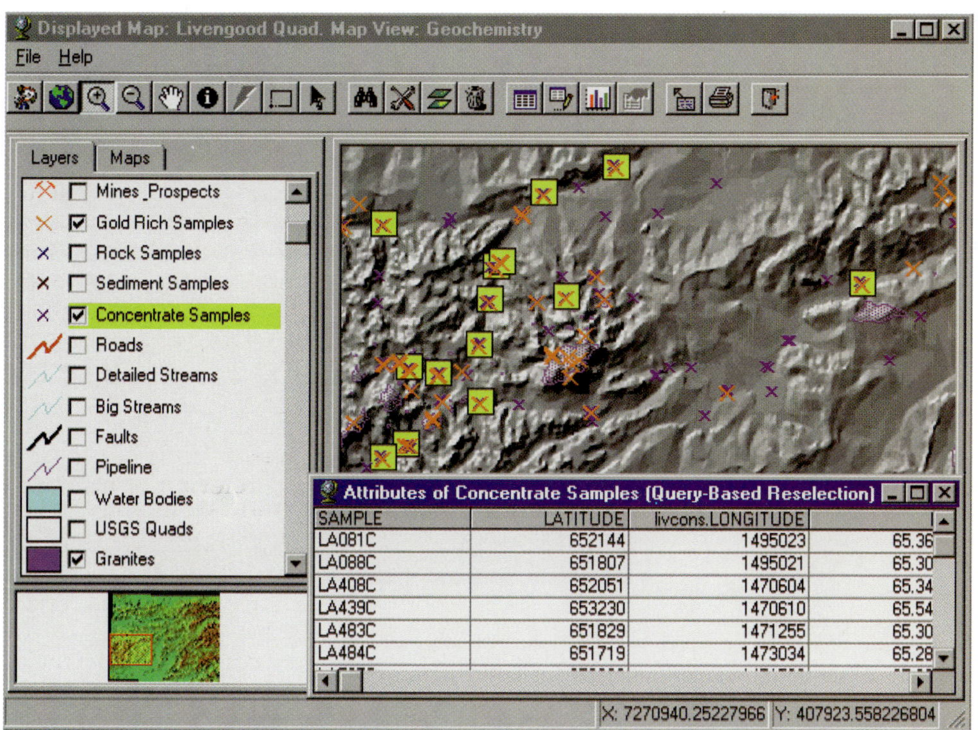

Figure 10—Geochemistry view. This is a "zoomed-in" view showing the geochemical samples on a hill-shaded digital elevation model (DEM) base. Shown are the results of a query on concentrate samples using gold greater than 100 ppb. A subset of the data matching this expression is displayed in a separate window and can be saved to an ASCII file.

Figure 11—This view shows an underlying synthetic aperture radar (SAR) image with land status layered on top.

In recent years, GUI standards have rapidly evolved, and almost all desktop computers now use a standard GUI format. Consequently, interfaces for software running on such computers that are constructed using these same standards will enhance the software's user-friendliness (Egenhofer and Frank, 1992). GeoExplorer's interface uses these common GUI standards as much as possible in the Windows-based environment.

Beyond consideration of common GUI standards, GeoExplorer was constructed in such a way as to make sense to those dealing with geoscience data. For example, a traditional geologic map interface consists of the following major components.

- Elements: heading (topic, scale, author)
- Legend
- Map face or views
- Index maps
- Cross-sections
- Scale bars

All of these basic elements have been incorporated into GeoExplorer and dynamically linked together.

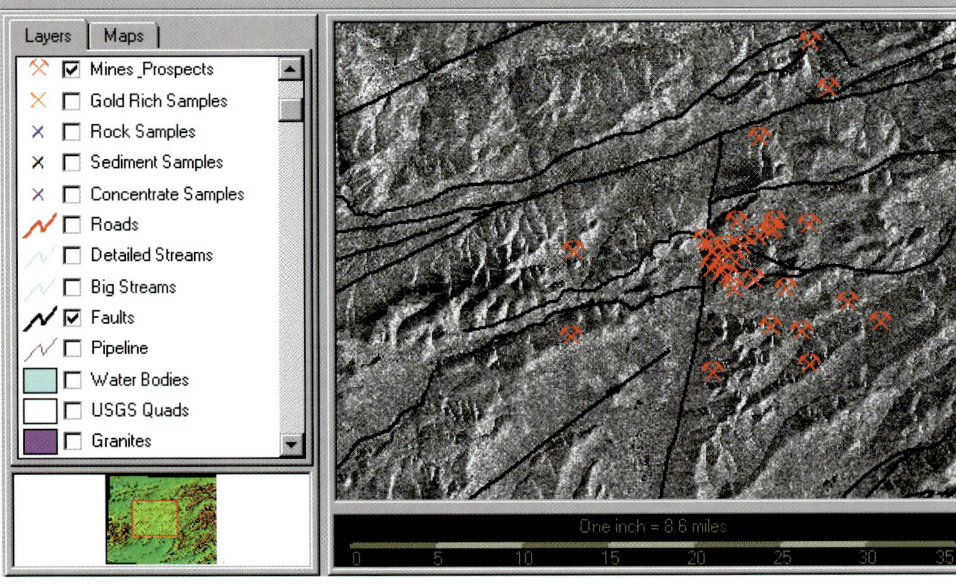

Figure 12—A synthetic aperture radar (SAR) image displaying layers of faults and mines. There appears to be a coincidence of faults and mines in this particular view.

When one zooms in on a given view, the index map reflects that change. The scale bar also changes dynamically with the change in view; nonetheless, the ability to provide all of these capabilities in traditional form was somewhat limited. For example, the legends typically included on geologic maps are often large, complex, and intricate. These are difficult to see clearly on a small computer screen. Multi-well cross-sections and long seismic traces are also difficult to display because of screen size. In contrast, some highly useful features can be included in interactive maps that static maps cannot incorporate. For example, GeoExplorer provides a dynamic index map that highlights the current map or view (see Figure 2). This feature allows instantaneous identity of location within a broad study area. It also facilitates panning within a study area.

CONTEXT

Although certainly not the last point to be considered in GIS development, the final issue to be discussed here is context. Context on maps consists of spatial information that the user does not explicitly request, but which is necessary to interpret the results of a spatial query. Typical GIS approaches provide only mechanisms to view or analyze information, relying on the user to select the appropriate data layers to add context. For the Livengood Quadrangle, context is a critical concern. The area contains a highly disparate collection of geoscience and base-level data that are difficult to interpret, even for geologists familiar with the region; consequently, for this particular project it was assumed that users would know little about the study area, or about Alaska for that matter.

A unique aspect of GeoExplorer is its ability to provide context for all levels of the data. GeoExplorer was designed such that each particular view presents an accompanying number of layers that together provide context for the one specific layer of interest. At startup, GeoExplorer presents an opening view of the state of Alaska on which a quadrangle grid is superimposed (refer to Figure 1). The addition of layers, such as a quadrangle index grid, provides context for the study area. Contrasted with this view is the state view showing only the populated places with no context. Thus, the user is given an opening view of the state of Alaska and a choice of data layers. The specific kinds of information that can be used to provide context within GeoExplorer are referred to as "context data layers." When used as a foundation on which to superimpose geoscience information, these same kinds of information are often referred to as "base map data layers."

As has been described, context data layers can be used to further enhance GeoExplorer's opening view. An example of one of these kinds of information is the airport layer (full-scale airports, landing strips, etc.), which is shown in Figure 13. Airports are an important contextual information source in Alaska because much of the state is accessible only by air. Note the index map at the lower left of the figure showing what part of the state is being displayed in the main view. Other context or base map layers contain information such as the following:

- Generalized Alaska coastline
- USGS quadrangle index (various scales)
- Town locations and outlines
- Mines and mineral prospect locations
- Hydrography (various scales)
- Digital elevation model (DEM) that can be used as an image (both raw and hillside versions are included)
- Township-range-section grid
- Photos (via hotlinking)
- SAR imagery
- Land status

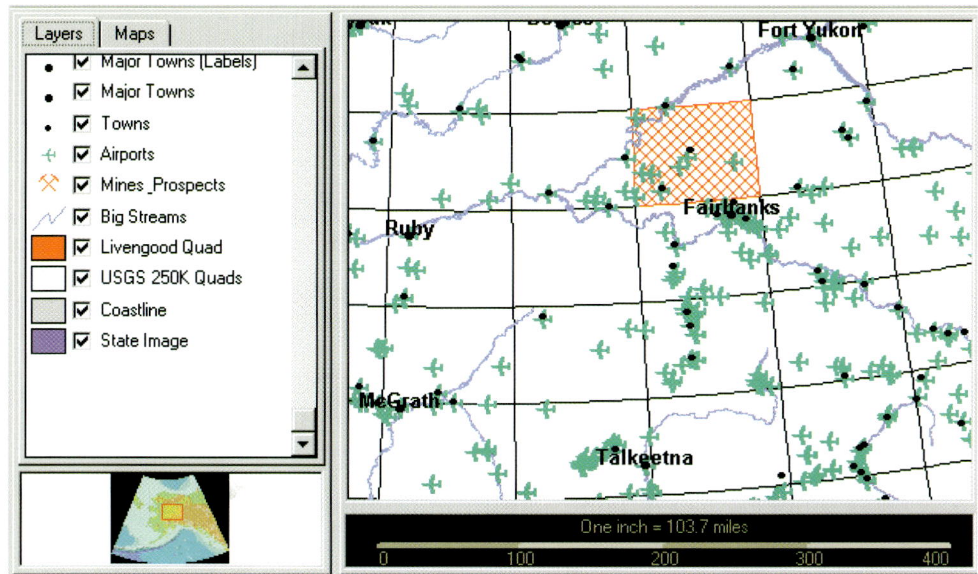

Figure 13—A view showing the "zoomed-in" area at the center of the state.

PRODUCTS

Finally, GIS applications are capable of generating a number of interesting information products. GeoExplorer, which is provided on the CD along with the Livengood Quadrangle data, supplies the ability to print letter-size maps containing titles, scale bars, and simple legends. These maps can be produced for any view selected by the user. For those who desire more sophisticated output, GeoExplorer can generate predesigned maps in the form of postscript files that can be photocopied or otherwise reproduced. It can also create Arc/Info plot files and ArcView layouts that can easily be printed using most other GIS software products.

SUMMARY

GeoExplorer is a prototype desktop GIS application that takes advantage of the CD-ROM as a delivery mechanism for complex GIS data. New GIS software technology that incorporates Microsoft's COM standards was used to construct a user-friendly GUI, quickly develop an application, and maximize system performance. The GUI, in particular, is the result of extensive collaboration with the USGS and feedback from prospective users, including geologists, landpersons, and regulatory professionals. The technology was successfully employed to rapidly and efficiently accomplish the project objectives, adhering to programming standards for the Windows-based computing environment. In doing so, it was still possible to provide access to data in a typical geoscience context, and to display the information in a typical geoscience fashion. The traditional elements of a geologic map can be digitally represented and can be dynamically linked so that interpreting data becomes easier and more powerful.

Use of GIS applications such as GeoExplorer can be a valuable means of rapidly packaging spatial data in a wide range of geoscience settings. These applications offer a particularly useful way to package and present to field personnel complex sets of data that are critical to petroleum exploration, development, and production.

REFERENCES CITED

Andrews, S. K., and D. W. Tilton, 1993, How multimedia and hypermedia are changing the look of maps, *in* Proceedings, 11th International Symposium on Computer-Assisted Cartography (Auto/Carto 11). Bethesda, MD: American Society for Photogrammetry and Remote Sensing (ASPRS), p. 348–366.

Asche, H., and C. M. Herrmann, 1994, Designing interactive maps for planning and education, *in* Visualization in Modern Cartography (A. M. MacEachren and D. R. F. Taylor, eds.). New York: Pergamon Press, p. 215–242.

Bernhardsen, T., 1999, Geographic Information Systems: An Introduction, 2nd ed. New York: John Wiley & Sons, 372 p.

Chung, M. Y., 1995, An object-oriented approach for handling topology in VPF products, *in* Proceedings, GIS/LIS '95, vol. 1. Bethesda, MD: American Society for Photogrammetry and Remote Sensing (ASPRS), p. 138–174.

Egenhofer, M. J., and A. U. Frank, 1992, User interfaces for spatial information systems: Manipulating the graphical representation, *in* From Geoscientific Map Series to Geo-Information Systems (Renier Vinken, comp.). Anscchrift des Herausgebers: Alfred-Bentz-Haus, Postfach, Hanover, p. 60–69.

Kang, S. B., 1995, Object-relational data modeling for GIS, *in* Proceedings, GIS/LIS '95, vol. 1. Bethesda, MD: American Society for Photogrammetry and Remote Sensing (ASPRS), p. 535–544.

Orfali, R., D. Harkey, and J. Edwards, 1996, The Essential Distributed Objects Survival Guide. New York: John Wiley & Sons, 604 p.

Somers, R., 1997, Framework: Introduction and Guide: National Spatial Data Infrastructure. Reston, VA: U.S. Geological Survey, Final Report for Contract 96-5000-0810, 106 p.

McCammon, R. B., 1997, Map showing mineral resource potential of the Livengood Quadrangle, Alaska. Reston, VA: U.S. Geological Survey, Open File Report 97-484-B.

GIS Resources

The technology of geographic information systems (GIS) has a lively and active following beyond the realm of petroleum exploration and development. Consequently, the amount of general information available is rather vast; and to the novice, it would perhaps even appear overwhelming. It seems as though there are always new software programs to consider, new innovations to try, and new papers to read.

In an effort to reduce this information to a more manageable size, two chapters have been prepared for inclusion in this section. In the first chapter, entitled **GIS—Public Domain and Inexpensive Software**, Betty Gibbs of Gibbs Associates provides an excellent review and assessment of no- or low-cost software resources that would be satisfactory for individuals just wanting to try out the technology and to learn more about it. The examples provided may also be appropriate for use by independent operators who require or desire some of the GIS functionality, but who do not have the financial resources to purchase the more comprehensive packages that are commercially available. In the context of her review, Gibbs also points the reader to numerous Internet resources that can be accessed to obtain extensive additional information on an as-needed basis. In the second chapter, Tim Coburn and Paul Yarka, who was previously Senior Scientist and GIS Specialist at Marathon Oil Company's Petroleum Technology Center, provide a comprehensive, though not exhaustive, bibliography of references that would be particularly useful for those working in petroleum exploration and development. This chapter, entitled **Bibliographic Resources for GIS in Petroleum Exploration and Development**, contains sections specifically devoted to technological developments, to case studies and applications, and to various other topics, all of which have a specific geoscientific context or manifest some level of importance to geoscience activities. Information about the availability of glossaries of GIS terminology is also provided.

For additional information and material, readers are encouraged to contact professional organizations which play a significant role in the GIS community. Such organizations include the International Association for Mathematical Geology (IAMG), the National Center for Geographic Information and Analysis (NCGIA), the Geological Survey of Canada (GSC), the U.S. Geological Survey (USGS), and various state geological surveys (such as those in Kansas, Illinois, and West Virginia). GISWorld also distributes a number of excellent trade publications pertaining to GIS technology and applications. Finally, readers may want to visit the Internet sites of various commercial entities to learn more about available GIS products and services, as well as those of the many academic institutions who are engaged in ongoing GIS research activities.

Gibbs, B., 2000, GIS—public domain and inexpensive software, *in* T. C. Coburn and J. M. Yarus, eds., Geographic information systems in petroleum exploration and development: AAPG Computer Applications in Geology, No. 4, p. 263–280.

Chapter 20

GIS—Public Domain and Inexpensive Software

Betty Gibbs
Gibbs Associates
Boulder, Colorado, U.S.A.

ABSTRACT

Depending on the specific situation, public domain and inexpensive geographic information system (GIS) software packages can be effectively applied to petroleum exploration and development problems. Several of these packages are reviewed and compared here. Descriptions of the packages are provided, along with examples of some of the graphic output that can be produced. In addition, other sources of information pertaining to GIS, particularly those available on-line, are listed.

INTRODUCTION

Public domain software is an important resource for the petroleum professional, and this is especially true in the case of geographic information systems (GIS) software. One of the hallmarks of public domain software is that it is often available free of charge; however, this is not always the case. Unless a program can be downloaded from the Internet, the provider often charges a media fee to cover copying and shipping costs. Source code for a public domain program is often available, although technical support for using the software or to decipher the code usually is not.

"Public domain" also means a program can be used freely and the code can be included in any software whether privately or commercially developed. Contouring algorithms produced by the National Oceanographic and Atmospheric Administration (NOAA) are examples of code from public domain software that have been incorporated into many commercial mapping programs.

Public domain software has a reputation for being hard to use, having poor user interfaces, and having little or no technical support. This is often the case, especially when the original developer cannot be located; however, software that has a broad audience appeal does not always follow the usual pattern.

Public domain software is often developed by someone in a government agency who needs a specialized calculation that cannot be done any other way. The developer is often more interested in getting answers to a problem than in producing a software package that is easy for other people to use. If a program has broad appeal, such as the GSMAP program developed at the U.S. Geological Survey for geological mapping, either the developing agency or a commercial software developer will continue to improve the program.

As suggested, a commercial software development company can use public domain code as the basis for developing its products. The company will acquire a useful public domain program and prepare it for general consumption by adding a more palatable user interface (such as Windows), graphics, and standard compatibility with other software. A price tag will also be added, which can range from low cost (less than $1000) to tens of thousands of dollars. The commercial cost usually depends on the effort required of the developer to bring the software to market and whether or not the software has a broad audience. Users must decide if they want to commit the effort to learn the public domain version or pay the price to obtain the commercially available program.

Sometimes public domain software has a wide audience and attracts subsidiary developers who share enhancements. An example of a GIS public domain program that has become a successful mainstream product is the geographic resources analysis support system (GRASS), a product developed by the

U.S. Army Corps of Engineers. GRASS has attracted a community of users who have added features and menus, and it has become a popular low-cost alternative. The public domain version of the software has also been used by a commercial developer to produce a Windows-based package called GRASSLands. GRASS is described in more detail in the following sections.

Public Domain Software Versus Inexpensive Commercial Products

As suggested previously, a primary incentive for using public domain software is its low cost. Ten years ago, the cost factor was significant because commercial GIS products were priced in the tens of thousands of dollars, and substantial training was required to learn to use them. At such high prices, a company could easily justify spending the time and resources to learn public domain software.

The software scene has changed dramatically. Conversion of software to Windows, the open database environment, the increased the power of PC hardware, and the development of new, easier-to-use GIS software have changed the marketplace. Windows-based or graphically oriented commercial GIS software can be purchased for as little as $500, and several very popular programs such as MapInfo (see description below) are priced at around $1000. These programs are easy to learn and use, and it is difficult to justify spending the time to learn a command-based product unless specific applications not available elsewhere are needed.

The user population has also changed during the past 10 to 15 years. A decade or so ago, someone who wanted to use a computer needed to know at least a little about programming to effectively use software. The development of so much software for the Windows environment has changed the need for users to know how the programs are written. Few technical professionals now know much about programming. Their skills are more oriented toward using software instead of writing it. This is as it should be. Professionals can do the work they are paid to do and not spend a major portion of upfront time writing code to accomplish their tasks.

Another effect of the change in the software environment is that many professionals do not have the time or desire to learn and use public domain software when they can obtain inexpensive and easier to use commercial products for a small investment in cost and learning time. However, someone who has a need for more complex or demanding calculations than available in these low-cost commercial packages would likely prefer public domain software if it has more flexibility and more extensive capabilities. Such individuals will also likely have a high level of computing skill and programming ability.

Several years ago, the U.S. Bureau of Land Management, the developer of MOSS (map overlay and statistical system) realized it could no longer support a continuing programming effort and made the transition to Arc/Info, a popular product developed by the Environmental Systems Research Institute (ESRI). The U.S. Army Corps of Engineers Construction Engineering Research Lab (CERL), developer of the popular GRASS software, made a similar decision and no longer enhances GRASS. The user community and commercial developers have assumed the software development tasks for GRASS.

Chapter Focus

This chapter focuses on the overall capabilities of public domain GIS software programs and how to locate or obtain them via the Internet. All the products discussed here were developed for applications other than petroleum exploration and development, but they are nonetheless directly applicable and have been used in various oil and gas settings.

Three public domain products are highlighted and described. In addition, information resources are given for free versions of commercially available software and low-cost products. The Internet provides access to free software that can be directly downloaded, as well as access to shareware and information about commercially available GIS packages. Addresses of contacts are given for those readers who do not have Internet access. A wealth of data resources for use with GIS software is also available on the Internet; however, it would not be possible to even begin to list all the data resources here because of the enormous amount of material available.

Following is a list of several public domain and inexpensive packages and how to find them. The three most readily available and widely used public domain programs are:

- GRASS (developed by the U.S. Corps of Engineers)
- LandView (a U.S. Census product provided with CDs containing extensive surface features, cartographic, and demographic information)
- MOSS/PC-MOSS (developed by the U.S. Bureau of Land Management)

Popular inexpensive or free versions of commercially available programs include:

- IDRISI, an inexpensive GIS software package available from Clark University
- MapInfo, a commercially available Windows-based software package with extensive coordinate conversion options
- Atlas GIS for Windows, a PC-based data management and integrated spreadsheet program
- ArcView 1.0, a free GIS viewing program from ESRI, makers of ArcInfo, and ArcView 3.2, which is a low-cost newer version
- TNTLite, a product available from MicroImages that has all the functions of the full TNTmips program, with the restrictions that it accommodates only a limited size grid and data cannot be exported
- ILWIS (integrated land and water information

system), a package that provides image processing, raster-based spatial modeling, and access to tabular databases
- OSUMAP, a free raster-based package for PCs running DOS. It is an instructional package and a good GIS starter program. Data sets and workbooks are available
- GMT (generic mapping tools), a free collection of 60 UNIX tools used to manipulate two- and three-dimensional data sets
- pMAP (professional map analysis package), a PC-based program for analyzing spatial data
- MicroDEM, a free program for analyzing and displaying digital elevation models (DEMs) and other gridded data sets

Information on these packages (and many more) is available on the Internet or directly from the source agency or vendor. Source or executable code can also be downloaded directly from the Internet or ordered from the developer or vendor.

Discussion groups aimed at the use and workings of GIS software packages can also be found on the Internet. Anyone can read the discussion group messages or contribute. For software users who need help, these groups are an excellent networking resource for discovering what to try next.

"Collector sites" are another Internet resource that can lead to sources of software, data, and a broad selection of other GIS-related information. These sites are usually compiled by an individual who has searched the Internet for any information on a selected topic (GIS, for example). The site information is usually categorized, and sometimes a sentence or two of description is added. Some of the collector sites for GIS software and other GIS information are listed in the following sections.

GRASS 4.1.2: GEOGRAPHIC RESOURCES ANALYSIS SUPPORT SYSTEM

GRASS is a UNIX-based GIS package developed at CERL. The program principally provides GIS management tools for the Army's environmental planners and land managers and includes applications for civil works project planning and design. It is versatile enough to accommodate a wide range of applications, such as resource management, image processing, and satellite data analysis.

The complete source code is available for download from the Internet. GRASS has many capabilities, including the handling of raster, vector, and point data and image processing. An extensive network of GRASS supporters has developed add-on programs, a mailing list, and other useful information about this software. GRASS can be downloaded from several Internet ftp sites.

Audience

GRASS is a high-end public domain GIS for use in sophisticated multi-variable analysis; however, it is flexible enough to be used by beginners as well as experts. Custom code can be integrated into the main program, and the extensive user network has developed a wide selection of software for specific applications that use GRASS as a base.

Commercial vendors have developed graphical user interfaces (GUIs) so novice GIS users can run the system. The GUI shell programs turn the command-oriented program into software that is accessible to users at all levels.

The program was developed in a UNIX environment (it runs under X-Windows or Open Windows); however, in-house programming expertise is needed to effectively employ the program as it exists in the downloadable format. A worldwide user community consisting of universities, government agencies, and individuals has also developed special modules and applications, many of which are also available on the Internet. A wide variety of modules has been created for specialized operations such as hydrological modeling and erosion prediction.

Baylor University offers a commercial Windows PC version of GRASS and supports the software. Baylor has an extensive Internet Web site containing downloadable PC versions of GRASS as well as a Windows 95/NT version.

Capabilities and Functions

GRASS is essentially a command-based program with around 390 different commands provided for GIS functions. Primary GRASS capabilities include the following.

- Display, geo-referencing, comparison classification of satellite and aerial photographic imagery
- Inputting of raster, vector and point map data, and production of hard copies on printers and HPGL pen plotters
- Linking to a database management system

Command line input gives the program instructions for running the processes. The commands are an internal language that allows users and programmers to create applications and link them to other software packages.

Data are captured and accessed through digitization, scanning, using a screen-pointing device, mounting a floppy disk, or reading computer tapes. Subsets of data can also be created with the software by selecting data elements from existing files for analysis. Output includes statistical tables, text files, or maps that can be printed or displayed on a color monitor.

GRASS is an open system and can be used for developing new software. New modules can be created, existing modules can be modified, and links can be established to other packages. ANSI C and Fortran 77 compilers are needed for developing such modules and for compiling existing modules from source code. A programmer's guide is available for help with software development.

User's Manual and Example Data Sets

A complete GRASS user's manual is available on-line at the following World-Wide Web address: *<http://www.baylor.edu/~grass/userman/>*.

The CERL documentation for GRASS 4.1 can be found at : *<ftp://129.229.103.1/pub/grass/grass4.1/documents/>*.

A beginner GRASS/GIS tutorial is available on-line at *<http://www.geog.le.ac.uk/assist/grass/seeds/gs_html.html>*.

The tutorial consists of four sessions and uses an agronomic data set. The data sets are available on-line in zipped, UNIX, PostScript, and WordPerfect formats at *<http://www.geog.le.ac.uk/assist/grass/seeds/>*.

A hard-copy beginner's manual can be ordered. It contains an introduction to the system, as well as examples showing how GRASS commands are prepared and used. Other information concerning terminology, data storage, and basic data processing is also included. The manual costs $18.00 per copy for 1–9 copies and $16.00 for 10 or more copies. First-class postage in the U.S. is $2.00. The cost of fourth-class U.S. postage is included in the price. Additional postage is required for international shipping.

The Beginners GRASS Manual can be ordered from Evergreen Software Services, 2208 N. Water Street, Ellensburg, WA 98926-2053 [telephone: (509) 925-2513; e-mail: jeanh@ellensburg.com].

Data File Formats Accepted

GRASS accepts the following types of data formats:

- Raster data (PPM, TIFF, TGA, USGS DEM, DMA DTED, PPM/3, GIF, SUN, ASCII)
- Vector data (ARC/INFO, DLG, IDRISI, TIGER, ASCII, DXF, MOSS)
- Images (ERDAS, NHAP, BIL, SPOT, LANDSAT MSS and TM, BSQ)

Command File Example

Following are a few examples of command setups:

- g.remove muledeer1 (remove a raster layer)
- g.rename muledeer2,muledeer1 (rename a raster layer)
- g.region rast = muledeer1 (make a raster layer active)

The r.mapcalc module is used for performing arithmetic calculations on gridded raster map data. Expressions can be stated in integer or floating point format and can include functions. To change raster values to real negative numbers, instead of unsigned integers, use the r.mapcalc command module as follows: r.mapcalc muledeer1a="if(muledeer1>32768,muledeer1-65536,muledeer1)"

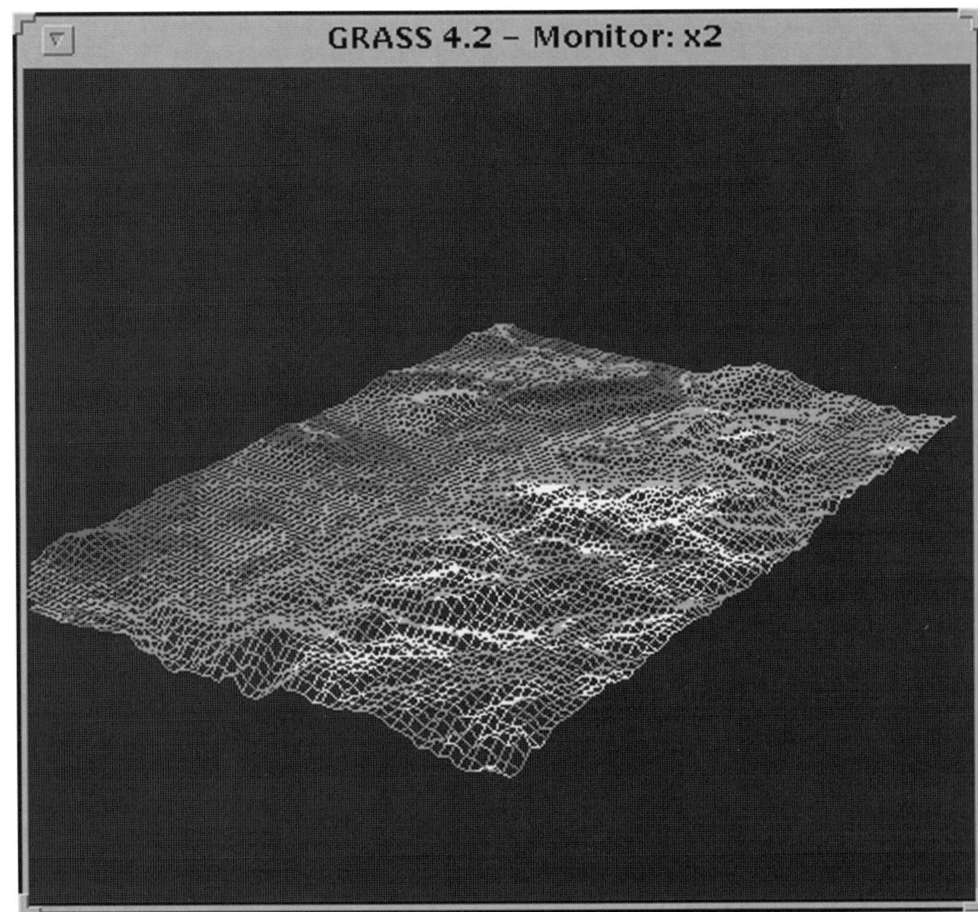

Figure 1—A GRASS 3-D image of an area around Spearfish, South Dakota, showing geology draped over an elevation wire frame. A digital file of this image is available at <http://www.baylor.edu/grass/>.

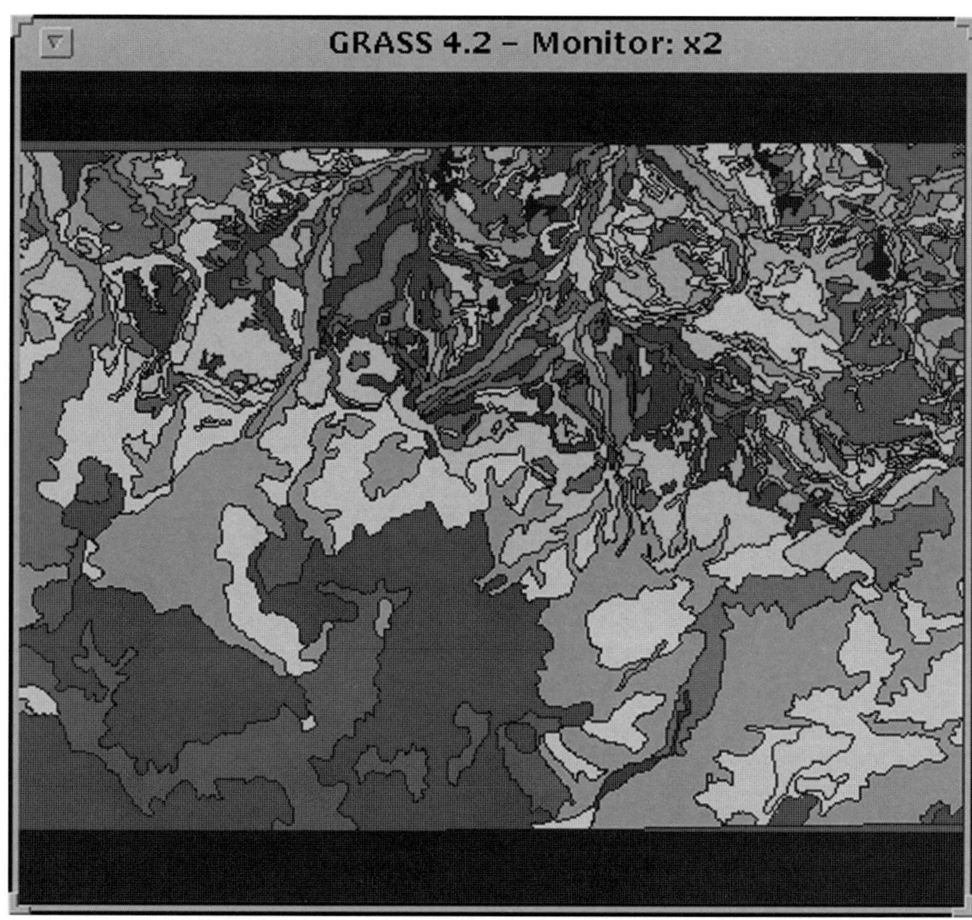

Figure 2—A GRASS 3-D image of an area around Spearfish, South Dakota, showing a soil layer map. A digital file of this image is available at <http://www.baylor.edu/~grass/>.

Example Graphics

Elaborate maps can be produced with GRASS, such as shaded-relief, color-coded contours, 3-D topographic maps, color raster maps, thematic maps, and many more. Figures 1 and 2 illustrate the kind of graphic output that can be produced with GRASS.

Hardware Requirements

Grass on UNIX

GRASS is compatible with a variety of hardware platforms running UNIX operating systems. UNIX operating systems on workstations that will run GRASS include (but are not limited to) DEC Alpha OSF/1, HP UX 90, Linux, SCO UNIX, SGI Irix, and SUN Solaris. The C source code compiles most easily on a SUN system.

The minimum hardware configuration includes a monitor that can display 256 colors, a processor running UNIX or a similar operating system, at least 8 MB of system memory, at least 140 MB of disk space (300–600 MB is recommended), a line printer, a 1/4″ or 1/2″ tape drive, and a mouse pointing device. Other hardware options include a digitizer for map input, any of several color printers for hard-copy output, a plotter that runs HPGL, and modems or network connections. In addition, a graphics library must be loaded with the system.

GRASS on PCs

GRASS will also run on a PC under Windows 95/NT using the functionality of Windows drawing tools. The software will also work on PCs running some version of UNIX (SCO, Linux). A PC must have a minimum of 8 MB of memory, but 32 MB or more is recommended for better performance. At least 300 MB of disk space is needed for the code and executables, and more for data. A C compiler must be loaded on the system to use or access the source code.

Availability

To find how and where to download the GRASS software, visit the following Internet site: *<http://www.baylor.edu/~grass/acquire.html>*.

This site contains source code programs and the Spearfish sample database but not contributed programs. Compiled binaries are also available.

Other information about GRASS is available from the GRASS Development Team, U. S. Army Corps of Engineers, Construction Engineering Research Laboratories (CERL), 2902 Newmark Dr., Champaign, IL 61821-1076 [telephone: (217) 352-6511].

Also visit the Hannover (Germany) University site: *<http://www.geog.uni-hannover.de/grass/>*. Access to the latest versions of GRASS can be found on this site. Other resources and contacts are also available.

Versions of GRASS for SCO UNIX and Macintosh are available from Central Washington University. The cost for a set of GRASS binaries (distributed via ftp) is $350. The source code is available free with the binaries. GRASS for Macintosh and Apple is available by special arrangement. The Mac and Apple versions are limited in capabilities. For more information, contact James Hinthorne, Director, GIS Laboratory, Central Washington University, Ellensburg, WA 98926-7420 [telephone: (509) 963-2826; e-mail: geol1@cwu.edu].

Source code is also available on the GRASS 4.1.2 CD-ROM distributed by CRSSA-Rutgers University. The CD-ROM and the source code tape (including source code and executables) are $100.00 each, plus a U.S. shipping cost of $10.00 ($50.00 outside the U.S.) per order.

To order the CD, contact Center for Remote Sensing and Spatial Analysis (CRSSA), Environmental and Natural Resource Sciences Building, 14 College Farm Rd., Rutgers University, New Brunswick, NJ 08901-8551 [telephone: (732) 932-1582; fax: (732) 932-2587; e-mail: grass-sales@crssa.rutgers.edu].

The GRASS Research Group at Baylor University has assumed the task of updating GRASS to version 5.0 (most recent: 4.2) and enhancing the documentation. CERL is no longer able to support GRASS for budgetary reasons.

GRASS information, software (PC and UNIX), and resources are available on *<http://www.baylor.edu/~grass/>*.

This site includes a complete on-line GRASS Users Manual, information about support, frequently asked questions (FAQ), a mailing list, ftp sites, and data sources. The latest GRASS releases for UNIX and Windows can be downloaded. Also included are archived past releases, data sets (such as the Spearfish data), manuals, and other software-related items. For other information, contact the GRASS Research Group, Center for Applied Geographic and Spatial Research, Baylor University, P.O. Box 97351, Waco, TX 76798-7351.

Other GRASS Resources

Listed below are resources or programs originally developed by individuals and organizations other than CERL. Companies or organizations use the GRASS command language as a foundation and develop their own user interfaces based on Windows or X-Windows that are more easily handled than the original command language.

- *GRASS on DEC*. GRASS 4.1 is available for a Digital Equipment Corporation (DEC) ULTRIX workstation from Central Washington University, Office of Continuing Education [contact Phillip Richins at telephone: (509) 963-1504; fax: (509) 963-1558].
- *GRASS on Macintosh*. Several commercial enterprises have ported GRASS to A/US, the Macintosh version of UNIX. This version of GRASS can be purchased from Central Washington University or Logiciels et Applications Scientifiques (L.A.S.), 1570 Chabanel West, Montreal, Quebec, Canada H4N 1H4 [telephone: (514) 858-1104; fax: (514) 389-9373].
- *XGRASS*. This is an X/Motif GUI version of GRASS. XGRASS was compiled to run only on an X11/Motif window manager. It can be found on the CERL ftp site. Support for this enhancement has been discontinued.
- *GRASSXMI* (GRASS with X/Motif Interface). This is a product developed by GPZ Technology that can be customized for clients. For further information, visit the GPZ Web site at *<http://www.gpz.com>*, or send an e-mail message to <support@gpz.com>.

GRASS Tutorials

CERL provides GRASS tutorials that can be downloaded from its ftp site: *<ftp://129.229.103.1/pub/grass/grass4.0/documents/tutorials>*.

Introduction to GRASS GIS Software by Markus Neteler (University of Hannover, Germany) can be found at *<http://www.geog.uni-hannover.de/grass/gdp/neteler/>*.

Project ASSIST (University of Leicester, UK—Academic Support for Spatial Information SysTems) provides education and training materials for GRASS, as well as for two other products (Arc/INFO and IDRISI); see *<http://www.geog.le.ac.uk/assist>*.

GRASS Seeds Beginner's Tutorial is available with sample data at *<http://www.geog.le.ac.uk/assist/grass/seeds/>*.

GRASSLinks 2.0: a Public Access GIS

GRASSLinks, a Web interface to GIS, offering public access to spatial information such as spatial databases and GIS software resources. The purpose of the Links is to facilitate easy sharing of data and applications among users. The following site describes how to acquire GRASSLinks software and install it on a computer: *<http://www.regis.berkeley.edu/gldist/>*.

GRASSLAND

GRASSLAND is a commercial, Windows-based (95/NT) GIS software package built on GRASS that provides a graphically oriented environment for presentation and manipulation of vector and raster data. Windows provides a familiar interface for PC users. The program employs the full capabilities of the open geospatial datastore interface (OGDI) to read almost any geospatial file format, including DIGEST (VRF, DTED, ADRG), Grass, DXF, and many more. The program uses Tcl/Tk, an X-based scripting language.

The price of GRASSLAND 1.1 for Windows/95/NT is $849 per license. Special educational pricing is available. A free UNIX release is offered, which is usually at least one version behind the PC product. For more information contact Global Geomatics Inc., 1600 St. Martin Blvd. East, Tower A, Suite 650, Laval, Quebec, Canada H7G 4R8 [telephone: (800) 584-0345 or (450) 668-4949; fax: (450) 668-2822; e-mail: dj@globalgeo.com; Web: *<http://www.globalgeo.com>*].

GRASSLAND on the Web

The GRASSLAND Web site contains extensive information about the software and how to participate in the user community. Frequently asked questions can be viewed on *<http://www.globalgeo.com/grassland/technical/faq/faqfrm.html>*.

This is a good place to find information about tutorials, database structure, new releases, useful related sites, and mailing lists. Subscribe to the Grassland Users mailing list by sending an e-mail message to majordomo@globalgeo.com and include the following two lines in the message text:

subscribe g_linfo
end

The GRASSLAND tutorial on the Global Geomatics Web site helps new users understand the demonstration version of the program. Users can contribute scripts and programs that can be accessed by the user community. Tasks can be automated and analysis routines developed using TCLAND the basic GRASSLAND C programs. An animated demonstration (+3MB) can be downloaded from<*http://www.globalgeo.com/grassland/wtsn.html*>.

LANDVIEW III

The NOAA and the U.S. Environmental Protection Agency (EPA) jointly developed LandView. MARPLOT for Windows is the mapping module that includes U.S. map boundaries, such as states, counties, congressional districts, census tracts, and block groups. Database search and query tools are provided for navigating and scaling, creating new map boundaries, and inserting raster images. Other capabilities include map layer management and raster image display.

Specific data about the United States are available and include roads, rivers, and railroads (based on TIGER/Line 95), jurisdictional and statistical boundaries, EPA-regulated sites, and key geographic features of the United States provided by the U.S. Geological Survey and other federal agencies. It also includes demographic and economic data from the Census Bureau, as well as information about miscellaneous public structures and facilities. Information is presented in a geographic context that includes:

- Jurisdictional entities (states, counties, cities and towns, and congressional districts) and other geographic entities (such as ZIP code reference points)
- A detailed network of major and minor roads, rivers, and railroads
- Census block groups and tracts
- Selected landmarks

The software provides:

- Desktop mapping capabilities for displaying, searching, and identifying map objects
- Thematic mapping
- Choosing display attributes based on database information or relationships to other map objects
- Printed maps and reports

"Buffer" queries are not supported. Only a limited set of symbols is available, and there is no capability to create or import new symbols.

Audience

LandView integrates companion data available on 11 CDs and is principally designed for use by individuals who need to generate maps based on U.S. Census data. Information is also included for Puerto Rico, Virgin Islands, American Samoa, Guam, and Northern Mariana Islands. The CDs are an excellent resource for surface feature details that can be used to create thematic and customized maps.

The software represents an inexpensive entry into the world of GIS. It is useful for showing surface features but does not satisfy highly technical needs.

Hardware Requirements

The program runs under Windows 3.1 or Windows 95. For optimal performance under Windows, the following configuration is suggested:

- *PC systems*: Pentium or 486/DX2 66 MHz IBM-compatible PC, with 8 MB of random access memory (RAM) running Microsoft Windows 3.1/95/NT or Windows for Workgroups 3.1.

A Macintosh version of this program is also available. The configurations for Macintosh are:

- *Macintosh 68K system*: a 68,020 processor or higher running Mac/OS System 7 or later, with 8 MB of RAM (12 MB of RAM recommended)
- *Macintosh Power PC system*: Mac/OS System 7.1.2 or later, with 8 MB of RAM (16 MB of RAM recommended)

All three systems need:

- 10 MB of free space on a hard disk drive
- CD-ROM drive
- VGA monitor, or better (or Macintosh-compatible monitor)
- Mouse or other pointing device

User's Manual

A user's manual is available with the software. The on-line address to download a PDF version is <*http://www.census.gov/geo/www/tiger/*>.

Sample data sets for two counties are also available at this location. A tutorial can be browsed on-line or downloaded.

Data

As suggested above, 11 CDs are available that provide well-researched and tested data from several U.S. government agencies. The first ten discs of the CD set contain information grouped by state with detailed networks of roads, rivers, and railroads based on the TIGER/Line 1995 files. Each of the ten discs provides the same functions as on the national disc, but only for the states contained on that disc. The U.S. summary (Disc 11) is designed for creating thematic maps or performing radius calculations that extend over a number of states.

The data on the CDs include the following:

- *Census Data:* selected 1990 demographic and economic census information from the Census Bureau, including economic and housing characteristics

- *Air Facilities database:* contains information maintained by the EPA's office of air and radiation in its aerometric information retrieval system (AIRS). AIRS stores information on air quality, point source emissions, and area/mobile source data required by federal regulation from the 50 states
- *Hazardous Waste Facilities data* from the resource conservation and recovery information system (RCRIS) and the biennial reporting system (BRS): BRS contains information on facilities that treat, store, or dispose Resource Conservation and Recovery Act (RCRA) hazardous wastes or are classified as large quantity generators (LQGs) under RCRA
- *Superfund* site information: data from the comprehensive environmental response, compensation, and liability information system (CERCLIS)
- *Toxic Release Inventory Facilities System* (TRIS): contains information about releases and transfers of more than 300 toxic chemicals and compounds to the environment
- *Wastewater Discharge Facilities:* the permit compliance system (PCS) is a national computerized management information system that automates entry, updating, and retrieval of national pollutant discharge elimination system (NPDES) data and tracks permit issuance, permit limits and monitoring data, and other data pertaining to facilities regulated under NPDES. PCS records water-discharge permit data on more than 75,000 facilities nationwide
- *Watersheds:* over 2,000 watershed boundaries and an EPA database from the Index of Watershed Indicators (IWI). Also see <http://www.epa.gov/surf>
- *Ozone Non-attainment Areas:* this database indicates which U. S. counties have been in an ozone non-attainment status
- *Dams:* 1995–96 update of the National Inventory of Dams CD-ROM, prepared under contract for the United States Federal Emergency Management Agency (FEMA)
- *Airports:* listed on the National Transportation Atlas Databases, Bureau of Transportation Statistics
- *Nuclear Sites:* information from the Nuclear Regulatory Commission
- *U. S. Highways and Water*—information that is included as layers on the MARPLOT map
- *Canadian Places:* this database is from the Digital Chart of the World (DCW), Edition 1. The MARPLOT map layers for Canadian Province boundaries, roads, and railroads also come from the DCW
- *Mexican Places:* this database also comes from the DCW, as do the MARPLOT map layers for Mexican State boundaries, roads, and railroads
- *Schools, Hospitals, Religious Institutions, and Cemeteries:* contains information from the U. S. Geological Survey's geographic names information system (GNIS)

Example Graphics

An example of the kind of graphics that can be produced with Landview III can be found at *<http://www.census.gov/apsd/pp98/pp.html>*.

Availability

A free copy of LandView III can be downloaded from *<http://www.census.gov/geo/www/tiger/>*.

The program and individual county data can also be downloaded from the Internet. The software and data for one county at a time can also be retrieved from the following Web site: *<http://www.RTK.NET:80/landview/>*.

As noted, LandView is also distributed on a set of 11 compact discs containing data that can be used with the software. The ten local discs each contain a grouping of states. The 11th CD (U.S. summary) covers the entire U.S. and includes all data except the detailed feature network derived from TIGER/Line 1995 files. Each CD costs $99. The complete set of 11 discs is available for $549. For more information, contact the U.S. Department of Commerce, Bureau of the Census (MS 1921), P.O. Box 277943, Atlanta, GA 30384-7943 [telephone: (301) 457-4100]. The following Web site contains a good description of the software with data screen and map examples: *<http://www.census.gov/apsd/pp98/pp.html>*.

MOSS AND PC-MOSS

MOSS—Map Overlay and Statistical System

MOSS is the data analysis component of a GIS originally developed by the Western Energy and Land Use Team (WELUT) of the U.S. Bureau of Land Management (BLM). It is part of a more complete GIS that includes capabilities such as data entry (automated digitizing systems, ADS), raster data processing (map analysis package, MAP), and cartographic output (cartographic output systems, or COS).

PC-MOSS Version 1.0 consists of a subset of 40 of the MOSS vector processing commands. The elements of map digitizing, projection changes, and map plotting are not provided in PC-MOSS. The MOSS IMPORT command is available in PC-MOSS and provides one mechanism for entering map data into the system.

Data from other PC-based digitizing programs can be converted to the MOSS IMPORT format for use in MOSS. Any data transfer capability between a minicomputer running MOSS and a PC running PC-MOSS would permit transfer via the IMPORT format. A tape drive interface for the PC or a serial communications link would provide the necessary capability. The MOSS EXPORT command is not available in PC-MOSS, but would help with moving MOSS data into other PC systems (such as CAD systems) for cartographic output. PC-MOSS was developed to address a particular need and has limitations that preclude its use as a stand-alone, full-function GIS.

The BLM has moved away from using MOSS and is now using Arc/Info, a commercially available product. GIS technology has improved since MOSS was developed, and the burden of maintaining the program can no longer be justified. BLM support efforts currently focus on collection of metadata and transition of existing GIS data to the ARC/Info format. MOSS is no longer being updated, and only two people are available to provide technical support.

Audience

MOSS is a UNIX-based, raster GIS software program. As noted, the PC version contains selected commands from the larger program. Prospective users should be prepared to spend more time learning to use this software on either a PC or UNIX machine than would be required with other products. An attraction of the software is that it is a free public domain product and provides a good selection of GIS analysis tools.

Even though MOSS is no longer supported by the BLM, a user base still exists. Although MOSS databases are available and still useful, and the file format can be read by several GIS packages, the analytic capabilities of the program are available in other, more recently developed products; so the need to actually use the software itself is diminishing even for users who are looking for an inexpensive package.

A discussion of the program is included here because it is available for free from the BLM ftp site, and the PC-MOSS program is available for free from the Computer Oriented Geological Society (COGS) ftp site. An understanding of the complete MOSS system is useful for applying the PC version and for understanding the formats of existing MOSS databases.

Both of these programs are more suitable for the experienced computer and GIS user than for beginners. Some knowledge of programming is also helpful.

Capabilities and Functions

The PC-MOSS commands provide the following capabilities:

- Map file management
- Attribute import, creation, and editing
- Data manipulation—feature selection, buffers, mathematics and statistics on attributes, measurement and display of distance between points; perimeter measurement, frequency analysis, overlay of polygons, joining and intersection of maps, and proximity calculation
- Generation of maps—creation and editing of map features, legend creation, cross-hatching of areas, and plotting of maps
- Production of attribute and session reports

Hardware Requirements

MOSS

Data General and other UNIX workstations are required for running MOSS.

PC-MOSS

Minimum configuration: IBM-compatible PC/XT or better, running DOS, 640 kB memory, math co-processor, 600 × 200 graphics card and monitor.

PC-MOSS requires the DOS operating system and a Tektronix graphics-terminal-compatible device driver for the PC. The device driver software interprets all instructions being written to the PC console. The driver correctly interprets Tektronix PLOT10 instructions for the PC monitor and draws the appropriate vectors on the screen. Because the device drivers are the same for both the UNIX and PC versions of MOSS, the UNIX version moves to the PC environment without changes to the graphics output routines.

Note that the modeling and analysis portions of the program can be run without graphics. In fact, prospective users will likely be more satisfied with transporting analysis results to more graphically oriented software. See the section below about data formats for an additional caution about transferring MOSS databases to other software.

Language and User Interface

PC-MOSS was developed using IBM Professional FORTRAN 77.

User's Manual

The documentation for PC-MOSS is available on the COGS ftp site (see following). Documentation for MOSS is included with the source code. Example data sets are available with both programs.

Data Formats Accepted

The MOSS/PC-MOSS data formats are not compatible with all other GIS software. Work would be required by the user to convert models developed with these programs into a format that could be imported by other products, particularly the commercially available packages currently in vogue. However, several GIS programs, such as GRASS, do accept the MOSS format.

Availability

As noted, MOSS source code can be downloaded from *<ftp://ftp.blm.gov/pub/gis/moss_src>*.

The files are compacted (.gz extension). Documentation and sample data are also available at this location.

PC-MOSS can be found on the Internet at:*<ftp://ftp.csn.org/COGS/MOSS>*. This site contains only the binary version of the code. Documentation is also included at this location, along with sample data.

For more information about MOSS, contact Bureau of Land Management, Division of Advanced Data Technology, Branch of Technology Applications and Assistance, Building 50, Denver Federal Center, P.O. Box 25047, Denver, Colorado 80225-0047 [telephone: (303) 236-0990].

INEXPENSIVE, COMMERCIALLY AVAILABLE GIS PROGRAMS

Four inexpensive commercially available GIS software packages are listed and briefly described. These four are only a small sample of all the low-cost GIS programs available.

ArcView 3.2 is a widely used program available from ESRI.

The IDRISI program developed at Clark University is not public domain software, but is widely used and was one of the first inexpensive commercial GIS packages available. The Graduate School of Geography maintains, enhances, and supports this software as a commercial product. IDRIS1 is the name of an ancient mapper.

MapInfo is another widely used inexpensive program. It runs in Windows on a PC and is easy to learn and use. A useful feature of MapInfo is its provision for easily converting between coordinate systems and mapping projections.

Atlas GIS is another inexpensive commercially available software program. It has a comprehensive selection of processes, and it will read spreadsheet and DBMS files.

ArcView GIS 3.2

ArcView 3.2 is an inexpensive commercial program that is available from ESRI. It is Windows-based and is compatible with the Arc/Info full professional GIS package. The software capabilities include:

- Creation of publication-quality maps from existing data and addition of data in a tabular format, display of data on maps, selection of layers, colors, type styles and sizes, and much more
- Linking of charts, tables, drawings, photographs, and other files
- Display, query, summarization, and organization of data geographically
- Analysis of trends in data using attributes
- Developing custom tools, interfaces, and complete applications using the built-in object-oriented scripting language

Audience

ArcView was developed as a GIS tool for the casual user. Training is recommended, but it is possible to learn to use the software by reading the manual and practicing with the aid of tutorials. The software was developed for people who have a need to use a GIS, but are not necessarily GIS professionals. It is oriented more towards analysis of data rather than creation of extensive databases.

Data can be created with ArcView 3.2, but the program performs more efficiently when accessing information from other sources. While several extensions are available that have direct applications to geological mapping, ArcView 3.2 only provides two-dimensional analyses (which would not be optimum for geological investigations).

Capabilities and Functions

Though not as robust as the full Arc/Info package, a complete GIS analysis can still be performed with ArcView 3.2. It includes a report writer, the ability to manipulate themes (data layers), extensive map building graphics, a wide selection of geoprocessing operations (such as the ability to buffer, dissolve, merge, clip, and perform intersections and unions), address matching, and geocoding. A few of the software's other features include:

- Cartographic wizards for developing high-quality map composition
- Thousands of industry/application-specific symbols
- An extensive range of data projections and datum transformations (e.g., NAD27 to NAD83)
- Many data classifications (equal area, equal interval, natural breaks, standard deviation, data normalization, etc.)
- The ability to create layouts that contain maps, tabular data, charts, and other graphics
- Spatial querying
- A wizard that creates buffers around objects in a theme
- The ability to select features in one map based on those in another map
- Tabular joining of data based on the location of features in a map
- The ability to overlay levels of data to create new data
- Data aggregation to build new information
- Creation of custom legends

User's Manual and Example Data Sets

A full-color manual is available that steps the user through key tasks. The manual details the steps necessary to use various features of the software. A tutorial is provided that can be accessed on-line.

Five CDs of data are included with the software. The data sets consist of shapefiles that include geographic information about the United States, such as boundaries, census tracts, census block populations, zip codes, landforms and land features (lakes, rivers, etc.), major roads, railroads, airports, landmarks, and much more. Other data files are included that contain similar information from Canada, Mexico, Europe, and a selection of other countries.

Data File Formats Accepted

Arc View 3.2 accepts data in the following formats:

- dBASE, INFO, Spatial Database Engine (SDE), and ASCII
- ODBC data access, including Access and Paradox
- FoxPro, Oracle, Sybase, Information, DB/2, INGRES, or any ODBC/SQL-computer database
- Spatial data in an RDBMS with SDE as a client
- SAP R/3 data
- MapInfo (MIF), S-57, SDTS (raster, point)

Data can be exported to the following: ESRI shapefiles, DXF, Placeable WMF, Windows Metafile, Windows Bit Map, PostScript, CGM (Binary, Character, Clear Text), JPEG.

Optional Extensions

As noted above, a selection of extensions is available that adds additional capabilities to the ArcView 3.2 basic software. Those that are directly applicable to geological mapping include:

- Spatial Analyst (advanced spatial analysis and data modeling)

- 3-D Analyst (tools to create, analyze, and display three-dimensional surface data)
- Image Analysis (enhanced image integration, display, and analysis)
- Internet Map Server (real-time mapping and GIS applications for the Web)

Hardware Requirements

ArcView 3.2 runs on industry-standard personal computers with at least a Pentium or better Intel-based microprocessor, a hard disk, and Microsoft Windows. 32 MB RAM is recommended. It will also run on several UNIX machines, including Hewlett Packard, Silicon Graphics, IBM, and Sun workstations. ArcView GIS 3.0a will run on a Macintosh with MacOS System 7.1 or later. 12 MB RAM is required for the Mac version.

Availability

A free CD is available from ESRI which contains a 30-day evaluation copy of ArcView GIS. ArcView GIS 3.2 and optional extensions are available to purchase from ESRI, ESRI resellers, and ESRI international distributors. Contact ESRI offices at ESRI, 380 New York Street, Redlands, CA 92373-8100 [telephone: (800) 447-9778; e-mail: info@esri.com; Web: *<http//www.esri.com/software.arcview/index.html>*].

IDRISI

Government agencies and individuals all over the world use the IDRISI program. It was developed by the Graduate School of Geography at Clark University and is supported by a full-time staff of development, technical support, and customer service professionals within the Clark Labs for Cartographic Technology and Geographic Analysis. The IDRISI project is a nonprofit organization, but the software is not public domain.

IDRISI is a raster-based microcomputer GIS and image processing system. Both Windows and DOS versions are available. Analytical operations support full multi-tasking. The system also supports multiple map windows with full multi-layer composition capability. A context-sensitive help system is incorporated. The Windows version can access all raster graphic displays and printers (color and monochrome) supported by the Windows system.

Capabilities and Functions

IDRISI provides raster analytical functionality covering the full spectrum of GIS needs from database query, to spatial modeling, to image enhancement and classification. Capabilities are included for GIS applications ranging from environmental monitoring to natural resource management. The program can perform statistical analysis, change and time series analysis, multi-criteria and multi-objective decision support, uncertainty analysis (including Bayesian and fuzzy set analysis), and simulation modeling (including force modeling and anisotropic friction analysis).

Images can be up to 32,000 rows × 32,000 columns. Full geodetic-quality geo-referencing includes reference ellipsoid, datum, projection, and false origin specifications. Rubber-sheet transformation and forward and backward projection uses ellipsoidal formulas and geodetic datum adjustment.

Hardware Requirements

The minimum requirements for the DOS version are 80486 DX processor or better, a monitor with SuperVGA graphics, 8 MB or more of RAM, a mouse, and an appropriate printer.

The Windows version has the following minimum requirements: 4 MB RAM (8 MB or more recommended); an 80386 or higher processor is required, with math coprocessor support (DX series chip) strongly recommended; Windows 3.1 or a compatible operating system running in 386-enhanced mode, including Windows 95, Windows for Workgroups 3.11, Windows NT 3.5/4.0, OS/2 2.2 or 3.0; an 8-bit (256 color) or better graphics card with a minimum resolution of 640 × 480 (1024 × 768 or better recommended).

Audience

Students and GIS professionals all over the world use the software. It is in active use in over 130 countries by a wide range of research, government, local planning, resource management, and educational institutions. The special facilities included for environmental monitoring and natural resource management make it a good tool for technical professionals; and the graphically oriented system is easy enough for beginners to use.

User's Manual and Example Data Sets

Sample data sets and a user's manual with tutorials are provided with the Windows and DOS versions of the software as commercial products.

Data Formats Accepted

The following file types can be imported or exported into/from IDRISI: Arc/Info shapefiles, ERDAS, MAP, DLG, DEM, TIFF, BMP, DXF, Map Info, GRASS, Atlas, GIS, GEOSPOT, and many others. The program can directly use database files from Access, dBase, and FoxPro, three popular commercial database management packages.

Availability

A variety of licensing options are available for both the Windows and DOS versions, with special discounts available for nonprofit, academic, and government institutions. Prices range from $95 for a student starter version to $990 for a commercial user.

The IDRISI software package is distributed solely by the IDRISI Project. To obtain information or to place an order, contact the Project at: The IDRISI Project, Graduate School of Geography, Clark University, 950 Main Street, Worcester, MA 01610-1477 [telephone: (508) 793-7526; fax: (508) 793-8842; e-mail: idrisi@clarku.bitnet or idrisi@vax.clarku.edu].

Other IDRISI Resources

A free demonstration product is available from *<http://www.clarklabs.org/13-Dnlds/13-Dnlds.htm>*.

The demonstration provides an overview of the "look and feel" and functionality of the full IDRISI system for

Windows. It also includes an on-line five-exercise tutorial, complete with sample data sets; but there are limitations on the sizes of images that can be processed. The tutorial can be accessed through the Help menu.

A tutorial in HTML format is also available on-line via the Internet. The full tutorial can be downloaded from *<http://www.sbg.ac.at/geo/idrisi/wwwtutor/tuthome.htm>*.

An Internet e-mail forum is set up for users at mailserv@toe.towson.edu. To subscribe, send a one-line e-mail containing the message: subscribe idrisi-l.

MapInfo

MapInfo is a GIS package that provides some data management capabilities, but whose real strength is an ability to combine data tables with graphic representations of spatial data (integrating database information with maps). Within the MapInfo system, it is possible to display, zoom, edit, and print maps. Foxpro, dBase, and other DBMS files can be used directly or imported in ASCII format. MapInfo also comes with a dBase compatible database manager that includes all the ZIP codes in the U.S., along with state outlines, a world map, and a sample street map.

Maps can be displayed of any area at any scale with an absolute accuracy of 4 in. It is also possible to display multiple layers of maps, boundaries, and points, and to have different levels of detail displayed at different scales. The user controls annotations, along with the colors, line types, and symbols to be drawn on the maps. A map can be positioned by entering an address, a latitude/longitude, or a boundary, and any of these locations can be pinpointed.

Audience

MapInfo is available in multiple languages and is sold worldwide through direct sales, resellers, and distributors. The system is relatively easy to learn and can be effectively used by beginners who are familiar with Windows. Experienced GIS professionals, as well, will find a good selection of analysis tools and an excellent set of coordinate conversion capabilities.

Capabilities and Functions

Raster Image Support. MapInfo allows an array of vector maps to be displayed in rich detail with a new raster underlay feature. Users can bring satellite images, aerial photographs, and scanned paper maps into MapInfo to incorporate remote sensing information.

Thematic Map Enhancements. Color, shading, symbols, graphs, legends, and other graphic elements can be used to visually represent data on thematic maps. These features include:

- Logical grouping of data for precise analysis and accurate presentation
- Insertion of pie, half-pie, and bar charts that show and compare multiple variables simultaneously
- Enhanced color spreads, making it easier to select a series of colors for a map display

Data Connectivity. The SQL DataLink connectivity module provides client/server remote data access to standard corporate databases for MapInfo applications. The SQL DataLink is extended for purposes of accessing information maintained in the UNIX and Macintosh environments, in addition to Windows, allowing connections to Oracle and Sybase databases. For UNIX, this connection is direct.

MapBasic Development Environment. MapBasic is a programming language for developing custom desktop mapping applications for GUI systems. Specific features of the system include the following:

- Information retrieval related to a selected point on the screen; for example, it is possible to show a map of the United States, click the label button on the toolbox, and have the state appear. An information button on the toolbox reveals information contained in the database related to the spot clicked
- Data query to extract information and show results on a graphic display as well as in a table
- Information query using a simple "select" by defining the field to search, operators (mathematical and logical), and functions
- A full suite of map coordinate systems and projections (Mercator, Albers, Lambert, State Plane, and many more)
- "Buffering" of any object on a map, including points, polygons, lines, and polylines; a buffer defines a zone of a specified distance around objects
- Polygon analysis, including union and intersection
- Digitization of single or multiple control points
- Geocoding for purposes of attaching spatial geographic coordinates to a table of information
- Multiple map output
- A wide selection of fill patterns, line types, and symbols
- A custom symbol generator
- Map layer combination. Maps can be added, removed, and reordered. Settings for layers include turning the layer on or off, and changing line color/style, fill color/pattern, symbol type/color, or text font/color. The "settings" capability allows layers to be edited and/or selected, establishes the zoom layering range (for display at particular zoom scales), and defines columns for entering labels on a map.

Hardware Requirements

MapInfo runs on PCs, as well as on Macintoshes, and on Sun and Hewlett Packard workstations. Minimum requirements for the PC Windows version include 486 or higher processor, Windows 3.1/95/NT, and a super VGA monitor/graphics card or better.

The cost of the software for Windows and for UNIX is $1,295 and $2,495, respectively. MapInfo for Macintosh costs $1,295. Digital data and other special modules are priced individually.

User's Manual and Example Data Sets

Documentation and sample data sets come with the program as part of the commercial product.

Data Formats Accepted

MapInfo reads the following formats: .GIF, .JPEG, .TIFF, .PCS, .BMP, .TGA, .BIL. Other file types accepted include DXF, Arc/Info, Atlas GIS, and Intergraph MGE. The software can read data directly from the Access, DB/2, Informix 5, Ingres, Oracle7, Gupta SQLBase, Microsoft SQL Server, and Sybase 10.x database management software.

Example Graphics

An example of the kind of graphics that can be produced with MapInfo can be found at *<http://www.mapinfo.com/solutions_and_services/mapinfo_in_action/telecom/step1.html.>*.

Availability

A free demonstration of MapInfo is available at *<http://testdrive.mapinfo.com/tdc/miprodemo.nsf/signup?openform>*. Free information, documentation, demos, and utilities are found at *<http://www.mapinfo.com/solutions_and_services/free/index.html>*.

For a list of worldwide distributors, see the MapInfo Web site at *<http://www.mapinfo.com/partners_and_alliances/reseller_locator/index.html>*. For other information, contact Mapping Information Systems Corporation, One Global View, Troy, NY 12180-9399 [telephone: (518) 285-6000; fax: (518) 285-6066; (800) FAST-MAP].

Atlas GIS for Windows

This Windows-based program provides mapping and basic spatial analysis functions. Statistical and geographic data are turned into meaningful information for decision-making and presentation purposes. Convenience features include point and click, a button bar, and a floating toolbox. Titles, legends, and scale bars update automatically as views on a map are changed.

Audience

This software works well for producing professional-quality maps of surface features. For the petroleum professional, Atlas GIS is appropriate for creating facilities maps and maps of surface geology. Computer and GIS beginners will be able to run this software after completing the tutorial. The script-writing capability also makes the software feasible for more advanced users who have more complex requirements.

Capabilities and Functions

A built-in database manager directly accepts Lotus-Excel- or dBase-compatible files directly and links them to a map. Client-server SQL databases can be queried using the built-in Q+E query builder, or SQL queries can be performed on Atlas or other dBase tables.

A page layout system facilitates the display of a complete "document;" that is, page layout templates help in creating professional-looking maps. The program automatically adds latitude and longitude coordinates, census tracts, and block or block group codes to geocode information from anywhere in the U.S. Geographic and database analysis can be done using "what if" and "show me where" scenarios. Map features such as points, lines, or areas can be combined, or "buffers" can be created around map features.

Multiple database tables can be opened within any mapping project. Having separate data tables for each layer makes it easier to view, query, and organize data. Every data table is a fully integrated spreadsheet, linked directly to a map. Atlas GIS analysis tools are included, such as buffering of points, lines, and areas, combining regions and their attribute data, and calculating area-weighted data aggregation.

Hardware Requirements

Atlas GIS runs on 80486 PCs and faster computers with a minimum 8MB RAM; however, 16 MB or more is recommended for better performance. Standard monitors are acceptable, although a fast VLB video card will increase performance. Atlas GIS will run with all versions of Windows and on DOS 5.0. The cost of the system is $495.

User's Manual and Example Data Sets

Documentation and sample data are provided with the purchase of the program as part of the commercial product.

A separate tutorial is available for $50. The tutorial leads the user through actual examples. All tutorial data are also included with each copy of the software.

Example Graphics

An example of the kind of graphics that can be produced with Atlas GIS can be found at *<http://www.esri.com/software/atlas/index.html.>*.

Availability

The Atlas GIS software, originally developed by Strategic Mapping, Inc., was acquired and is marketed by ESRI. Atlas users can upgrade to the current version of ArcView, an ESRI product, for $249. Information is available on-line at *<http://www.esri.com/software/atlas/index.html>*.

The software and tutorials can be ordered from the Web site (U.S. customers only). For more information, contact ESRI, 380 New York Street, Redlands, CA 92373-8100 [telephone: (800) 447-9778].

FREE VERSIONS OF COMMERCIAL SOFTWARE

Several commercial vendors offer a free version (usually an early one) of their software. The software is provided as executable code, and source code is not available. Two such free programs are listed here.

ArcView 1.0

ArcView is an ESRI desktop mapping and GIS tool available for free. The free version is an early incarnation of the software (Version 1). As previously noted, later versions have more capabilities but have a price attached. The program is designed for creative display of spatial information and can be used to show any combination of data.

ArcView is available for PC-compatibles, Macintosh computers, Digital Equipment VAX computers, and a variety of UNIX-based workstations. The program works directly with Arc/Info, ArcCAD, and PC Arc/Info databases (all of which are ESRI products).

To download Version 1.0 go to *<ftp://greenwood.cr.usgs.gov/pub/arcview1/>*. To download other versions of ArcView and other ESRI software, go to *<http://www.esri.com/software/arcview/avsoftware.html>*.

User support and help can be obtained through the mailing lists ArcView-L and ESRI-L. To subscribe to any of the ESRI mailing lists, go to the Web site and fill out the form *<http://www.esri.com/usersupport/support/selfhelp/lists-sub.html>*.

For more information, contact Environmental Systems Research Institute, Inc., 380 New York Street, Redlands, CA 92373 [telephone: (909) 793-2853, ext. 1375; fax: (909) 793-2853; e-mail: support@esri.com].

TNTLite/TNTMips

MicroImages, Inc. provides TNTLite as a free version of TNTMips, its professional software for geospatial data analysis. The free TNTLite product has all the features of the professional version, except the size of project file objects is limited, and data can be shared only with other copies of TNTLite (export processes are disabled.)

Features include raster, vector, CAD, and TIN display, editing, and interconversion; pin mapping (symbolic map overlays from database query); symbol, line, and fill pattern design tools; automated map legend generation; database query for selection and styling of points, lines, and polygons; interactive tabular selection and viewing for vector, CAD, and TIN objects; 3-D display; video capture; mosaic preparation; georeferencing and warping to many map projections; elevation mapping and orthoimage creation; classification and interpretation; statistical measurements; smart line-following; buffer zone creation; polygon fitting; watershed and viewshed analysis; screen and page layout, including scale bars, map grids, and legends; outline annotation fonts and tools; and customization with the Spatial Manipulation Language.

The program is freely available for downloading at *<http://www.microimages.com/tntlite/>*. It can also be ordered on CD-ROM for $35 to cover the cost of shipping and handling. The CD also comes with the current set of *Getting Started* tutorial booklets for learning geospatial analysis. For more specific information, contact TNTMips, MicroImages, Inc., 201 N. 8th Street Suite 15, Lincoln, NE 68508-1347 [telephone: (402) 477-9562; fax: (402) 477-9559; e-mail: info@microimages.com].

OTHER GIS AND MAPPING PROGRAMS AND RESOURCES

The following GIS and mapping programs are listed because they are inexpensive and have a variety of uses and capabilities. The list is by no means comprehensive, but it is intended to provide an entry point for finding just the right software for just the right application. Searching the Internet turns up many more offerings than are listed here, and even these programs appear on many lists. To find more information about each offering, use the Internet addresses listed or just perform a search on the program name.

ILWIS: Integrated Land and Water Information System

ILWIS was produced by the International Institute for Aerospace Survey and Earth Sciences (ITC). This PC-based package combines the benefits of image processing capabilities, raster-based spatial modeling, and tabular databases. A primary feature is the ability to interactively overlay multiple raster maps and remotely sensed images and integrate them with attribute data analysis. Version 2.2 of the ILWIS program runs under Windows. Earlier versions ran under DOS.

ILWIS offers data capture, data import and export, raster and vector data management tools (querying, data structure conversion, table management), visualization, analysis (spatial modeling), and output/presentation options. Multiple maps and annotation layers can be displayed in one map window. Three-dimensional views include interactive editing of viewing parameters. Extensive image processing facilities is another feature. Comprehensive calculations can be performed on both maps and tables. Several types of geostatistical analysis are provided.

The cost of ILWIS is higher than for some of the other programs listed in this group, but the software is still affordable. The price for commercial use is $2,650; the educational price is $1,590, and for students it is $344.

For more information, contact PCI-ILWIS, Nederland B.V. Hengelosestraat 99, P.O. Box 6, 7500 AA Enschede, The Netherlands [telephone: +31 53 4874337; fax: +31 53 4874484; e-mail: ilwis@itc.nl; Web: *<http://www.itc.nl/ilwis/ilwis.html>*].

OSUMAP

A good introductory package is OSUMAP, which was developed at Ohio State University (OSU) and is maintained by the Department of Geography. This is a raster-based package for PCs running DOS. It is designed as an instructional program but is also useful for small GIS applications. It was released in 1994.

OSUMAP supports databases of up to 50,000 grid cells and operates on any DOS-based PC with a graphics display and hard disk. The graphics driver distributed with the program is royalty-free and device independent.

A free copy of the program plus a variety of data sets and other documentation can be downloaded from the OSU Department of Geography Internet home page.

Data sets are specifically available at <*ftp://bastet.sbs.ohio-state.edu/osumap/data_sets/*>.

Copies of the software may be distributed to students and others only on a no-charge or on a cost-recovery basis. For more information, contact Duane F. Marble, Department of Geography, Ohio State University Columbus, Ohio 43210 [telephone: (614) 292-2250; fax: (614) 292-6213; e-mail: marble.1@osu.edu]. The home page for OSUMAP is <*http://thoth.sbs.ohio-state.edu/osumap*>.

GMT—Generic Mapping Tools Software

GMT is a free, public-domain collection of approximately 60 UNIX tools that allows users to manipulate two- and three-dimensional data sets (operations include filtering, trend fitting, gridding, projecting, etc.). The tools produce Encapsulated PostScript (EPS) file illustrations. Capabilities include the ability to produce simple *x–y* plots, contour maps, artificially illuminated surfaces, and 3-D perspective views. Graphics appear in black and white, gray tone, or 24-bit color and can include hachure patterns. Twenty common map projections are supported, as well as linear, log, and power scaling. Support data such as coastlines, rivers, and political boundaries are included.

GMT is used by 6000 scientists and engineers worldwide and most users are earth scientists. The software has been installed on a variety of UNIX systems, including (but not limited to) computers from Cray, Sun, IBM, DEC, HP, SGI, Data General, Apple, and Next. It will run on PCs and Macintoshes running LINUX or Windows/NT.

The software requires the netCDF library from the National Center for Atmospheric Research (NCAR), which is available at <*ftp://unidata.ucar.edu/pub/netcdf*> or <*http://www.unidata.ucar.edu/packages/netcdf*>. The GMT source code can be found at several locations including <*ftp://pahoehoe.soest.hawaii.edu*>.

For more information, contact Paul Wessel, School of Ocean and Earth Science and Technology, University of Hawaii, 1680 East-West Road, POST 802, Honolulu, HI, USA 96822 [e-mail: wessel@soest.hawaii.edu].

The Professional Map Analysis Package (pMAP) and the Academic Map Analysis Package (aMAP)

pMAP is a low-cost PC software system for analyzing spatial data. Spatial data is obtained from a variety of sources, such as map sheets, surveys, digital maps, and other GIS systems. The program consists of a database of base maps and derived information layers, internally referenced to a common geographic location. A command language is used to create, store, analyze, and display maps. More than 50 processing operations are provided, including:

- Optimal path determination
- Visual exposure
- Edge characterization
- Spatial interpolation
- Slope/aspect determination
- Effective proximity
- Coincidence statistics
- Roving window summaries
- Map overlay
- Searches
- Thematic mapping
- Contouring
- Shape/pattern analysis
- Contiguity analysis

The program also provides macro file capabilities, pop-up windows, and contextual help. Tutorials are accessed through a menu interface.

pMAP uses a raster data structure, and input can be supplied in the form of gridded data, or digitized points, lines, or polygons. Output includes summary tables, character-based or raster pattern maps, monochrome or color displays, and standard files for enhanced color and line graphics.

The Academic Map Analysis Package (aMAP) is the educational version of pMAP. pMAP allows maps containing up to 1,000,000-grid cells (1000 × 1000) to be produced, but aMAP is constrained to 6400 grid cells (80 × 80). An extensive set of "digital slide shows" and a video series support classroom instruction.

Hardware system requirements include DOS 5.0 or higher (runs as a DOS application under Windows and NT), VGA graphics, and 16 MB RAM. Both software packages are being enhanced and upgraded for Windows/NT, with revisions expected to be available soon.

The price for a single-user license of pMAP is $495. A multi-user license costs $895. The price of an aMAP multi-user license is $199. Prices do not include shipping. A 10% educational discount on pMAP is available for colleges and universities.

The software is discussed in detail at <*http://blaze.innovativegis.com/basis/basis/pmap.htm*>.

For more information, contact Spatial Information Systems, 2000 South College Avenue, Suite 300, Fort Collins, CO 80525 [telephone: (970) 490-2155; fax: (970) 490-2300; e-mail: sis@innovativegis.com; Home page: <*http://blaze.innovativegis.com/basis/basis/sis.html*>].

MicroDEM

MicroDEM is a free program that can be found on the Bureau of Land Management's ftp Internet site. It analyzes and displays DEMs and other gridded digital data sets. It can manipulate scanned satellite data or air photos and overlays DEM contours on the images. The program is an outgrowth of terrain analysis work done at the U.S. Naval Academy.

General terrain analysis functions include colored tinted maps, contour maps, 3-D views, and histograms of elevation distribution. The program can perform some, but not extensive, coordinate transformations. GIS functions allow overlays of information and comparisons of models. The program can be downloaded from <*ftp://ftp.blm.gov/pub/gis/microdem*>.

INTERNET DISCUSSION GROUPS, MAILING LISTS, AND OTHER GIS INFORMATION RESOURCES

Thousands of discussion groups and mailing lists are available on the Internet. The following apply specifically to using GIS. Discussion groups are an excellent way to find out how other people are using GIS software, the problems they have, and how they answer questions pertaining to specific situations.

GIS Mailing Lists

Following are several GIS-specific mailing lists.

- ESRI-L—This group is dedicated to discussions about ESRI products (Arc/Info, ArcView, etc.). The e-mail contact is esri-l-human@esri.com. An archive of the ESRI-L mailing list is available at *<gopher://gopher.gis.umn.edu/11/rsgis/lists/esri-l>*.
- AGIS-L—This mailing list is for discussion of Atlas-GIS and desktop mapping issues. Join the AGIS mailing list by sending e-mail to majordomo@ciesin.org. Type in the following message: subscribe agis. Archived messages from AGIS-L are accessible at *<http://listhost.ciesin.org>*.
- MAPINFO-L—To subscribe to this list, send e-mail to majordomo@csn.net. For the message, type: subscribe MAPINFO-L. Frequently asked questions about MapInfo are addressed at *<http://www.inmap.co.nz/mi_faq/contents.html>*.

Other Resources

A compilation of GIS mailing lists can be found at *<http://www.csn.net/~bthoen/ores/gis/mail.html>*.

The comp.infosystems.gis newsgroup covers all GIS topics and all aspects of geographic information systems. The Web site *<http://www.dejanews.com>* is an easy way to access the messages in newsgroups and to post your own. Finally, a general list of frequently asked questions about GIS is available at *<http://www.census.gov/ftp/pub/geo/www/faq-index.html>*.

GeoDirectory

An excellent source for information about GIS software for all price levels is GeoDirectory, published by Adams Business Media. GeoDirectory is available in hard copy, or it can also be found on the Internet at *<http://www.GeoPlace.com>*.

A subscription to one of the GIS magazines, such as *GIS World*, *Mapping Awareness*, or several others, includes a free copy of GeoDirectory as a supplement. The information listed encompasses services, data, hardware, software, and resellers. For additional information, contact Adams Business Media Company, 2101 South Arlington Heights Road, Arlington Heights, IL 60005-4142 [telephone: (847) 427-9512; fax: (847) 427-2006].

GIS COLLECTOR SITES

Following is a list of a few "collector sites" for GIS information, software, data, tutorials, and other resources. These sites provide an excellent starting place for finding out about the latest in GIS developments.

- *<http://www.geo.ed.ac.uk/home/GISwww.html>*

Bruce M. Gittings and Anup Pradhan at the Department of Geography, University of Edinburgh, maintain this list of GIS resources, in collaboration with the Association for Geographic Information.

- *<http://www.sni.net/gis/>*

Bill Thoen's extensive list of GIS resources.

- *<http://wwwinweh.uwaterloo.ca/unuinweh/gis1.htm>*

This is a list of resources developed for the United Nations by the University of Waterloo in Canada. Hundreds of references are included.

- *<http://www.geocities.com/SiliconValley/2682/gis.html>*

This site identifies other Internet sites where GIS data, software, library and publications information, GIS events on-line, mailing lists, jobs, and maps can be located.

- *<http://www.ciesin.org/>*

The Center for International Earth Science Information Network at Columbia University (CIESIN) specializes in global and regional network development, science data management, decision support, and training, education and technical consultation services. CIESIN is the World Data Center A (WDC-A) for Human Interactions in the Environment. This is a good place to start if you are looking for data to use with a GIS software package.

- *<ftp://ftp.blm.gov/pub/gis>*

This area on the U.S. Bureau of Land Management's ftp site is listed because of the extensive resources contained, including software, data sets, images, and other information. Some of the software contained on the site is provided as source code, whereas some is only available as executable code. Everything is free, but programming skills may be required to use some of the programs. Many of the programs listed are command-line driven.

ON-LINE GIS TUTORIALS

Individuals who have only a vague idea of what GIS is about should consider spending time with one of the online GIS tutorials. This approach will provide an introduction and a good start at understanding the technology.

Two kinds of GIS tutorials are available on the Internet. Many of the public domain and commercial software providers provide tutorials about how to use their software, but other tutorials about using GIS in general are also available. As a starting point, the location for the U.S. Geological Survey GIS tutorial is listed below, along with a brief description of its contents.

To find other tutorials on-line, use any of the search programs and search for the words "GIS" and "tutorial." A Web search on AltaVista resulted in over 13,000 hits. Many of the hits were from collector sites that provide tutorials as well as other GIS learning resources.

U.S. GEOLOGICAL SURVEY

The Internet site of the U.S. Geological Survey, *<http://www.usgs.gov/research/gis/title.html>*, provides specific links to information about how a GIS works, what is special about a GIS, and GIS applications.

- *How a GIS works* includes instructions on relating information from different sources, data capture, data integration, projection and registration, data structures, and data modeling
- *What is special about a GIS* includes discussion about how maps and other data are stored or filed as layers of information, and how to perform complex analyses. Some specific topics are information retrieval, topological modeling, networks, overlay, and data output.
- *Applications of GIS* includes GIS through history; using GIS (mapmaking, site selection, emergency response planning, simulating environmental effects, graphic display techniques, etc.), and the future of GIS.

In addition to the information specific to GIS, the U.S. Geological Survey Web site contains data sources, actual data available for downloading, and software for related tasks such as coordinate conversion. For more information, contact U.S. Geological Survey, 804 National Center, Reston, VA 20192.

FINAL WORDS

The software described here represents only a small selection of programs available for GIS and related utilities. Besides software, there is a plethora of information about data sources, consultants who can help with generating GIS databases, utilities for coordinate conversion, graphical visualization software, Internet map publishing, and much, much more.

Internet searching turns up almost too many options for places to look for software, data, and other GIS information. Familiarity with GIS technology and the resources available can be achieved with only a small investment of time in using some of the information listed in this article. The collector sites are a good place to start because someone has already performed an extensive search and categorized the results. Individuals who peruse the collector sites will find many places to return to, and will build a specialized bookmark file for future reference.

Finally, readers are referred to Clarke (1999) and Guptill (1988) which present comparative information about software products similar to the material discussed here. Other good textual sources can be found in most good public and private libraries.

DISCLAIMER

All factual material presented here in the form of telephone numbers, e-mail addresses, URL addresses, capabilities, and prices was current at the time of writing. Because computer technology changes rapidly, the reader is encouraged to pursue the latest information.

REFERENCES CITED

Clarke, K. C., 1999, Getting Started with Geographic Information Systems, 2nd ed. Upper Saddle River, N.J.: Prentice Hall, p. 220–226.

Guptill, S. C., 1988, A process for evaluating geographic information systems (abs.). Abstracts, GIS Symposium: Integrating Technology and Geoscience Applications (Hugo F. Thomas, ed.). Washington, D.C.: National Academy of Sciences, p. 88.

Yarka, P. J., and T. C. Coburn, 2000, Bibliographic resources for GIS in petroleum exploration and development, *in* T. C. Coburn and J. M. Yarus, eds., Geographic information systems in petroleum exploration and development: AAPG Computer Applications in Geology, No. 4, p. 281–292.

Chapter 21

Bibliographic Resources for GIS in Petroleum Exploration and Development

Paul J. Yarka[1]
The Convergent Group
Englewood, Colorado, U.S.A.

Timothy C. Coburn
Department of Mathematics and Computer Science, Abilene Christian University
Abilene, Texas, U.S.A.
and
Mathematical Geology Section, Kansas Geological Survey
Lawrence, Kansas, U.S.A.

ABSTRACT

The literature on geographic information systems (GIS) is a rapidly expanding body of knowledge that now represents developments and applications from almost every scientific arena. During the past decade, GIS technology has attracted extensive interest from within the petroleum industry and its related geoscience and engineering disciplines. With over 250 citations, this bibliography details much of what has been reported and written during this time period concerning the use of GIS in petroleum exploration and development. Citations are organized into five general categories: defining concepts and critical discussion, available data, issues in data management and integration, technological developments, and case studies and applications.

INTRODUCTION

Over the past 10 years, there has been a virtual explosion of interest and applications in the use of geographic information systems (GIS) in geology and other geoscientific disciplines. One of the segments of the GIS industry that is experiencing the most rapid development, and one in which some of the most dynamic and exciting applications of GIS technology are arising, is petroleum GIS. Within the petroleum industry, the power and potential of GIS is nowhere more evident than in upstream exploration and development.

The citations contained in the following bibliography provide extensive evidence of the growing influence of GIS in the collection, integration, management, analysis, and interpretation of geoscience data, and its particular application in the decision-making aspects of petroleum exploration and development. Although the list is not exhaustive, these resources provide the reader with an excellent cross-section of information that has been disseminated over approximately the last decade through active research, scholarly literature, project reviews, discussions and presentations, and other professional activities.

[1]Formerly, Petroleum Technology Center, Marathon Oil Company, Littleton, Colorado, U.S.A.

To assist readers in identifying the most appropriate bibliographic resources for their use, citations are organized into five broad categories: defining concepts and critical discussion, data availability, data integration and data management issues, technological developments, and case studies and applications. Every attempt has been made to categorize the citations into the best group, but the boundaries, by their very nature, are somewhat fuzzy.

Although the general GIS literature is quite vast, only those citations judged to be most pertinent to petroleum exploration and development have been included; nonetheless, not all citations are specifically petroleum related. Each one, however, involves concepts and ideas that are directly applicable to the search for oil and gas or to the management of oil and gas resources. In addition, within some of the more narrowly defined specialties of petroleum exploration and development, such as remote sensing, numerous additional applications of GIS technology are not specifically included here.

A conscious effort has been made to exclude many references that generally pertain to defining GIS, the basics of GIS, how to get started with GIS, and hardware/software system needs. Plentiful sources of information on these topics exist in the popular literature, and in all good private and public libraries. Also, not all references cited elsewhere in this text are included here.

Finally, excellent glossaries of GIS terminology are also available elsewhere. Some of these glossaries are more generally oriented toward geoscience than others. Four excellent examples are Clarke (1999), Bonham-Carter (1994), Prescott (1994), and McDonell and Kemp (1995), cited here. Readers are also encouraged to check the Internet, where additional information can be found.

DEFINING CONCEPTS AND CRITICAL DISCUSSION

Alderman, Willis, Charles Iglehart, and Marvin Hollenbeck, 1994, The petroleum and computer industries share an amazing history of growth. Earth Observation Magazine, vol. 3, no. 5, p. 42–44.

Balcom, M., and D. Brockelbank, 1995, Decision support using GIS in an oil and gas implementation, *in* Proceedings, 9th Annual Symposium on Geographic Information Systems in Natural Resources Management, vol. 1. (P. O'Reilly, ed.). Ft. Collins, CO: GIS World, p. 235–239.

Berry, Joseph K., 1993, Beyond Mapping: Concepts, Algorithms, and Issues in GIS. Ft. Collins, CO: GIS World, Inc., 246 p.

Bilden, Dean M., 1999, GIS pumps profits into frac jobs. Oil and Gas World, vol. 91, no. 1, p. 33–34.

Chirico, Peter, 1997, GIS: the bedrock of geologic mapping. Geo Info Systems, vol. 7, no. 10, p. 26–31.

Clarke, K. C., 1986, Recent trends in geographic information system research. Geo-Processing, vol. 3, p. 1–15.

Clarke, Keith C., 1999, Getting Started with Geographic Information Systems, 2nd ed. Upper Saddle River, NJ: Prentice Hall, 338 p.

Coshey, Robert J., 1990, The explorationist's GIS (abs.). AAPG Bulletin, vol. 74, p. 1320.

Cowen, David J., 1987, GIS vs. CAD vs. DBMS: What are the differences, *in* Proceedings of the 7th International Symposium on Computer-Assisted Cartography (Auto/Carto 7). Bethesda, MD: American Society of Photogrammetry and Remote Sensing (ASPRS) and American Congress on Surveying and Mapping (ACSM), p. 46–56.

Dangermond, Jack, 1986, CAD vs. GIS: Automatic mapping includes two technologies, and it's important to understand the difference. Computer Graphics World, vol. 9, p. 73–74.

Fisher, P. F., and R. E. Lindenberg, 1989, On distinctions among cartography, remote sensing, and geographic information systems. Photogrammetric Engineering and Remote Sensing, vol. 55, p. 1431–1434.

Fogotson, James M., Jr., and John L. Stout, 1969, Future of well-data information systems, *in* Computer Applications in the Earth Sciences (D. F. Merriam, ed.). Lawrence, KS: Kansas Geological Survey, p. 61–72.

Forster, John R., and John C. Horne, 1996, GIS below the earth's surface. Earth Observation Magazine, vol. 5, no. 5, p. 27–30.

Fried, Charles, 1989, Operation database: petroleum geographic information systems project summary (abs.). Computer Contributions of the Computer Oriented Geological Society (COGS), vol. 5, no. 3, p. 111.

Fried, Charles C., and Jay E. Leonard, 1990, Operation database petroleum GIS results: evaluation and application of geographic information systems to exploration and production (abs.). AAPG Bulletin, vol. 74, p. 658.

Fried, Charles C., and Jay E. Leonard, 1990, Petroleum 3-D models come in many flavors. Geobyte, vol. 5, no. 1, p. 27–30.

Giles, J. R. A., and K. A. Bain, 1995, The nature of data on a geological map, *in* Geological Data Management (J. R. A. Giles, ed.). London: Geological Society of London, Special Publication 97, p. 33–40.

GIS in the oil patch, 1990. GIS World, vol. 3, no. 1, p. 56–58.

Gray, C., 1993, Geographic information systems provide benefits in data access and analysis in exploitation (abs.). Extended Abstracts of Papers, 5th Conference and Technical Exhibition, European Association of Petroleum Geoscientists and Engineers, Stavanger, Norway, p. E052.

Guptill, S. D. (ed.), 1988, A Process for Evaluating Geographic Information Systems. Reston, VA: U.S. Geological Survey, Federal Interagency Coordinating Committee on Digital Cartography, Technology Exchange Working Group, Open-File Report 88–105, 57 p.

Harlow, C., 1989, The future of geographic information systems—18 important trends, *in* Proceedings of the 9th International Symposium on Computer-Assisted Cartography (Auto/Carto 9). Falls Church, VA: American Congress on Surveying and Mapping and

American Society of Photogrammetry and Remote Sensing, p. 16–25.

Hastings, David, and Stanton H. Moll, 1986, On geographic information systems; the new explorers, a one-act play. Geotimes, vol. 31, no. 7, p. 8–10.

Hodson, Mark W., 1990, GIS and petroleum geology: a typical Hollywood production. Bulletin of the Houston Geological Society, vol. 32, no. 8, p. 48–49.

Hough, Harold, 1992, Field geologists take high road to prospecting. Earth Observation Magazine, vol. 1, no. 4, p. 26–28.

Huang, X., and B. Chen, 1989, A discussion on some trends of geographic information systems. Acta Geographica Sinica, vol. 44, p. 230–236.

Janzen, A., and K. McKay, 1995, Issues in corporate implementation of GIS at PanCanadian, *in* Proceedings, 9th Annual Symposium on Geographic Information Systems in Natural Resources Management, vol. 1 (P. O'Reilly, ed.). Ft. Collins, CO: GIS World, p. 240–243.

Juracek, Kyle E., 1996, U.S. Geological Survey's role in state GIS. Geo Info Systems, vol. 6, no. 7, p. 36–40.

Katz, S. S., 1985, Introduction to geographic information systems. COGSletter (newsletter of the Computer-Oriented Geological Society, Denver, CO), September issue.

Katz, Solomon S., 1989, Geographic information systems; more than just another computer mapping program, *in* 21st Application of Computers and Operations Research in the Mineral Industry (Alfred Weiss, ed.). Littleton, CO: Society of Mining Engineers of the American Institute of Mining, Metallurgical, and Petroleum Engineers, p. 597–602.

Lang, Laura, 1988, Exploring related technologies, Part II: Are GIS and CAD evolving toward the same answer? Geobyte, vol. 3, no. 4, p. 12–15.

Lang, Laura, 1989, GIS goes 3-D: As the GIS sheds its 2-D mold, exciting 3-D applications begin to emerge. Computer Graphics World, vol. 12, no. 3, p. 38–46.

Leonard, J. E., 1986, Geographic information systems (GIS). Geobyte, vol. 1, no. 6, p. 64.

Leonard, J. E., 1988, Integration, Part I: PC products—Windows & GIS revisited. Geobyte, vol. 3, no. 2, p. 87–88.

Leonard, Jay E., 1989, Operation database update; workstations, integrated applications, GIS, and more. Geobyte, vol. 4, no. 3, p. 13–14.

Leonard, Jay E., and Charles C. Fried, 1989, Petroleum problem sets challenge geographic information systems. Geobyte, vol. 4, no. 4, p. 46–51.

Leonard, Jay E., and Charles C. Fried, 1989, Conference preview & participant profiles: 18 GIS vendors match product capabilities against oil industry needs. Geobyte, vol. 4, no. 5, p. 26–59.

Leonard, Jay E., Charles C. Fried, and Ken Milam, 1989, GIS results: 15 vendors rise to the challenge. Geobyte, vol. 4, no. 6, p. 26–48.

Longley, Paul A., Michael F. Goodchild, David J. Maguire, and David W. Rhind, 1999, Geographical Information Systems, vol. 1 (Principles and Technical Issues). New York: John Wiley & Sons, 580 p.

Longley, Paul A., Michael F. Goodchild, David J. Maguire, and David W. Rhind, 1999, Geographical Information Systems, vol. 2 (Management Issues and Applications). New York: John Wiley & Sons, 1101 p.

McDonnell, Rachael, and Karen Kemp, 1995, International GIS Dictionary. Cambridge: Geoinformation International, 111 p.

McGee, David, 1994, GIS offers solutions to challenges of integrating petroleum data. Earth Observation Magazine, vol. 3, no. 8, p. 45–47.

Miller, Betty M., 1994, The role of geographic information systems and expert systems technology in natural resource management, *in* Proceedings of the Canadian Conference on GIS (Y. C. Lee, ed.). Ottawa, ON: Canadian Institute of Geomatics, p. 723–732.

Morgan, Nina, 1991, Exploration: Plotting out the future. Petroleum Economist, vol. 58, no. 3, p. 31–32.

New geographic information system speeds exploration mapping, 1995. World Oil, vol. 216, no. 2, p. 96–98.

Prescott, George W. (ed.), 1994, A Practitioner's Guide to GIS Terminology: A Glossary of Geographic Information System Terms. Federal Way, WA: Data West Research Agency, 408 p.

Rathfeeder, Jeffrey, 1989, These maps can find oil—or sell burgers. Business Week, no. 3095 (March 13), p. 134.

Reinemund, John A., John S. Pallister, and Paul P. Hearn, 1997, Geologic mapping in the global marketplace (abs.), *in* Abstracts with Programs. Boulder, CO: Geological Society of America, vol. 29, no. 6, p. 423.

Ruta, T., 1995, An analysis of GIS in oil and gas: Problems and prospects, *in* Proceedings, 9th Annual Symposium on Geographic Information Systems in Natural Resources Management, vol. 1 (P. O'Reilly, ed.). Ft. Collins, CO: GIS World, p. 219–221.

Ruta, T., 1995, GIS in the petroleum industry: Lessons for vendors, *in* Proceedings, 9th Annual Symposium on Geographic Information Systems in Natural Resources Management, vol. 1 (P. O'Reilly, ed.). Ft. Collins, CO: GIS World, p. 251–253.

Simmons, Reid G., 1982, Spatial and temporal reasoning in geologic map interpretation (abs.), *in* Proceedings, National Conference on Artificial Intelligence. Pittsburg, PA: Carnegie-Mellon University.

Smith, Ted, 1998, Geographic information systems. Geotimes, vol. 43 (June), p. 27.

Turner, A. Keith, 1997, The best of times or the worst of times. GIS World, vol. 10, no. 1, p. 64–65.

Turner, A. Keith, 1997, Where are we and where are we going? GIS World, vol. 10, no. 11, p. 40.

Turner, A. Keith, 1990, Three-dimensional GIS: Possibilities attract geoscientists in many industries, worldwide. Geobyte, vol. 5, no. 1, p. 31–32.

Wilson, H., and E. Bonino, 1993, The role of geographic information systems in oil exploration and production (abs.), *in* Extended Abstracts of Papers, 5th Conference and Technical Exhibition. Stavanger, Norway: European Association of Petroleum Geoscientists and Engineers, p. E051.

Yarka, Paul J., 1993, GIS: An evolving exploration and exploitation technology, World Oil, vol. 214, no. 9, p. 45–47.

AVAILABLE DATA

Ferderer, Dave, and Craig Wandrey, 1999, Digital data available from the U.S. Geological Survey (abs.). Redlands, CA: Environmental Systems Research Institute, Inc., Annual Meeting, Petroleum Users Group, Houston, TX.

Larsgaard, Mary, 1996, Providing access to earth-sciences spatial data: Metadata and the Alexandria digital library (abs.), *in* Abstracts with Programs. Boulder, CO: Geological Society of America, Annual Meeting, vol. 28, no. 7, p. A-66.

Laxton, John, 1998, Data index: An on-line index to European geoscience data. Earthwise, vol. 11, p. 9.

Scholz, Donna K., Dale A. Johnson, Kevin Lowell, and Karla K. Sprenger, 1998, Internet access to U.S. Geological Survey geospatial data and images, *in* Abstracts and Program, U.S. Geological Survey National Computer Technology Meeting (C. A. Babcock and S. B. Mathey, eds.). Reston, VA: U.S. Geological Survey, Open-File Report, 98-0280, p. 13.

Turner, A. Keith, 1998, U. S. Geological Survey releases free oil and gas datasets. GIS World, vol. 11, no. 9, p. 40–41.

ISSUES IN DATA MANAGEMENT AND INTEGRATION

Albert, T. M., 1988, Knowledge-based geographic information systems (KBGIS): New analytic and data management tools. Mathematical Geology, vol. 20, p. 1021–1035.

Bonham-Carter, G. F., 1991, Integration of geoscientific data using GIS, *in* Geographical Information Systems, vol. 2 (D. J. Maguire, M. F. Goodchild, and D. F. Rhind, eds.). New York: Wiley, p. 171–184.

Breunig, Martin, and Axel Perkhoff, 1992, Data system integration for geoscientific data, *in* Proceedings, 5th International Symposium on Spatial Data Handling, vol. 1 (P. Bresnahan, E. Corwin, and D. Cowen, eds.). Columbia, SC: University of South Carolina, College of Humanities and Social Sciences (in conjunction with the International Geographical Union, Commission on GIS), p. 272–281.

Casar Gonzalez, Ricardo, 1994, Modelo conceptual de un sistema de informacion aplicado a la exploracion petrolera. Boletin de la Asociacion Mexicana de Geologos Petroleros, vol. 44, no. 1, p. 40–60.

Chung, C. F., and Fabbri, A. G., 1993, The representation of geoscience information for data integration. Nonrenewable Resources, vol. 2, p. 122–139.

Collins, D. R., R. Brownrigg, and J. Ross, 1995, Management, analysis and display of digital well-log data for presentation of geologic information in the thrid dimension, *in* Proceedings, 9th Annual Symposium on Geographic Information Systems in Natural Resources Management, vol. 1 (P. O'Reilly, ed.). Ft. Collins, CO: GIS World, p. 266–268.

Crow, Bill W., 1996, Digital technologies correct and integrate data for more precise mapping, World Oil, v. 217 (September), p. 45–48.

Denhart, Sally L., 1993, Geologic database design: The foundation of a successful project, *in* Proceedings of the 13th Annual ESRI User Conference, vol. 1. Redlands, CA: Environmental Systems Research Institute, Inc., p. 467–475.

Ehlers, M., G. Edwards, and Y. Bedard, 1989, Integration of remote sensing with geographic information systems: a necessary evolution. Photogrammetry Engineering and Remote Sensing, vol. 55, p. 1619–1627.

Etak, Inc., 1995, CRADA factsheet: Digital geographic data collection and revision. Reston, VA: U.S. Geological Survey, Fact Sheet 0143–95.

Fowler, Robert A., 1997, Map accuracy specifications, Part 1. Earth Observation Magazine, vol. 6, no. 10, p. 33–36.

Fowler, Robert A., 1997, Map accuracy specifications, Part 2. Earth Observation Magazine, vol. 6, no. 11, p. 33–35.

Goodchild, M. F., 1992, Geographical data modeling. Computers & Geosciences, vol. 18, p. 401–408.

Hamilton, Steven D., 1995, Applications and limitations of raster data in geographic information systems. Earth Observation Magazine, vol. 4, no. 6, p. 43–45.

Johnson, A. I., C. B. Pettersson, and J. L. Fulton, 1992, Geographic Information Systems (GIS) and Mapping—Practices and Standards. West Conshohocken, PA: American Society for Testing and Materials (ASTM), Special Technical Publication 1126.

Jones, C. B., 1989, Data structures for three-dimensional spatial information systems in geology. International Journal of Geographical Information Systems, vol. 3, p. 15–31.

King, R., 1995, GIS as an information integrator in an oil and gas company, *in* Proceedings, 90th Annual Symposium on Geographic Information Systems in Natural Resources Management, vol. 1 (P. O'Reilly, ed.). Ft. Collins, CO: GIS World, p. 222–225.

McDonald, J., and L. H. Wickstrom, 1997, Initial data capture for Ohio's oil-and-gas-well geographic information system, GIS/LIS '97. Bethesda, MD: American Society of Photogrammetry and Remote Sensing (ASPRS), p. 214–218.

O'Connor, J. T., 1994, Analytical data-management systems. 1: Data management strategy for geological databases (microform). Denver, CO: U.S. Geological Survey, Open-File Report 94–905.

Oldow, John S., Carlos L. V. Aiken, Xueming Xu, Karl H. Fleischmann, Nahs G. AveLallement, Marino Ostos, and Franklin Yoris, 1996, Digital geological and geophysical database of Northern Venezuela (abs.), *in* Abstracts with Programs. Boulder, CO: Geological Society of America, Annual Meeting, vol. 28, no. 7, p. A-264.

Peters, Douglas C., 1997, Data quality and documentation: What you don't know about your GIS data can come back to haunt you (abs.), *in* Official Program with Abstracts. Tulsa, OK: AAPG, Annual Convention, vol. 6, p. A-92.

Raines, Gary L., Boyan Brodaric, and Bruce R. Johnson, 1997, Progress report: digital geologic map data model, *in* Proceedings of a Workshop on Digital Mapping Techniques: Methods for Geologic Map

Data Capture, Management, and Publication (D. R. Soller, ed.). Reston, VA: U.S. Geological Survey, Open-File Report 97-0269, p. 43–46.

Saunders, Michael R., John A. Shields, and Michael R. Taylor, 1995, Improving the value of geological data: A standardized data model for industry. Geological Society Special Publications, vol. 97, p. 41–53.

Shaw, Brian R., and Harry Waller, 1988, AAPG exchange format for the transfer of geologic and petroleum data (abs.), *in* Abstracts, GIS Symposium: Integrating Technology and Geoscience Applications (H. F. Thomas, ed.). Washington, D.C.: National Academy of Sciences, p. 59.

Stark, P. H., 1991, Large petroleum data bases: The three I's—integrity, integration, interface (abs.). AAPG Bulletin, vol. 75, p. 676.

Turner, A. Keith, 1997, Location is everything in GIS. GIS World, vol. 10, no. 9, p. 38.

Viljoen, David, 1997, Topological and thematic layering of geological map information: improving efficiency of digital data capture and management, *in* Proceedings of a Workshop on Digital Mapping Techniques: Methods for Geologic Map Data Capture, Management, and Publication (David R. Soller, ed.). Reston, VA: U.S. Geological Survey, Open-File Report 97-0269, p. 15–21.

Walker, J. D., R. A. Black, J. K. Linn, A. J. Thomas, R. Wiseman, and M. G. D'Attilio, 1996, Development of geographic information systems–oriented databases for integrated geological and geophysical applications. GSA Today, vol. 6, no. 3, p. 1–7.

Waller, H. O., D. O. Guinn, M. A. Herkommer, and B. R. Shaw, 1990, Exchange format for transfer of geologic and petroleum data (abs.). AAPG Bulletin, vol. 74, p. 787.

Winczewski, L. M., 1990, Out of Babel: Emerging standards for data management in the petroleum industry (abs.), *in* Official Program with Abstracts. Tulsa, OK: AAPG, Annual Convention.

Youngmann, Carl, 1988, Spatial data structures for modeling subsurface features, *in* Abstracts, GIS Symposium: Integrating Technology and Geoscience Applications (H. F. Thomas, ed.). Washington, D.C.: National Academy of Sciences, p. 85.

TECHNOLOGICAL DEVELOPMENTS

Abel, Roger D., and W. Mark Tuttle, 1996, Geographical information system (GIS) and remote sensing in the exploration environment. GeoArabia (Manama), vol. 1, no. 1, p. 96–97.

Abert, Curtis C., 1996, Utilizing GIS and DEMS for landscape and geological modeling (abs.), *in* Abstracts with Programs. Boulder, CO: Geological Society of America, Annual Meeting, vol. 28, no. 7, p. A-525.

Aiken, Carlos L. V., Qing Ming Yang, Mamadou Balde, Karl H. Fleischmann, Xueming Xu, Manish Bhatt, Mauricio F. de la Fuente, and Luis Miguel Mitre, 1996, A GIS program for the geology and geophysics and related data of Mexico (abs.), *in* Abstracts with Programs. Boulder, CO: Geological Society of America, Annual Meeting, vol. 28, no. 7, p. A-264.

An, P., 1992, Spatial Reasoning and Integration Techniques for Geophysical and Geological Exploration Data, Unpublished Ph.D. Thesis, University of Manitoba, 280 p.

An, P., W. M. Mood, and A. Rencz, 1991, Application of fuzzy set theory for integration of geological, geophysical and remote sensing data. Canadian Journal of Exploration Geophysics, vol. 27, p. 1–11.

An, P., W. M. Moon, and G. F. Bonham-Carter, 1992, On a knowledge-based approach of integrating remote-sensing, geophysical and geological information, *in* Proceedings IGARSS '92, vol. 1. Piscataway, NJ: Institute of Electrical and Electonics Engineers (IEEE), Inc., International Geoscience and Remote Sensing Symposium, p. 34–38.

Applying GIS to geology, 1992. Geotimes, vol. 37 (August), p. 69–70.

Bedell, R. L., Jr., 1994, GIS for the Geosciences. Boulder, CO: Geological Society of America, Short course notes, Annual Meeting, 193 p.

Berquist, C. R., Jr., 1996, Techniques for conversion of paper maps to digital and the creation of new digital geologic maps in Virginia (abs.), *in* Abstracts with Programs. Boulder, CO: Geological Society of America, Annual Meeting, vol. 28, no. 7, p. A-264.

Black, Ross A., Chris Schmeissner, and J. Douglas Walker, 1996, Geophysical model development within a GIS environment (abs.), *in* Abstracts with Programs. Boulder, CO: Geological Society of America, Annual Meeting, vol. 28, no. 7, p. A-265.

Bonham-Carter, G. F., 1994, Geographic Information Systems for Geoscientists: Modeling with GIS. Oxford: Pergamon Press, 398 p.

Brockhaus, J. A., J. E. Fucella, and S. Loomer, 1996, Incorporation of GIS techniques in a geology field course, *in* GIS/LIS '96. Bethesda, MD: American Society of Photogrammetry and Remote Sensing (ASPRS), p. 656–662.

Brownrigg, R., D. R. Collins, and J. Ross, 1995, Changing the face of geologic mapping: technology and data source integration for improved geologic map production, *in* Proceedings, 9th Annual Symposium on Geographic Information Systems in Natural Resources Management, vol. 1 (P. O'Reilly, ed). Ft. Collins, CO: GIS World, p. 257.

Burroughs, W. A., 1988, GIS in oil and gas exploration: Automating geological interpretation (abs.) *in* Abstracts, GIS Symposium: Integrating Technology and Geoscience Applications (H. F. Thomas, ed.). Washington, D. C.: National Academy of Sciences, p. 52.

Carlson, E., 1987, Three-dimensional conceptual modeling of subsurface structures, *in* Technical Papers, vol. 4. Bethesda, MD: American Society of Photogrammetry and Remote Sensing (ASPRS) and American Congress on Surveying and Mapping (ACSM), Annual Convention.

Cesar Gonzalez, Ricardo, 1994, Conceptual model of an information system applied to petroleum exploration. Boletin de la Asociacion Mexicana de Geologos Petroleros, vol. 44, p. 40–60.

Colman-Sadd, S. P., J. S. Ash, J. P. Hayes, and L. W. Nolan, 1996, Management of Geological Map Units in a Geographic Information System. St. Johns, NF: Newfoundland and Labrador Geological Survey, Current Research Report 96-1.

Colman-Sadd, S. P., J. S. Ash, and L. W. Nolan, 1997, GeoLegend, a database system for managing geological map units in a geographic information system. Computers & Geosciences, vol. 23, p. 715–724.

Connell, Joseph F., 1994, Conversion of geologic quadrangle maps to geologic coverages (microform). Denver, CO: U.S. Geological Survey, Open-File Report 94-359, 23 p.

Feineman, D. R., 1994, GIS and petroleum geology in retrospect. Papers and Extended Abstracts for Technical Programs, IAMG '94. International Association for Mathematical Geology, p. 122–127.

Gomez-Herrera, J. E., O. Rodriguez-Moran, and V. Cintas-Sanchez, 1996, Diseno de un sistema informatico para petroleo: INFOPET.GIS (abs.). AAPG Bulletin, vol. 80, p. 1295.

Goodwin, P. B., M. R. Choiniere, F. W. Harris, and B. P. Dean, 1996, Improving exploration with geographical information system (GIS) technology (abs.). AAPG Bulletin, vol. 80, p. 1297.

Guptill, S., and D. Nystrom, 1986, Geographic information systems: an important tool for spatial analysis U.S. Geological Survey Yearbook. Reston, VA: U.S. Geological Survey, Fiscal Year 1985, p. 9–16.

Hall, Robert J., 1984, Remote sensing applications to the development of an integrated data base for oil and gas exploration, *in* Proceedings, 1983 National Conference on Resource Management Applications—Energy and Environment, vol. IV: Remotely Sensed Geographic Information Systems in Geologic Applications (Y. M. Schiffman, N. E. Schiffman, J. O. Brumfield, and V. B. Robinson, eds.). Center for Earth Resource Management Applications (CERMA), p. 74–81.

Hamilton, D. E., J. A. Downing, and T. A. Jones, 1987, Graphic interface for a geologic mapping system, *in* Proceedings, NCGA Mapping and Geographic Information Systems '87. Dunkirk, MD: National Computer Graphics Association.

Harbaugh, J. W., and P. A. Martinez, 1992, Two major problems in representing geological well data and seismic data in petroleum-bearing regions via 3-D geographic information systems, *in* Three-dimensional modeling with geoscientific information systems (A. K. Turner, ed.). Boston: D. Reidel Publishing Company, NATO ASI Series C—Mathematical and Physical Sciences, vol. 354, p. 291–302.

Herring, John R., 1989, A fully integrated geographic information system, *in* Proceedings of the 9th International Symposium on Computer-Assisted Cartography (Auto/Carto 9). Bethesda, MD: American Society of Photogrammetry and Remote Sensing (ASPRS) and American Congress on Surveying and Mapping (ACSM), p. 828–837.

Hinze, William J., and Brian R. Shaw, 1989, Geographic information systems symposium—integrating technology and geoscience applications. Geophysics: The Leading Edge of Exploration, vol. 8, p. 52–53.

Holzman, Bud, 1991, Geographic information systems for exploration decisions. Bulletin of the South Texas Geological Society, vol. 32, no. 4, p. 7–8.

Jones, T. A., and M. M. Bush, 1987, Graphical display of three-dimensional geologic models, *in* Proceedings, NCGA Mapping and Geographic Information Systems '87. Dunkirk, MD: National Computer Graphics Association.

Katz, S. S., 1985, Using the Map Overlay and Statistics System (MOSS) as a geologic tool, *in* Proceedings of the Denver Geotech '85 (Micro Computer Applications for Energy and Mineral Exploration). Westminster, CO: American Institute of Petroleum Geologists (AIPG), Special Issue No. 7.

Katz, Solomon S., 1988, Emulating the PROSPECTOR expert system with a raster GIS. Abstracts, GIS Symposium: Integrating Technology and Geoscience Applications (H. F. Thomas, ed.). Washington, D.C.: National Academy of Sciences, p. 27.

Kavouris, M., and S. E. Masry, 1987, An information system for geosciences: design considerations, *in* Proceedings of the 8th International Symposium on Computer-Assisted Cartography (Auto/Carto 8). Bethesda, MD: American Society of Photogrammetry and Remote Sensing (ASPRS) and American Congress on Surveying and Mapping (ACSM), p. 336–345.

Lee, Y. C., 1997, Conceptual Models for Geographic Information Systems. Unpublished Ph.D. dissertation, University of New Brunswick.

Linn, Jonathan K., Douglas J. Walker, and Ross A. Black, 1996, Computer demonstration utilizing ArcView for display and analysis of geologic and geophysical data (abs.), *in* Abstracts with Programs. Boulder, CO: Geological Society of America, Annual Meeting, vol. 28, no. 7, p. A-264.

List, Franz K., and Coy H. Squyres, 1996, Geologic mapping from satellite data: contribution of geologic remote sensing and GIS technology to hydrocarbon and mineral exploration. Proceedings of the Thematic Conference on Geologic Remote Sensing, vol. 11, p. I.309–I.310.

Loudon, T. V., 1986, Digital spatial models and geologic maps, *in* Proceedings, Auto/Carto London, vol. 2. (M. Blakemore, ed.). London: International Cartographic Association, p. 60–75.

Lukatela, Hrvoje, 1989, GIS future: Automated cartography or geo-relational solid modeling?, *in* Proceedings of the 9th International Symposium on Computer-Assisted Cartography (Auto/Carto 9). Bethesda, MD: American Society of Photogrammetry and Remote Sensing (ASPRS) and American Congress on Surveying and Mapping (ACSM), p. 341–347.

Madigan, Michael E., James A. Sturdevant, and Clark A. Roberts, 1988, Using GIS to evaluate map data. Geotimes, vol. 33, no. 4, p. 14–15.

Maling, D. H., 1991, Coordinate systems and map projections for GIS, *in* Geographical Information Systems: Principles and Applications, vol. 1 (D. J. Maguire, M. F. Goodchild, and D. W. Rhind, eds.). London: Longman Group UK Ltd.

McCabe, Bernie, 1992, Using Arc/Info to predict the

presence of oil or gas using our tool called EUREKA!, *in* Proceedings, 12th Annual ESRI User Conference, Part II. Redlands, CA: Environmental Systems Research Institute, Inc., p. 99–100.

McColloch, Gayle H., Jr., 1997, GIS data models for geology (abs.), *in* Official Program with Abstracts. Tulsa, OK: AAPG, Annual Convention, vol. 6, p. A-80.

McCulloch, Thomas M., Charles E. Ogrosky, and Benjamin S. Ramey, 1995, Technical transition in cartography at the U.S. Geological Survey. Cartography and Geographic Information Systems, vol. 22, p. 140–148.

McGee, David, 1994, Interface merges analysis programs. The American Oil & Gas Reporter, vol. 37, no. 11, p. 68–74.

McLeod, Ronald G., Larry J. Simon, and Steven L. Adams, 1984, Development of information data bases for nonrenewable resource assessment, *in* Proceedings, 1983 National Conference on Resource Management Applications—Energy and Environment, vol. IV: Remotely Sensed Geographic Information Systems in Geologic Applications (Y. M. Schiffman, N. E. Schiffman, J. O. Brumfield, and V. B. Robinson, eds.). Center for Earth Resource Management Applications (CERMA), p. 48–64.

Mennim, K., 1986, Raster techniques in geological map production, *in* Proceedings, Auto/Carto London, vol. 1 (M. Blakemore, ed.). London: International Cartographic Association, p. 267–271.

Miller, B. M., 1991, Integration of geographic information systems and expert systems technology for resource management, *in* Proceedings, Resource Technology 90 (G. J. Buhyoff, ed.). Bethesda, MD: American Society of Photogrammetry and Remote Sensing (ASPRS), 2nd International Symposium on Advanced Technology in Natural Resources, p. 727–736.

Moll, S., and D. Hastings., 1986, Geographic information systems in the earth sciences, *in* Proceedings: Geographic Information Systems Workshop. Bethesda, MD: American Society of Photogrammetry and Remote Sensing (ASPRS), p. 357–364.

Moore, R. F., 1991, GIS in the minerals and oil exploration industry (abs.), *in* Exploration in a Changing Environment: Abstracts of the 8th Conference and Exhibition. Sydney: Australian Society of Exploration Geophysicists, vol. 30, p. 81–82.

Park, Larry A., 1997, GIS as a basin, play, and acquisition evaluation tool (abs.), *in* Official Program with Abstracts. Tulsa, OK: American Association of Petroleum Geologists, Annual Convention, vol. 6, p. A-90.

Perez, G., J. M. Ellis, and H. H. Davis, 1996, Using GIS technology to integrate remote sensing and field work data with templates (abs.), *in* Official Program with Abstracts. Tulsa, OK: AAPG, Annual Convention, vol. 5, p. A–111.

Petroleum and GIS industries: Tapping into each other's potential, 1994. Earth Observation Magazine, vol. 3, no. 5, p. 23–24.

Pfeil, Russell W., and James M. Ellis, 1995, Evaluating GIS for establishing and monitoring environmental conditions of oil fields (abs.). AAPG Bulletin, vol. 79, p. 595.

Pinney, De Anne S., Gina R. Cruse, Katherine L. Queens, and Gail Davidson, 1995, GIS applications in geological mapping and the production of derivative maps (abs.), *in* Abstracts with Programs. Boulder, CO: Geological Society of America, Cordilleran Section, 91st Annual Meeting, vol. 27, no. 5, p. 71.

Raleigh, Michael P., 1993, The integrated solution. Western Oil and Gas World, vol. 50, no. 7 (suppl.), p. 16–18.

Raper, J. F., 1988, The three-dimensional geographical information system: a design manifesto, *in* Abstracts, GIS Symposium: Integrating Technology and Geoscience Applications (H. F. Thomas, ed.). Washington, D.C.: National Academy of Sciences, p. 81.

Raper, J. F., 1989, The 3-dimensional geoscientific mapping and modelling system: a conceptual design, *in* Three-Dimensional Applications in Geographic Information Systems (J. Raper, ed.). London: Taylor and Francis, p. 11–20.

Raper, J. (ed.), 1989, Three-Dimensional Applications in Geographical Information Systems. London: Taylor and Francis, 189 p.

Robinove, Charles J., Principles of Logic and the Use of Digital Geographic Information Systems. Reston, VA: U.S. Geological Survey, Circular No. 977, 10 p.

Scharek, P., T. Tullner, and G. Turczi, 1994, The use of geographic information systems in regional and applied geology. Zeitschrift fur Angewandte Geologie, vol. 40, no. 2, p. 87–91 (in German).

Schau, M., 1995, Synthetic assembly of digital geological maps: from data-driven to knowledge-driven construction, *in* Proceedings, 9th Annual Symposium on Geographic Information Systems in Natural Resources Management, vol. 1 (P. O'Reilly, ed.). Ft. Collins, CO: GIS World, p. 258–265.

Schetselaar, E. M., P. M. Van Dijk, and Y. A. Al Fasatwi, 1990, Digital images processing of geophysical data using a raster-based GIS. ITC Journal, vol. 1990, no. 3, p. 248–252.

Sheath, Nigel J., 1986, Development of an integrated map plotting system for hydrocarbon exploration, in Proceedings, Auto/Carto London, vol. 2 (M. Blakemore, ed.). London: International Cartographic Association, p. 66–75.

Shirley, Kathy, 1995, Applied high tech tackles problems. AAPG Explorer, vol. 16, no. 8, p. 20–21.

Smith, D. R., and A. R. Paradis, 1989, Three-dimensional GIS for the earth sciences, *in* Three-Dimensional Applications in Geographical Information Systems (J. Raper, ed.). London: Taylor and Francis, p. 149–154.

Stark, Phillip H., 1989, GIS applications for petroleum mapping (abs.). Computer Contributions of the Computer-Oriented Geological Society (COGS), vol. 5, no. 3, p. 118.

Starr, L. E., D. A. Nystrom, and J. T. O'Kelley, Jr., 1987, Geographic information system developments within the U.S. Geological Survey, *in* Proceedings: 13th Cartographic Conference, vol. 1. Reston, VA: U.S. Geological Survey, p. 223–226.

Stoa, Rodney S., 1995, PCs ease geographic oil and gas

data base visualization. Oil & Gas Journal, vol. 93 (September 25), p. 79–80.

Tarleton, John, 1995, GIS in hydrocarbon exploration and production. Geodetical Information Magazine, vol. 9, no. 11, p. 33, 35, 37–38.

Thomas, H. F. (ed.), 1988, Abstracts, GIS Symposium: Integrating Technology and Geoscience Applications. Washington, D.C.: National Research Council, 179 p.

Thorleifson, Tracy, 1999, Using ArcView GIS and 3-D shapes to model complex geoscience data (abs.). Redlands, CA: Environmental Systems Research Institute (ESRI), Annual Meeting, Petroleum Users Group.

Turner, A. K., ed., 1991, Three-Dimensional Modeling with Geographic Information Systems. Dordrecht: Kluwer Academic Publishers, 443 p.

Turner, A. Keith, 1998, Help define 3-D GIS' future. GIS World, vol. 11, no. 3, p. 40–41.

Van Driel, J. N., 1989, Three-dimensional display of geologic data, *in* Three-Dimensional Applications in Geographic Information Systems (J. Raper, ed.). London: Taylor and Francis, p. 1–9.

Van Driel, J. N., and J. C. Davis (eds.). Digital Geologic and Geographic Information Systems; Short Course in Geology. Washington, D.C.: American Geophysical Union, 1989, 62 p.

Williams, Van, 1997, Using the GSMCAD program with GPS for data collection in the field and as a quick and efficient way of creating Arc/Info geologic map coverages, *in* Proceedings of a Workshop on Digital Mapping Techniques: Methods for Geologic Map Data Capture, Management, and Publication (D. R. Soller, ed.). Reston, VA: U.S. Geological Survey, Open-File Report 97-0269, p. 83–87.

Ye, Deliao, 1997, Geographic information system—a new technique for petroleum exploration and development. Experimental Petroleum Geology, vol. 19, p. 179–182.

CASE STUDIES AND APPLICATIONS

Bannon, Jackie, and Paul Batey, 1995, BP turns to image processing for visualization of seismic data. Earth Observation Magazine, vol. 4, no. 2, p. 23–25.

Barringer, Jennifer, Michael Schlegel, Sarah Terry, and Leslie Zilm, 1998, Characterization of regional landforms and landscape analysis in support of an oil field development in the remote Arctic Timan Pechora region of Russia (abs.), *in* Official Program with Abstracts. Tulsa, OK: AAPG, Annual Convention, vol. 7.

Bentz, Cristina Maria, and Fernando P. Miranda, 1998, Data integration using GIS in the Tucano Basin, Brazil, as an aid to hydrocarbon exploration (abs.). AAPG Bulletin, vol. 82, p. 1892.

Best, Steven Eric, 1995, Landsat Thematic Mapper and GIS Special Applications to the Mapping of Fractures in the Austin Chalk of South Central Texas. Unpublished M.S. Thesis, Texas A & M University, 169 p.

Blundell, Stuart, David O. Baldwin, and Natalie J. Anderson, 1996, Application of geographic information systems (GIS) in decision analysis for monitoring acquifer systems during oilfield development projects (abs.). AAPG Bulletin, vol. 80, p. 967.

Boltze, Uta, and Michael S. Rosenbaum, 1996, The application of GIS to analyzing 2-D models of porous systems in the context of gas mobility. Marine and Petroleum Geology, vol. 13, p. 365–372.

Bonham-Carter, G. F., and F. P. Agterberg, 1990, Application of a microcomputer-based geographic information system to mineral-potential mapping, *in* Microcomputer Applications in Geology II (J. T. Hanley and D. F. Merriam, eds.). New York: Pergamon Press, p. 49–74.

Byars, Bruce W., Bethany D. Rinard, and Thomas T. Goforth, 1995, A GIS link for first order seismic source assessment. Geo Info Systems, vol. 5, no. 2, p. 34–37.

Carlton, R. Bradley, 1994, Evaluating the Antrim Shale formation using a geographic information system (abs.). AAPG Bulletin, vol. 78, p. 1328.

Carter, Melvan D., and John S. Fischer, 1988, GIS and reservoir boundary determination: Defining a distributary front reservoir with seismic and well logs on an interactive workstation (abs.), *in* Abstracts, GIS Symposium: Integrating Technology and Geoscience Applications (H. F. Carter, ed.). Washington, D.C.: National Academy of Sciences, p. 78.

Clark, Michael, 1995, Application of Geographic Information Systems (GIS) Technology and Remote Sensing to a Geological Analysis of the Regions, West Texas Solitario Dome and Terlingua Uplift. Unpublished M.S. Thesis, Stephen F. Austin State University, 50 p.

Clark, M., 1996, Integration of geographic information systems and remote sensing to a geologic analysis of the Llano Uplift, Texas. Geographical Bulletin, vol. 38, no. 1, p. 7–13.

Coe, J. A., and E. P. Nelson, 1997, Characterization of fracture networks using close-range photogrammetric mapping and GIS analysis, *in* Fractured Reservoirs: Characterization and Modeling (Thomas E. Hoak, Alan L. Klawitter, and Peter K. Blomquist, eds.). Denver, CO: Rocky Mountain Association of Geologists, p. 43–47.

Comet, Paul A., Christopher R. King, and K. F. M. Thompson, 1996, Geographic information systems applied to petroleum geology (abs.), *in* Abstracts with Programs. Boulder, CO: Geological Society of American, Annual Meeting, vol. 28, no. 7, p. A-527.

Corbley, Kevin P., 1995, EOCAP project targets Antrim Shale. Earth Observation Magazine, vol. 4, no. 12, p. 36–39.

Craigg, Steven D., 1988, Analysis of spatial geologic data using surface modeling and geographic information system software—applications to regional geologic mapping, San Juan structural basin, New Mexico, Colorado, Arizona, and Utah, *in* Abstracts, GIS Symposium: Integrating Technology and Geoscience Applications (H. F. Thomas, ed.). Washington, D.C.: National Academy of Sciences, p. 9.

Donaldson, Kurt A., 1993, An integrated approach to reservoir modeling—Granny Creek Field, West Virginia, *in* Studies in Eastern Energy and the Environment (A. P. Schultz et al., ed.). Tulsa, OK: Virginia Division of Mineral Resources, Special Publication 132, p. 16–19.

Donaldson, Kurt A., 1994, An Integrated Approach to Reservoir Modeling. Unpublished M.S. thesis, University of West Virginia, 259 p.

Dong, Feng, 1996, Three-Dimensional Models and Applications in Subsurface Modeling. Unpublished M.S. Thesis, University of Calgary, 109 p.

Dwyer, John L., and J. Thomas Nash, 1991, Spatial Analysis of Geochemical and Geologic Information from the Tonopah 1° × 2° Quadrangle, Nevada. Reston, VA: U.S. Geological Survey, Open-File Report.

Elkington, Gary, R. Malcolm Lansley, Federico Martin, and Randy Utech, 1995, Uses of GIS data in 3-D seismic designs and acquisition, *in* Expanded Technical Program Abstracts with Biographies. Tulsa, OK: Society of Exploration Geophysicists, Annual Meeting, vol. 65, p. 957–959.

Ellis, J. M., W. Narr, P. B. Goodwin, and G. Perez, 1995, Geotechnologies facilitate geologic mapping and field operations in Columbia. Earth Observation Magazine, vol. 4, no. 2, p. 18–21.

Everham, W. D., and J. E. Huatoon, 1996, Subsurface databases: graphical display and error detection for stratigraphic interpretation in the Michigan Basin (abs.), *in* Abstracts with Programs. Boulder, CO: Geological Society of America, Annual Meeting, vol. 28, no. 7, p. A-265.

Ferrier, G., and G. Wadge, 1997, An integrated GIS and knowledge-based system as an aid for the geological analysis of sedimentary basins. International Journal of Geographical Information Science, vol. 11, p. 281–297.

Fishman, Jeremy, M. Bhatt, Carlos Aiken, Karl H. Fleischmann, Mamadou Balde, and Mauricio F. de la Fuente, 1996, The integration of remote sensing, GIS converted geology and GPS controlled gravity in an analysis of the structure of Ecuador (abs.), *in* Abstracts with Programs. Boulder, CO: Geological Society of America, Annual Meeting, vol. 28, no. 7, p. A-265.

Flewelling, Douglas M., Andrew U. Frank, and Max Egenhofer, 1992, Constructing geological cross sections with a chronology of geologic events, *in* Proceedings, 5th International Symposium on Spatial Data Handling, vol. 2 (P. Bresnahan, E. Corwin, and D. Cowen, eds.). Columbia, SC: University of South Carolina, College of Humanities and Social Sciences (in conjunction with the International Geographical Union, Commission on GIS), p. 544–553.

Foley, Donald C., 1995, How Conoco uses GIS technology to map geology, geography through time. Oil & Gas Journal, vol. 93 (May 8), p. 75–77.

Froomer, Norman, Robert Hughes, and John Young, 1993, The use of GIS for the environmental review of oil and gas exploration and development plans on the outer continental shelf of the Gulf of Mexico, *in* Proceedings of the 13th Annual ESRI User Conference, Vol. 1. Redlands, CA: Environmental Systems Research Institute, Inc., p. 463–465.

Gallaher, David W., 1992, A GIS case study of a petroleum production problem in Lea and Eddy counties, New Mexico (abs.), *in* Official Program with Abstracts. Tulsa, OK: AAPG, Annual Convention, vol. 1, p. A-44.

Gao, Changlin, and Deliao Ye, 1997, Petroleum geology of the Tarim Basin, northwest China; recent advances. Journal of Petroleum Geology, vol. 20, p. 293–244.

Ghiselin, Dick, 1997, Optimizing well analysis and design—birth of a system. Petroleum Engineer International, vol. 20 (October), p. 75–76.

Grace, John D., and David B. Walsh, 1997, Development of a GIS on domestic oil and gas resources (abs.), *in* Official Program with Abstracts. Tulsa, OK: AAPG Annual Convention, vol. 6, p. A-42.

Hughes, J. D., 1996, Optimizing resource analysis of stratified sequences with 3-D modeling, GIS, remote sensing and 3-D visualization (abs.), *in* Annual Meeting Abstracts. Sydney: Geological Society of Australia, vol. 41, p. 203.

Jackson, Alan, 1991, Resource assessment in Western Austalia using a geographic information system (abs.). AAPG Bulletin, vol. 75, p. 600–601.

Jacobsen, R. B., 1993, Application of digital geographic information system in compilation and analysis of surficial geologic maps, *in* Abstracts with Programs. Boulder, CO: Geological Society of America, North Central Section, 27th Annual Meeting, vol. 25, no. 3, p. A-28.

Jones, S., 1988, Subsurface geological mapping in the Blanding Basin of Southeast Utah using *MOSS, in* Proceedings of the 5th National MOSS Users Conference. Lakewood, CO: U.S. Bureau of Land Management.

Kacewicz, Marek, 1994, An integration of secondary hydrocarbon migration modeling and GIS into resource assessment (abs.), *in* Official Program with Abstracts. Tulsa, OK: AAPG, Annual Convention, vol. 3, p. A-182.

Keighin, C. William, and Christopher J. Schenk, 1992, The Arc/Info geographic information system applied to geologic investigations of the Norphlet Formation, Alabama and Mississippi (abs.), *in* USGS Research on Energy Resources 1992: Program and Abstracts (L. M. H. Carter, ed.). Reston, VA: U.S. Geological Survey, Report C-1074, p. 41.

Kettler, David J., and Robert C. Howe, 1993, Comparison of geologic data sets with petroleum production utilizing GIS technology. Proceedings of the Thematic Conference on Remote Sensing for Exploration Geology, vol. 9, p. 445–456.

Krumm, R. J., R. R. Pool, E. D. McKay, and L. R. Smith, 1988, Geologic applications of an extensive GIS database: examples from Illinois. Abstracts, GIS Symposium: Integrating Technology and Geoscience Applications (H. F. Thomas, ed.). Washington, D. C.: National Academy of Sciences, p. 70.

Lang, Laura, 1994, Using GIS for overseas exploration. PetroSystems World, July–August issue, p. 16–18.

Lang, Laura, 1998, GIS coordinates Chevron Nigeria. Hart's Oil and Gas World, vol. 90, no. 3, p. 53–56.

Lavrent'yeva, N. I., and Yu. V. Polyushkin, 1993, A dendochronology database within an integrated geographic information system. Mapping Sciences and Remote Sensing, vol. 30, p. 198–204.

Lengyel, Patrick, and John Fairs, 1995, Remote sensing and GIS technology applied to hydrocarbon exploration. Earth Observation Magazine, vol. 4, no. 10, p. 25–28.

Liu, Jia Y., William R. Bryant, and William W. Sager, 1997, GIS-based study of the Alaminos Canyon, Gulf of Mexico, and its linkage with WWW (abs.), *in* Official Program with Abstracts. Tulsa, OK: AAPG, Annual Convention, vol. 6, p. A-70.

Luneau, Barbara A., Susan M. Landon, and Mark W. Longman, 1994, Using an integrated database to determine regional facies variations and petroleum potential; Niobrara Formation, Rocky Mountain region, USA, *in* Official Program with Abstracts. Tulsa, OK: AAPG, Annual Convention, vol. 3, p. 202.

Mannino, Antonio, and Paul A. Montagna, 1996, Fine-scale spatial variation of sediment composition and salinity in Nueces Bay of South Texas. The Texas Journal of Science, vol. 48, p. 35–47.

McKenna, Thomas E., 1994, Mapping the hydrogeology of sedimentary basins using a geographic information system and data from the petroleum industry (abs.), *in* Abstracts with Programs. Boulder, CO: Geological Society of America, Annual Meeting, vol. 26, no. 7, p. A-415.

McKenna, T. E., and X. Li, 1997, Development of a geographic information system for sedimentary basin analysis, *in* Proceedings of the 12th International Conference on Applied Geologic Remote Sensing, vol. 2. Ann Arbor, MI: ERIM International, Inc. (Environmental Research Institute of Michigan), p. 340–343.

McKnight, Cleavy L., 1997, Using GIS to improve structural interpretation: Junggar and Turpan Basin margins, NW China (abs.), *in* Official Program with Abstracts. Tulsa, OK: AAPG, Annual Convention, vol. 6, p. A-80–81.

Meehan, Kenneth T., 1991, Integrating geologic and economic data using a geographic information system (GIS): Offshore Gulf of Mexico (abs.). AAPG Bulletin, vol. 75, p. 633.

Miller, B. M., 1988, Application of geographic information systems technology to resource assessment within sedimentary basins, *in* Abstracts, GIS Symposium: Integrating Technology and Geoscience Applications (H. F. Thomas, ed.). Washington, D.C.: National Academy of Science/National Research Council, p. 79–80.

Miller, B. M., 1989, Applications of geographic information systems to exploration studies in the San Juan Basin, New Mexico, *in* Geoscientific Information Systems Applied to Exploration and Research. Lakewood, CO: Denver GeoTech '89, p. 121–123.

Miller, B. M., 1990, Applications of geographic information systems (GIS) to exploration studies in the San Juan Basin, New Mexico (abs.). AAPG Bulletin, vol. 74, p. 719.

Miller, B. M., 1992, Guide to the development and application of geographic information systems for sedimentary basin analysis—case study for the San Juan Basin, New Mexico and Colorado. Reston, VA: U.S. Geological Survey, Bulletin 2025, 37 p.

Miller, B. M., P. C. Latzke, and B. R. Schachte, 1990, Videocassette demonstrating the applications of geographic information systems techniques to basin characterization—case study for the San Juan Basin, New Mexico. Denver, CO: U.S. Geological Survey, Open-File Report 90-275-A, 6 p.

Miller, B.M., P. C. Latzke, and B. R. Schachte, 1990, Videocassette demonstrating the applications of geographic information systems techniques to basin characterization—case study for the San Juan Basin, New Mexico. Denver, CO: U.S. Geological Survey, Open-File Report 90-275-B, VHS videocassette tape, 20 minutes.

Monson, Lawrence M., 1994, Shallow gas well site selection on Fort Peck Reservation, Montana, *in* GIS/LIS '94. Bethesda, MD: American Society of Photogrammetry and Remote Sensing (ASPRS), p. 590–599.

Morris, Lance G., 1994, An Application of Remote Sensing and Geographic Information Systems to Hydrocarbon Exploration in South-Central Kentucky. Unpublished M.S. Thesis, Murray State University, 112 p.

Nagihara, S. S., R. L. Arbo, and A. J. Bertagne, 1999, An integrated approach using GIS for evaluating exploration and production potential in the Gulf of Mexico (abs.), *in* Official Program with Abstracts. Tulsa, OK: AAPG, Annual Convention, vol. 8, p. A-99.

Orange, D. L., M. Angell, and D. Lapp, 1999, Seafloor mapping for exploration and production: seep detection, geohazards, decreased cycle time, and GIS to manage it all (abs.), *in* Official Program with Abstracts. Tulsa, OK: AAPG, Annual Convention, vol. 8, p. A-102–103.

Panetta, Brian J., Ernest A. Mancini, Ed Blair, and Robert Crumb, 1998, GIS-based petroleum reservoir data management system: key to effective reservoir management of the Lower Cretaceous Donovan Sandstone reservoir, Citronelle field, southwestern Alabama (abs.), *in* Abstracts with Programs. Boulder, CO: Geological Society of America, Annual Meeting, vol. 30, no. 4, p. 54.

Prasetyo, Hardi, 1993, The application of geographic information system (GIS) for compiling digital "tectonic map" of the Indo-Pacific offshore region: Prototype. Bulletin of the Marine Geological Institute of Indonesia, vol. 8, no. 4, p. 21–39.

Riebau, A., D. Fox, D. Dietrich, D. Mussard, and W. Marlatt., 1986, Regional risk identification analysis applicable to resource development of H_2S-contaminated natural gas fields in southwest Wyoming, *in* Proceedings: Geographic Information Systems Workshop. Bethesda, MD: American Society of Photogrammetry and Remote Sensing (ASPRS), p. 345–356.

Roessner, Sigrid, 1997, GIS in quantitative geomorphology—assessment of denudation rates in the Central Kenya Rift, *in* Geographic Information Research: Bridging the Atlantic (M. Craglia and H. Couclelis, eds.). London, Taylor & Francis, p. 441–455.

Romie, J. E., 1985, Use of a geographic information system for geologic site characterization studies in the Paradox Basin, Southeastern Utah, *in* Geographic Information Systems in Government (B. K. Opitx, ed.), p. 379–402.

Ross, J., D. R. Collins, and R. Brownrigg, 1995, GIS management of petroleum production data for estimation and mapping of regional reserves, future production rates and discovery potential, *in* Proceedings, 9th Annual Symposium on Geographic Information Systems in Natural Resources Management, vol. 1 (P. O'Reilly, ed.). Ft. Collins, CO: GIS World, p. 244–250.

Scotese, Christopher R., and Malcolm I. Ross, 1989, Paleogeographic reconstructions using interactive three-dimensional computer graphics (abs.). AAPG Bulletin, vol. 73, p. 410.

Seber, D., M. Vallve, E. Sandvol, D. Steer, and M. Barazangi, 1997, Middle East tectonics: applications of geographic information systems. GSA Today, vol. 7, no. 2, p. 1–6.

Silliphant, Laura J., Cartherine Hanks, Daniel D. Naber, and Sharon W. George, 1995, Fracture analysis of selected areas of the northeastern Brooks Range using digital elevation models and a geographic information system (abs.), *in* Abstracts with Programs. Boulder, CO: Geological Society of America, Cordilleran Section, 91st Annual Meeting, vol. 27, no. 5, p. 78.

Silvaji, S., 1995, Evaluating some applications of GIS technology in geological data management and processing methods: a case study from part of the Northwest Borneo Basin, Sabah. Newsletter of the Geological Society of Malaysia, vol. 21, no. 3, p. 217.

Slutz, Gregory S., and J. Marc Young, 1995, Petroleum system modeling with a geographic information system in the Gulf of Thailand (abs.), *in* Official Program with Abstracts. Tulsa, OK: AAPG, Annual Convention, vol. 4, p. A-89.

Slutz, Greg S., 1999, Exploration GIS: Concept to well site (abs.). Redlands, CA: Environmental Systems Research Institute (ESRI), Petroleum Users Group, Annual Meeting.

Stark, Philip H., 1978, Computer information systems and mapping in petroleum resource analysis, *in* Proceedings, Computer Mapping for Resource Analysis (J. C. Davis and Silvana Levi de Lopez, eds.). Lawrence, KS: Kansas Geological Survey, p. 96–109.

Takahashi, Kenneth I., and Debra K. Higley, 1990, Animated geographic information system of petroleum exploration through time in the Denver Basin of Colorado, Nebraska, and Wyoming (abs.). AAPG Bulletin, vol. 74, p. 1347.

Taylor, David J., 1988, Geographical information systems as an integrated tool in geophysical data interpretation (abs.), *in* Abstracts, GIS Symposium: Integrating Technology and Geoscience Applications (H. F. Thomas, ed.). Washington, D.C.: National Academy of Sciences, p. 43.

Thayer, Ruth M., 1993, GIS operations at the Wind River Agency, Bureau of Indian Affairs, Wyoming, *in* Wyoming Geological Association Special Symposium on Oil and Gas and Other Resources of the Wind River Basin, Wyoming (William R. Keefer, William J. Metzer, and Larry H. Godwin, eds.). Casper, WY: Wyoming Geological Association, p. 25–30.

Tracey, Thomas, 1992, Oil and Gas Well Drill Site Location Analysis in the Targhee National Forest Using a Geographic Information System. Unpublished M.S. thesis, University of Idaho, 63 p.

Trent, Victoria P., 1990, Using a geographic information system for production of a well drilling favorability map (abs.), *in* Abstracts with Programs. Boulder, CO: Geological Society of America, Southeastern Section, Annual Meeting.

Treworgy, Colin, and Margaret Bargh, 1988, Use of a GIS as a tool for geologic mapping and analysis. Abstracts, GIS Symposium: Integrating Technology and Geoscience Applications (H. F. Thomas, ed.). Washington, D.C.: National Academy of Sciences, p. 29.

Turner, A. K., 1989, The role of three-dimensional geographic information systems in subsurface characterization for hydrogeological applications, *in* Three-Dimensional Applications in Geographic Information Systems (J. Raper, ed.). London: Taylor and Francis, p. 115–128

Turner, A. Keith, 1998, GIS enlightens the dark continent. GIS World, vol. 11, no. 1, p. 36–37.

Ulmasvai, F., 1996, Forecast for oil and gas deposits based on GIS technologies (abs.), *in* Abstracts, 30th International Geological Congress. Trondheim, Norway: International Union of Geological Sciences, p. 472.

Usery, E. L., R. R. P. Deister, and D. J. Barr., 1988, A geological engineering application of a knowledge-based geographic information system, *in* GIS/LIS '88. Bethesda, MD: American Society of Photogrammetry and Remote Sensing (ASPRS).

Vallve, Marisa, Dogan Seber, Eric Sandvol, and Muawia Barazangi, 1996, A menu-driven GIS application to store, access, manipulate, and plot geologic/geophysical databases: an example from the Middle East and North Africa region (abs.), *in* Abstracts with Programs. Boulder, CO: Geological Society of America, Annual Meeting, vol. 28, no. 7, p. A-264.

Walsh, D. B., and J. D. Grace, 1996, Visualizing petroleum systems with a combination of GIS and multimedia technologies: an example from the West Siberia Basin (abs.), *in* Official Program with Abstracts. Tulsa, OK: AAPG, Annual Convention, vol. 5, p. A-146.

Walsh, Jim, 1997, Assessing field remediation economics with integrated data analysis. Earth Observation Magazine, vol. 6, no. 4, p. 16–19.

Watson, Richard L., and Kermit G. Witherbee, 1988, Application of oil and gas well database management, GIS, and geologic mapping at the Colorado State Office, Bureau of Land Management, *in* Abstracts, GIS Symposium: Integrating Technology and Geoscience Applications (H. F. Thomas, ed.). Washington, D.C.: National Academy of Sciences, p. 76.

Webster, T. L., J. B. Murphy, and S. M. Barr, 1998, Anatomy of a terrane boundary: an integrated structural, geographic information system, and remote sensing study of the late Paleozoic Avalon-Meguma terrane boundary, mainland Nova Scotia, Canada. Canadian Journal of Earth Sciences, vol. 35, p. 787–801.

Wilsher, W. A., and M. J. Dewit, 1989, A GIS solution for Gondwana geoscientific data. Journal of African Earth Science, vol. 9, p. 271–274.

Wolfe, J. D., 1997, Using GIS to locate high potential zones of hydrocarbon migration, *in* GIS/LIS '97. Bethesda, MD: American Society of Photogrammetry and Remote Sensing, p. 610–619.

Xu, Xueming, Carlos Aiken, Kent Neilsen, Gregory Lyman, and Mamadou Balde, 1997, Digital geological mapping in three dimensions in the Austin Chalk, Plano, Texas (abs.), *in* Abstracts with Programs. Boulder, CO: Geological Society of America, vol. 29, no. 6, p. A-293.

Yang, Qing Ming, Xueming Xu, Carlos L. V. Aiken, Karl H. Fleischmann, Mamadou Balde, Gregory D. Lyman, Tingdong Li, Shufang Geng, and Xun Liu, 1996, GIS conversion of the geoscientific information of Southeast Asia (including China and adjacent areas) and along a geotransect across the Tarim Basin, Xinjiang Province (abs.), *in* Abstracts with Programs. Boulder, CO: Geological Society of America, Annual Meeting, vol. 28, no. 7, p. A-266.

Yarka, P. J., 1994, Progress on Tunisia GIS. Technical memorandum, Marathon Oil Company.

Yun, S., and Y. Sun, 1999, A development of ArcGeology, GIS-based geological D/B, and analysis of geological structure, Eoil area, Korea (abs.), *in* Official Program with Abstracts. Tulsa, OK: AAPG, Annual Convention, vol. 8.

Zech, Robert S., Jennie L. Ridgley, Kimberly I. Cunningham, G. Michael Reimer, A. Curtis Huffman, Jr., Steven M. Condon, and Bill S. Larsen, 1995, Energy studies on native American lands funded by U. S. Department of Energy—near surface methane and hydrocarbon occurrence, Southern Ute Indian Reservation, Colorado, and Jicarillo Apache Indian Reservation, New Mexico, *in* Energy and the Environment—Application of Geosciences to Decision-Making (L. M. H. Carter, ed.). Reston, VA: U. S. Geological Survey, Circular 1108, p. 66–67.

Looking Ahead

Like all developing technologies, the future of GIS—particularly in petroleum exploration and development—is really unknown. However, the era of high-tech computing is advancing at a rapid clip, and capabilities that were only dreams a few years ago are rapidly becoming reality. It seems almost certain that the processes used to discover and exploit oil and gas resources will become more computer intensive with the passage of time, and that various individual computational tools will become parts of larger, more integrated, and more comprehensive computer-based systems. In this final section, the editors propose and consider just one of the many possible ways in which GIS could be combined with technologies of similar focus to yield more effect decision making in the petroleum exploration and development arena. This chapter is entitled **GIS and Geostatistics—A Marriage of Technologies**.

Coburn, T. C., and J. M. Yarus, 2000, GIS and geostatistics: a marriage of technologies, *in* T. C. Coburn and J. M. Yarus, eds., Geographic information systems in petroleum exploration and development: AAPG Computer Applications in Geology, No. 4, p. 295–298.

Chapter 22

GIS and Geostatistics: A Marriage of Technologies

Timothy C. Coburn
Department of Mathematics and Computer Science, Abilene Christian University
Abilene, Texas, U.S.A.
and
Mathematical Geology Section, Kansas Geological Survey
Lawrence, Kansas, U.S.A.

Jeffrey M. Yarus
Roxar, Inc.
Houston, Texas, U.S.A.

ABSTRACT

Together, the technologies of geographic information systems (GIS) and geostatistics have revolutionized the way spatial information is characterized, analyzed, and interpreted. Because the two are so closely related in their objectives, a merger of their functionalities is particularly intriguing, having the potential to further recast the nature of spatial data assessment. A brief discussion of the individual capabilities of both GIS and geostatistics is presented here, along with some considerations and implications for combining them, specifically in the context of petroleum exploration and development. Such a blending of approaches holds special promise in this arena because many situations involve limited data that have resulted from restricted subsurface sampling, and more effective decision making may therefore be achieved.

INTRODUCTION

In recent years there has been an explosion of interest and activity surrounding the concepts of spatial data analysis. It is as if there were some pent-up demand for new and better ways to look at data that is spatially oriented. Ironically, spatial data analysis has been around for decades. Geographers, for example, have been analyzing spatially referenced information essentially for as long as their discipline has been in existence. Nonetheless, at no time in the past has the scientific community witnessed anything close to the kind of flurry of activity pertaining to spatial information that is going on right now.

In the petroleum industry, one does not have to look any further than the annual conventions of the professional societies to discover the heavy and growing emphasis on spatial data and spatial data analysis. Of course, the principal reason for this phenomenon is the computer, along with the recent, rapid explosion in computational power. Never before have there been these kinds of computational resources, both in terms of speed and algorithms, that are available today; and never before has it been possible to handle and treat the kinds of interrelationships that spatially oriented data present.

Two particular technologies that are experiencing phenomenal growth in terms of interest, use, and technological development are geostatistics and geographic information systems (GIS). Although these two technologies have evolved from distinctly different beginnings, their ultimate objectives are quite similar. Geostatistics is a collection of non-deterministic methodologies designed to estimate or predict the values of naturally occurring phenomena, those values being related in some way by distance, area, volume, or space. Geostatistical methods provide the ability to characterize and assess uncertainty in the information used to construct models and to evaluate multivariable relationships. In the geoscience context, geostatistics can be used, for example, to estimate rock and fluid properties away from drill-hole locations. Geographic information systems, in contrast, are computer-based processes for co-registering and displaying the commonality of occurrence of multiple factors in a plane, area, or volume of interest, and for assisting users in conceptualizing interpretations through the visualization of data. GIS provides the ability to simultaneously observe multiple data types and volumes of information of varying sizes, thereby facilitating the discovery of relationships that might otherwise go unnoticed or that might require more extensive time and resources to detect.

In their own individual camps, both within and without the petroleum industry, much has recently been made about the benefits and potential of geostatistics and GIS (Rathfeeder, 1989; Shirley, 1996). Clearly, both technologies fall squarely in the realm of spatial analysis. In fact, in 1996, the second in a series of major international symposia on spatial accuracy assessment (Mowrer et al., 1996) explicitly linked the two together, along with remote sensing, to formally establish a triumvirate of techniques for use in addressing classes of natural resource problems associated with spatial variability.

Certainly, there is already a fairly substantial literature on the more general relationship between spatial statistics and GIS (e.g., Thomas, 1991; Haining and Wise, 1991; Goodchild et al., 1992; Anselin and Gettis, 1992; Griffith, 1993a, b; Anselin et al., 1993; Fotheringham and Rogerson, 1994; Journel, 1996; Unwin, 1997). In fact, because literature is becoming so extensive, there is a need to distinguish what exactly is meant by the term "spatial statistics." The various ideas range from map accuracy or positional uncertainty, to database accuracy (or completeness and correctness of the numbers), to propagation of error, to the assessment of variability or components of error in spatial data. The proceedings of the *International Symposium on Spatial Data Quality* conducted in Hong Kong in July 1999 is an excellent example of the degree to which GIS has become focused on spatial statistics. Although these notions all pertain to important statistical aspects of spatially oriented data, none of them really conveys the definition commonly attributed to geostatistics, and none of them accurately communicates the ideas that are envisioned by a marriage of geostatistics with GIS.

Individually, the two technologies have lengthening histories, most developments spanning roughly the last 25 years; however, it is only recently that the power and potential of a merger of the two has come to be appreciated (Burrough, 1999). Their integration, both theoretically and in applied ventures, consequently, has not yet come to full fruition. Exciting new avenues of research lie on the horizon that blend ideas and concepts from GIS and geostatistics that will provide the next leap in spatial problem-solving capability. In the coupling of GIS with geostatistics, what is envisioned is the use of geostatistical constructs to analyze point data contained in GIS coverages, or conversely, the use of GIS to improve geostatistical estimates from various information and/or to portray them as GIS coverages.

Some work has already been accomplished in other disciplines (Liebhold et al., 1993; Rhodes and Myers, 1993; Nelson et al., 1999), but the petroleum industry, particularly the upstream segment, is poised to make significant contributions to this marriage of technologies. In fact, in the context of petroleum exploration and development, there is already evidence that the boundaries of the two technologies are becoming blurred as the industry moves to a more umbrella-like approach to basin analysis, reservoir characterization, reserves assessment, and field development (Witherbee and Watson, 1988; Turner, this volume; Bonham-Carter, this volume). Such a blended approach is expected to alleviate the inadequacies of employing individual technologies alone.

As a case in point, stochastic models of the subsurface constructed with geostatistical techniques and software typically fail to take advantage of all the data available to the geoscientists. This may result from an inability to access the data or, more commonly, from an inability to access it in a usable form. Many subsurface relationships are difficult to characterize and quantify with geostatistics alone, particularly in the presence of limited data, and it would be extremely beneficial to be able to use information external to the geostatistical modeling process as a guide. To borrow a notion from computer science, what is needed is the ability to "co-process" or "parallel process" the geostatistical analysis results along with other concomitant data, and perhaps to do so in an interactive fashion, in much the same way that the human brain simultaneously processes multiple sources of information. Employing GIS to visualize the "extra" information at scales and projections equivalent to those of the geostatistical model, perhaps as part of the same software environment, would provide the unique ability to interactively confirm assumptions and to augment or update the analytical results, essentially on-the-fly.

In geostatistical modeling of subsurface features, some common inputs are structural surfaces and faults derived from interpreted maps, lithology construed from seismic attributes, and point estimates of porosity, permeability, and fluid saturation determined from petrophysical logs or core samples. However, in many cases, adequate well control is lacking, and seismic data may even be absent or of poor quality. Consequently, there is a need for more information that can help identify trends affecting petrophysical properties, estimate variogram parameters, and determine flow direction or bedding orientation. Experienced practitioners will undoubtedly understand the difficulty that limited well control presents when attempting to construct horizontal variograms because the maximum direction of geological, geophysical, or petrophysical continuity cannot be determined with any degree of reliability. Having more wells may not even remedy the situation because they still may be too far apart to allow naturally occurring subsurface variation to be properly characterized with the variogram (i.e., the graph of the variogram may be ill-posed near its origin). In such cases, the geoscientist is left with having to model the variogram using a back-of-the-envelope approach, or searching for other evidence to support project hypotheses.

The sampling issues that give rise to this situation are somewhat peculiar to oil and gas exploration and development, and they underscore how important a link to GIS may be in this particular arena. With a direct connection to a GIS containing accessory information (whether in the form of maps, tables, images, or reports) that cannot be geostatistically modeled, it would be possible to immediately refine the variogram, thereby improving the overall subsurface characterization. The present-day topography of an area, the distribution of faults and fractures expressed at the surface, surface geochemistry, various attributes determined from satellite images and spectral analysis, and the size, shape, and proportional relationship of sedimentological features (rivers, bars, deltas, etc.) are all examples of the kinds of information that cannot be directly modeled using geostatistics, but which can be used to guide the modeling process.

In addition, GIS can be an extremely beneficial asset in the synthesis of geostatistical results. Today, considerable research is being undertaken to determine how to summarize equiprobable realizations and to select the best ones from among the many that can be produced. Although generating individual realizations is not particularly difficult, given sufficient computational resources, selecting the one, two, or three best ones with which to produce probability maps or to scale-up for reservoir simulation is a more challenging task. The difficulty arises from human inability to simultaneously compare numerous two-dimensional maps from a multidimensional perspective, the absence of reliable quantitative ranking or selection criteria, and insufficient time to physically produce all the graphics and to consider all possible scenarios without such objective guidelines. If GIS functionality is tied directly to geostatistical modeling routines in such a way that realizations can be immediately re-created as GIS coverages, then the visualization limitations are essentially eliminated. In addition, such coverages can be layered on other coverages representing auxiliary information (satellite images, surface lineaments, hydrocarbon seeps, ocean currents, pipeline infrastructure, etc.) to help determine their reasonableness relative to the actual physical system being described. Within the confines of GIS, geostatistical realizations can be stored and visualized as multiple layers, or even volumes, along with other kinds of data, and the functionality of GIS can be used to search for realizations that match a prescribed template or satisfy specific requirements.

From a complementary position, geostatistical methods are ideal tools with which to analyze and inform standard GIS coverages. All data that arise from naturally occurring phenomena exhibit variation; yet when presented and visualized as a GIS coverage, they take on an aura of certainty and determinism. In truth, many coverages that might be used in petroleum exploration and development scenarios possess little terrestrial variation (well and platform locations, pipeline infrastructure, wetland boundaries, etc.), yet they do vary in time, if only very slowly. Other coverages, such as those representing subsurface rock properties, are definitely uncertain because the data arise as a result of sampling. It is these kinds of coverages (porosity, effective permeability, water resistivity, etc.) that geostatistics can help to qualify through an assessment of observed variability.

A marriage of geostatistics and GIS has the potential to yield many other technical capabilities and benefits in the search for, and production of, oil and gas. Reservoir characterization, resource assessment, volumetric calculation, estimation of sedimentation rates, positioning of offshore platforms, siting of injector and development wells, and determination of reservoir communication and connectivity are just some of the many technical elements of petroleum geoscience and engineering that can benefit from a blending of these technologies.

From the standpoint of petroleum exploration and development, geostatistics and GIS are technologies that are here to stay. The degree to which they mature and become incorporated into the conventional geoscience toolbox, however, depends largely on how closely they become intertwined, both functionally and operationally. Effective blending of their technical capabilities may require extensive research and development, but the rewards are anticipated to be substantial. The petroleum industry has great incentive to lead and support such an enterprise, and it is hoped that the challenge will be accepted.

Obviously, the true significance of this integration of technologies can only be judged with the passage of time. Successful deployment, however, can only be accomplished through determined cooperation among all interested parties—the prospective users (petroleum oriented or otherwise), experts in the two technological camps, software developers and vendors, and academic research groups. Nonetheless, the prospects are extremely exciting to consider. The resulting outcome may be a whole new way in which the scientific community views the nature and character of spatial data. In the world of petroleum exploration and development, such an outcome could further revolutionize decision making—decision making that is, and will remain, so heavily dependent on the management, analysis, and interpretation of valuable information resources.

REFERENCES CITED

Anselin, L., and A. Getis, 1992, Spatial statistical analysis and geographic information systems. Annals of Regional Science, vol. 26, p. 19–33.

Anselin, L., R. F. Dodson, and S. Hudak, 1993, Linking GIS and spatial data analysis in practice. Geographical Systems, vol. 1, p. 2–23.

Burrough, Peter, 1999, GIS and geostatistics: essential partners for spatial analysis, *in* Proceedings, International Symposium on Spatial Data Quality, Hong Kong, China.

Fotheringham, A. S., and P. A. Rogerson (eds.), 1994, Spatial Analysis and GIS. London: Taylor & Francis, 1994.

Goodchild, M., R. Hanining, and S. Wise, 1992, Integrating GIS and spatial data analysis: problems and possibilities. International Journal of Geographical Information Systems, vol. 6, p. 407–423.

Griffith, D., 1993a, Advanced spatial statistics for analyzing and visualizing geo-referenced data. International Journal of Geographical Information Systems, vol. 7, p. 107–123.

Griffith, D., 1993b, Which spatial statistics techniques should be converted to GIS functions?, *in* Geographic Information Systems, Spatial Modeling and Policy Evaluation (M. Fischer and P. Nijkamp, eds.). Berlin: Springer-Verlag, p. 103–114.

Haining, R. and S. M. Wise, 1991, GIS and Spatial Data Analysis: Report on the Sheffield Workshop. Sheffield, UK: University of Sheffield, Regional Research Laboratory Initiative, Discussion Paper Number 11.

Journel, Andre G., 1996, Geostatistics—tools for advanced spatial modeling in GIS, *in* Applications of GIS to the Modeling of Non-Point Source Pollutants in the Vadose Zone (Dennis L. Corwin and Keith Loague, eds.). Madison, WI: Soil Science Society of America, SSSA Special Publication, vol. 48, p. 39–55.

Liebhold, Andrew M., Richard E. Rossi, and William P. Kemp, 1993, Geostatistics and geographic information systems in applied insect ecology. Annual Review of Entomology, vol. 38, p. 303–328.

Mowrer, H. Todd, Raymond L. Czaplewski, and R. H. Hamre, 1996, Spatial Accuracy Assessment in Natural Resources and Environmental Sciences: Second International Symposium. Ft. Collins, CO: U.S. Forest Service, Rocky Mountain Forest and Range Experiment Station, Report RM-GTR-277.

Nelson, Merritt R., Thomas V. Orum, Ramon Jaime-Garcia, and Athar Nadeem, 1999, Applications of geographic information systems and geostatistics in plant disease epidemiology and management. Plant Disease, vol. 83, p. 308.

Rathfeeder, Jeffrey, 1989, These maps can find oil—or sell burgers. Business Week, no. 3095 (March 13), p. 134.

Rhodes, Hannah R., and D. E. Myers, 1993, GRASS used in the geostatistical analysis of lakewater data from the ELS, Phase I, *in* Proceedings of the 1st International Conference/Workshop on Integrating GIS and Environmental Modeling (M. F. Goodchild, B. O. Parks, and L. T. Steyaert, eds.). New York: Oxford University Press, p. 438–436.

Shirley, Kathy, 1996, Geostats add up on bottom line. AAPG Explorer, vol. 17, no. 7, p. 1, 8–10.

Thomas, Daniel, 1991, Geostatistics for GIS. GIS World, vol. 4, no. 4, p. 113–115.

Unwin, David, 1997, GIS and spatial statistical analysis, *in* Geographic Information Research: Bridging the Atlantic (M. Craglia and H. Couclelis, eds.). London: Taylor & Francis, p. 399–411.

Witherbee, Kermit G., and R. L. Watson, 1988, Applied geostatistics, subsurface mapping, and reservoir management using geographic information system technology. Denver, CO: Computer-Oriented Geological Society (COGS), Computer Contributions 4.

Appendix A
Annotated Glossary of Acronyms Used by Authors

Acronym	Complete Name, Expression, or Concept	Extended Explanation or Description
AAPG	American Association of Petroleum Geologists	an international professional association
ADRG	Arc-digitized raster graphics	graphical displays that have been captured in digital form using software developed by Environmental Systems Research Institute
ADS	automated digitizing systems	electronic mechanisms for capturing non-numeric information contained in images
AGI	American Geological Institute	an international professional association
AGIS-L	Atlas GIS List	an internet mailing list for Atlas GIS users
AIRS	Aerometric Information Retrieval System	a record-keeping mechanism of the U.S. Environmental Protection Agency
aMAP	Academic Map Analysis Package	a GIS software package
ANSI	American National Standards Institute	an organization that promotes the use of U.S. industrial standards
ANWR	Arctic National Wildlife Refuge	a conservation region in northern Alaska
AOI	area of interest	a concept pertaining to petroleum exploration
API	American Petroleum Institute	an industry-supported advocacy and research organization
API	application programming interface	a commonly used computer programming protocol
ASCII	American Standard Code for Information Interchange	a data format commonly used for text files
ASPRS	American Society of Photogrammetry and Remote Sensing	an international professional association
ASSIST	Academic Support for Spatial Information SysTem	an entity of the University of Leicester
BIL	band interleaved by line	a format principally used for Geospot satellite raster images
BLM	Bureau of Land Management	an agency of the U.S. government
BLOB	binary large object	a coding regime used in object-oriented programming
BMP	bit map	a commonly-used image format
BRS	Biennial Reporting System	a record-keeping mechanism of the U.S. Environmental Protection Agency
BSQ	band sequential	a satellite imagery format
CAD	computer-aided drafting/drawing/design	also, a format for storing digital drafting files
CARP	Circum Arctic Regional Study	a project of Exxon Exploration Company
CD	compact disc	an information storage medium
CDC	Control Data Corporation	a computer technology company
CD-ROM	compact disc—read only media/memory	a common medium for storing data
CERCLIS	Comprehensive Environmental Response, Compensation, and Liability Information System	a record-keeping mechanism of the U.S. Environmental Protection Agency
CERL	Construction Engineering Research Laboratory	a facility of the U.S. Army Corps of Engineers

Acronym	Complete Name, Expression, or Concept	Extended Explanation or Description
CGM	computer graphics metafile	a file of programming commands used to produce graphical displays
CIESIN	Center for International Earth Science Information Network	an entity of Columbia University
COGS	Computer Oriented Geological Society	an international professional association
COM	component object model	a protocol used in object-oriented programming
CORBA	common object request broker	a protocol used in object-oriented programming
COS	cartographic output systems	computer protocols that invoke the drawing of maps
COTS	commercial-off-the-shelf	a self-contained software product
CRSSA	Center for Remote Sensing and Natural Resource Sciences	an entity of Rutgers University
DBMS	database management system(s)	computerized facilities for housing/accessing/processing data
DCOM	distributed component object model	a protocol used in object-oriented programming
DCW	Digital Chart of the World	a set of maps in electronic format produced by the U.S. government
DDE	dynamic data exchange	a database programming protocol
DEC	Digital Equipment Corporation	a computer technology company
DEM	digital elevation model	a 3-D numerical representation of the earth's topography
DGI	Dynamic Graphics, Inc.	a 3-D visualization software company
DIGEST	Digital Geographic Information Exchange Standard	a spatial data format
DLG	digital land grid	a 2-D numerical representation of a physical earth property
DMA	Defense Mapping Agency	an agency of the U.S. government
DOE	Department of Energy	an agency of the U.S. government
DOS	disc operating system	the electronic instructions used to make some computers work
DRG	digital raster graphics	electronic representations of images produced by non-electronic means
DSS	decision support system(s)	computer-based mechanisms for information synthesis
DTED	digital terrain elevation data	a numerical 3-D representation of the topography of the earth
DXF	drawing exchange format	a commonly used format used to maintain images
E&P	exploration and production	upstream activities of the petrolem industry
ECU	Experimental Cartography Unit	an entity of the Royal College of Art, London
EDT	extended object data types	categories of information used in object-oriented programming
EPA	Environmental Protection Agency	an agency of the U.S. government
EPS	encapsulated postcript	a commonly used graphics format
ERDAS	Earth Resources Data Analysis System	a company that produces software for managing satellite imagery data
ESRI	Environmental Systems Research Institute	a GIS software company
ESRI-L	ESRI List	an internet mailing list for users of ESRI products
FAQ	frequently asked questions	a commonly used method to express users' concerns about products
FEMA	Federal Emergency Management Agency	an agency of the U.S. government
FGDC	Federal Government Data Committee	an entity of the U.S. government

Acronym	Complete Name, Expression, or Concept	Extended Explanation or Description
FSD	field size distribution	a term commonly used in hydrocarbon resource assessment
GCDB	geographic coordinate database	digital records of locations on the earth
GES	Geographix Exploration System	a GIS-like computer software product
GIF	graphics image format	a format for maintaining/transferring images
GIS	geographic information system(s)	a computer-based data management and visualization mechanism
GKS	Graphical Kernel System	a graphical programming protocol
GMMS	geoscientific mapping and modeling system(s)	computer-based mechanisms for describing/analyzing earth processes
GMT	generic mapping tools	a collection of data manipulation tools available from the University of Hawaii
GNIS	Geographic Names Information System	a record-keeping system of the U.S. Geological Survey
GOCAD	Geological Computer Aided Drawing	a 3-D geoscience modeling system
GPS	Global Positioning System	a satellite-based system for establishing exact ground locations
GRASS	Geographic Resources Analysis Support System	a GIS software package
GRI	Gas Research Institute	a non-profit organization engaged in support of natural gas research
GSA	Geological Society of America	an international professional association
GSC	Geological Survey of Canada	an agency of the Canadian government
GSIS	geoscience information system(s)	a computer-based data management and visualization mechanism
GUI	graphical user interface	the physical presentation of information displayed on a computer monitor
HP	Hewlett Packard	a computer technology corporation
HPGL	Hewlett Packard graphics language	a protocol commonly used to produce graphical displays or to store images
HTML	Hypertext Markup Language	a commonly used graphics format
IAMG	International Association for Mathematical Geology	an international professional association
IBM	International Business Machines	a computer technology corporation
IGES	initial graphics exchange specification	a programming protocol for managing images
IGS	integrated geo system(s)	a collection of GIS procedures developed by Oil and Gas Development Company, Ltd.
IIASER	International Institute for Aerospace Survey and Earth Sciences	an international research organization located in The Netherlands
IIS	Internet Information Server	a data management protocol
ILWIS	Integrated Land and Water Information System	a GIS software package
IMS	Internet Map Server	a product of Microsoft Corporation
IP	image processing	the general idea of capturing and using information contained in images
IS	information system(s)	mechanisms (usually computer based) for processsing/managing/storing data
IT	information technology	the processes and procedures of electronic communication
ITC	International Institute for Aerospace Survey and Earth Sciences	the Dutch acronym for this scientific entity

Acronym	Complete Name, Expression, or Concept	Extended Explanation or Description
IWI	Index of Watershed Indicators	a record-keeping system of the U.S. Environmental Protection Agency
JPEG	Joint Photographic Expert Group	a still photographic image compression method
LAS	Log ASCII Standard	a data format commonly used for well logs
L.A.S	Logiciels et Applications Scientifiques	a Canadian computer technology company
LFM	large field minimum	a concept pertaining to petroleum resource assessment
LOM	limit of migration	a concept petaining to petroleum resource assessment
LQC	large quantity generator	a concept pertaining to environmental pollution
LTM	large trap minimum	a concept pertaining to petroleum resource asessment
MAC, Mac	Macintosh	a desktop computer developed by Apple Corporation
MAP	Map Analysis Package	a GIS software package
MARPLOT	Mapping Application for Response Planning and Local Operation Tasks	a mapping package developed by the U.S. Environmental Protection Agency
MGE	Microstation Graphics Engine	an automated drawing system developed by Intergraph Corporation
MIF	MapInfo format	a specific data storage/transfer format used by the MapInfo GIS software
MIPS	millions of instructions per second	a unit in which computational speed is measured
MIS	management information system(s)	a term commonly used to describe various computational aspects of managing/processing data
MMS	Minerals Management Service	an agency of the U.S. government
MOEB	millions of equivalent barrels	a measurement unit in which hydrocarbon resources are commonly expressed
MOSS	Map Overlay and Statistical System	a GIS software package
MSS	Multispectral Scanner	a satellite which produces space-based imagery of the earth; a type of satellite image
NAD27	North American Data of 1927	a commonly used map projection based on Clarke's 1866 reference ellipsoid
NAD83	North American Data of 1983	a commonly used map projection based on the geocentric ellipsoid
NCAR	National Center for Atmospheric Research	an intercollegiate research organization support by the U.S. government
NCGIA	National Center for Geographic Information and Analysis	an entity of the U.S. government
netCDF	network common data form	a data format promoted by the National Center for Atmospheric Research
NHAP	National High Altitude Photography	an image format; a kind of image reposited by the U.S. government
NOAA	National Oceanic and Atmospheric Administration	an agency of the U.S. government
NPDES	National Pollutant Discharge Elimination System	a record-keeping mechanism of the U.S. Environmental Protection Agency
NT	New Technology	a version of Microsoft Wiindows
OGDC	Oil and Gas Development Company, Ltd.	a petroleum company in Pakistan
OGDI	open geospatial datastore interface	a method of transferring and sharing spatial data

Acronym	Complete Name, Expression, or Concept	Extended Explanation or Description
OGIS	Open GIS Consortium	an industry-supported entity that promotes standardization and common usage of GIS data
OLE	object linking and embedding	an aspect of object-oriented programming
OSF	Open Software Foundation	an industry-supported entity that promotes common access to software functionality and data
OSUMAP	Ohio State University Mapping Package	a GIS software package
PC	personal computer	desktop computing hardware and software
PCS	Permit Compliance System	a record-keeping mechanism of the U.S. Environmental Protection Agency
PCS	Paintbrush Computer Software	an early version of Paintshop PCX graphics file format
PDF	Portable Document Format	a document storage and transport format
PI	Petroleum Information, Inc.	a data collection and distribution company
pMAP	Professional Map Analysis Package	a GIS software package
POSC	Petrotechnical Open Software Corporation	an industry-supported organization that promotes the development of a common petroleum data model
PPDM	Public Petroleum Data Model	a standard method for storing, maintaining, and updating petroleum data
PPM	principal period of maintenance	a term pertaining to raster images; a raster image format
RAM	random access memory	the temporary internal information storage capacity of a computer
RCRA	Resource Conservation and Recovery Act	a regulation of the U.S. government
RCRIS	Resource Conservation and Recovery Information System	a mechanism for recording data pertaining to environmental cleanup
RDBMS	relational database management system(s)	a commonly used method of storing/accessing/retrieving/processing data
ROIP	recoverable oil in place	a concept pertaining to petroleum resource assessment
SAP	Systems Applications and Products	SAP America, developer of the R/3 internet application protocol
SAR	synthetic aperture radar	a type of space-based image of the earth
SCO	The Santa Cruz Operation, Inc.	a company specializing in Unix-based computer technology
SDBMS	spatial database management system(s)	mechanisms and techniques for storing/accessing/retrieving/processing spatial information
SDE	spatial data extension(s)	a database management and programming protocol
SDK	software development kit	the computational tools used in constructing computer applications
SDTS	Spatial Data Transfer Standard	a format used to share spatial information among software applications
SDVC	Spatial Data and Visualization Center	an entity of the University of Wyoming
SEG	Society of Geophysicists	an international professional association
SEG-P	Society of Geophysicists-P	a data format commonly used for positioning seismic shot point locations
SEG-Y	Society of Geophysicists-Y	a data format commonly used with seismic trace information
SGI	Silicon Graphics, Inc.	a computer technology company
SNV	standard normal variable	a concept pertaining to the discipline of statistics
SPE	Society of Petroleum Engineers	an international professional association

Acronym	Complete Name, Expression, or Concept	Extended Explanation or Description
SPOT	Spot Image, Inc.	a French company engaged in the production of satellite imagery
SQL	Structured Query Language	a common method of retrieving/accessing data stored in an RDMS
TCL	Tool Command Language	a programming language
TCL-Tk	Tool Command Language toolkit	the functionality of the Tool Command Language used in developing graphical user interfaces
TDS	total dissolved solids	a term pertaining to the compositional makeup of liquids
TGA	Targa	a raster image format
TIFF	tagged image file format	a data format frequently used with scanned images
TIGER	Topologically Integrated Geocoding and Referencing	a kind of data produced and distributed by the U.S. Census Bureau
TIN	triangulated irregular network	a mathematical process for producing a map
TM	Thematic Mapper	a satellite which produces space-based imagery of the earth; a type of satellite image
TNTMips	The New Thing—Map and Image Processing System	a GIS software package
TRIS	Toxic Release Inventory System	a mechanism used by the U.S. Environmental Protection Agency to record incidences of environmental pollution
TSD	treat, store, dispose	a concept pertaining to environmental pollution
TST	true stratigraphic thickness	a concept pertaining to petroleum exploration
TVT	true vertical thickness	a concept pertaining to petroleum exploration
UKOOA	United Kingdom Offshore Oil Scouts Association	a data format commonly used with siesmic navigation information
USGS	United States Geological Survey	an agency of the U.S. government
UTM	universal transverse mercator	a common map projection technique
VAX, Vax	virtual address extension	the operating system of computers produced by Digital Equipment Corporation
VESA	Video Electronics Standards Association	an international professional association
VGA	Video Graphics Adapter	a resolution, color depth, and scan rate spcification for video displays (computer monitors)
VLB	VESA Local Bus	a type of computer video card; a type of computer architecture
VRF	vector relational format	a spatial data format
VRML	virtual reality modeling language	a methodology for producing 3-D graphics that incorporate motion
WDC	World Data Center	an international repository for data about the earth
WELUT	Western Energy and Land Use Team	an entity of the U.S. Bureau of Land Management
WIMS	Wyoming Internet Map Server	a computing system that supports ready access to map products
WMF	Windows metafile	a file about information files that can be processed by Microsoft Windows
WWW	World Wide Web	a name frequently given to the Internet, or part of an Internet address
YPF	Yacimientos Petroliferos Fiscales	the national oil company of Argentina

Appendix B
Vendors and Products Cited by Authors

Vendor/Manufacturer/Developer	Product
Adobe Systems, Inc.	PostScript
Apple Corporation	A/UX general computer equipment
AutoDesk, Inc.	AutoCAD
Baylor University	GRASS (Windows version)
Borland International, Inc.	dBase (formerly a product of Ashton Tate Corporation) Delphi
Central Washington University	GRASS (SCO UNIX and Macintosh versions)
Centura Software Corporation	Gupta SQLBase
Cerro Verde Geotech	CVGMap
Clark University, The IDRISI Project	IDRISI
Computer Associates Inc.	Ingres
Computer Corporation of America	Model 204 (or M204)
Control Data Corporation	general computer equipment
Corel Corporation	Paradox Wordperfect
Cray Research, Inc.	computers and equipment
Data General Corporation	general computer equipment
Datalink Software, Inc.	SQL Datalink
Digital Equipment Corporation (now part of Compaq, Inc.)	ULTRIX workstations and other computer equipment
Dynamic Graphics, Inc.	EarthVision Geologic Structure Builder
Environmental Systems Research Institute	ArcCAD Arc/Info ArcView 1.0, 3.2 ArcX Atlas GIS (formerly a product of Strategic Mapping, Inc.) MapObjects
ERDAS, Inc.	Imagine
Geological Survey of Canada	FieldLog
Geotech Computer Systems	PetroSpase Spase
Global Geomatics	GRASSLand
GPZ Technology	GRASSXMI (X/Motif interface version of GRASS)
Hewlett-Packard Corporation	UX 90 workstations and other computer equipment
IBM Corporation	DB/2 general computer equipment Professional Fortran 77
IHS Energy Group	Edge

Vendor/Manufacturer/Developer	Product
Informix, Inc.	Informix
Intel Corporation	Pentium processors
Intergraph Corporation	MGE
International Institute for Aerospace Survey and Earth Sciences	ILWIS
Landmark Graphics Corporation	Geographix, or GES (formerly a product of Geographix, Inc.) OpenExplorer SeisWorks Stratamodel (formerly a product of Stratamodel, Inc.)
Lotus Development Corporation	Lotus 1-2-3
Mapping Information Systems Corporation	MapBasic MapInfo
Massachusetts Institute of Technology	X-Windows
Micro Images, Inc.	Spatial Manipulation Language TNTLite TNTMips
Microsoft Corporation	Access ActiveX Component Object Model Excel FoxPro Internet Explorer Internet Information Server PowerPoint SQL Server Visual Basic Visual C++ Visual FoxPro Windows 95, 3.1, NT Windows for Workgroups Word
National School of Geology, Nancy, France	GOCAD
Netscape Communications	Navigator
Next Corporation	workstations and other computer equipment
Ohio State University, Department of Geography	OSUMAP
Oil and Gas Development Company, Ltd. (Pakistan)	IGS
Open Software Foundation	LINUX (originally developed by Linus Torvalds) Motif Open Windows
Oracle Corporation	Oracle Spatial Cartridge
Petroconsultants, S. A. (a part of IHS Energy Group)	Petroconsultants International Oil and Gas Databases
Petroleum Information/Dwights (a part of IHS Energy Group)	GRID
Schlumberger	CPS-1 (formerly a product of Radian Corporation)
Silicon Graphics, Inc.	IRIX workstations and other computer equipment
Spatial Information Systems, Inc.	aMAP pMAP

Vendor/Manufacturer/Developer	Product
Spot Image, Inc.	general satellite imagery
Sun Microsystems, Inc.	Solaris SparcStation and related computer equipment
Sybase, Inc.	PowerBuilder Sybase
Tektronix Inc.	PLOT10
Tobin International	general mapping products and databases
University of Hawaii, School of Ocean and Earth Sciences and Technology	GMT
U.S. Bureau of Land Management	MicroDEM MOSS
U.S. Bureau of the Census	LandView (formerly a product of NOAA/EPA) Marplot for Windows
U.S. Army Corps of Engineers, Construction Engineering Research Laboratory	GRASS
U.S. Environmental Protection Agency	MARPLOT
U.S. Geological Survey	GSMAP
X/Open International Consortium	UNIX (originally developed at Bell Laboratories)

Index

Access, 251, 253, 254, 255
ActiveX, 251, 253
Aeromagnetic map data, 141
AMAP (Academic Map Analysis Package), 277
Analysis with GIS, 18, 20
  Arc/Info, 140
  Atlas GIS, 275
  EarthVision, 140, 206, 209
  Gap Mountain prospect (Nevada), 141-143
  Gulf of Suez, 206-210
  maps, 23-24
  MOSS/PC-MOSS, 264, 270-271
  PetroSpase, 153, 154-160, 163-166, 167-169
  spatial, 11, 21, 50, 275-277
Apache Web server, 232
Application programming interface (API), 39-40
Applications of GIS, 43-45, 69, 213
  Amoco Venezuela, 230-235
  Austral Basin (Argentina), 161-166
  CARP (Circum Arctic Regional Study), 71-83
  Citronelle field (Alabama), 195-203
  evaluating international acquisitions, 151-170
  exploitation, 245-246
  exploration, 239-245
  fault density and subsurface fluid flow, 121-135
  fluid migration in sedimentary basins, 121-134
  functionally integrated, 40
  Gabon (West Africa), 154-161, 162, 163, 164, 165
  Gap Mountain prospect (Nevada), 137-148
  Gulf of Suez, 205-212
  hydrocarbon resource assessment, 173-184
  image processing, 41
  integrated management, 41, 215-224, 227-235
  integrated spatial analysis, 41
  integration with World Wide Web, 230-235
  Molasse basin (Upper Austria), 121-135
  North Sea, 166-169
  Port Hudson field (Louisiana), 187-193
  regional study in GIS, 71-83
  social issues, 12-13
  spatial relationships, 121-135
  waterflood optimization, 195-203
Arc/Info, 38, 50, 51, 53, 56, 140, 251
ArcView, 18, 52, 53, 232, 251, 272-273, 276
Assessment, hydrocarbon resource, 173-186
Atlas GIS, 272, 275
Austral Basin (Argentina), 161-166
AutoCAD, 52, 53
Automated cartography, 28
Automated feature counting, 182
Automated mapping functions, 228

Bedrock geology, registration and digitization, 141
Benefits of GIS, 73, 76-77, 237-238
  accessible to larger user base, 188, 237
  accuracy greater, 188
  agreement on common data elements, 81
  better quality control, 176
  continuous digital workflow, 122
  cost effectiveness, 77, 188, 238
  data visualization, 237
  database flexibility, 252
  decision support, 123
  digital analysis of multi-component processes, 122
  direct ties between maps and databases, 175
  dynamic mapping, 237
  early understanding of hydrocarbon occurrence, 77
  easier interaction with institutional partners, 82
  enhanced portability, 238
  facilitates hydrocarbon system approach, 175
  greater productivity, 23
  high degree of geological integration, 81, 197
  integration with digital communications systems, 122
  integration of interdisciplinary contributions, 237
  interactive query and analysis of results, 176
  linkage of multiple software applications, 237
  multiple presentation formats, 237
  multiple products easily "zippered," 77
  new maps from old, 76-77
  performance enhanced, 238
  projects completed in timely manner, 77
  proliferation of new tools, 238
  self-documenting system, 77
  spatial analysis of large datasets, 122-123
  spatial information in one system, 76
  supports multiple computer platforms, 237
  uses variety of input formats, 126
Bibliographic resources, 281-292
Binary large object data-type (BLOB), 39
Boundaries, ambiguous or gradational, 29
Bureau of Land Management (BLM), 140, 270, 277

CAD (computer-aided design) systems, 39, 41, 56, 220
CARP (Circum Arctic Regional Study), 71-83
Cartographic output systems (COS), 270

Cartography. *See* Mapping
Case studies. *See* Applications
Center for Remote Sensing and Spatial Analysis (CRSSA), 268
CERL (Construction Engineering Research Laboratories), 267
CGM (computer graphics metafile format), 234
Chance of adequacy maps, 181-182, 183
Circum Arctic Regional Study (CARP), 71-83
Citronelle field (Alabama), 195-203
Closure database, 178, 179-181
COGS (Computer Oriented Geological Society), 271
COM/DCOM (Component object model/ distributed component object model), 40
Commercial applications of GIS, 28-29
Commodity ratios, 178, 180
Common object request broker architecture (CORBA), 40
Communications, 213, 217, 218
Component object model/distributed component object model (DOM/DCOM), 40
Computer graphics metafile format (CGM), 234
Computer graphics. *See* Graphics
Computer mapping. *See* Mapping
Computer Oriented Geological Society (COGS), 271
Conceptual geological modeling, 137-150
Conflict resolution, 12-13
Construction Engineering Research Laboratories (CERL), 267
Context data layers, 258-259
Contour mapping, 220
CORBA (common object request broker architecture), 40
COS (cartographic output systems), 270
Cost of GIS, 54, 65, 66, 263, 264, 271-275
  cost effectiveness, 32, 77, 188, 238
  GIS vs. other techniques, 151-152
  See also Economic potential; Hardware; Software; Staffing
Cost surface modeling, 131-133
Cross-section, digitization, 141
CRSSA (Center for Remote Sensing and Spatial Analysis), 268
CVGMap, 153

Data acquisition, 8-19, 21, 22, 213
  3-D conceptual model, 140
  automated digitizing systems (ADS), 270
  Citronelle field (Alabama), 198-199, 206
  Gulf of Suez, 206
  Molasse basin (Austria), 126-128
  Port Huron field, 190-191
Data analysis. *See* Analysis
Data combination, 20, 21. *See also* Integration of GIS data
Data communications bridge, 217, 218
DataLink connectivity module, 274
Data models, 22, 57
Data organization, 19, 29. *See also* Data acquisition; Database management; Modeling
Data prediction. *See* Modeling; Prediction
Data processing, 140-141, 217, 218, 231
  transformation, 21, 23, 53-54, 141
Data publishing, 251
Data quality, 230-231
Data search, *See* Search
Data storage, 30-31, 80, 230-231
Data symbology, 238
Data vendors, 43
Data visualization. *See* Visualization
Database connectivity, 41
Database management, 213, 245
  CVGMap, 153
  data types, 141, 192
  database management systems, 18
  Gap Mountain prospect (Nevada), 141
  international acquisitions, 152-153
  PetroSpase, 153-160, 167-169
  RDBMS, 216-217
  "self-serve" concept, 231-232
  storage, 230-231
  web-based, 227-235
  See also Software programs for GIS
Database search. *See* Search
DCW. *See* Digital Chart of the World
DDE (dynamic data exchange), 199
Decision support system (DSS), 222
Delphi, 251, 253
DEM (Digital elevation model), 23, 140
Desktop GIS, 18, 39, 41, 251, 276
Determining large trap minimum and large field minimum, 181
DGI. *See* Dynamic Graphics Inc.
Digital cartography, 41
Digital Chart of the World (DCW), 57, 153, 154
Digital raster graphics (DRG), 140
Digital vs. non-digital data, 63-64
Discrete volume elements (voxels), 30
Discussion groups, 265, 277-278
DRG (Digital raster graphics), 140
DSS (decision support system), 222
Dwights Energydata. *See* PI/Dwights
Dynamic data exchange (DDE), 199
Dynamic Graphics Inc. (DGI), 140, 206
Dynamic simulation, 14

E&P. *See* Exploration and production
EarthVision, 140, 206, 209
Economic potential of GIS, 174, 184
  cost effectiveness, 77, 188, 238
  evaluating international acquisitions, 151-170
  international E&P, 152

EDTs (Extended object data types), 39
Egyptian General Petroleum Company, 205
Environmental Systems Research Institute (ESRI), 4, 24, 38, 52, 135, 140, 153, 232, 250
Exploitation, 203, 245-247
Exploration, 239-245
Exploration plays assessment, 173-184
Exploration and production (E&P), 37, 40
Extended object data types (EDTs), 39
Exxon templates, 84-120
  bedrock and tectonics, 85, 86-93
  closures, 120
  contours, 94, 95
  faults, 96, 97
  miscellaneous polygons and lines, 98, 99
  non oracle point data, 100, 101
  oracle point data, 102
  polygonized values, 103, 104
  sedimentary lithofacies, 105, 106-119

Faults, 127-128, 130-134, 139, 141-143
Feature counts, 178, 180, 182
Federal Government Data Committee (FGDC), 251
Field size distributions (FSD), 78, 179
FieldLog, 18
Fieldwide analysis, 199
Flat files, 56, 57
Flow units, 201
Fluid flow systems, 32, 121-134
Focus area analysis, 199-200
Format engine, IGS applications, 217, 218
FSD (field size distributions), 78, 179
Future of GIS, 32-34, 45, 58, 293, 295-299
  applications, 41-42
  GIS and geostatistics, 295-298
  vendors view, 41-43

Gabon (West Africa), 154-161, 162, 163, 164, 165
Gap Mountain prospect (Nevada), 137-148, 137-149
GCDB (Geographic Coordinate Database), 140
Geo-objects, 30
Geo-rectification, 54, 63-64
Geo-referencing, 21, 22
Geo-spatial locations, 140
Geodetic system, 218
GeoDirectory, 278
GeoExplorer, 213, 253-254, 258-259
Geographic Coordinate Database (GCDB), 140
Geographic-to-UTM transformation, 141
Geographix, 52, 53, 56, 198
Geologic play assessment, 173-184
Geologic risk, 177-178
Geologic subplays, 177, 179
Geoscientific information systems (GSIS), 27, 29-33
Geoscientific mapping and modeling systems (GMMS), 30
Geospatial analysis. *See* Analysis
Geostatistics, 4, 11, 19, 295-298
Geotech Computer Systems, 153
GIS (Geographic Information Systems), 56, 64
  basic elements, 138
  and CAD convergence, 56
  character and composition, 3-5, 196–197
  combination of software and database, 63
  components, 21-24
  cost, 54, 65, 66
  definition, 5, 122
  effective use, 43-45
  evaluating international acquisitions, 151-170
  integration with physical database, 38-39
  interoperability, 40
  issues, 63-67
  modeling. *See* Modeling
  onion metaphor, 64
  as part of corporate database, 33
  in petroleum exploration and development, 61-67
  realities, 47
  reasons for using, 122
  recent advances, 39, 228-230
  as software package, 63
  spatial data extensions, 227, 228
  spatial database, 63, 71
  support and maintenance, 58
  users, 43, 238
  vendor perspective, 37-45
  see also Benefits of GIS; Cost of GIS; Problems and limitations of GIS; Requirements of GIS
Global Geomatics Inc., 268
GMMS (geoscientific mapping and modeling systems), 30
GMT (generic mapping tools), 277
GOCAD, 18
Graphics with GIS
  computer graphics metafile (CGM) format, 234
  digital raster graphics (DRG), 140
  IGS capabilities, 218-220
  image processing, 39
  integrated geo systems (IGS), 218-220
  integrated graphics presentation, 220, 221
  Landview III, 269-270
  tools, 30, 217-218, 218-219
  see also Mapping for GIS; Software for GIS; Software programs for GIS
GRASS (geographic resources analysis support system), 264, 265-269
GRASSLands, 264, 268-269
GRASSLinks 2.0, 268
GSIS (geoscientific information systems), 27, 29-33
GSMAP, 263

Hardware requirements of GIS, 66
  ArcView, 3.2, 272-273

Atlas GIS, 275
GeoExplorer, 253
GRASS, 267
IDRISI, 273
LANDVIEW III, 269
MapInfo, 274
MOSS, 271
History of GIS, 9-10, 28-29, 38
computer mapping, 9-10
data structure, 10
historical perspective, 3-5
spatial database management, 10
vendor perspective, 38-39
Hotlinking, 252
HTML (HyperText Markup Language), 252-253
Hydrocarbon systems approach, 173, 177
HyperText Markup Language (HTML), 252-253

IDRISI, 271-272, 273-274
IGS (integrated geo systems), 218-220
IHS Energy Group, 43, 49, 153
IHSEnergy's Edge, 43
ILWIS (Integrated Land and Water Information System), 276
Image processing (IP) systems, 18
IMS (Internet map server), 227, 228-229
Information systems (IS), 38
Institute of Geological Sciences, 28
Integrated geo systems (IGS), 218-220
Integrated Land and Water Information System (ILWIS), 276
Integration of GIS data, 4, 5, 18, 20, 231
data management, RDBMS, 216-217
data management and analysis, 32, 215-224
digital communications systems, 122
geological, 81, 197
interdisciplinary contributions, 4, 237
interpretation, 4
with LandSat imagery, 245
and overlay of multiple maps, 245, 246
with physical database, 38-39
with World Wide Web, 230-235
Interactive seismic survey planning, 220, 223
Intergraph Corporation, 134
International Institute for Aerospace Survey and Earth Sciences, 276
International projects, 57-58
Internet discussion groups, 277-278
Internet GIS applications, 33
Internet map server (IMS), 227, 228-229
Internet. *See* Web-based applications
Investment opportunities evaluation, 174, 184
IS (Information systems), 38

Java, 41

Landmark Graphics Corporation, 206
Landscape analysis, 12
LandView III, 264, 269-270
Large field minimum (LFM), 178, 181
Large trap minimum (LTM), 178, 181
LFM (large field minimum), 178, 181
Limitations of GIS. *See* Problems and limitations
Linking assessment results to maps, 183
LTM (large trap minimum), 178, 181

M204 database, 57
Map algebra, 11
MapBasic, 274
Map overlay techniques, 28, 264, 270-271
MapInfo, 18, 272, 274-275
MapObjects, 52, 249, 251-253, 255
Mapping with GIS, 9
automated cartography, 28
average porosity, 191, 192
average water saturation, 191, 192
base map generator, IGS, 220, 221
base maps, 239
basin maps, 245
chance of adequacy, 181-182, 183
color management, 254-255, 256
cumulative gas production maps, 242-244
desktop cartography, 39
drainage buffer maps, 246, 247
fault plane maps, 246
gas field size, 244, 245
geochemical data maps, 255, 257
geologic structure maps, 241-242
geology maps, 254-255, 256
initial oil potential map, 243, 245
integration with LandSat imagery, 245
integration of multiple maps, 245, 246
land status, 255-256, 257
map feature generalization, 28
net effective sand maps, 191, 192
net pay, 191, 193
net pay contour maps, 246, 247
operator maps, 239-240, 241
overlay techniques, 28, 133
petroleum, 50-59
production maps, 240-241
purpose of, 9
real-time, 222, 224
reconnaissance maps, 239
residual-aeromagnetic, 141
sand isopach map, 245-246
spud date maps, 239, 240
structure, 191
synthetic aperture radar (SAR) image, 255, 257, 258
Web-based, 33-34
well classification map, 244, 245
See also Software for GIS; Software programs for GIS
Metadata, 251, 252-253

Methodology
  creation GIS database and RDBMS, 126
  PetroSpase, 153, 154-160, 163-166, 167-169
  play assessment, 173-184
MicroDEM, 277
Modeling with GIS, 27-34, 138
  2-D surface model, 142
  3-D conceptual geologic, 137-150
  3-D fault geometry, 141-142
  3-D Gulf of Suez, 206-212
  automated cartography, 28
  categories of GIS, 29
  commercial geographic information systems, 28-29
  conceptual geological, 137-150, 139-140
  deformation across fault zone, 130-131
  DEM (digital elevation model), 140
  digital elevation model and DRG image, 140-141
  fault surface, 206, 207, 208
  fluid migration using cost surfaces, 131-133
  future for GIS in petroleum exploration and production, 32-34
  Gap Mountain prospect (Nevada), 141-143
  geochemical data, 147-148
  geologic structure and stratigraphy, 144-145
  geoscientific information systems (GSIS), 29-32
  geostatistical, 296-297
  holistic approach, 14-15
  limitations of GIS, 29
  normal faults, 142
  Paleozoic section (Gap Mountain prospect), 145-146
  porosity and permeability, 208, 209
  predictive, 13
  Quaternary alluvium, 147
  raster vs. vector data structure, 29, 56
  rock properties, 147
  spatial data management, 28
  surface representation methods, 30, 32
  tertiary volcanics, 147
  thrust faults, 142-143
  trends in subsurface characterization, 32
  volume representation methods, 30
  volumetic analysis, 147
  See also Software for GIS; Software programs for GIS
Molasse basin (Austria), 124-135
MOSS/PC-MOSS (map overlay and statistical system), 264, 270-271
Multimedia, 252

Natural Resources Data Clearinghouse of Spatial Data and Visualization Center (SDVC), 33
Net effective sand maps, 191, 192
Net pay maps, 191, 193
Net-pay thickness, 147
NetCDF library, 277
Nevada Division of Mines and Geology, 140
North Sea, 166-169
Object linking and embedding (OLE), 40
Object-oriented programming, 253-254
Octrees, 30
OGIS (Open GIS Consortium), 40, 134
OLE (Object linking and embedding), 40
Open GIS Consortium (OGIS), 40, 134
Open standards, 39-40
OpenExplorer, 43
Opportunity analysis, 174, 184
Oracle Corporation, 135
OSUMAP, 276-277

Paleomap Project, 4
Permeability distribution and fluid flow, 130-131, 132
Petro-technical Open Software Corporation (POSC), 37, 40
Petroconsultants, 57, 153
Petroleum Information. *See* PI/Dwights
PetroSpase, 153, 154-160, 163-166, 167-169
PI/Dwights, 49-60, 50
Pipeline data, 53
Play assessment, 173-184
Play chance, 176, 177
Play elements, 173, 176, 177, 179
Play summary map, 173-174, 175, 176, 179, 182
PMAP (professional Map Analysis Package), 277
Polygon analysis, 274
Porosity and permeability, 208, 209
Port Hudson field (Louisiana), 187-193
POSC (Petro-technical Open Software Corporation), 37, 40
PowerBuilder, 251, 253
PPDM. *See* Public Petroleum Data Model Association
Prediction, 18, 20. *See also* Modeling
Problems and limitations of GIS, 5, 122-123, 227-230
  complexity, 19, 229
  dimensional limitations, 29
  implementation, 77, 79, 81
  international acquisitions, 152
  lack of corroborating scientific knowledge, 14
  language barrier, 58
  limitations, 29, 32-33
  separate from corporate database, 38-39, 228
  specialized spatial analysis functions, 228
  staff, 62, 63, 64-65, 228
  stand-alone technology, 227-228
  support burdens, 228
Product generation capabilities, 148
Productivity, Web-based GIS, 234-235
Project ASSIST, 268
Prospect counts, 176
Proximity functions, 128
Public domain and inexpensive software, 263-282
Public Land Survey System, 140

Public Petroleum Data Model Association (PPDM), 37, 40, 57
QC Data, 43

Raster data structure, 18, 22, 23, 29, 56, 270
RDBMS. *See* Relational database management system
Real-time geographic mapping, 222, 224
Recoverable oil-in-place (ROIP), 200, 201
Regional study, circum Arctic (CARP), 71-83
Relational database management system (RBDMS), 38-39, 122, 216-217, 228
Requirements of GIS systems, 62
- application of GIS data standards, 82
- change in operating paradigm, 62, 82
- continual upgrading, 62, 65
- data handling flexibility, 216
- data management, 62, 216-217
- development of 3-D mentality, 62
- financial costs, 54, 65, 66, 263, 264, 271-275
- graphical tools, 216
- hardware, 66
- integrated data processing, analysis and interpretation, 216
- need for increased reliance on computer, 62
- networking capability, 66
- product-based planning, 82
- for regional study, 71, 73
- staff, 62, 63, 64-65, 82, 228
- user interface, 64

Resampling of attributes of polygons, 23
Reservoir assessment, 187-193, 195-203, 205-212
Reservoir volume calculation, 209-210
Resources on GIS
- bibliographic, 281-292
- GeoDirectory, 278
- software, 263-279
- Web-based, 278-279

Risk analysis, 32, 177-178
Risked mean resource, 174, 176
Rock property modeling, 147
ROIP (Recoverable oil-in-place), 200, 201

Sampling, 297
Scheduling, 75
SDE (spatial database engine databases), 253
SDVC (Natural Resources Data Clearinghouse of Spatial Data and Visualization Center), 33
Search, 18, 20, 22-23, 274
- 3-D, 222, 224
- data filtering, 238
- database, 18, 20, 222, 224, 238
- IGS, 222
- innovative, 222
- spatial queries, 32-33
- web-based, 229, 232, 233

Seismic interpretation, 4, 234-235
SeisWorks, 206
Social solutions, 12-13
Software for GIS
- 3-D geological data, 18
- collector sites, 265
- constructing a GIS, 66, 67
- COTS (commercial-off-the-shelf) desktop applications, 41
- CVGMap, 153
- desktop GIS applications, 18, 39, 41, 251
- discussion groups, 265
- free versions, 275-276
- geoscience application, 217
- incorporation of GIS concept, 4
- inexpensive, 271-272
- off-the-shelf GIS systems, 63, 65-66
- public domain, 263-271
- seismic line cross-section digitizing program, 141
- spatial analysis, 19
- surface modeling, 18, 19
- systems issues, 40-41

Software programs for GIS, 264-277
- Access, 251, 253, 254, 255
- ActiveX, 251, 253
- aMAP (Academic Map Analysis Package), 277
- ArcView, 3.2, 18, 52, 53, 232, 251, 272-273, 276
- ArcView (Version 1), 276
- Atlas GIS, 272, 275
- AutoCAD, 52, 53
- Delphi, 251, 253
- EarthVision, 140, 206, 209
- GMT (Generic Mapping Tools), 277
- GRASS (geographic resources analysis support system), 264, 265-269
- GRASSLands, 264, 268-269
- GRASSXMI, 268
- GSMAP, 263
- IDRISI, 271-272, 273-274
- ILWIS (Integrated Land and Water Information System), 276
- Java, 41
- LandView III, 264, 269-270
- MapInfo, 18, 272, 274-275
- MapObjects, 52, 249, 251-253, 255
- MicroDEM, 277
- Microsoft, 41, 42
- Microsoft Internet Information Server (IIS), 232
- MOSS/PC-MOSS (map overlay and statistical system), 264, 270-271
- netCDF library, 277
- OSUMAP, 276-277
- PetroSpase, 153, 154-160, 163-166, 167-169
- pMAP (professional Map Analysis Package), 277
- PowerBuilder, 251, 253
- SeisWorks, 206
- TNTLite/TNTMips, 276
- Visual C++, 251, 253
- Visual FoxPro, 251, 253

XGRASS, 268
Spatial analysis. *See* Analysis
Spatial Cartridge, 134
Spatial data extensions, 228
Spatial data transfer standard (STDS), 122
Spatial database, 21, 22
Spatial database engine databases (SDE), 253
Spatial Information Systems, 277
Spatial object types, 41
Spatial queries, 32-33. *See also* Search
Spatial reasoning, 12-13
Spatial relationships, 128-130
SQL (structured query language), 40, 41, 153
SQL2 (structured query language), 40
SQL3 (structured query language), 40
Stacked pay potential, 174
Staff, 63, 64-65, 82, 228
Statistics, 11, 19, 295-298
SDTS (spatial data transfer standard), 122
Storage. *See* database management
Stratamodel, 18
Subsurface characterization, 30, 32
Subsurface fluid migration, 121-134
Sun Microsystems, 134
Surface modeling and contouring, 18, 30, 31
Symbology, 238
System architecture, 38-39, 40-41, 216-220

Tagged image file format (TIFF), 140
Team approach, 188-190
Techniques
GIF vs. other, 151-152
map overlay, 28
spatial analysis, 121
voxel, 30-31
Three-dimensional technology, 30-33, 138, 220-222
Three-tier architecture, 38-39, 40-41
3-D GIS, 30
TIFF (tagged image file format), 140
TINs (triangulated irregular network methods), 31
TNTLite/TNTMips, 276
Tobin International, 43
Topographical relationships, 22
Topology, 56
Transformations, 23. *See also* Analysis with GIS
Triangulated irregular network methods (TINs), 31
Tunnel method, 130

Umbrella technology, 122
Undiscovered resource potential, 173-184
UNIX, 232, 268, 277
U.S. Defense Mapping Agency, 153
U.S. Department of Energy (DOE), 196
U.S. Geological Survey, 52, 53, 213, 249-260
digital elevation model (DEM), 140
Global Land Information System, 140
Web resources, 250, 279

Vector data structure, 18, 22, 23, 29, 56
Vendor perspective, 37-45
Veritas, 43
Vertical data, 55
Virtual Reality Modeling Language (VRML), 33
Visual Basic, 251, 253
Visual C++, 251, 253
Visual development languages, 251
Visual FoxPro, 251, 253
Visualization, 18, 19-21, 22-23, 33, 50, 237
Volume representation methods, 30
Volumetric analysis, 147
Voxel methods, 30
VRML (Virtual Reality Modeling Language), 33

Waterflood optimization, 195-203
Web resources, 267-268, 278
Web-based applications, 33, 41, 229-230, 278
data management, 227-235
integration with GIS, 227-235
mapping, 33-34, 58
public access website, 268
Wyoming Internet Map Server, 33
Well data, Dwights, 55
Western Energy and Land Use Team (WELUT), 270
WIMS (Wyoming Internet Map Server), 33
Windows NT, 4.0, 54
Windows NT server, 232
Work process overview, 74
World Wide Web, 33-34, 41, 58, 227-235, 278, 279
Wyoming Internet Map Server (WIMS), 33

XGRASS, 268